IAASTD

International Assessment of Agricultural Knowledge, Science and Technology for Development

East and South Asia and the Pacific (ESAP) Report

IAASTD

International Assessment of Agricultural Knowledge, Science and Technology for Development

UNDP

UNEP

UNESCO

THE WORLD BANK

W.H.O. O.M.S.

WHO

GLOBAL ENVIRONMENT FACILITY

IAASTD

International Assessment of Agricultural Knowledge, Science and Technology for Development

East and South Asia and the Pacific (ESAP) Report

Edited by

Beverly D. McIntyre	Hans R. Herren	Judi Wakhungu	Robert T. Watson
IAASTD Secretariat	Millennium Institute	African Centre for Technology Studies	University of East Anglia

Library of Congress Cataloging-in-Publication data.

International assessment of agricultural knowledge, science and technology for development (IAASTD) : East and South Asia and the Pacific (ESAP) report / edited by Beverly D. McIntyre . . . [et al.].
p. cm.
Includes bibliographical references and index.
ISBN 978-1-59726-544-7 (cloth : alk. paper) —
ISBN 978-1-59726-545-4 (pbk. : alk. paper)
1. Agriculture—East Asia—International cooperation.
2. Agriculture—South Asia—International cooperation.
3. Agriculture—Pacific Area—International cooperation. 4. Sustainable development—East Asia. 5. Sustainable development—South Asia.
6. Sustainable development—Pacific Area. I. McIntyre, Beverly D.
II. Title: East and South Asia and the Pacific (ESAP) report.
HD1428.I543 2008
338.95′07—dc22 2008046046

British Cataloguing-in-Publication data available.

Printed on recycled, acid-free paper

Interior and cover designs by Linda McKnight, McKnight Design, LLC.

Manufactured in the United States of America

10 9 8 7 6 5 4 3 2 1

Contents

Statement by Governments

All countries present at the final intergovernmental plenary session held in Johannesburg, South Africa in April 2008 welcome the work of the IAASTD and the uniqueness of this independent multistakeholder and multidisciplinary process, and the scale of the challenge of covering a broad range of complex issues. The Governments present recognize that the Global and Sub-Global Reports are the conclusions of studies by a wide range of scientific authors, experts and development specialists and while presenting an overall consensus on the importance of agricultural knowledge, science and technology for development also provide a diversity of views on some issues.

All countries see these Reports as a valuable and important contribution to our understanding on agricultural knowledge, science and technology for development, recognizing the need to further deepen our understanding of the challenges ahead. This Assessment is a constructive initiative and important contribution that all governments need to take forward to ensure that agricultural knowledge, science and technology fulfills its potential to meet the development and sustainability goals of the reduction of hunger and poverty, the improvement of rural livelihoods and human health, and facilitating equitable, socially, environmentally and economically sustainable development. In accordance with the above statement, the following governments accept the East and South Asia and Pacific Report:

Bangladesh, Bhutan, China, (People's Republic of), India, Lao People's Democratic Republic, Maldives, Philippines, Republic of Palau, Solomon Islands, Timor-Leste, Viet Nam (11 countries).

While approving the above statement Australia did not fully accept the East and South Asia and Pacific Report. Australia recognizes the IAASTD initiative and reports as a timely and important multistakeholder and multidisciplinary exercise designed to assess and enhance the role of AKST in meeting the global development challenges. The wide range of observations and views presented however, are such that Australia cannot agree with all assertions and options in the report. The report is therefore noted as a useful contribution which will be used for considering the future priorities and scope of AKST in securing economic growth and the alleviation of hunger and poverty.

Foreword

The objective of the International Assessment of Agricultural Knowledge, Science and Technology for Development (IAASTD) was to assess the impacts of past, present and future agricultural knowledge, science and technology on the

- reduction of hunger and poverty,
- improvement of rural livelihoods and human health, and
- equitable, socially, environmentally and economically sustainable development.

The IAASTD was initiated in 2002 by the World Bank and the Food and Agriculture Organization of the United Nations (FAO) as a global consultative process to determine whether an international assessment of agricultural knowledge, science and technology was needed. Mr. Klaus Töepfer, Executive Director of the United Nations Environment Programme (UNEP) opened the first Intergovernmental Plenary (30 August-3 September 2004) in Nairobi, Kenya, during which participants initiated a detailed scoping, preparation, drafting and peer review process.

The outputs from this assessment are a Global and five Sub-Global reports; a Global and five Sub-Global Summaries for Decision Makers; and a cross-cutting Synthesis Report with an Executive Summary. The Summaries for Decision Makers and the Synthesis Report specifically provide options for action to governments, international agencies, academia, research organizations and other decision makers around the world.

The reports draw on the work of hundreds of experts from all regions of the world who have participated in the preparation and peer review process. As has been customary in many such global assessments, success depended first and foremost on the dedication, enthusiasm and cooperation of these experts in many different but related disciplines. It is the synergy of these inter-related disciplines that permitted IAASTD to create a unique, interdisciplinary regional and global process.

We take this opportunity to express our deep gratitude to the authors and reviewers of all of the reports—their dedication and tireless efforts made the process a success. We thank the Steering Committee for distilling the outputs of the consultative process into recommendations to the Plenary, the IAASTD Bureau for their advisory role during the assessment and the work of those in the extended Secretariat. We would specifically like to thank the cosponsoring organizations of the Global Environment Facility (GEF) and the World Bank for their financial contributions as well as the FAO, UNEP, and the United Nations Educational, Scientific and Cultural Organization (UNESCO) for their continued support of this process through allocation of staff resources.

We acknowledge with gratitude the governments and organizations that contributed to the Multidonor Trust Fund (Australia, Canada, the European Commission, France, Ireland, Sweden, Switzerland, and the United Kingdom) and the United States Trust Fund. We also thank the governments who provided support to Bureau members, authors and reviewers in other ways. In addition, Finland provided direct support to the Secretariat. The IAASTD was especially successful in engaging a large number of experts from developing countries and countries with economies in transition in its work; the Trust Funds enabled financial assistance for their travel to the IAASTD meetings.

We would also like to make special mention of the Regional Organizations who hosted the regional coordinators and staff and provided assistance in management and time to ensure success of this enterprise: the African Center for Technology Studies (ACTS) in Kenya, the Inter-American Institute for Cooperation on Agriculture (IICA) in Costa Rica, the International Center for Agricultural Research in the Dry Areas (ICARDA) in Syria, and the WorldFish Center in Malaysia.

The final Intergovernmental Plenary in Johannesburg, South Africa was opened on 7 April 2008 by Achim Steiner, Executive Director of UNEP. This Plenary saw the acceptance of the Reports and the approval of the Summaries for Decision Makers and the Executive Summary of the Synthesis Report by an overwhelming majority of governments.

Signed:

Co-chairs
Hans H. Herren,
Judi Wakhungu

Director
Robert T. Watson

Preface

In August 2002, the World Bank and the Food and Agriculture Organization (FAO) of the United Nations initiated a global consultative process to determine whether an international assessment of agricultural knowledge, science and technology (AKST) was needed. This was stimulated by discussions at the World Bank with the private sector and nongovernmental organizations (NGOs) on the state of scientific understanding of biotechnology and more specifically transgenics. During 2003, eleven consultations were held, overseen by an international multistakeholder steering committee and involving over 800 participants from all relevant stakeholder groups, e.g., governments, the private sector and civil society. Based on these consultations the steering committee recommended to an Intergovernmental Plenary meeting in Nairobi in September 2004 that an international assessment of the role of AKST in reducing hunger and poverty, improving rural livelihoods and facilitating environmentally, socially and economically sustainable development was needed. The concept of an International Assessment of Agricultural Knowledge, Science and Technology for Development (IAASTD) was endorsed as a multi-thematic, multi-spatial, multi-temporal intergovernmental process with a multistakeholder Bureau cosponsored by the FAO, the Global Environment Facility (GEF), United Nations Development Programme (UNDP), United Nations Environment Programme (UNEP), United Nations Educational, Scientific and Cultural Organization (UNESCO), the World Bank and World Health Organization (WHO).

The IAASTD's governance structure is a unique hybrid of the Intergovernmental Panel on Climate Change (IPCC) and the nongovernmental Millennium Ecosystem Assessment (MA). The stakeholder composition of the Bureau was agreed at the Intergovernmental Plenary meeting in Nairobi; it is geographically balanced and multistakeholder with 30 government and 30 civil society representatives (NGOs, producer and consumer groups, private sector entities and international organizations) in order to ensure ownership of the process and findings by a range of stakeholders.

About 400 of the world's experts were selected by the Bureau, following nominations by stakeholder groups, to prepare the IAASTD Report (composed of a Global and five Sub-Global assessments). These experts worked in their own capacity and did not represent any particular stakeholder group. Additional individuals, organizations and governments were involved in the peer review process.

The IAASTD development and sustainability goals were endorsed at the first Intergovernmental Plenary and are consistent with a subset of the UN Millennium Development Goals (MDGs): the reduction of hunger and poverty, the improvement of rural livelihoods and human health, and facilitating equitable, socially, environmentally and economically sustainable development. Realizing these goals requires acknowledging the multifunctionality of agriculture: the challenge is to simultaneously meet development and sustainability goals while increasing agricultural production.

Meeting these goals has to be placed in the context of a rapidly changing world of urbanization, growing inequities, human migration, globalization, changing dietary preferences, climate change, environmental degradation, a trend toward biofuels and an increasing population. These conditions are affecting local and global food security and putting pressure on productive capacity and ecosystems. Hence there are unprecedented challenges ahead in providing food within a global trading system where there are other competing uses for agricultural and other natural resources. AKST alone cannot solve these problems, which are caused by complex political and social dynamics, but it can make a major contribution to meeting development and sustainability goals. Never before has it been more important for the world to generate and use AKST.

Given the focus on hunger, poverty and livelihoods, the IAASTD pays special attention to the current situation, issues and potential opportunities to redirect the current AKST system to improve the situation for poor rural people, especially small-scale farmers, rural laborers and others with limited resources. It addresses issues critical to formulating policy and provides information for decision makers confronting conflicting views on contentious issues such as the environmental consequences of productivity increases, environmental and human health impacts of transgenic crops, the consequences of bioenergy development on the environment and on the long-term availability and price of food, and the implications of climate change on agricultural production. The Bureau agreed that the scope of the assessment needed to go beyond the narrow confines of S&T and should encompass other types of relevant knowledge (e.g., knowledge held by agricultural producers, consumers and end users) and that it should also assess the role of institutions, organizations, governance, markets and trade.

The IAASTD is a multidisciplinary and multistakeholder enterprise requiring the use and integration of information, tools and models from different knowledge paradigms including local and traditional knowledge. The IAASTD does not advocate specific policies or practices; it assesses the major issues facing AKST and points towards a range of AKST options for action that meet development and sustainability goals. It is policy relevant, but not policy prescriptive. It integrates scientific information on a range of topics that are critically interlinked, but often addressed independently, i.e., agriculture, poverty, hunger, human health, natural resources, environment, development and innovation. It will enable decision makers to bring a richer base of knowledge to bear on policy and management decisions on issues previously viewed in isolation. Knowledge gained from historical analysis (typically the past 50 years) and an analysis of some future development alternatives to 2050 form the basis for assessing options for action on science and technology, capacity development, institutions and policies, and investments.

The IAASTD is conducted according to an open, transparent, representative and legitimate process; is evidence-based; presents options rather than recommendations; assesses different local, regional and global perspectives; presents different views, acknowledging that there can be more than one interpretation of the same evidence based on different world views; and identifies the key scientific uncertainties and areas on which research could be focused to advance development and sustainability goals.

The IAASTD is composed of a Global assessment and five Sub-Global assessments: Central and West Asia and North Africa – CWANA; East and South Asia and the Pacific – ESAP; Latin America and the Caribbean – LAC; North America and Europe – NAE; and Sub-Saharan Africa – SSA. It (1) assesses the generation, access, dissemination and use of public and private sector AKST in relation to the goals, using local, traditional and formal knowledge; (2) analyzes existing and emerging technologies, practices, policies and institutions and their impact on the goals; (3) provides information for decision makers in different civil society, private and public organizations on options for improving policies, practices, institutional and organizational arrangements to enable AKST to meet the goals; (4) brings together a range of stakeholders (consumers, governments, international agencies and research organizations, NGOs, private sector, producers, the scientific community) involved in the agricultural sector and rural development to share their experiences, views, understanding and vision for the future; and (5) identifies options for future public and private investments in AKST. In addition, the IAASTD will enhance local and regional capacity to design, implement and utilize similar assessments.

In this assessment agriculture is used to include production of food, feed, fuel, fiber and other products and to include all sectors from production of inputs (e.g., seeds and fertilizer) to consumption of products. However, as in all assessments, some topics were covered less extensively than others (e.g., livestock, forestry, fisheries and the agricultural sector of small island countries, and agricultural engineering), largely due to the expertise of the selected authors.

The IAASTD draft Report was subjected to two rounds of peer review by governments, organizations and individuals. These drafts were placed on an open access web site and open to comments by anyone. The authors revised the drafts based on numerous peer review comments, with the assistance of review editors who were responsible for ensuring the comments were appropriately taken into account. One of the most difficult issues authors had to address was criticisms that the report was too negative. In a scientific review based on empirical evidence, this is always a difficult comment to handle, as criteria are needed in order to say whether something is negative or positive. Another difficulty was responding to the conflicting views expressed by reviewers. The difference in views was not surprising given the range of stakeholder interests and perspectives. Thus one of the key findings of the IAASTD is that there are diverse and conflicting interpretations of past and current events, which need to be acknowledged and respected.

The Global and Sub-Global Summaries for Decision Makers and the Executive Summary of the Synthesis Report were approved at an Intergovernmental Plenary in April 2008. The Synthesis Report integrates the key findings from the Global and Sub-Global assessments, and focuses on eight Bureau-approved topics: bioenergy; biotechnology; climate change; human health; natural resource management; traditional knowledge and community based innovation; trade and markets; and women in agriculture.

The IAASTD builds on and adds value to a number of recent assessments and reports that have provided valuable information relevant to the agricultural sector, but have not specifically focused on the future role of AKST, the institutional dimensions and the multifunctionality of agriculture. These include: FAO State of Food Insecurity in the World (yearly); InterAcademy Council Report: Realizing the Promise and Potential of African Agriculture (2004); UN Millennium Project Task Force on Hunger (2005); Millennium Ecosystem Assessment (2005); CGIAR Science Council Strategy and Priority Setting Exercise (2006); Comprehensive Assessment of Water Management in Agriculture: Guiding Policy Investments in Water, Food, Livelihoods and Environment (2007); Intergovernmental Panel on Climate Change Reports (2001 and 2007); UNEP Fourth Global Environmental Outlook (2007); World Bank World Development Report: Agriculture for Development (2008); IFPRI Global Hunger Indices (yearly); and World Bank Internal Report of Investments in SSA (2007).

Financial support was provided to the IAASTD by the cosponsoring agencies, the governments of Australia, Canada, Finland, France, Ireland, Sweden, Switzerland, US and UK, and the European Commission. In addition, many organizations have provided in-kind support. The authors and review editors have given freely of their time, largely without compensation.

The Global and Sub-Global Summaries for Decision Makers and the Synthesis Report are written for a range of stakeholders, i.e., government policy makers, private sector, NGOs, producer and consumer groups, international organizations and the scientific community. There are no recommendations, only options for action. The options for action are not prioritized because different options are actionable by different stakeholders, each of whom has a different set of priorities and responsibilities and operates in different socioeconomic and political circumstances.

1

Contextual Realities

Coordinating Lead Authors:
Arturo S. Arganosa (Philippines) and Revathi Balakrishnan (USA)

Lead Authors:
Li Xiande (China) and Fu Qin (China)

Contributing Authors:
Zhu Xiaoman (China), M. Monirul Qader Mirza (Canada)

Review Editors:
Indu Grover (India), Rajendra Shrestha (Nepal)

Key Messages

Key Messages

1. East and South Asia and Pacific (ESAP) has considerable diversity in agroecological zones, which affects the resources available for production systems. Changes in resources have implications for productivity and sustainability of production systems. This region was rich in natural resources and biodiversity but is vulnerable to disasters and faces losing biodiversity. Decreasing farm size poses a major constraint to agricultural productivity and adoption of AKST (agricultural knowledge, science and technology). Wide variations in agriculture demand diverse AKST to ensure productivity and sustainable development.

2. People are the wealth of ESAP. Since this region is home to three of the world's most populous countries, investing in people will yield development dividends. The demographics of the region are changing rapidly, due to a remarkable decline in fertility and increase in life expectancy. Developing countries in the region have a relatively large youthful population, with labor supply advantages compared to Organisation for Economic Cooperation and Development (OECD) countries with rapidly aging populations.

3. Urbanization in the region has accelerated in the last five decades, with implications for demand for food and a significant rural population lacking basic services and education demanding attention on rural development. The agricultural labor force is changing, with increased participation of mostly underpaid and unpaid females, children and family. The region has made significant gains in education and literacy. But educational attainment has been marked by gender and rural disparities, with uneven gains in human capital. The region also has had high international and internal migration, leading to labor flight but with remarkable growth in remittances received.

4. Human welfare in the region has improved overall, but South Asia continues to have a high concentration of poverty and poor nutrition. Significant disparities in well-being have been observed between urban and rural areas. Rural communities have experienced increasingly greater poverty, with many women among the rural poor. Persisting poverty and food insecurity within ESAP developing countries require public assistance programs to provide safety nets for the marginalized population. Two current threats to human well-being in the region are HIV and AIDS and the highly pathogenic avian influenza, both of which have adverse effects on the rural economy.

5. Stringent trade barriers adopted by industrial countries have constrained international trade in the region. The AKST system has been challenged by the task of assisting farmers in ESAP to adopt good farming practices and improve the quality of exportable produce and commodities to overcome import constraints, such as food safety standards. As multinational negotiations, such as with the World Trade Organization (WTO), have achieved little progress, regional free trade agreements have been promoted to develop regional trade blocks and strengthen intraregional trade, like the free trade agreement between China and the Association of Southeast Asian Nations (ASEAN).

6. Domestic trade in agriculture has played a sizeable role in national economies of the region, although it has not often been explicitly addressed in discussions on the effects of trade on agriculture. As urbanization increases and economic conditions improve, the demand for high-quality and high-value agriculture commodities should expand. This domestic demand would affect trade in agriculture.

1.1 IAASTD Framework

An assessment is a critical, objective evaluation and analysis of information, including local knowledge, designed to meet user needs and support decision making. It is an application of experts' judgment in providing scientific answers to policy questions, quantifying the level of confidence wherever possible.

Agriculture in this report is defined broadly to include crops, livestock and pastoralism, fisheries, biomass, and agricultural goods and services, and land management, such as forestry and agroforestry.

Variations in grouping of countries adopted by different UN agencies under ESAP affect using United Nations (UN) data to arrive at generalizations on regional trends. The countries that make up ESAP (Table 1-1) are different in size, geography, agroecological systems, production systems, culture, religion and political systems, economic performance and social development. The complex regional realities are shaped by historical trends, agroecological environments, farming practices, contradictions surrounding agriculture trade and aid to farmers, and investment in agriculture knowledge, science and technology. Collectively these affect AKST generation and application with significant variation in processes and outcomes in achieving the goals of development and sustainability in ESAP countries.

The conceptual framework (Figure 1-1) provides guidance on common concepts and terminology and enables systematic analysis and appraisal of the primary goals of the assessment. It illustrates links among several components and the process, methods and tools for addressing them. Components include direct drivers of change: availability and management of natural resources, climate change, labor, energy and AKST use; and indirect drivers: economic change, demographic change, changes in level and availability of education, sociopolitical changes, changes in infrastructure, agricultural knowledge, science and technology. The assessment focuses on interactions among the drivers to understand how to facilitate development and sustainability goals.

1.2 ESAP Agroecological Production Systems

The ESAP region covers South and East Asia and the Pacific. The countries are diverse in population, size, economy and agroecological zones and the resource base varies. This resource base, among and within countries, determines the prevalent production system. Under each production system, crops, livestock, fisheries, forestry or in any combination, there is a set of appropriate AKST, which may come

Table 1-1. *East and South Asia and Pacific (ESAP) countries*

East Asia	South Asia	Pacific
Brunei Darussalam	Bangladesh	Australia
Cambodia	Bhutan	Fiji
China, People's Republic of	India	Kiribati
Indonesia	Maldives, Republic of	Marshall Islands, Republic of
Japan	Nepal	Micronesia, Federated States of
Korea, Republic of	Sri Lanka	New Zealand
Lao, People's Democratic Republic		Palau, Republic of
Malaysia		Papua New Guinea
Mongolia		Samoa
Myanmar		Solomon Islands
Philippines		Tonga
Singapore, Republic of		Vanuatu
Thailand		
Timor-Leste		
Democratic Republic of Vietnam		

from external or internal sources—including traditional ones.

1.2.1 Agroecology, climate and natural resources

The agroecological zones in ESAP countries range from warm arid tropics to cool subtropics. This diversity is important because agricultural production systems are sensitive to local climate, soil and other biophysical attributes, making them less transferable (Pardey et al., 2006). The agroecological zone determines the vegetation and the length of the growing period. ESAP has eight major agroecological zone classifications.

ESAP has very divergent climatic zones, from temperate to arid. Monsoon, the region's most important climate feature, is the wind system that dominates the climate of South Asia and the area around the Indian Ocean. Differential heating and cooling of landmass and oceans between summer and winter creates seasonal reversals of direction. The wind blows from the northeast, toward the sea, in winter—the dry monsoon—and from the southwest, toward the land, in summer—the wet monsoon (Banglapedia, 2006).

Annual rainfall varies from over 10,000 mm in parts of the Central Highlands of Papua New Guinea to almost zero in the Gobi and Australian deserts. Likewise, across

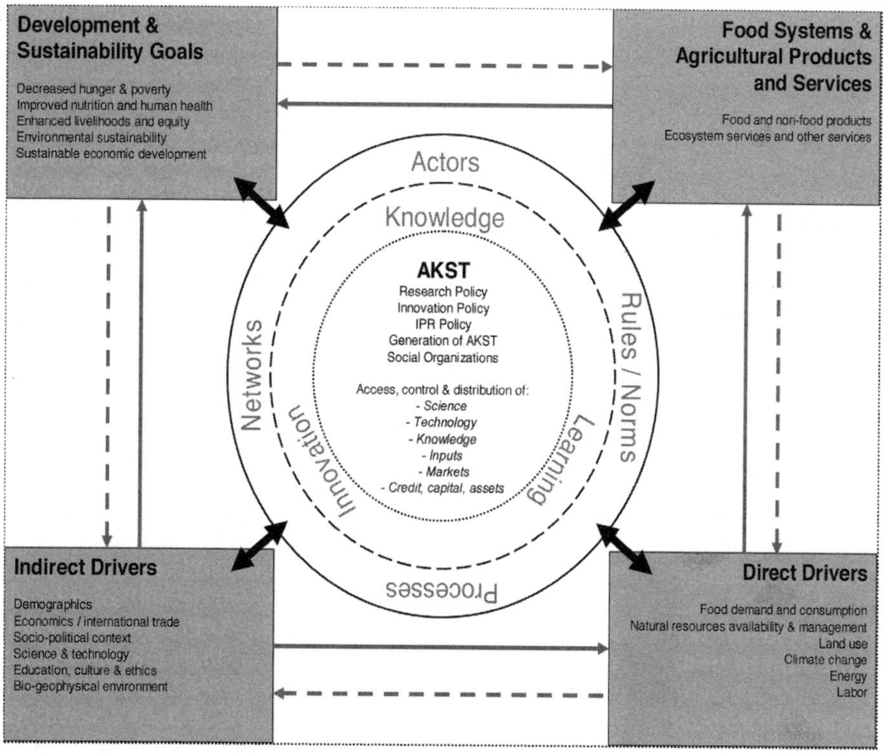

Figure 1-1. *Conceptual framework for IAASTD.*

the region there is considerable variation in recorded temperatures. During the winter months in Mongolia the temperature commonly falls below zero, whereas in the summer months in the arid regions of Pakistan and Australia daytime temperatures can rise above 50°C.

Rainfed agriculture is restricted in many countries to that period coinciding with the monsoon. However, in some of the more humid parts of the region rainfall occurs throughout the year, giving a 12-month growing season. In the northern and high-altitude parts of the region the length of growing season is curtailed by low temperatures, below 5°C. Tropical cyclones and typhoons are a feature in much of the region and result in heavy downpours, with the risk of high runoff and flooding. The worst effects of the strong winds, tidal surges and heavy rainfall are mostly felt in coastal and island areas; the influence of some cyclones may extend into the interior of the Asian continent. Islands in the Pacific and the Philippine archipelago are especially vulnerable. The smallest islands cannot deflect typhoons and cyclones and are not large enough to moderate climatic circulation patterns. They are vulnerable to drought and other climate events, which can destroy complete ecosystems. The effects of climate variability and change on agriculture are projected to steadily manifest directly in changes in land and water regimes. Changes in the frequency and intensity of drought and flooding and the amount of storm damage are expected. Climate change is expected to result in long-term water and resource shortages, worsening soil conditions, drought and desertification, disease and pest outbreaks on crops and livestock and sea level rise. Vulnerable areas are expected to experience losses in agricultural productivity, primarily from reduction in crop yields (Rosenzweig et al., 2002).

In contrast, climate change is also expected to result in some beneficial effects, particularly in temperate regions. Lengthening growing seasons, carbon fertilization and improved conditions for crop growth are forecast and should stimulate gains in agricultural productivity in high-altitude regions, such as Northern China and Mongolia (Mendelsohn et al., 2004).

In ESAP agriculture land is the primary resource. Land categories are arable permanent crops, permanent pasture, and forests and woodlands. East Asia has the biggest total land area among the major regions in Asia, with more than 1 billion ha, followed by Southeast Asia and South Asia. China leads all the ESAP countries with a 932,743,000 ha, followed by Australia, India and Indonesia. Countries in the Pacific Islands and Maldives have the least land area.

In arable and permanent croplands, South Asia has the largest area because India tops all ESAP countries, with 202,835,000 ha. East Asia comes next with 160,796,000 ha, followed by Southeast Asia with 95,361,000 ha. Industrial countries Australia, Japan and New Zealand have a combined 56,043,000 ha of arable and permanent cropland. China has 400,001,000 ha and Mongolia 129,300,000 ha permanent pasture, giving East Asia the highest potential for livestock production for ruminants. Australia has 391,565,000 ha of permanent pasture area, which also provides the country the opportunity to produce a lot of ruminants.

1.2.1.1 Forest resources

Forests cover about 25% of Asia and the Pacific. The Pacific Islands, with 65% forest cover, and insular Southeast Asia, with 53%, have the highest proportion of land-user forest. Papua New Guinea has the largest rainforest coverage in the Pacific region and accounts for the third largest block of tropical rainforest in the world (Chatterton et al., 2000). South Asia has relatively little forest cover.

Although ESAP contains only about 5% of the world's forests, it accounts for about 25% of forest loss over the last decade. The Philippines has had the highest rate of deforestation, followed by Pakistan, Thailand and Malaysia. However, the largest losses have occurred in Indonesia and Myanmar (Waggener, 2001). Between 1990 and 2000 the region experienced considerable decline in forest cover, with the greatest decline in the islands of Southeast Asia, followed by continental Southeast Asia and the Pacific Islands (Waggener and Lane, 1997). Forests in the South Pacific are being removed at an unsustainable rate (ESCAP, 2000).

The Asia and Pacific region is also home to the world's greatest concentration of mangroves. Once thought of as coastal wasteland, mangroves have been destroyed at alarming rates for agriculture, aquaculture and firewood. Up to half of mangrove destruction in recent years has been prompted by the desire to create shrimp farms (UN Atlas of the Sea, 2002). Over the last 20 to 30 years, with the help of the UNESCO Mangrove Programme and other international initiatives, government planners and fisheries experts have become more aware of the many roles that mangroves play as a nursery for many coastal and aquaculture fish species, as a key buffer that reduces the impact of sediment flows onto offshore reefs and as a barrier to protect against storm surges and tsunamis (Vannucci, 1997). About 90% of all marine organisms spend some portion of their life cycle within mangrove systems (Adeel and Pomeroy, 2002). Mangroves have some commonality with open access natural forests in management (see 2.2.5).

ESAP has over 552 million ha of forests, of which 477 million ha are natural. However, only about 249 million ha have been available and suitable for harvesting (Waggener, 2001). The natural forests throughout ESAP up until very recently have been seen mostly as a vast natural source of raw timber for export income. However, there is general agreement on the need to change from a focus on timber exploitation to an emphasis on management for sustainable, multiple-use natural forests (Enters, 1997). In the face of increasing deforestation, many countries across ESAP, including China, New Zealand, Philippines, Sri Lanka, Thailand, and Viet Nam, imposed several partial, temporary or selective bans on logging in natural and old growth forests. The results of these restrictions have been mixed and a number of case studies have indicated that bans can have unanticipated effects on timber supply, forest harvesting, transport, processing and consumption of forest products and on forest residents and those who depend on forestry for their livelihoods (Waggener, 2001).

Plantation forestry is another form of management in the region. In 2000, ESAP accounted for 61% of global forest plantations. Five ESAP countries accounted for 55% of the world's forest plantations and 91% within Asia and the

Pacific (Brown and Durst, 2003). This is a rather new phenomenon, with the average age of Asia's industrial plantations less than 15 years (FAO, 2001).

With diminishing availability of large-diameter timber from natural forests in the region, plantation forestry is expected to become the dominant source for wood in ESAP. The region has more than 80% of forest plantations in the tropics. At present, most legally produced industrial wood in the region is sourced from plantations. Most plantation forestry in the region can be described as intensive management of monocultures for producing a relatively narrow range of products and species; the main species are pine, teak, poplar, acacia and eucalyptus (Enters, 1997).

Because of the extent of plantations in China and their short rotation, most of Asia's plantation forests are aged less than 15 years. This has come mostly from a rapid acceleration in plantation establishment in China and the short rotation generally used. This sector has considerable diversity in ownership, management, scale of operation and products. Plantations were established to meet the need for several different products, including fuelwood, poles, wood chips and furniture wood, and various estate crops including rubber, oil palm and coconut. Until 25 years ago, forest plantations were largely smallholder or government operated. The trend now is for increasing private investment and management of forest plantations to meet an increasing demand for wood for pulp, furniture and particleboard. Smallholder plantations have sprung up to meet this market in the Philippines (Garrity and Mercado, 1994; Pasicolan et al., 1997).

Agroforestry has come to mean many different things, but in its simplest form it refers to incorporating and using trees in farming. The focus has been primarily on smallholders. The practice gained widespread attention by government agencies and nongovernmental organizations as a way to address a range of soil conservation objectives and meet livelihood needs. Because of its potential for increased food security, poverty reduction and environmentally sound land management, a CGIAR-supported international research center is now devoted to agroforestry research and development. Agroforestry is defined as a dynamic, ecologically based, natural resources management system that, by integrating trees with other crops and enterprises on farms and in the agricultural landscape, diversifies and sustains production. Tree farms and nut plantations managed as a monocrop are not considered agroforestry (Beetz, 2002).

1.2.1.2 Water resources

Except for Australia and some of the Pacific Islands, ESAP is relatively well endowed with water resources: for a total area with 21% of the world's land surface, it has 28% of its water resources. Water endowments vary widely among the countries. The figure of 2,000 m^3 for each person annually is usually used as an indicator of water scarcity; China was reaching this limit, while India had 1,700 m^3 and the Republic of Korea only 1,450 m^3. For Asia as a whole, about 80% of the water withdrawals are for agriculture—the range is from 95-96% for Bangladesh, Bhutan and Sri Lanka to 50-60% in Japan, Republic of Korea, Malaysia and Viet Nam. For the Pacific, the water withdrawals varied from about 1% in Papua New Guinea to 75-78% in Australia and Fiji.

The hydrology of ESAP is dominated by the monsoon climate, which induces large interseasonal variations in river flows. In this situation, average annual river flow is a poor indicator of the water resources available. In the absence of regulation, most of the water flows during a short season when it is usually less needed. In Bangladesh, for example, the surface flow of the driest month was only 18% of the annual average; in Indonesia, it was 17 percent. In India, the flow distribution of some rivers in the monsoon period represents 75 to 95% of the annual flow. In North China, 70 to 80% of the annual runoff is concentrated between May and September (FAO, 2006a).

Water in shallow underground aquifers has been significant in developing and diversifying agricultural production in the region, particularly in China and India. Groundwater offers a primary buffer against the vagaries of climate and surface water. Because groundwater is available on demand, crops irrigated with it are often more productive than those irrigated with surface water.

1.2.1.3 Aquatic resources

ESAP countries contributed 64% to the total global fishery production in 2004 (FAO, 2007). ESAP fisheries are vital for food security, supplying valuable animal protein, minerals and vitamins. Fisheries generate employment, reduce poverty and earn revenue through domestic and export trade. People use all sizes and types of fish, discarding little. Fisheries products come from two sources: capture of wild fish, shrimp and other aquatic organisms from the sea and inland open water bodies, and aquaculture in fresh water or in brackish or marine waters.

The increase in ESAP fish production in recent years has largely been attributed to the significant development of aquaculture. As opposed to the stagnation or decline in capture fisheries, aquaculture production has increased at a rapid rate. As a result of aquaculture knowledge, science and technology have been constantly generated and refined.

A significant increase in the global human population, reduced supply of food fish, high prices for exportable aquatic species from the open water and increased demand for them has stimulated aquaculture to quickly develop and flourish. Many rural farmers urgently need increased income from their limited and gradually shrinking agricultural landholdings to meet the minimum necessities of life. Farming aquatic organisms is a profitable proposition; this activity has been rapidly gaining importance for producing food, creating employment, reducing poverty and increasing earnings through domestic and export trade.

Within ESAP in 2004, seven countries, China, India, Indonesia, Japan, Philippines, Thailand and Viet Nam, produced the most by volume from aquaculture, including aquatic plants. China alone produced 41,661,660 tonnes, or 78 percent. The next six countries accounted for 17%, the remaining countries 5 percent.

1.2.1.4 Livestock

Millions of rural households in ESAP countries depend on domesticated animals for food, farm power and income. The region is home to 30% of the world's livestock species. Developing Asian countries had the world's highest growth rates of production and consumption of food from livestock (Steinfeld et al., 2006ab). The dynamic Asian livestock sector is growing at a rate between 3.5 and 5% annually—more rapidly than crops such as cereals, vegetables and pulses—driven partly by increasing population, rising incomes and changes in consumer lifestyles.

Livestock production in Asia and the Pacific grew rapidly from 1992-1994 to 2002-2004, with the most rapid growth occurring in China, 93%; Viet Nam, 93%; and the Philippines, 79% (FAO, 2006b). For the region as a whole, all categories of livestock products grew rapidly. The most rapid production growth was in poultry, 83% and eggs, 78%. Rapid growth in poultry and egg production was widespread throughout the region. Production of milk exhibited strong growth in East Asia, 136%; Southeast Asia, 65%; South Southwest Asia, 52%; and the industrial economies, 33%. Pork production grew 50% for the region, with strong growth in Southeast Asia, 55%; East Asia, 53%; and the Pacific Islands, 44%. Although most ESAP countries were technically capable of increasing production in meat, milk and eggs, most faced shortages of key feed ingredients. As a result, there was a large and burgeoning world trade in feed crops. On the other hand, the drive by livestock growers to serve urban markets led to intensive production, with problems of livestock waste, land management and distribution of meat products. Awareness increased of the potential for transmitting diseases from animals to humans, particularly with the bird flu, or avian influenza, crisis. Diseases affecting animals and humans could spread rapidly across the region, creating transboundary epidemics. Concerns remained about the rising demand for livestock feed, increased need for veterinary services and training, loss of genetic resources and need for extension for profitable livestock opportunities for small-scale producers (FAO, 2006b). More than half the small-scale farmers in Asia rely on livestock as a major source of income and nutrition. However, small-scale producers have mostly not been a part of the rapid rise in intensive animal production (FAO, 2006b).

In population, the figures from 1994 to 2004 for all the livestock and poultry species showed no definite trends. Ruminant numbers seemed not to have increased, except goats, which increased by almost 100 million over ten years. The number of buffalo declined by more than 10 million in ten years. Pigs and poultry, nonruminants, increased, with chickens increasing by almost 10 billion from 1994 to 2004. Growth in production of poultry and pork resulted in a growing shift away from pasture systems. As livestock production became more intensified, feed shifted from locally available resources to commercial feed concentrates, particularly in pig and poultry production (Steinfield et al., 2006b).

1.2.1.5 *Plant biodiversity*

ESAP encompasses parts of three of the world's eight biogeographic realms and includes the world's highest mountain system, the second largest rainforest complex and more than half the world's coral reefs. The rainforests of Southeast Asia contain more than 25,000 species of flowering plants, equivalent to about 10% of the flora of the world. The region as a whole encompasses two-thirds of the world's flora. Almost all nations in the region, except Singapore and Brunei Darussalam, depend heavily on harvesting natural products directly.

Flora and fauna of the region are increasingly threatened, but only a few countries have designated more than 15% of their land as protected areas. The drive for increased agricultural production has resulted in loss of genetic diversity. The area of land under rice cultivation rose by only 25% between 1960 and 1970, although production rose by 77% by replacing traditional varieties with higher-yielding, semidwarf varieties. More than 100,000 varieties of rice were found in Asia early in the 20th century. In 2002 there were less than a dozen modern rice varieties being planted on 70% of land being cultivated for rice (Dano, 2002). In Indonesia, 1,500 varieties of rice disappeared from 1975 to 1990 (see subchapter 2.4.1).

The Indo-West Pacific is the key area for shallow-water marine biodiversity. Coastal habitat loss and degradation, combined with increased sediment, nutrient and pollutant discharge into coastal areas, is a major cause of concern, particularly for the island countries. The rates of loss of coral reef and mangrove habitat in this region are among the highest in the world. Thailand alone lost about 0.2 million ha of mangrove forest from 1961 to 1993. Conversion of mangrove forest to shrimp aquaculture and the use of unsustainable fishing practices, such as blast fishing, were widespread. However, the effects of such unsustainable practices on regional biodiversity are difficult to quantify.

Although terrestrial biodiversity loss has been a major concern, actual losses still have to be quantified. As much as 70% of major vegetation types in Indo-Malaya have been lost, with a possible associated loss of up to 15% of terrestrial species. Dry forests suffered 73% loss and moist forests 69%, while wetlands, marsh and mangroves were reduced by 55%. Overall habitat losses were most acute in the countries of the Indian subcontinent, the People's Republic of China, Thailand and Viet Nam (ESCAP, 1995b).

The underlying causes of biological diversity loss in the region include international trade, particularly the trade in timber, which results in habitat loss; population growth, leading to accelerated rates of change in land use; poverty and demand for common access resources, leading to their unsustainable consumption; introduction of nonnative species, leading to destruction of predator and prey equilibrium; and improper use of agrochemicals, leading to loss of aquatic species. Other major reasons include loss of keystone species, extensive deforestation and habitat loss, increased trafficking in animals and animal body parts, widespread conversion of land to agriculture and construction of large-scale dams.

In response, national governments have implemented conventions related to biodiversity and are taking measures to protect biologically rich areas. Twenty-nine ESAP countries had ratified the Convention on Biological Diversity by 1 May 1996. Several regional conventions covering parts of ESAP dealt with specific aspects of biological diversity; the most significant were the Convention on Conservation of Nature in the South Pacific (Apia Convention), the ASEAN Agreement on the Conservation of Nature and Natural Resources (ASEAN Agreement), and the Convention on the Protection of the Natural Resources and the Environment of the South Pacific (SPREP Convention).

Progress in designating protected areas has generally been positive. It is clear that almost all countries in the region understand the importance of establishing terrestrial and aquatic natural reserves by creating national parks, wildlife sanctuaries and gene pool reserves. The number and area of protected areas in both South and Southeastern Asia has increased dramatically. The Pacific region has also shown a major increase in the number of protected areas, although the increases have been less dramatic.

Biological diversity has finally been accepted as a legitimate issue nationally and internationally in ESAP, with conventions on biological diversity and designation of protected areas. However, patterns of unsustainable use and conflicting policies contribute to continued losses throughout the region. With only 10 to 30% of natural habitats remaining in many countries, any further decrease will have serious consequences for biodiversity (ESCAP, 1995a). High rates of population and economic growth in most countries of the region suggest even greater losses will occur, unless decisive action is taken. Such action could include intensifying protected-area systems and zoological parks, botanical gardens, gene resource centers, seed banks and tissue culture techniques.

1.2.2 Production potential

1.2.2.1 Farm size

In general, the diversity in concepts used to define small farms makes definition difficult. By applying a common approach of size of landholdings or livestock numbers the overwhelming majority of these farms (87%) were in Asia. In Asia, China alone had almost half of the world's small farms, followed by India with 23%. Other leaders in the region, in descending order include Indonesia, Bangladesh and Viet Nam. Despite steady economic growth in many Asian countries over the decades, small farms still dominate in rural areas (Nagayets, 2005). Small farms characterize agriculture in Asia and small Pacific Island countries, while extremely large farms dominate in Australia. In wealthier countries such as Japan and the Republic of Korea, average farm size has been increasing, but at a slow pace. For example, between 1956 and 2003, average farm size in Japan increased just 0.60 ha. The increase in the Republic of Korea from 1969 to 2002 was 0.58 ha (Fan and Chan-Kang, 2003). In contrast, national average farm size is still decreasing in most Asian developing countries. For example, average farm size in Nepal decreased from 0.95 ha in 1992 to 0.79 ha in 2002. Similar trends occurred in Pakistan and the Philippines during the 1990s (FAO, 2006b). In India research demonstrated an association between decreasing farm size and more hunger and poverty. The study documented that the incidence of hunger among farmers with landholdings less than 0.5 ha was 32% and the incidence of poverty was 38%; the likelihood of being affected by hunger dropped to 12% and poverty dropped to 13% for farmers who cultivated more than 4 ha (Singh, 2004). But farms are becoming larger in dynamic agricultural areas close to large cities, such as Suphan Buri province near Bangkok, Thailand. From 1993 to 2003, total agricultural land in Suphan Buri declined, but the number of agricultural households declined even more rapidly. Families migrated to Bangkok or assumed nonagricultural rural jobs. Active land rental markets have been important in the land consolidation in Suphan Buri (Dawe, 2005).

Urbanization created pressure on maintaining agricultural land and production. In 2005, the net loss of arable land was 361,600 ha, about 0.3%, of which 138,700 ha was used for construction. From 1998 to 2005, farmland decreased by 7.6 million ha, about 6.2% of the total amount of arable land. The per capita area of cropland in China was only 0.93 ha in 2005, 40% of the world average. To achieve a higher production rate from the small remaining area of cultivatable land, China became the world's largest consumer of fertilizers and the second largest of pesticides. Consequently, much cultivated land and farm produce have been contaminated, especially with pesticide residues (Fu et al., 2007).

1.2.2.2 Farming systems

Within the diverse agroecological systems and variations in natural resources, the region has developed unique farming systems. Rice–wheat and rainfed mixed farming cover about half of the land in South Asia. Rice–wheat farming is characterized by a summer paddy crop followed by an irrigated winter wheat crop, sometimes with a short spring vegetable crop. Rice–wheat farming covers a broad swathe across India and Pakistan, from the Indus irrigation area in Sindh and Punjab and across the Indo-Gangetic plain to the northeast of Bangladesh. About 60% of rice–wheat land is cultivated, about three-quarters irrigated. The system integrates crops and livestock significantly; about 119 million cattle are used for draft power, milk and manure for composting. About 73 million small ruminants are kept, principally for meat. The area has 484 million people, 254 million in agriculture.

Rainfed mixed farming covers the largest area within the subcontinent and, with the exception of a small area in northern Sri Lanka, is confined to India. This system covers 147 million ha, with about 59% under cultivation. Rice, wheat, pearl millet, sorghum, a wide variety of pulses, many oilseeds, sugarcane, vegetables and fruit are grown. About 16% of the cultivated area is irrigated. About 126 million bovines and 64 million small ruminants are partially integrated with cropping. In many instances, relatively small areas are irrigated from reservoirs. In recent decades, tube wells have contributed to stable cereal production. Vulnerability stems from substantial climatic and economic variability. Poverty is extensive and its severity increases markedly after droughts.

Three farming systems predominate in the land area in East Asia and the Pacific: upland intensive mixed 19%, pastoral 20%, arid 20%. These can be further classified, depending on the production systems.

Upland intensive mixed farming is found in uplands and hills of moderate altitude and slope in humid and subhumid agroecological zones. The total area of the system is 314 million ha, with an agricultural population of 310 million—the second most populous system, after lowland rice, in the region. The cultivated area is 75 million ha, of which less than one-quarter is irrigated. This is the most widespread and most heterogeneous farming system in the region, including some remnant shifting cultivation, with major areas in all countries of East and Southeast Asia. The system is the cultivation of a wide range of mostly permanent crops, but the specific crops preferred depend on geography, climate, slope, terracing and water regime. A significant crop area, mainly rice, is irrigated from local streams and rivers. Livestock production is important in most farm livelihoods. The area has 52 million large and 49 million small ruminants. Livestock contribute draft power, meat, cash income and savings. Off-farm work is an important source of income for many poor households.

Pastoral farming is found in semiarid and arid temperate plains and hills, with fewer than 120 growing days annually. The system is extensive in western China and much of central and northern Mongolia. It covers 321 million ha but has no more than 42 million agricultural people. The cultivated area is just over 12 million ha, with about 20% irrigated in dispersed zones. The system is dominated by transhumant pastoralism, characterized by mixed herds of camels, cattle, sheep and goats extensively grazing native pasture. Irrigated crops include cotton, barley, wheat, pulses, peas, broad beans, potatoes and grapes; sericulture is sometimes practiced. Severe poverty, often triggered by drought or severe winters, with consequent loss of livestock, is common in both pastoral and irrigated areas.

The area of arid farming in western China and southern Mongolia covers about 322 million ha, supporting about 9 million cattle and 59 million small ruminants. Only a little over 1%, less than 4 million ha, is cultivated, of which about two-thirds are irrigated. Some large-scale irrigation is concentrated in the west; pastoralists use scattered, small-area irrigation to supplement their livelihoods. The area has about 24 million people, 17 million of whom are pastoral or agricultural. Apart from these arable areas, the dominant arid areas are used for opportunistic grazing. Poverty is extensive and, especially after droughts, severe.

Except for Australia, most nations in the Pacific are relatively small islands and atolls. On the small islands as on most other small islands, traditional agriculture is agroforestry, where trees are planted and protected for their great variety of functions and products, including food. Food or fruit trees and shrubs are most common in permanent village tree groves and intercropped in home gardens. They included a wide range of coconut palms, banana and plantain cultivars, breadfruit, edible pandanus (screw pine) varieties (especially on atolls), fruit trees, nut and seed trees, and kava (a root used for a traditional alkaloid social beverage). Most of these species are aboriginal, pre-European introductions, but some are indigenous.

Atoll islands have among the most infertile soils in the world and almost no surface freshwater sources. Despite inadequate land, soil and water and relatively high populations, atoll societies have developed sophisticated subsistence agroforestry systems based on coconut, breadfruit, pandanus (screw pine), native fig, bananas (on the wetter islands) and giant swamp taro. This pit cultivation uses leaves of salt-tolerant coastal trees and plants as mulch and fertilizer. It is also used for important staple tree crops to ensure their survival in the atoll.

1.2.3 Production constraints

ESAP has rich and diverse natural resources and has assimilated agricultural science and technology to achieve remarkable agricultural productivity, although many production constraints have presented risks.

1.2.3.1 Degradation of natural resources

Environmental degradation can increase the impact of floods and landslides, just as disasters such as wildfires, droughts and floods can cause serious damage to forests, farmland and livestock. Small-scale measures to increase environmental resilience include social forestry, fish farming, drought-resistant crops and rainwater harvesting. In India, local knowledge of indigenous, hardy seeds has helped farmers recover from the loss of cash crops devastated by drought and pests (IFRCRC, 2004).

Overextraction of groundwater can result in water declining beyond the economic reach of pumping technology. Groundwater depletion is a widespread problem in many areas in the region, especially in the semiarid areas. Poorer farmers are hit the most. When near the sea or in proximity to saline groundwater, overpumped aquifers are prone to saline intrusion. Groundwater quality is also threatened by the application of fertilizers, herbicides and pesticides that percolate into aquifers. These nonpoint sources of pollution from agricultural activity often take time to become apparent, but their effects can be long lasting, particularly with persistent organic pollutants.

Capture fisheries stagnated or dwindled in most ESAP countries and other world regions. Historically, the vast sea and the lakes, rivers and canals were rich sources of fish. As the human population increased, fish and other fisheries organisms have been heavily exploited for human food. In addition, fishery products have been used as industrial raw materials for producing fish meal.

Unscrupulous application of technology eventually resulted in overfishing and depletion of ocean fish stocks. Despite caution from scientists, many rich marine fishing grounds all over the world have been excessively exploited for years. Aquatic habitat change or destruction from massive construction of embankments for flood control, drainage, irrigation, temporary damming of rivers, excessive surface water withdrawal, aquatic pollution from pesticides, indiscriminate release of industrial effluent and unplanned construction of rural roads and culverts that obstructed fish movement have all contributed to the destruction of fisheries.

1.2.3.2 Natural hazards

Natural disasters are grouped in three specific categories: hydrometeorological disasters, including floods, wave surges, storms, droughts, extreme temperatures, forest and scrub fires, landslides and avalanches; geophysical disasters, divided into earthquakes, tsunamis and volcanic eruptions; and biological disasters, covering epidemics and insect infestations. ESAP suffers frequent natural disasters with considerable human and economic loss. The most recent and dramatic natural disaster, which caught the world's attention and empathy, was the 2004 tsunami. Since 2000, the region has suffered major earthquakes, floods, tsunami and pestilence. "Both hydrometeorological and geophysical disasters have become more common, becoming respectively 68 and 62% more frequent over the decade. This reflects longer-term trends. However, weather-related disasters still outnumber geophysical disasters by nine to one over the past decade. Among natural disasters, floods are the most reported events in Africa, Asia and Europe, while windstorms are most frequent in the Americas and Oceania" (IFRCRC, 2004). Among the top 50 countries with major economic loss from natural disasters are 14 countries from ESAP, with Japan ranked second, China third, India sixth and Indonesia eighth (Table 1-2) (Guha-Sapir et al., 2004).

Frequent disasters make agriculture and land-based production in ESAP a high-risk venture. The livelihoods of communities dependent on agriculture and natural resources and with limited diversification are vulnerable. Landslides across the southern Philippines in December 2003 killed 200 people and left thousands homeless, reigniting the disaster prevention debate. From 1971 to 2000, natural disasters killed 34,000 Filipinos. From 1990 to 2000, 35 million people were severely affected by natural disasters (IFRCRC, 2004). A windstorm in 2002 led to considerable land and crop loss, affecting 100 million people in China (Guha-Sapir et al., 2004). For many countries in South and East Asia floods have become annual, alternating with drought. In the Pacific, cyclones present constant threats to livelihoods. In Thailand, the 2004 tsunami had a devastating effect on the livelihoods of villagers in over 400 fishing and faming communities along the Andaman coast. Many of the communities' livelihood assets were lost (FAO, 2006d). Lost livelihoods and basic productive assets were similar in other countries affected by the tsunami, such as India, Indonesia, the Maldives and Sri Lanka. Since 2004 Indonesia has been affected by many disasters—tsunami, avian influenza, volcanic eruption, haze and floods—that have taxed the capacity of government to manage disaster and tested people's resilience.

Disasters are further obstacles to overcome in trying to reduce poverty and achieve sustainability. In the region, increasingly emphasis has been placed on early warning systems for disaster, information access for local disaster-prone communities, community approaches in disaster management and risk reduction, on exploration of strategies to improve agriculture extension, and on local government support for community approaches.

1.2.3.3 Pests and pathogens

ESAP agricultural communities, as those in every other region, face risk to productivity from pests and crop and from livestock diseases. The region is recognized for its integrated pest management programs with community participation and farmer field school training methods. Yet in recent days the region has been the focus of global attention because of avian influenza.

Highly pathogenic avian influenza. Since 2004, the highly pathogenic avian influenza epidemic presents a high risk to small-scale farmers in ESAP who practice mixed farming, combining crops and livestock. The emergency officially began in December 2003, when a highly contagious avian influenza struck chickens on a farm near Seoul, Republic of Korea, and spread rapidly across the country. Within weeks, simultaneous outbreaks in Cambodia, China, Indonesia, Japan, Lao PDR, Thailand and Viet Nam had devastated domestic fowl. The impact has been distributed within the entire poultry market chain, affecting producers, consumers and employees in the retail industry. In some areas, farmers lost more than half their poultry (FAO, 2005a).

Table 1-2. *Incidence of natural disasters and reported economic damage: 1974-2003.*

UN region and subregion	Total number of natural disasters with economic damage reported	Economic damages reported (2003 US$ millions)
	1974-2003	1974-2003
	Asia	
Eastern	302	425,502
South-Central	217	102,353
South-East	306	43,867
Western	48	68,015
Total	873	639,736
	Oceania	
Australia–New Zealand	120	38,382
Melanesia	32	1,940
Micronesia	8	1,416
Polynesia	22	1,489
Total	182	43,227

Source: Guha-Sapir et al., 2004.

1.3 Demographics

People in ESAP are both producers and consumers of AKST. Only a few population indicators with immense and immediate implications for AKST were explored: male and female population, aging of population, urban and rural population trend, agriculture labor disaggregated by male and female workers, child labor in agriculture, unpaid work in farming, literacy and education among men and women, migration realties and contributions of migrants to capital.

1.3.1 Regional demographic trends

People are the wealth of East and South Asia and the Pacific. The region encompasses three of the world's most populous countries and developing countries that have a relatively large youthful population. China, India, Indonesia, Bangladesh and Japan are among the top ten in population size (U.S. Census Bureau, 2008). From 2000 to 2005, three countries in ESAP were among the six countries in the world that had half the world's estimated 77 million annual increase in population. These countries and their rate of increase were India, 21%; China, 12%; and Bangladesh, about 4%. India is expected to overtake China as the most populous country in the world by 2035 (ECOSOC, 2004).

The population of the Pacific Islands reached about 8.6 million in 2004, an increase of approximately 1.7 million people over the past ten years. Population distribution remained largely unchanged: the five largest countries and territories that comprise Melanesia had the vast majority, 86.4%, of the regional population, followed by much of the smaller island countries and territories of Polynesia, 7.4%, and Micronesia, 6.2%. Two out of every three Pacific Islanders live in Papua New Guinea. Fiji's population is 25% larger than ten Polynesian island countries and territories combined. The fertility rate in the Pacific Islands is still moderately high, while mortality is declining, contributing to increased population (Haberkorn, 2004).

In ESAP, since people are the fundamental resource for sustainable development, investment in people would bear development dividends. Human resource-centered strategies present opportunity for sustainable development but also present enormous challenges to ensure equitable access to education, productive assets, goods and services to the billions of people. Transforming a large reserve of human resources to human capital and driving development will be the core challenge for achieving development with social sustainability.

In Asia fertility declined remarkably. The average number of children born to Asian women declined by more than half, from 5.4 in 1970 to 2.4 in 2003. Average life expectancy of Asian men and women increased about 15 years over the same period. Life expectancy for males increased from 52 years in 1970 to 66 years in 2003; for females, from 54 to 70 years (Hugo, 2005), overtaking men's life expectancy in nearly every country. In some Asian countries, however, girls were more likely than boys to die during early childhood and in others an unusual preponderance of male births pointed to sex selectiveness (Westley, 2002). Between 1950 and 2005 in most of the region's countries women gained and improved the sex ratio trend of the number of males per 100 females. Sex ratio also indicates gender equity by reflecting women's chances of survival. The population sex ratio improved either with decrease in female to male difference or with female gains over males. A few exceptions were Brunei Darussalam, India, Samoa and Tonga (UNDESA, 2004).

The region recorded increases in its aged population and female-headed households. The elderly population grew rapidly, in both numbers and percentage. The aging population proportion in industrial ESAP countries was greater than in the less industrial ones. Asia is one of the world's fastest aging regions; the percentage of elderly is projected to double between 2000 and 2030, but with differences among the countries (Kaneda, 2006). Industrial economies Australia, Japan and New Zealand had a rapid rate of aging; by 2050, 25% of their population will be over 60. From 1950 to 2005, all but a few countries in the region, Bangladesh, Maldives, Nepal and Papua New Guinea, increased their population aged over 60 years (UNDESA, 2004). An aging population challenges productivity and innovation in agriculture, and the potential for saving and investment. It increases poverty among the rural elderly.

The decrease in fertility and the aging population in mostly industrial countries in the region contrasts with a growing youthful population in developing countries. The outlook for the future in Asia is that the youth population will increase to 685 million by 2040, when they will comprise 14%. While the young adult population will continue to grow over the next two decades in developing countries, their numbers will decrease in most OECD nations in the region (Hugo, 2005). Yet while a large youth population presents developing countries with a labor pool advantage, the lack of appropriate skills will form a barrier to using human resources effectively.

1.3.2 Accelerated urbanization with a significant rural population

Since 1950, Australia, China, Fiji, Indonesia, Japan, Korea DPR, the Republic of Korea and the Philippines have lost rural population. This has applied to most industrial countries in the region. In most developing countries, however, the urban population is less than 50% of the total. Countries that depend on agriculture as the economic driver have an urban population of less than 30%; these include Bangladesh, Bhutan, Cambodia, India, Lao PDR, Nepal, Papua New Guinea, Samoa, Solomon Islands, Sri Lanka, Vanuatu and Viet Nam. The projection for China is that the urban population will be 60% by 2030 (UNDESA, 2003).

While the decrease in rural population will be minimal in Asia, the Pacific will gain rural population between 2010 and 2030 (UNDESA, 2004). The Asian Development Bank estimates there will be 2.2 billion rural Asians by 2020 and that this population will have much lower access to health and education and have less general well-being (ADB, 2000). By 2030, this region still will have a substantial rural population, demanding attention to agriculture, rural livelihood strategies and investment in rural physical and social service infrastructure.

Asia is expected to experience rapid urbanization from 2005 to 2030; by 2030, 55% of Asian inhabitants are projected to live in urban areas. Although economic growth and prices are closely monitored drivers of food demand, demographic changes—urbanization, growth in population and changes population age—likely will have more profound long-term effects on the region's food system. It will be affected by migration, the aging population and urban demand for a more varied diet, with a premium on convenience (Coyle et al., 2004).

1.3.3 Agricultural labor: Feminization, child labor and unpaid work

The overall share of agricultural employment decreased between 1995 and 2005 from 44.4 to 40.1%. This decline was seen in all regions, except East Asia, where the share in agriculture remained stable. With a few exceptions, from 1979 to 2002 the percentage of agricultural labor in the total labor force decreased (Figure 1-2). The decline was remarkable among the wealth creators, such as Japan and Republic of Korea, Australia and New Zealand. For poorer wealth producers, such as Bangladesh, Bhutan, Cambodia, India, Lao PDR, Nepal, Papua New Guinea, and the Solomon Islands, however, agriculture employed a large proportion of people and the rate of decrease was less. Thailand and China still illustrated the dominance of agriculture in employment, although they were high-growth countries. In general, for the poorer countries in the region, agriculture continues to be important for employment and livelihoods.

The World Employment Report for 2004/2005 contended that rural nonfarm activities were important for household income; this also applied to poor households engaged in agriculture (ILO, 2004). In Asia, various estimates suggested that one-third of rural labor participated

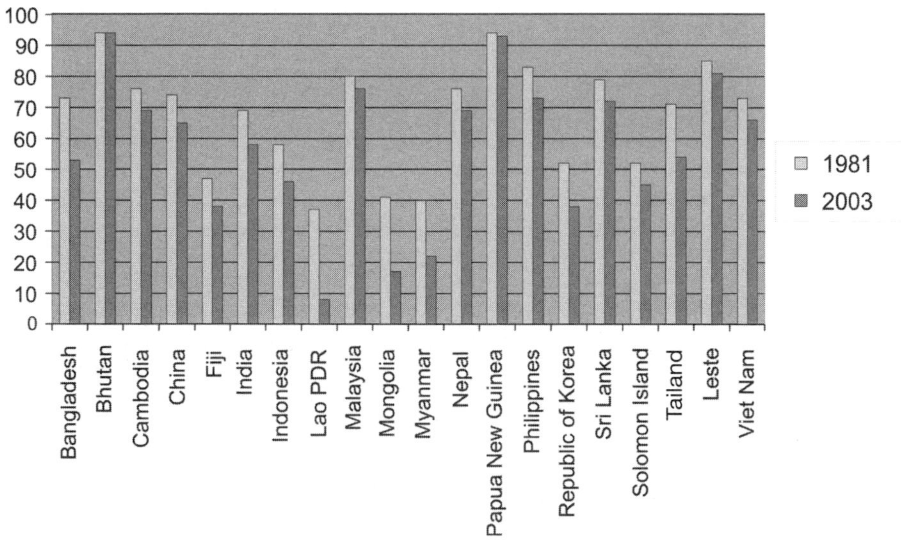

Figure 1-2. *ESAP agriculture labor 1981-2003. Sources: FAO, 2005c; FAO STAT; World Bank, 2006*

in nonfarm activity. When agriculture stagnated, nonfarm employment offered a way out; these workers were pushed into this employment, not pulled by dynamic nonfarm activities (Islam, 1997).

The region's females participated in agriculture at 33.1% for South Asia and 47.2% for East Asia and the Pacific (ILO, 2006). From 1995 to 2002 female labor continued to be important in the region's agriculture labor force. If women's contribution as family workers also was considered, then women were critical in the region's agriculture (UNDP, 2005). In recent decades the debate on agricultural labor in the region focused on "feminization of agriculture," the predominance of women in the sector. Evidence has shown that more women than men participate in agriculture. Indeed, women's participation in agriculture is substantial and increasing (UNDP, 2004).

Asia was the world's most densely populated region and also had the most child labor, approximately 61% of the total. About one in five children in Asia worked, 21% (ILO, 1998). ESAP had the most child workers in the 5-to-14 age group, about 127 million. Not all these children were classified as child laborers, although they were below the minimum working age. Within ESAP child labor in agriculture was common among boys and girls, and most child laborers live in rural areas (ILO, 2005). Families in poverty consider child labor an asset to improve access to income and food, but a productive, literate and educated asset is lost.

Unpaid women and men contribute much to economic activity but are not credited appropriately in the United Nations System of National Accounts. The unpaid work within the System of National Accounts boundary included unpaid work in a family enterprise or agricultural holding (ESCAP and UNDP, 2003). Estimates of women in the labor force were not comparable internationally because in many countries large numbers of women assist on farms or in other family enterprises without pay and countries differed in the criteria to determine the extent such workers were counted as part of the labor force (World Bank, n.d.). In South Asia, female official employment rates usually were low because of arbitrary definitions. If definitions were revised and all activities for which women are traditionally responsible were incorporated, a huge difference in activity rate would be noted (Mahabub ul Hag Development Centre, 2003). Until and unless the unpaid work by men, women and children is measured as labor contribution to agriculture and rural economic production, the labor contribution of rural households will not be fully accounted.

Linguistic, ethnic and religious diversity is the hallmark of the region. About 70% of the world's more than 250 million indigenous peoples live in Asia and the Pacific. Marginalization and poverty in many indigenous communities are closely linked to being deprived of the ability to lead lives they value (IFAD, 2002a). Ethnic diversity, while enriching the cultural heritage, in recent times has also caused ethnic conflict. Ethnic conflict adversely affects rural productivity and livelihood security (Wanasundera, 2006).

1.3.4 Education: Gender and rural disparities

South and West Asia had countries with a literacy rate of about 60%. Although the East Asia and Pacific region had 91% literacy, the highest rate among developing regions, its large population was home to 17% of illiterate adults worldwide. A considerable difference among regions in literacy gains was evident in UNESCO categories. In all regions, for both adults and youth, female literacy rates were lower than those of males. More remarkable was the poor literacy gain in South and West Asia, which included two of the most populous ESAP countries, India and Bangladesh (UNESCO, 2006). In South and West Asia, on average 93% of boys and 86% of girls of the relevant age were enrolled in primary education. This region had 38% of the world's out-of-school children, 56% girls.

In East Asia and the Pacific, on average 94% of boys and 94% of girls of the relevant age were enrolled in primary education. This region had 9% of the world's out-of-school children, 49% of whom were girls (UNESCO,

2004). Among the small Pacific Island countries, Papua New Guinea has the greatest gender gap.

In part, the high dropout rate in rural schools and among girls should explain the differences. Available data on the rural and urban differentials in adult literacy showed rural and urban disparity. More women than men were illiterate. Rural girls also had a higher school dropout rate. As the importance of agriculture employment decreases, changes in human capital will affect nonagricultural growth. From society's perspective, education provides a more adaptable and productive workforce, able to move with the times and adjust to technological change (Siamwalla, 2001).

In general, the region demonstrated gains in education and literacy, but with intraregional differences; it also showed improved gender parity in education, though gender gaps persist. The uneven educational achievement between genders and rural disparities present risks in transforming the large youth population into productive human capital—a workforce that could improve global competitiveness of the ESAP countries in providing trained labor.

1.3.5 Migration: Labor movements and capital gains

Asia provided half the world's international migrants and most of the international labor migrants; it became the primary source of migrants to most of the world's recipient countries. International migration in Asia in reached an unprecedented scale, diversity and significance, but inadequate data hamper understanding its extent and effect. Migration within the region was from poorer countries to more industrialized countries, seeking employment in agriculture and construction, while highly skilled labor sought employment in wealthier countries across the globe. Most international migration was nonpermanent labor. Movement involved mainly unskilled and semiskilled workers in low-paid, low-status, and dirty, dangerous and difficult (3D) jobs eschewed by local workers in the fast-growing, labor-short nations of Asia and the Middle East (Hugo, 2005). The capital gains through remittances from emigration of females were so high for some Asian governments that female labor export targets were included in government development programs. These migrants went out to improve economic returns, in spite of experiencing social and economic discrimination and personal risk. Sri Lanka had in its expatriate labor force more women than men (IOM, 2005).

Patterns of internal migration varied among countries of Asia, partly from variations in economic and cultural structures (Guest, 2003). Rural-to-urban flows still dominated migration in most Asian countries because of the high rural population. Women were increasingly involved, and their temporary emigration, economically motivated, continued to be important. The rural-to-urban migration in China was obvious. Social factors profoundly changed the system and the society (Asia-Pacific Migration Research Network, n.d.). In China, Southeast Asia and India temporary migration increased. Studies in India indicate that rural households migrated and improved their economic returns, in spite of the risks and family disruptions. Rural women also migrated with adult males or in groups of women (Deshingkar and Grimm, 2005).

Migration in the Pacific Island countries is seen as a way to improve economic and professional opportunities. "The currently widely perceived disparities in economic development and welfare between the Pacific states, especially the smallest countries and territories of Polynesia and Micronesia, and the fringing metropolitan countries, have contributed to substantial migration but also increasing pressures for further international migration. Migration remains, in different forms, a time honored strategy from a poor area to a richer one in the search for social and economic mobility at home and abroad" (Connell, n.d.).

Migrants are key contributors to wealth in their home countries. Migrant remittances are an economic benefit that reduce the incidence and severity of poverty in origin countries. The funds from migrants directly increase recipients' income and improve household consumption. Remittances reduce household economic shock in adverse times, such as crop failure and natural disaster (World Bank, 2006). Over the last decade China, India and the Philippines have received the highest remittance flows. In small economies remittances contribute significantly to foreign exchange funds in the receiving countries (World Bank, 2006). In small Pacific Island countries remittances augment household and national economies. Remittances to rural households supply capital for investment in small-scale agriculture or off-farm enterprises. Human flight turns into financial benefit for the migrants' families and the origin countries.

1.4 Human Well-being

Human well-being in ESAP improved over the last five decades, as measured in life expectancy. But persisting gaps remain in poverty reduction, general human health, food security and nutrition. Agriculture is prominent in human well-being, which includes health, nutrition, poverty and rural livelihood. These components are also related to the development goals. Country data in the region varies greatly for indicators used to measure them. Common indicators for human health include life expectancy, infant mortality and access to safe water and sanitation.

Life expectancy for babies born in 2000-2005 is 67 years for all Asia, 75 years for the Pacific and 65 years for the world. This compares with industrial country life expectancy of 75 years and developing country of 65 years. In Bangladesh, Bhutan and India life expectancy has increased by a decade or more. The extreme is Bhutan, where life expectancy for babies born from 1980 to 1985 was 48 years, compared with 63 years for those born from 2000 to 2005 (see 1.3).

While the region gained remarkably in economic growth and trade links, poverty is still common. Poverty perpetuates a cluster of insecurities in health, food and nutrition. In many ESAP countries, although economic growth has led to substantial reduction in poverty, income increased unequally. In ESAP, between 1990 and 2001, the number of people living on less than US$1 a day dropped by nearly one-quarter of a billion. In developing countries of the region the proportion of the population living below the $1-a-day poverty line was 22%, although for least developed countries it was 38% (UNDP, 2006). In ESAP the least developed countries are Afghanistan, Bangladesh, Bhutan, Cambodia, Kiribati, Lao People's Democratic Republic, Maldives, Myanmar, Nepal, Samoa, Solomon Islands, Timor-Leste, and Vanuatu. It has become necessary to distinguish among "poverty", "extreme poverty" and "deprivation", since the region still has extreme deprivation. In 2002, at the $1-a-day measure, extreme poverty estimates revealed that almost 690 million Asian were poor. Using a more "generous" poverty threshold of US$2 a day, 1.9 billion Asians were poor, with South Asia home to the most.

As measured by FAO, 16% of the Asia and Pacific region was undernourished, but this comprised 64% of the world's undernourished population. Since 1995-1997, Asia and the Pacific have reduced overall the number and prevalence of undernourished people. However, recently the number of undernourished again increased. From 2001 to 2003, India had the most, 212 million, undernourished, followed by China with 150 million. The other Asia and Pacific countries together had 162 million undernourished (FAO, 2006c). Girls and women in poor households were included in the "hunger vulnerable" group. Poor maternal nutrition and health can be the hub of the vicious cycle that passes hunger from one generation to another, with reduced capacity among children with low birth weights to be productive adults (FAO, 2005b). Nutritional deficiency among women and children in South Asia is a major crisis in the making.

On average, two out of every three malnourished children in the world live in South and Southeast Asian countries (FAO, n.d.). Malnutrition and underweight prevalence in children under 5 years in developing ESAP countries is about 31%, compared with 28% for all developing countries. In Bangladesh, India and Nepal nearly 50% of children less than 5 years are underweight. The estimated deaths per 1000 live births of children younger than 5 years in 2002 was less than ten for Australia, Japan, Malaysia, New Zealand and Singapore, and around 100 for India and Myanmar. Gains in nutrition have not been equitable in the same country nor in the region, with many poor and landless not benefiting as much as the rest of the population. In some cases, the use of chemicals and irrigation often associated with the Green Revolution have had a negative effect on human health, through vector- and waterborne diseases, pollution of water supplies and direct exposure to pesticides.

Sanitation, access to potable water and nutrition could be improved for about three-quarters of people in urban areas and one-third of the people in rural Asia and some of the developing Pacific Island nations. Most countries in Asia improved their sources of drinking water, but about 1 billion people still did not have access to safe drinking water (WHO, 2004; WRI, 2005).

In recent times ESAP faced threats to human well-being from economic turmoil, epidemics, and ethnic and political conflicts. Notable were the East Asian economic crisis of the late 1990s, severe acute respiratory syndrome (SARS) in early 2000, increasing HIV incidence, and currently, the highly pathogenic avian influenza. ESAP countries demonstrated resilience by coping or recovering from these crises to achieve well-being targets, but the effects remain.

The East Asian economic crisis threatened sustainable growth that undermined economic, health, food security and educational opportunities. In 1997/1998 millions of people fell below the poverty line and created concerns over labor rights. Women in low-paid labor were most severely affected (Jones, 1998; Heller, 1999). In 2003 a major threat came from the SARS outbreak. It affected the regional economy and was contained within a year, illustrating the resiliency of the countries. Recently, HIV and avian influenza have been significant threats to human well-being, with both health risks and economic consequences.

In 2006 about 8.6 million people lived with HIV in Asia, including 960,000 people who became newly infected in 2005. In Oceania, about 7,100 people acquired HIV in 2006, bringing to 81,000 the number of people living with the virus, three-quarters of whom lived in Papua New Guinea (UNAIDS and WHO, 2006). Women comprised 13% of adults with HIV in East Asia, the Pacific, and South and Southeast Asia. In Cambodia, India, Japan, Papua New Guinea, Sri Lanka and Thailand, infection rates among women aged 15 to 24 were higher than among their male counterparts.

The human welfare effect of HIV and AIDS on economic output was likely to be felt hardest in the household. The economic impact was predicted to be severe where millions lived under US$1 a day. Two major causes for financial and material burden were a drastic increase in healthcare expenditures and severe reduction in income of patients and caregivers. In a rampant epidemic, local economic loss can accumulate and drag down national economic growth. Epidemics significantly affect the labor supply (ESCAP, 2003). The labor loss in agriculture from HIV and its economic burden on rural households affect rural economic viability and capacity to innovate. Another threat to human well-being has been the avian influenza epidemic, a human health crisis that could have adverse global economic effects.

Agricultural Livelihoods and Poverty

Across the region, far more poor lived in rural than urban areas (ADB, 2004). Gender-differentiated poverty and poverty among children were frequently observed and often cited. Asia and the Pacific has nearly two-thirds of the world's poor, and two-thirds of the region's poor are women. Poverty is particularly acute for rural women (ADB, 2004). The proportion of the poor in the region varied by country and within countries. In all countries, the major groups of rural poor were the landless, marginal farmers, tenants and indigenous people. The region's less favored areas were home to about 40% of the rural poor. They were rainfed farmers, forest dwellers, highlanders and indigenous peoples. Agriculture productivity was low in upland areas, where ethnic minority groups dominated (IFAD, 2002a).

In South Asia rice farming, arid and rainfed mixed farming demonstrate extensive poverty. In East Asia and the Pacific there is extensive poverty in upland intensive mixed, pastoral and arid farming. In pastoral and arid farming, drought brings on poverty. In all farming systems economic alternatives to farming that are undertaken include off-farm work.

Alternative Systems to Access Food

Most of the developing countries in the region have a large population dependent on agriculture; the region also has many food-deficit, low-income countries. The region is home to many hungry people and rural poor. Farming systems include off-farm work as a livelihood strategy. Hence, the government and external agencies such as the UN and nongovernment organizations deployed programs to improve access to food and increase rural employment. The remarkable achievements in poverty reduction in China and India have come from public investment in rural areas. Public investments, particularly in human capital, physical capital, and science and technology, have been used to stimulate economic growth and reduce poverty. This investment in rural areas, where most of the poor reside, has been important in reducing rural poverty (Thorat and Fan, 2007). FAO recommended a twin-track approach to quickly reduce hunger and poverty. One track would create opportunities for the hungry to improve livelihoods and the second would require direct and immediate action to enhance access to food (FAO, 2002). These programs could take different forms, such as direct food assistance, food for work, and rural nonfarm employment. Rural public works programs generate nonfarm employment and reduce poverty. These programs would be complementary because they would mitigate income fluctuation (IFAD, 2002b).

Both food and cash transfers have increased household resources. The multidimensional nature of malnutrition and the nonlinear link between food consumption and nutrition make it difficult to attribute a nutritional outcome to either food or cash. A combination of food and cash transfers should be considered more widely, especially if done under a national social protection program (Gentilini, 2007). Policies to improve science and technology in rural areas, investment in rural areas to increase labor productivity, improved access to nonfarm work and direct food assistance all help decrease rural poverty and improve access to food. However, effective implementation and monitoring will require good collaboration among the stakeholders involved.

1.5 Trade

1.5.1 Agricultural GDP in ESAP

Trends in the agricultural share in national economies were not homogeneous across ESAP. Agriculture was very important in South Asia. It was important in South Asia trade. Though slowly declining in the past ten years, compared with industrial Europe, the share of agriculture in the gross domestic product (GDP) was high in East Asia and the Pacific and in South Asia (Table 1-3). Trade reform in export partners, particularly OECD countries, will affect a significant share of the population. East Asia and the Pacific has been a net agricultural exporter for most of the past two decades. The region's trade position after the WTO was created, however, fluctuated. The region became a net importer in 1996, followed by rapid growth in net exports in 1998. South Asia is a net agricultural importer in a region in which India is the only country that is a net agricultural exporter. It is also dominant in the region's exports.

The agricultural share in GDP and in total trade has declined over the last decades in many ESAP countries, but it remains a significant source of employment, income and economic activity. The share of agriculture in the GDP ranges from 14 to 57%, from Kiribati to Myanmar; agriculture and agricultural products represent a large share of regional exports. Products include natural rubber, palm oil, rice, fruits and vegetables, mainly exported to the United States, Europe and Japan. Imports are primarily cereals and dairy products, mainly from the United States and Europe.

Table 1-3. *Agricultural share of GDP in ESAP region, 2004.*

ESAP countries	GNI per capita (US$)	Agricultural GDP (% of total GDP)
East Asia and the Pacific		
Australia	27,070	3
Brunei	estim > 10,066	n/a
Cambodia	350	33
China	1,500	13
Fiji	2,720	15
Indonesia	1,140	15
Japan	37,050	1
Kiribati	970	14
Korea, Republic of	14,000	4
Lao, People's Democratic Rep. of	390	47
Malaysia	4,520	9
Marshall Islands	2,320	n/a
Micronesia, Federated States of	2,300	n/a
Mongolia	600	21
Myanmar	estim < 826	57
New Zealand	19,990	10
Palau	6,870	n/a
Papua New Guinea	560	29
Philippines	1,170	14
Samoa	1,840	15
Singapore	24,760	n/a
Solomon Islands	560	n/a
Thailand	2,490	10
Timor-Leste	550	26
Tonga	1,860	28
Vanuatu	1,390	15
Vietnam	540	22
South Asia		
Bangladesh	440	21
Bhutan	760	33
India	620	21
Maldives	2,410	n/a
Nepal	250	40
Sri Lanka	1,010	18

n/a = not available.

Source: World Bank, 2006.

Many countries in this region trade a large share of their GDP, mostly in primary or processed primary products. Tariffs and market access are important to East Asian exporters, but in the region, agricultural protection remains considerably higher than industrial protection.

The economy of East Asia and the Pacific has grown rapidly and poverty has fallen. The GDP of this region grew 8.5% in 2004. The number of East Asians living on less than US$2 a day declined by about 250 million between 1999 and 2004. Countries in the region were on track to meet the Millennium Development Goal for poverty reduction, although there was wide variation in progress across and within countries. China exerted strong economic influence through trade and cross-border production networks. China's growth helped strengthen economic integration within East Asia and increased the region's integration into the global economy. Many countries were considering how to maximize the opportunity China presented, while managing the challenges. High prices for natural resources, especially oil, likely will slow growth in the years ahead. Several other risks also threaten to reduce the rate of growth.

Most of the population in South Asia depend on agriculture and related activities for their livelihood. Despite more than five decades of policy commitment to industrialization, agriculture still is important for most of the countries. All countries in South Asia are low and middle income (Table 1-3). The share of agriculture in total GDP ranges from 18 to 40%, from Sri Lanka to Nepal.

The share of agricultural products in total exports has declined significantly over the past two decades. However, in net foreign exchange earnings, agriculture is much more important than it appears in gross export earnings. The decline in agriculture's share in total exports in these countries cannot be explained solely by the rapid growth in exports of manufactured products. There is considerable evidence that the region lost market share in several agricultural products in which they had comparative advantage, because some countries hold a significant antiexport bias in their incentive structures. South Asian agricultural exports have a significant share of world trade in only five products: spices, rice, tea, oilseeds and jute. In all other major internationally traded agricultural goods, South Asia has less than 4% of the market share.

1.5.2 Trade flows: Main players, commodities and partners

ESAP countries' trade dynamics are vibrant and marked by complex and growing bilateral and multilateral trade agreements. The Asian trade has gone global, with ESAP countries emerging as exporters and importers. Cross-border agriculture trade has increased. In these countries agriculture trade is also important in domestic economies and is driven by the increasing purchasing power of a growing middle class. But aggregate data on agricultural domestic market effects are difficult to obtain and analyze, since some trade happens informally and in rural and urban links. Hence, this subchapter focuses on international trade in ESAP.

In import and export trade value, Japan, China, Australia, Thailand and South Korea are the top five countries in ESAP, followed by Malaysia, Indonesia, India, New Zealand and Singapore. Japan with US$71 billion and China with US$66 billion were also the leading traders in the world in 2004.

As for exports, China ranks fifth and Australia sixth in the world. They are the biggest exporting countries in ESAP, followed by Thailand, Malaysia, Indonesia, New Zealand and India.

Japan is the biggest importer in the region, also the second biggest importer in the world, just behind the United States. China is the biggest exporter and also one of the biggest importers in the region, ranking fourth in the world. Other big agricultural importers are South Korea, India and Malaysia.

If the ASEAN countries are regarded as a group, the large traders in the region are China, Japan, ASEAN, Australia, New Zealand, South Korea and India. The Pacific countries, even though copra and cocoa beans are important, occupy only a marginal place in total trade value.

In products traded, Australia and New Zealand export mainly livestock products, especially mutton and lamb, beef, milk products and wool. Indonesia and Malaysia export palm oil and rubber. Thailand, Viet Nam, Cambodia and India export a large amount of rice and fisheries products. The main world rice exporters come from ESAP, especially from ASEAN countries. China exports mainly vegetables, fruits and maize. India, Sri Lanka and China are the major world exporters of teas. The Pacific countries export copra, cocoa beans and, to a lesser extent, raw sugar.

Japan, China and South Korea are the three biggest importers; Japan and South Korea import most of their agricultural products, mainly cereals and meat products. Japan is low in food self-sufficiency, importing about 60% of its supply. China and India import mainly land-intensive products, such as soybeans, wheat, cotton and edible oils. Since Singapore has almost no agriculture, it relies almost entirely on food imports.

Apart from internal trade among countries in the region, USA, Brazil and Europe were the main providers of agricultural products to ESAP. The USA, Europe and Russia were the main destinations for ESAP agricultural exports. However, trade within the region is important. For example, 66% of China's exports go to Asia, Japan and Korea alone accounting for more than 40% of China's total export. ASEAN is also an important trade bloc, with strong trade relations among members.

Australia and New Zealand had close trade relationships with Pacific countries; they were the major exporters to these countries and the main importers from them. Australia has long been the major source of imports for many of the Pacific Island economies and its importance has increased significantly, except in Tonga and Vanuatu. For agricultural products as a whole, New Zealand had a relatively small import share, except in Fiji, Samoa and Tonga, and that share declined in recent years.

Asian economies are more important as suppliers of imports than as markets for exports for Pacific Island countries, except for Papua New Guinea, the Solomon Islands and Vanuatu. They had a significant import share in many Pacific Island countries, which increased quite sharply in Fiji and Papua New Guinea. However, their share in the imports of Samoa and Tonga eroded considerably and the United States became a much more important import source.

References

ADB. 2000. A study of rural Asia: An overview—beyond the Green Revolution. Asian Dev. Bank, Manila and Oxford Univ. Press, Hong Kong.

ADB. 2004. Key indicators 2004: Poverty in Asia. Measurement, estimates, and prospects. Available at http://www.adb.org/Documents/Books/Key_Indicators/2004/default.asp. Asian Dev. Bank, Manila.

Adeel, Z., and R. Pomeroy. 2002. Assessment and management of mangrove ecosystems in developing countries. Trees 16:235-238.

Asia Pacific Migration Research Network. n.d. Migration issues in the Asia-Pacific—China. Available at http://www.unesco.org?most/apmrnw10.htm.

Banglapedia. 2006. National Encyclopedia of Bangladesh. Available at http://banglapedia.org/.

Beetz, A. 2002. Agroforestry overview: horticulture systems guide. Available at www.attra.ncat.org./attra-pub/agroforestry. National Sustainable Agric. Inform. Service.

Brown, C., and P.B. Durst. 2003. State of forestry in Asia and the Pacific, 2003. Status, changes and trends. Asia-Pacific Forestry Commission, Rap Publ. 2003/22. FAO Reg. Off. Asia Pacific, Bangkok.

Chatterton, P., Y. Bun, C. Hung, K. Whimp, and P. Eddowes. 2000. A future of our forest. Strategies for community forestry and conservation in Papua New Guinea. World Wide Fund for Nature and the World Bank, Suva.

Connell, J. n.d. International migration from the Pacific Islands. Available at http://www.unescap.org/esid/psis/population/popseries/apss160/apss160chap2.pdf

Coyle, W., B. Gilmour, and W.J. Armbruster. 2004. Where will demographics take the Asia-Pacific food system? Amber Waves 2:14-21. Available at http://www.ers.usda.gov/AmberWaves/june04/features/WhereWillDemographics.htm.

Dano, E. 2002. Biodiversity, biopiracy and ecological debt. Available at http://www.jubileesouth.org/journal/preface.htm. Jubilee South, Philippines.

Dawe, D. 2005. Economic growth and small farms in Suphan Buri, Thailand. Presentation at the FAO Symposium on Agricultural Commercialization and Small Farms, Rome, 4-5 May 2005.

Deshingkar, P., and S. Grimm. 2005. Internal migration and development: a global perspective. IOM, Geneva.

ECOSOC (Econ. Social Council of the UN). 2004. World demographic trends. Thirty-Eighth Session. E/CN.9/2005/8. Comm. Popul. Dev., NY.

Enters, T. 1997. Asia-Pacific forestry sector outlook study: Technology scenarios in the Asia-Pacific forestry sector. Asia-Pacific Forestry towards 2010. Working Pap. APFSOS/WP25. FAO, Rome.

ESCAP. 1995a. State of environment in the Asia Pacific. Econ. Social Commission for Asia and the Pacific (ESCAP), Bangkok.

ESCAP. 1995b. Preparatory Meeting for the Ministerial Conference on Environment and Development in Asia-Pacific. ESCAP, Bangkok.

ESCAP. 2000. Developments in forestry over the past 10 years. Ministerial Conf. Environ. Dev. Asia and the Pacific, 31 Aug-5 Sept 2000, Kitakyushu, Japan.

ESCAP. 2003. Economic and social progress in jeopardy: HIV/AIDS in the Asian and Pacific region. UN, New York.

ESCAP and UNDP. 2003. Integrating unpaid work into national policies. UN, New York.

Evans, D., M. Gasiorek, P. Holmes, L. Lacovone, K. Jackson, S. Robinson, and J. Rollo. 2005a. A framework for evaluating regional trade agreements involving developing countries. Draft report to Dfid. DFID, London.

Evans, D., R. Kaplinsky, and S. Robinson. 2005b. The emergence of India, China, and an East and South-East Asia Region on the world economy: A typology of drivers and impacts. IDS Bulletin, Oct 2005. Available at http://www.kent.ac.uk/economics/seminars/sempapers/Autumn0506/Asian_Drivers_IDS_Bulletin_Sept_05_DE_10.pdf

Fan, S., and C. Chan-Kang. 2003. Is small beautiful? Farm size, productivity and poverty in Asian agriculture. Plenary paper prepared for the 25th Int. Conf. Agric. Economists, Durban.

FAO. n.d. Fighting hunger in Asia-Pacific. FAO Reg. Off. Asia Pacific, Bangkok.

FAO. 2001. State of the world's forests: 2001. Available at http://www.fao.org/DOCREP/003/Y0900E. FAO, Rome.

FAO. 2002. Anti-hunger programme. FAO, Rome.

FAO. 2005a. Avian influenza in Asia. Agriculture 21. Spotlight / 2005. Available at http://www.fao.org/AG/magazine/0502sp3.htm. Rome, FAO.

FAO. 2005b. The state of food insecurity in the World: 2005. FAO, Rome.

FAO. 2005c. Compendium of food and agriculture indicators. FAO, Rome.

FAO. 2006a. AQUASTAT FAO information system on water and agriculture. Available at http://www.fao.org/AG/agL/aglw/aquastat/aquastat.htm.

FAO. 2006b. The state of food and agriculture in Asia and the Pacific 2006. RAP Publ. 2006/03. FAO Reg. Off. Asia and Pacific, Bangkok.

FAO. 2006c. The state of food insecurity in the world: 2006. FAO, Rome.

FAO. 2006d. Waves of hope in Thailand: FAO response to tsunami disaster. FAO Reg. Off. Asia Pacific, Bangkok.

FAO. 2007. Fishery Information. Fishery Statistical Collections. FIGIS Data Collection. FAO, Rome.

Fu, B., X. Zhuang, G.B. Jiang, J. Shi, and Y. Lü. 2007. Environmental problems and challenges in China. Environ. Sci. Tech. 4(22):7597-7602. Available at http://pubs.acs.org/subscribe/journals/esthag/41/i22/html/111507feature_fu.html

Garrity, D., and A. Mercado. 1994. Reforestation through agroforestry: Market-driven small-holder timber production on the frontier. p. 239-263. *In* J.B. Raintree and H.A. Francisco (ed) Marketing of multipurpose tree products in Asia. Proceedings of an international workshop held in Baguio City, Philippines, 6-9 Dec 1993. Winrock Int., Bangkok.

Gentilini, U. 2007. Cash and food transfers: A primer. World Food Programme, Rome.

Guest, P. 2003. Bridging the gap: internal migration in Asia. Paper prepared for Conf. Comparative Perspective, Johannesburg, 4-7 June 2003.

Guha-Sapir. D., D. Hargitt, and P. Hoyois. 2004. Thirty years of natural disasters 1974-2004: the numbers. Center Res. Epidemiology for Disasters, Univ. Louvain, Belgium.

Haberkorn, G. 2004. Current Pacific population dynamics and recent trends. Available at www.spc.int/demog/en/stats/2004/2004%20Current%20Pacific%20population%20dynamics.doc.

Heller, P.S. 1999. Human dimensions of the Asian economic crisis. World Bank regional meeting in Bangkok, Social issues arising from the east Asia crisis, and policy implications for the future. Available at http://www.imf.org/external/np/speeches/1999/012199.htm. IMF, Washington DC.

Hugo, G. 2005. Migration in the Asia-Pacific region. Global Comm. Int. Migration. Available at http://www.us-asean.org/ASEANOverview/asean+3.asp.

IFAD. 2002a. Assessment of rural poverty in Asia and the Pacific. IFAD, Rome.

IFAD. 2002b. IFAD strategy for rural poverty reduction in Asia and the Pacific. Available at http://www.ifad.org/operations/regional/2002/pi/pi.htm. IFAD, Rome.

IFRCRC. 2004. World disaster report 2004. Available at http://www.ifrc.org/publicat/wdr2004/chapter8.asp. Int. Fed. Red Cross and Red Crescent Societies.

ILO. 1998. Statistics on working children and hazardous child labour in brief. Available at http://www.ilo.org/public/english/comp/child/stat/stats.htm. ILO, Geneva.

ILO. 2004. World employment report 2004-2005. ILO, Geneva.

ILO. 2005. Combating child labour in Asia and the Pacific: Progress and challenges. ILO, Geneva.

ILO. 2006. Global employment trends. Brief. January 2006. ILO, Geneva.

IOM (Int. Organization for Migration). 2005. World migration, 2005: Costs and benefits of international migration. ILO, Geneva.

Islam, N. 1997. Expanding rural non-farm employment. IFRI Rep. Vol. 19, no. 3. IFPRI, Washington DC.

Jones, S. 1998. Social cost of Asian crisis. The region's economic upheaval may lead to a human rights disaster. The Financial Times, London, Monday, 26 Jan 1998.

Kaneda, T. 2006. A critical window for policymaking on population ageing in developing countries. Population Reference Bureau, Washington DC.

Mahabub ul Hag Development Centre. 2003. Human development in South Asia 2002: agriculture and rural development. Oxford Univ. Press, Karachi.

Mendelsohn, R., A. Dinar, A. Basist, P. Kurukulasuriya, M.I. Ajwad, F. Kogan, and C. Williams. 2004. Cross-sectional analyses of climate change impacts. WPS 3350. World Bank, Washington DC.

Nagayets, O. 2005. Small farms: current status and key trends. Information brief. Prepared for the Future of Small Farms Research Workshop. Wye College, UK.

Pardey, P.G., N.M. Beintema, S. Dehmer, and S. Wood. 2006. Agricultural research: a growing global divide? IFPRI, Washington DC.

Pasadilla, O.G. 2006. Preferential trading agreements and agricultural liberalization in East and Southeast Asia. Asia-Pacific Research and Training Network on Trade (ARTNeT) Working Pap. Ser. 11. Available at http://econpapers.repec.org/paper/escwpaper/1106.htm.

Pasicolan, P.N.A., Udo de Haes, and P.E. Sajise. 1997. Farm forestry: an alternative to government-driven forestry in the Philippines. Forest Ecol. Manage. 99(102):261-274.

Rosenzweig, C.F.N., R. Tubieldo, E. Goldberg, E. Mills, and J. Bloomfield. 2002. Increased crop damage in the US from excess precipitation under climatic change. Global Environ. Change: Human Dimens. Policy 12:197-202.

Siamwalla, A. 2001. Study of rural Asia. Vol. 5. The evolving roles of the state, private and local actors in rural Asia. Available at http://www.adb.org./Documents/Books/Rural Asia. Asian Dev. Bank and Oxford Univ. Press.

Singh, P. 2004. Agriculture policy: Vision 2020 Background Pap. 24. Gov. India, New Delhi.

Steinfeld, H., P. Gerber, T. Wassenaar, V. Castel, M. Rosales, and C. de Haan. 2006a. Livestock's long shadow: Environmental issues and options. FAO, Rome.

Steinfield, H., T. Wassenaar, and S. Jutzi. 2006b. Livestock production systems in developing countries: Status, drivers, trends. Sci. Tech. Rev. Off. Int. Epizooties 25(2):505-516.

Thorat, S., and S. Fan. 2007. Public investment and poverty reduction. Lessons from China and India. Econ. Polit. Weekly Feb:704-710.

UNAIDS and WHO. 2006. AIDS epidemic update. UNAIDS and WHO, Geneva.

UN Atlas of the Sea. 2001. Human settlements on the coast: Types of habitats at risk, Mangroves. Available at http://www.oceanatlas.org. UNEP, Nairobi.

UNDESA. 2003. World urbanization prospects. The 2003 revision. Available at http://esa.un.org/unup. UN Dep. Econ. Social Affairs, NY.

UNDESA. 2004. World population prospects. Available at http://www.un.org/esa/population/publications/WPP2004/2004Highlights_finalrevised.pdf. UN Dep. Econ. Social Affairs, NY.

UNDP. 2004. UNDP Human development report 2004. Table 27, Gender equality in economic activity: human development indicators. Available at http://hdr.undp.org/reports/global.

UNDP. 2005. HDR global reports 2004. Table 27. Gender inequality in economic activity, and HDR Global reports 2005. Table 28. Gender inequality in economic activity. Available at http://hdr.undp.org/reports/global. UNDP, NY.

UNDP. 2006. Asia-Pacific Human Development Report 2006. Trade on human terms. Available at http://www.undprcc.lk/rdhr2006/rdhr2006_report.asp. UNDP Reg. Centre, Colombo.

UNESCO. 2006. Education for all global monitoring report, 2006. Available at http://www.unesco.org/education/GMR2006/full chapt2_eng.pdf. UNESCO, Paris.

UNESCO Institute for Statistics. 2004. UNESCO-EFA. Global monitoring report 2003/4. UNESCO, Paris.

U.S. Census Bureau. 2008. Countries and areas ranked by population. Available at http://www.census.gov/cgi-bin/ipc/idbrank.pl. US Census Bureau, Washington DC.

Vannucci, M. 1997. The UNESCO Mangrove Programme (1982-1990). p. 173-179. *In* Science and Technology in Asia and the Pacific. UNESCO, Paris.

Waggener, T.R. 2001. Logging bans in Asia and the Pacific: An overview. P.B. Durst, (ed) Forests out of bounds: Impacts and effectiveness of logging bans in natural forests in Asia-Pacific. RAP Publ. 2001/08. FAO Reg. Off. Asia Pacific, Bangkok.

Waggener, T.R., and C. Lane, 1997. Pacific Rim demand and supply situation, trends and prospects: implications for forest products trade in the Asia-Pacific region. Asia-Pacific Forestry Sector Outlook Study, Working Pap. APFSOS/WP/02. FAO, Rome.

Wanasundera, L. 2006. Rural women in Sri Lanka's post-conflict rural economy. Centre for Women's Research, Colombo and FAO Reg. Off. Asia Pacific, Bangkok.

Westley, S.B. 2002. Assessing women's wellbeing in Asia. Asia-Pacific Population and Policy o. 61. East-West Centre, Hawaii.

WHO. 2004. Water supply, sanitation and hygiene development. Available at http://www.who.int/water_sanitation_health/monitoring/jmp04_3.pdf

World Bank. n.d. Labour force structure. Available at http://devdata.worldbank.org/genderstats/home.asp

World Bank. 2006. World development indicators 2006. Available at http://web.worldbank.org/WBSITE/EXTERNAL/DATASTATISTICS/0,,contentMDK:20899413~pagePK:64133150~piPK:64133175~theSitePK:239419,00.html or http://siteresources.worldbank.org/DATASTATISTICS/Resources/GNIPC.pdf. World Bank, Washington DC.

WRI. 2005. The wealth of the poor. Part 2: Data and tables. Also at http://earthtrends.wri.org/pdf_library/data_tables/pop2_2005.pdf. UNDP, UNEP, World Bank, WRI, New York.

Other sources

http://www.aseansec.org/

http://www.apecsec.org.sg/content/apec/about_apec.html

http://www.bimstec.org/

http://www.bilaterals.org/rubrique.php3?id_rubrique=85

http://faostat.fao.org/

http://lnweb18.worldbank.org/ESSD/ardext.nsf/12ByDocName/RegionsSouthAsia

http://publications.worldbank.org/ecommerce/catalog/category_3914

2

History and Impact of AKST

Coordinating Lead Authors:
Danilo Cardenas (Philippines) and Hong Yang (Australia)

Lead Authors:
Duong Van Chin (Viet Nam), Mahmudul Karim (Bangladesh), Harold McArthur (USA), Charito Medina (Philippines), Vanaja Ramprasad (India), and Girija Shrestha (Nepal)

Contributing Authors:
Arturo Argañosa (Philippines), Richard Daite (Philippines), James B. Friday (USA), Fezoil Luz Decena (Philippines), Rasheed Sulaiman (India), Douglas Vincent (USA), Halina Zaleski (USA), Yuan Zhou (China)

Review Editors:
Satinder Bajaj (India), D.J. Connor (Australia)

Key Messages

1. Modern AKST has increased crop, livestock and fishery production over the last 50 years more rapidly than the population increased, improving food availability in the region. The increase has not, however, translated into complete availability, accessibility and affordability of food. Food insecurity, poverty and malnutrition remain widespread in some Southeast Asia countries.

2. AKST has improved availability of staple cereals and meat, but micronutrient deficiencies persist. Per capita consumption of cereals has increased in most countries in ESAP. Meat consumption has increased during the 1990s, particularly in China. However, underweight children are still prevalent, as are deficiencies in vitamin A and iron.

3. New irrigation AKST and expanded irrigated areas and inputs (chemical fertilizers, pesticides and herbicides), have made it profitable to grow high-yield varieties. However, increasing withdrawal of water for irrigation has intensified water scarcity in many areas. Irrigation was important in the Green Revolution in the 1960s through the 1980s. However, increasing water withdrawal has decreased discharge of many rivers, affecting aquatic, wetland and estuary ecosystems. The rapid decline in groundwater tables has reduced the resources available and increased costs of accessing water.

3. Inadequate attention has been paid to rainfed production and technology, despite its great importance in area, production and support of the rural poor. The paucity of rainfed technology has partly been from decisions by authorities to concentrate research and extension on irrigated areas because of the perception these areas are likely to yield greater return. For most crops current drought-resistant varieties are associated with low yield.

4. Grazing livestock production systems have generally shifted to mixed farming and intensive commercial production systems in most ESAP countries. The changes in livestock production systems have been related to increased urbanization, improved income and changes in dietary preference. In ESAP, the greatest growth has been in swine production and poultry production, which both depend heavily on commercially processed and traded feed and feed concentrates. Small-scale farmers have increasingly been marginalized.

5. Native forest cover continues to decline across the region as land is converted to agriculture and commercial logging. Although only 5% of the world's forests are in Southeast Asia, this region has had nearly 25% of global forest loss in the last decade. The greatest forest declines have been in the small island states, such as Micronesia, which lost half their forest cover. Increasing population pressure and illegal logging are the dominant drivers in Southeast Asia, East Asia and South Asia, driven mostly by China, a world leader in plantation forestry. Most plantation systems, except in Japan, are less than 15 years old.

6. AKST has enabled innovations in aquaculture, diversified the culture system, accelerated its productivity and improved its sustainability. The supply of aquaculture products to domestic and export markets, with valuable protein and other nutrients, has increased. AKST in aquaculture offset the stagnation and decreased productivity of marine and inland fisheries. Diversified aquaculture technology—pond, pen, cage, raft and raceway culture, monoculture, polyculture and integrated aqua-agriculture—have developed to suit the region's diverse aquatic environment.

7. Transgenic cultivars of some crops are increasingly grown in some countries, but not in others because of concerns of human and environmental safety and restrictions of export markets. Cotton, maize and wheat are the important crops. The most widespread use of transgenics is in China and India. The target characteristics are herbicide and insect resistance; research is seeking wider improvement, to stress resistance and nutritive quality.

8. Traditional knowledge and indigenous practices as part of AKST contribute to the welfare of many local communities in ESAP. However, many countries have had an evident trend of loss in traditional knowledge in agriculture because of historical neglect; fast demographic, economic, political and cultural changes; and the aggressive expansion of modern agriculture. Nongovernmental organizations (NGOs) and local communities have been active in using and developing traditional knowledge systems.

9. The funding of AKST in most ESAP countries remains inadequate, despite clear evidence linking productivity improvement to investment in agricultural research and development. The public sector continues to fund the bulk of agricultural research and development in many ESAP countries because of the "public good" character of many technologies that it generates. Private investment is largely concentrated on technology that can provide privileges to property rights, relevant to only a small portion of the needs of small-scale farmers. This observation has been prevalent in many developing countries, where limited and fluctuating funding has led to institutional instability.

10. Modern AKST, especially the intensive use of chemical fertilizer and pesticides, has had considerable adverse effect on sustaining soil, water, biodiversity and the ecology. Productivity in many food crops has increased. However, soils and water have deteriorated in many instances and some offsite effects have been observed. Technologies for improving soil fertility and water management are available but have not been widely adopted. Biodiversity conservation has not been fully integrated in major agricultural production systems.

11. The benefits of AKST have been inequitably distributed to the farmer community, particularly to women. Women's control over key economic resources, including land ownership, remains weak, despite women contributing the most time to agricultural production. Accrual of benefits by women has been limited by unequal access to education, information and capacity-building programs. Neither unpaid nor paid contributions to agricultural production by women and children have been fully recognized. Small-scale and group programs and special training, including in marketing and management, have helped some women benefit from AKST.

2.1 Agriculture and AKST in ESAP

In ESAP, the past half century witnessed a rapid population increase, from around 1.6 billion in 1960 to 3.4 billion in 2004. At the same time, remarkable economic growth took place in most countries, leading to significant increase in income and to demand for more and better food and other agricultural products. During this period, AKST experienced unprecedented progress, which has been the foundation underlying the growth of agricultural production. Despite the encouraging achievement, it was also evident in many ESAP countries that the benefits of economic development and the increased food supply were not equally distributed. Meanwhile, the exploitation of natural resources and intensive use of modern inputs caused serious environmental degradation, which undermined the sustainability of agricultural development.

Agriculture is traditionally important in the national economy in most countries in the region. However, the share of agriculture in the gross domestic product (GDP) has declined since World War II, especially in the last three decades. As of 2005, ESAP produced more than 80% of the world's rice, vegetables, jute, sweet potato and coconut. It provided more than half of the world's tea, tobacco and peanuts, while accounting for more than 25% of the wheat, maize, white potato, cassava, millet, melon and sugarcane. The region is also home to 30% of the world's livestock species. With the exception of the industrial countries in the region, most of the agriculture is in small holdings and diversified farming. Small farm size has limited the potential of employment and income from agriculture. Rural-to-urban migration, multiple occupations of the laborers and diversification of the rural economy have been evident in many countries.

Except in a few countries, the absolute size of the agricultural labor force has still been rising. Agricultural employment has been especially important for the livelihood of the poor. Agriculture has served as an employment buffer and safety net in the face of large economic shocks, such as the Asian financial crisis in 1997 and 1998.

During the last two to three decades, remarkable progress in rural economic development has taken place in many countries, particularly China and India. AKST has played an important role. The significant improvement in crop yields enabled a large increase in food production. The amount of rural population living under poverty has reduced substantially, although reduction rates vary greatly. However, evidence from India suggests that the augmentation of total food production has not always benefited the poor in increasing income and improving food security.

Population growth and economic development have generated greater demand for food. An increasing population combined with limited arable land means little agricultural area per capita—an average of 0.2 ha each person—in the developing countries in the region (FAO, 2006a). With most people on less than half of the total land, pressure on land, water, flora and fauna has been increasingly severe. Intensification of production with the support of modern AKST in almost all crop farming, animal husbandry, fishery and forestry has been evident in most ESAP countries.

Modern AKST, particularly that associated with the Green Revolution, has been developed to increase the quantity of agricultural products and enhance their resilience to physical stress. However, the effects of AKST have varied from positive to negative, depending on the wealth and farm size of the groups involved. Some aspects of AKST are applicable mainly to large and commercial farms; others are more suitable for small and subsistence farmers. Natural endowments, socioeconomics, culture and tradition all influence AKST innovation and adoption.

ESAP countries face new challenges for agricultural production. In many areas absolutely no more land is available for cultivation. Many areas, particularly parts of China and India, have endured water stress, threatening the sustainability of food production on irrigated land. At the same time, land degradation, environmental pollution, loss of biodiversity, and little or no investment in agricultural research and development have affected the agricultural potential of the region. In recent years, many ESAP countries have again become net food importers. A good understanding of past trends in agriculture and the effects of AKST in ESAP are useful in the search for appropriate AKST to meet the challenges of sustainable agricultural development and food security.

2.2 Trends in AKST: Agricultural Practices

2.2.1 Application of AKST to crop production

Over the past 50 years, the increase in land productivity has enabled farmers to feed twice as many people from less agricultural land. This productivity growth has been based mostly on generating, promoting, disseminating and adopting AKST from formal and informal agricultural extension organizations. The principal agricultural technologies adopted were mainly in water management, chemical fertilizer use, variety development and crop protection, mechanization, livestock feeding and disease control, and sustainable resource management.

2.2.1.1 Expanding irrigated areas and adopting irrigation technology

Irrigation is widely used in ESAP. In many countries, irrigation has had a long history, being closely linked to rice cultivation. Rapid population growth, limited arable land and continuous increase in demand for food in the past 50 years have driven an unprecedented expansion of the area under irrigation. Advances in modern dam construction, flow regulation and pumping equipment have provided the means to harness more water for irrigation.

Between 1961 and 2003, irrigated areas in ESAP more than doubled, with an annual growth rate of around 2.6%. In China and India, the pace surpassed the average of the rest of the EASP countries (Figure 2-1). However, since the late 1990s, the expansion of irrigated areas has slowed; even India and China appear to have a slight decrease. By 2003, about 28% of the cultivated land in the region had been brought under irrigation.

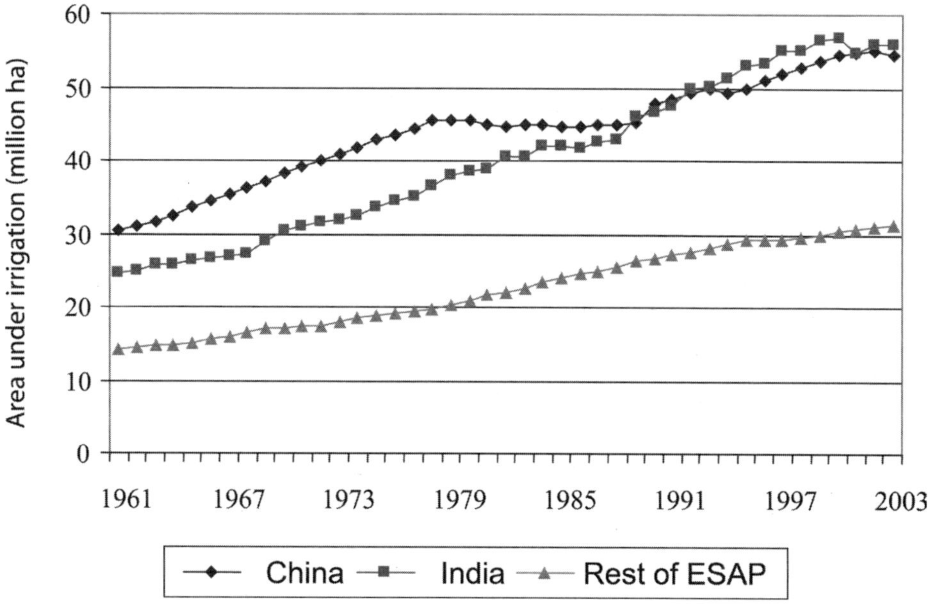

Figure 2-1. *Changes in irrigated area in ESAP, 1961-2003. Source: FAO, 2006a*

The pace of irrigation expansion and the share of irrigated total cropland have varied substantially across countries (Table 2-1). Japan is the only country that has experienced a decline in the amount of irrigated land, brought about by the shrinkage of the paddy rice area, which was nearly halved during 1961-2003 (FAO, 2006a).

Irrigation was important in the Green Revolution in the 1960s through the 1980s. Irrigation is crucial for stable and high yields and for increasing total food production. Crop yield on irrigated land is usually significantly higher than the yield on rainfed land.

Irrigation also enabled more intensive use of land. Double and triple cropping was able to spread to where the rainy season is long enough for only one crop a year. The high yield and increased intensity of land use resulted in a much larger proportion of food production on irrigated land than its share in total cultivated land. For the region as a whole, about 60% of the food production was from irrigated land, which was about 28% of total cropland. Irrigated areas produced about 70% in China and 50% in India of the total national food production (Lipton and Litchfield, 2003).

Traditionally, surface irrigation was the most widely used in ESAP. However, the past 50 years have seen a rapid expansion of groundwater irrigation. As electricity became more widespread and pumping and well-drilling technology improved, groundwater has become increasingly important, particularly in arid and semiarid areas. By the mid-1990s, half of China's irrigation water came from wells (Brown, 1994). In India, about 60% of the irrigated areas rely on groundwater (IWMI, 2002).

Groundwater is a primary buffer against the vagaries of climate and surface water. Because groundwater for crops can be used on time, it is often more productive than surface water. Some locations in India showed yield on crops irrigated by groundwater 2 to 3 t ha^{-1} higher than with canal irrigation (Shah et al., 2001; Lipton and Litchfield, 2003).

The proliferation of wells has owed a lot to inexpensive groundwater technology, such as treadle pumps. Low-cost pumps have helped fuel the groundwater boom, mostly with private investment from farmers. Groundwater is available in groundwater-rich areas to anyone who can afford a pump. It has been a boon to small-scale farmers, even poor ones. In China, the number of groundwater pumps owned per hundred farm households increased from 1.69 to 22.5 between 1985 and 2005 (China, 2006).

In surface irrigation systems, flood irrigation has been dominant in ESAP countries. Its generally low irrigation efficiency leaves ample room for saving water. However, in river basins, the water-saving potential may be lower than anticipated because part of the "lost water" upstream can recharge the groundwater and become available downstream (Molden and Fraiture, 2004). Commonly used water-saving technology included furrow and border irrigation, mulch and terracing. In the Ningxia Autonomous Region in China, official statistics showed that irrigation water withdrawal from the Yellow River decreased from 8.9 to 6.7 billion m^3 between 1998 and 2004, while the irrigated area increased from 387,000 to 406,000 ha, mostly by using various water-saving measures.

More advanced irrigation technology, such as sprinkler, microirrigation and laser leveling, was seen in the region, but on a rather small scale. In Japan, sprinkler irrigation and microirrigation were about 9.5% of the total irrigated area in the late 1990s. In India, it was 1.5%. Mongolia was the only country where sprinkler irrigation was significant, as large schemes were equipped with sprinkler irrigation in the 1980s (FAO, 1999). In China, sprinklers, drip irrigation and low pressure pipes were used in about 10% of the total irrigated area in 2004 (China, 2006). In other countries, national data for the application of irrigation techniques were generally not available.

By the mid-1970s, farmers had adopted drip irrigation in some ESAP countries, such as Australia and New Zealand. Drip irrigation is especially effective in arid and

Table 2-1. *Changes in irrigated area by country, 1961-2003.*

Country	Irrigated land (1000 ha)		Changes (%)	Area equipped for irrigation as % of cultivated land
	1961	2003		
Australia	1001	2545	154.2	5
Bangladesh	426	4725	1009.2	50
Bhutan	8	40	400.0	31
Cambodia	62	270	335.5	7
China	30411	54596	79.5	35
Fiji Islands	1	3	200.0	1
India	24685	55808	126.1	33
Indonesia	3900	4500	15.4	13
Japan	2940	2592	-11.8	55
Korea Dem People's Rep	500	1460	192.00	50
Korea Republic of	650	878	35.1	47
Laos	12	175	1358.3	17
Malaysia	228	365	60.1	5
Mongolia	5	84	1580.0	7
Myanmar	536	1870	248.9	17
Nepal	70	1170	1571.4	46
New Zealand	77	285	270.1	8
Philippines	690	1550	124.6	14
Sri Lanka	335	743	121.8	34
Thailand	1621	4986	207.6	26
Viet Nam	1000	3000	200.0	34
ESAP	69158	141645	104.8	28

Source: FAO, 2006a.

drought-prone areas, where water is scarce. It does not accumulate salt in the root zone and causes little soil erosion. Drip systems use 30 to 60% less water than furrow or sprinkler irrigation (Postel, 1996). In India, crop yields from drip irrigation were about 10 to 30% higher than those from surface irrigation (Postel, 1999). Despite the efficiency in water use, drip irrigation still is used on very little of the total irrigated area. In Australia, the percentage was about 7.8%. In China and India, it was below 0.1%. The large initial investment required for the equipment is a main constraint to its use in developing countries. Drip irrigation is used primarily to irrigate high-value horticultural crops.

Supplemental irrigation is the application of water at critical times. It can substantially improve yield and water productivity in arid and semiarid environments. In dry areas, supplemental irrigation can boost productivity of irrigation water by 10 to 20% (Oweis and Hachum, 2003). Technology for supplemental irrigation ranges from farm ponds to shallow groundwater pumped with treadle pumps (Barker and Molle, 2004). Supplemental irrigation could prevent total crop failure and stabilize and improve crop yields. However, it requires comprehensive knowledge and skills in crop management. Weather forecasts reduce risk and are an integral part of such a comprehensive strategy.

2.2.1.2 Water management in rainfed crop farming systems

Compared with irrigated agriculture, rainfed systems have been given little attention in most ESAP countries. The paucity of technology for rainfed areas partly relates to decisions by government authorities to concentrate research and extension in irrigated areas because they have been perceived more likely to yield a greater return on investment. Among the many constraints that limit rainfed agriculture, unreliable rainfall is probably the biggest. Water stress at the flowering stage of maize can reduce yield 60%, even if water is adequate during the rest of the season (Molden and de Fraiture, 2004). Recent years have seen increased biological science efforts to produce traits for drought tolerance and resistance to many pests and diseases.

Rainwater harvesting to retain water has been seen in some semiarid areas in ESAP. Rainwater harvesting has shown considerable potential in semiarid areas because it could supply limited irrigation at the key stages of crop growth by using stored rainwater. A number of cases in China and India have shown significant increase in crop productivity through rainwater harvesting. In Gansu Province in China, for example, yields of maize and wheat on the experimental sites increased over 50% (Liu et al., 2005).

Rainwater harvesting technology is simple for local people to install and operate. It is convenient because it provides water at the point of consumption and family members have full control of their own systems, reducing operating and maintenance problems. The disadvantage is the limited supply and uncertainty of rainfall. In addition, numerous small-scale water-harvesting and storage systems in a basin could have similar effects on river flows and aquatic ecosystems as a large dam and canal irrigation. For example, along the Yellow River, bunds and plugging gullies were effective in encouraging agriculture and in reducing erosion, but evidence showed these practices reduced river discharge (Zhu et al., 2003).

Improved land management techniques and agricultural production systems have received growing attention for improving water productivity of rainfed systems. Such technology has been referred to as "green [soil] water management." In some areas, minimum or zero tillage proved effective in improving soil moisture and crop yields in rainfed land (Hatibu and Rockström, 2005). Mulching, terracing, contouring and microbasins are also important in maximizing rainfall infiltration into the soil to increase yields. No-till and conservation agriculture maintains and improves crop yields and resilience against drought and other hazards, while protecting and stimulating the soil. The essential features of conservation agriculture are minimal soil disturbance and maintenance of a permanent cover of live or dead vegetative material. The cover protects the soil against erosion and water loss from runoff or evaporation. A major impediment to successfully introducing conservation agriculture is that management skills are complex. In many ESAP countries, any production system that includes crop rotation is complex because it calls for coherent management over more than one or two crop seasons. Farmers who have adopted these systems need to understand them and the reasons for the various procedures to be able to adapt them to their needs and conditions to balance crop rotation with market requirements (Box 2-1).

Recently, increasing emphasis has been on integrated rainwater and irrigation water management. Because obtaining additional water for irrigation is difficult and water in rainfed systems is unreliable, agricultural water management has shifted from pure rainfed or fully irrigated systems to emphasizing intricately connected soil conservation and supplemental, drip, ground and surface irrigation.

Box 2-1. Potential of rainfed agriculture.

There are different views on the potential of rainfed systems. Evidence exists for great potential for poverty reduction from new approaches to enhancing rainfed agricultural systems. New pro-poor small-scale, low-cost approaches such as treadle pumps, water bags and water harvesting are key to unlocking rainfed potential and reducing poverty on marginal rainfed lands. Although crop yields seem low considering the amounts of land, water, labor and capita required, new technologies are available to help farmers predict uncertain variables such as rainfall. This improved predictability can help increase the contribution of rainfed agriculture.

Others state that although rainfed agriculture has been the focus of research for many years, gains are not forthcoming. Dependence on approaches to enhancing rainfed agriculture involves high risk due to climate variability, particularly affecting small scale and poor farmers. As poor people often live in semiarid agricultural environments where the ability to cope with weather variation is very low, the failure of crop often means starvation or even death. A study in three semiarid watersheds in India showed that large scale investments in soil and water conservation did not have a significant impact on dryland yields, at least not under prolonged conditions of drought (Bouma and Scott, 2006).

2.2.2 Development and application of modern technology and inputs

2.2.2.1 High-yielding varieties—the Green Revolution
The historical focus by international and national research institutes has been food crop production technology, emphasizing improved yield varieties—the Green Revolution. Modern plant breeding and improved agronomy, including the use of inorganic fertilizer and pesticides, have been components of the strategy to increase production (Friedman, 1990). Nearly three-quarters, 71%, of production growth since 1961 has been from yield increases. Increased yields have contributed to greater food security within developing regions and contributed to declining real prices for food grains.

In the 1960s, when the International Rice Research Institute (IRRI) was formed, breeders found the main constraint to rice yield was the architecture of traditional tropical rice cultivars (Khush et al., 2001). Although tall cultivars responded positively to nitrogen fertilizers, competed well with weeds, and provided much straw for fodder, fuel and construction, they lodged and lost yield. The Japanese had realized the value of short-straw cultivars in the quest for high yield and introduced the trait into rice around 1900. By the 1950s, semidwarf rice could be found among the landraces in many Asian countries, including in subtropical China. Taichung Native 1 (TN1), a semidwarf cultivar from Taiwan (China), was first planted in the tropics in the late 1950s, but it was highly susceptible to major diseases and insects in the tropics (Peng and Khush, 2003). In 1962, IRRI introduced dwarfness into tropical rice by crossing the dwarf Taiwanese cultivar Dee-geo-woo-gen into the tall Indonesian cultivar Peta. The result was IR8 (the 8th cross), the first lodging-resistant and fertilizer-responsive cultivar. Farmers rapidly adopted it and it became the symbol of the Green Revolution in Asia. After the release of IR8, three more semidwarf cultivars, IR5, IR20 and IR22, were released during the 1960s, followed by another 17 in the 1970s and 13 during the 1980s (Peng and Khush, 2003). The release of IR8 increased the yield potential of tropical rice from 6 to 10 tonnes ha^{-1}. Its yield potential has hardly been surpassed in 40 years of breeding (Peng et al., 1999). The development of early-maturing varieties, particularly in rice, has enabled double and triple cropping in areas that previously produced only one or two crops a year. The dwarf varieties, less prone to lodging, could be grown more densely, using a smaller area (Robinson, 1996).

By 1970 almost all the area under high-yield seeds, about 94% of wheat and 98% of rice, was in Asia, of which nearly half was in India (Pearce, 1980). The maximum effect of the high-yield variety program in India was in wheat, where the coverage was 83% of the cropped area by 1985/1986. Rice was next, with about 57%. Coverage under cereals ranged between 30 and 46% (Groosman et al., 1991). Seed supply systems of new varieties replaced the traditional varieties.

High-yield wheat and rice were critically dependent on several inputs, so there was an increase in agroindustry. For example, in India, nitrogen fertilizer production increased from 0.37 million tonnes in 1967/1968 to 2.23 million tonnes in 1979/1980. Furthermore, production capacity had to be generated for tractors and other machines. Farmers had to invest their own capital to acquire these machines, which were often produced with the help of public financing agencies (Chaboussou, 2004).

Amid the wave of the Green Revolution, Chinese scientists led by Professor Yuan Longping bred the world's first rice hybrid in 1974. Hybrid rice yields about 15 to 20% more than the best of the improved or high-yielding varieties. By 2000, about half of China's total rice area was under hybrid rice cultivation. National average rice yields increased from 3.5 to 6.2 tonnes ha^{-1} between 1975 and 2000. Hybrid rice has particularly good potential to improve the food security of poor countries with scarce arable land, expanding populations and cheap labor. FAO, IRRI, the United Nations Development Programme (UNDP) and the Asian Development Bank supported improving national capacity in hybrid rice development and dissemination outside China (FAO, 2004).

2.2.2.2 *Mechanization*

Agricultural machinery is another modern technology that has contributed greatly to farming and crop production. Advances in farm machinery have changed the way people produce food worldwide. Agricultural machinery entails substantial cost to buy and operate but reduces labor considerably.

Rising wages and reduced availability of labor in many Asian countries forced farmers to mechanize, adjust cropping patterns and resort to migrant labor. In some cases, these changes were extraordinarily rapid. In the Central Plain of Thailand, the labor used in irrigated rice cultivation had declined from 57.5 person days per hectare in 1987 to just 8 person days by 1998, a decline of 86% in little more than a decade (Isvilanonda et al., 2000). The reduced labor was by mechanizing harvesting and switching from transplanting to direct seeding. Rapid changes occurred in southern China, where many farmers changed from triple cropping (rice–rice–winter crop) to a single rice crop to save labor.

Farmers often could not afford to buy agricultural machinery, so well-functioning rental markets were crucial. For example, while combine harvesting was widespread in the Central Plain of Thailand, only a small percentage of farmers owned a combine harvester. Use by owners, plus rental through cooperatives or government agencies, accounted for just 6% of use; the rest occurred in private rentals (Dawe, 2005). Rental markets often arose naturally in the absence of government restrictions.

Tractors were the most common machinery. The number of agricultural tractors in ESAP expanded rapidly, reaching 6.5 million in 2005 (Figure 2-2). It increased 14-fold from 1961 to 2005. Japan, with the most tractors in use, started mechanizing early. India's use of tractors increased rapidly, overtaking China in the 1980s and reaching a level

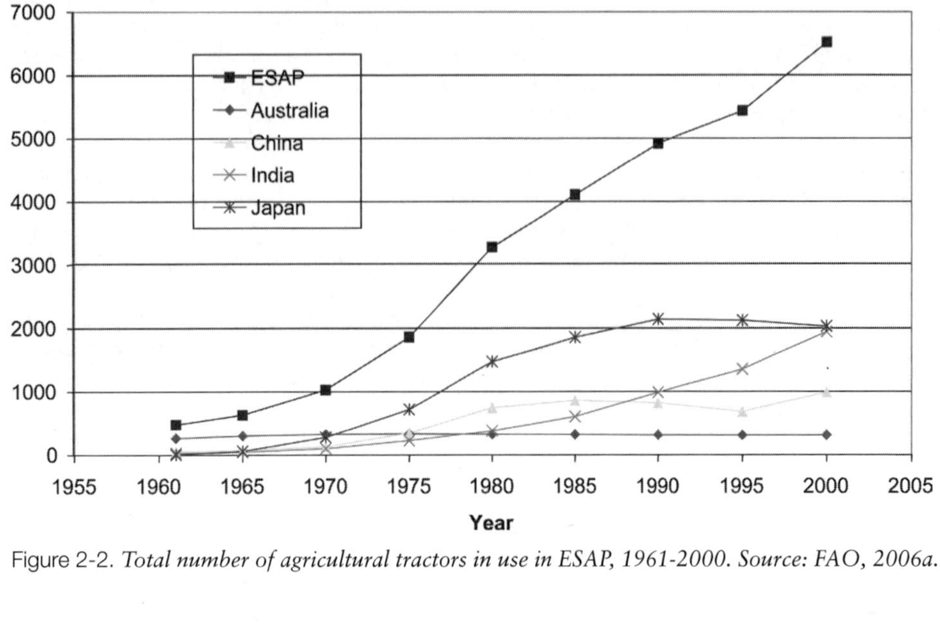

Figure 2-2. *Total number of agricultural tractors in use in ESAP, 1961-2000. Source: FAO, 2006a.*

similar to Japan's in 2000. In contrast, China used tractors moderately, which can be attributed to small-scale subsistence farming.

2.2.2.3 *Fertilization*

In ESAP, fertilizer use increased sharply and had reached 275 kg ha^{-1} by 2005 (Figure 2-3). The average annual growth rate was about 6.6%. About 61% of the fertilizers applied were nitrogen based; next was phosphorus (P) 24%, and potassium (K), 15%. Use of nitrogenous fertilizer increased 23-fold over this period.

Chemical fertilizer application varied significantly within ESAP. In East Asia, growth in usage was especially rapid, from 69 kg N ha^{-1} in 1978 to 155 kg N ha^{-1} in 2002. Growth was also rapid in Southeast Asia and South Southwest Asia, but much less than in East Asia. As a result, nitrogen use was much higher in East Asia than in Southeast Asia and South Southwest Asia. Over-reliance on this fertilizer led to nitrogen overdose in some high-yielding farmland in China. The adverse effects of excessive fertilizer use on the environment emerged as a serious concern (Zhu and Chen, 2002). Application of phosphate and potash fertilizer also grew rapidly, sometimes exceeding the growth in the use of nitrogen. On the other hand, many soil nutrients were mined, leaving many intensive rice systems exhibiting negative K balances (Dobermann et al., 2004). In some cases, reversing these imbalances would lead to higher profits for farmers.

2.2.2.4 *Crop protection*

Pesticide use in agriculture is on the rise in many developing countries in ESAP. Because data are often unavailable, it is difficult to paint a general picture of trends in ESAP or national pesticide use.

In China, the amount of pesticide used increased 1.8 times between 1991 and 2004. Pesticide use per hectare of sown area reached about 9 kg ha^{-1} in 2004 (Figure 2-4). Pesticide use was high in the wealthy and developed areas on the southeast coast, while poor areas, such as the northwest regions, used the least. Farmers growing grain in the North China Plain, who had used pesticides for many years, increased applications in response to pesticide resistance. Crops receiving the most applications were fruit, cotton, maize, and wheat. Pesticide use was high in greenhouses, where the chemicals were applied up to 10 times above the rate used in fields. Even in the field, it was not uncommon for farmers to double the recommended dose.

Importance needs to be placed on minimizing the negative health effects that pesticides, especially insecticides, have on farmers, who often spray with little or no protection. This can be done by educating farmers in integrated pest management (IPM) as promoted by FAO, through media campaigns and by strengthening regulatory enforcement. The perception of many farmers, extension service providers and even policy makers about the magnitude of crop losses caused by insect pests are often greatly exaggerated (Heong and Escalada, 1997) and probably contribute to pesticide overuse.

Plant breeding can also offer improved pest and disease resistance in new varieties. One example is the steady reduction in insecticide use on rice over the past 20 years in central Luzon, the rice bowl of the Philippines. Application rates are now lower than before the Green Revolution, but rice yields have increased. Another potential means of reducing insecticide use is Bt cotton, a genetically modified crop widely adopted in China and India, where collectively more than 70% of the region's cotton is produced. Reports indicate that insecticide use with Bt cotton has fallen dramatically and farmer health has improved (Pray et al., 2002).

2.2.3 Trend in crop production and application of AKST in major farming systems

2.2.3.1 *Growth of crop production and increase in food availability*

Over the past 50 years, food crop production has increased remarkably because of development in agricultural science, technology and modern inputs. The harvest area has

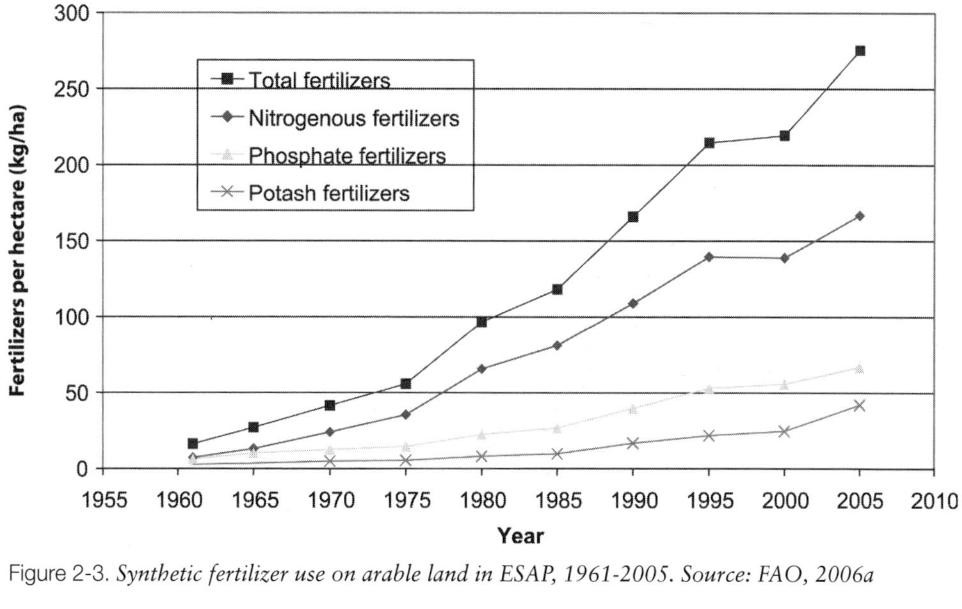

Figure 2-3. *Synthetic fertilizer use on arable land in ESAP, 1961-2005. Source: FAO, 2006a*

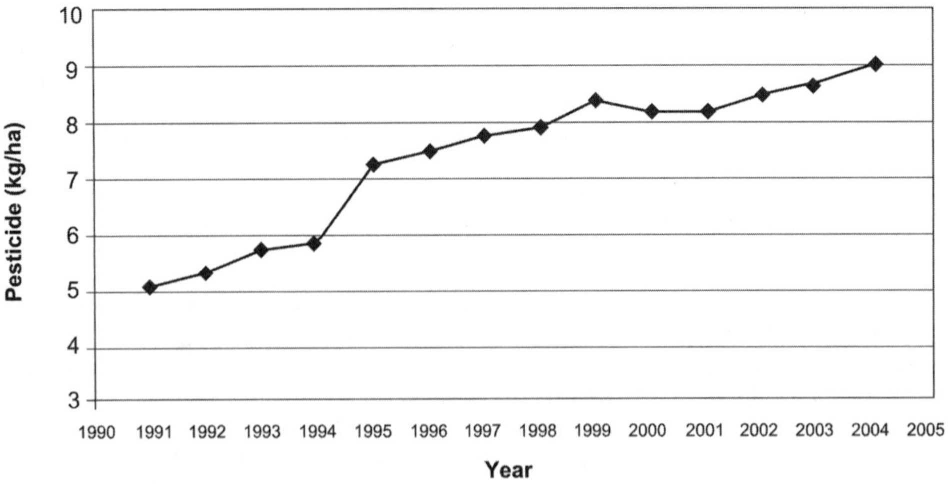

Figure 2-4. *Pesticide use in China, 1991-2004. Source: China Ministry of Agriculture, 2005.*

remained stable—288 million ha in 2005. However, cereal production increased threefold from 1961 to 2005, with a 2.7% annual growth rate (Figure 2-5). Among the cereal crops, paddy rice accounted for about 55% in 2005, followed by wheat 22%, and maize 19%. Among the ESAP countries, Australia, China, Indonesia, Laos, Pakistan, Philippines and Viet Nam experienced rapid growth in cereal production. For example, China's cereal production had an annual growth rate of 3.1%, mainly driven by maize production the last 20 years in response to the increased demand for animal feed.

With relatively stable crop areas, the growth in cereal production has come from increases in crop yields (Figure 2-6). From 1961 to 2005, the yields of maize, paddy rice and wheat increased remarkably. The yields for maize were 4.16 tonnes ha^{-1}, 4.16 tonnes rice ha^{-1} and 3.05 wheat tonnes ha^{-1} in 2005. However, agricultural performance varied substantially across countries. In maize production, Australia, Bangladesh, China and New Zealand had the highest yields in 2005. Bangladesh had the highest annual growth rate from 1961 to 2005, followed by the Republic of Korea. For paddy rice, Australia, China, Republic of Korea and Japan were the countries with the highest yields, while the small island countries had the lowest. Laos experienced the highest annual growth rate, followed by China. With wheat production, higher yields were in China, Japan, Republic of Korea and New Zealand; China had the highest annual growth rate.

In the ESAP countries, food production per capita increased steadily until 1990 and then began to level off (Figure 2-7). In 2005, the average cereal availability was 283 kg per capita. China had a rapid increase between the 1960s and the 1980s but tended to decrease after that, mainly from changes in the diet with the rise in income and more meat consumption. Indonesia also experienced steady increase in food provision. In contrast, India moderately improved domestic food production per capita.

2.2.3.2 *AKST in major crop farming systems—three cases*

Intensive irrigated rice production. Irrigation and the use

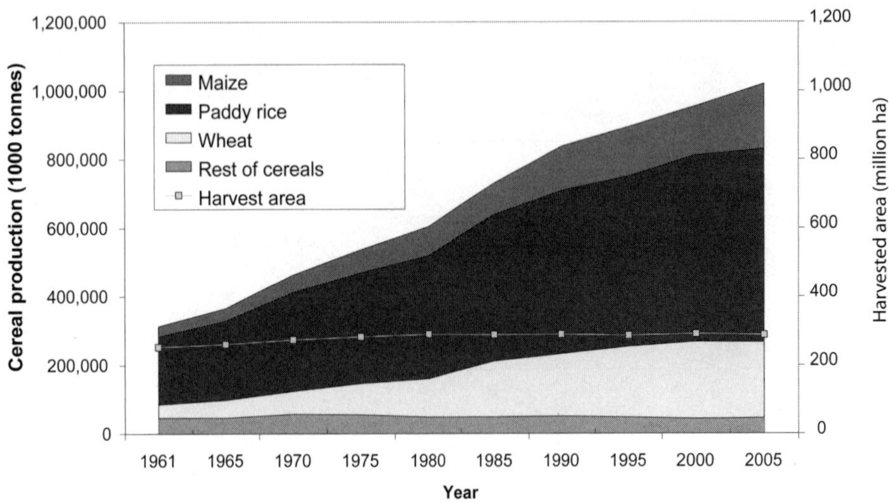

Figure 2-5. *Harvested area and cereal production in ESAP, 1961-2005. Source: FAO 2006a*

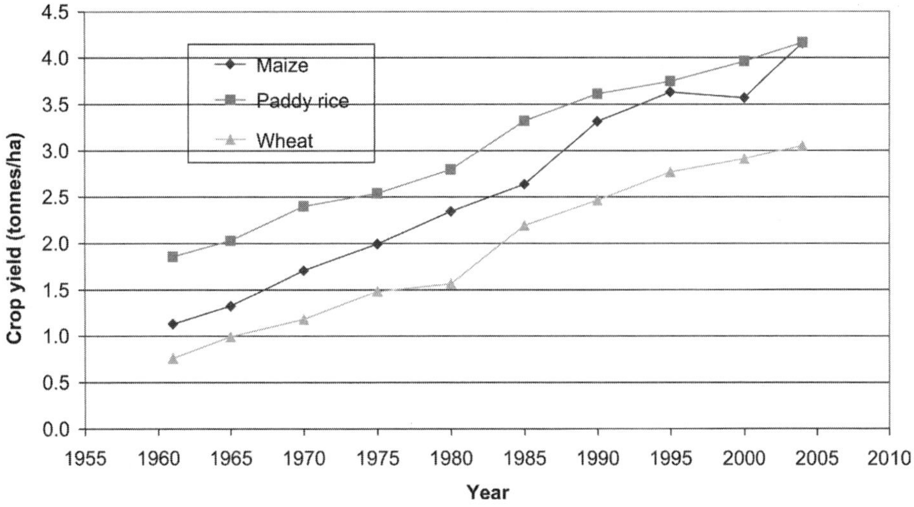

Figure 2-6. *Average crop yields in ESAP, 1961-2005. Source: FAO 2006a*

of dwarfing genes in tall cultivars led to great increases in rice yield. The release and widespread adoption of short-duration, high-yielding, semidwarf cultivars triggered investment in irrigation infrastructure and allowed more farmers to grow two or three rice crops each year. Tillage and intense management increased and soil remained submerged longer. Inorganic fertilizer and pesticide use increased, but the diversity of rice cultivars in the irrigated systems shrank. Higher yields resulted from the combination of increased yield of modern cultivars compared with the landraces they replaced. Improved crop nutrition was made possible by fertilizer application and improved plant resistance and pest management to minimize losses from weeds, insects and diseases (Cassman and Pingali, 1995). In the irrigated lowlands of Asia, which accounted for 75% of the world's rice production, average yield increased from 2 to 3 tonnes ha^{-1} in the 1950s to 5.3 tonnes ha^{-1} at the turn of the last century.

Although the quest to further increase the potential yield of inbred rice after the release of IR8 was not as successful as hoped, many new cultivars were better adapted to different environments and contributed to better nutritional quality and human health (Peng et al., 1999; Peng and Khush, 2003). Considerable progress was made, for example, in managing major rice diseases, such as bacterial blight, blast and tungro, through genetic improvement. The reason few disease outbreaks occurred in the past 15 years was the result of collaborative research between international research institutions and the national extension organizations in many developing countries. Combining resistance to insect pests with ecological crop management principles (Heong et al., 1995) greatly reduced the incidence and effect of pest outbreaks (IRRI, 2006).

Modern rice cultivars with origins in breeding research in the 1960s covered about 75% of Asian riceland. In irrigated areas, that proportion was often 80 to 100%. In Bangladesh, for example, 46 new cultivars developed from 1970 to 2001 had spread to 80% of the dry season irrigated rice area by the early 2000s. These modern cultivars have contributed to a 2.3% growth in rice yield each year

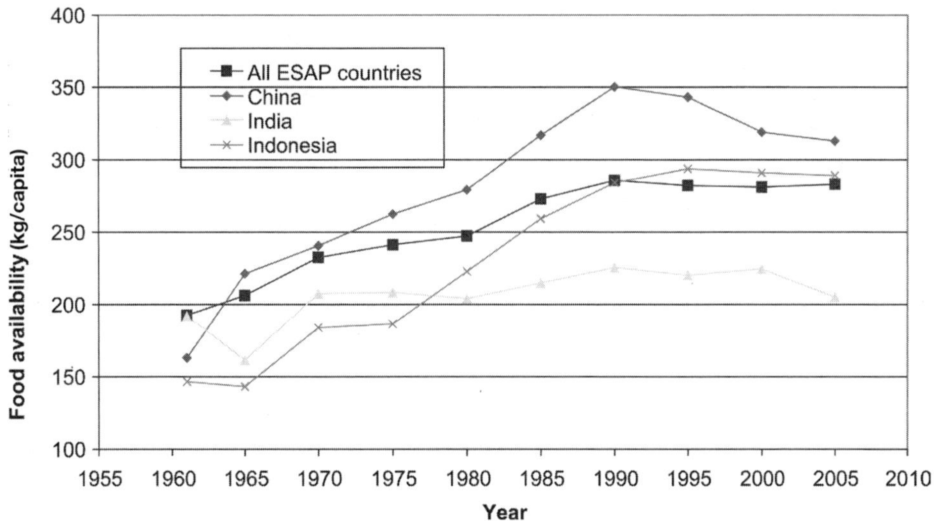

Figure 2-7. *Cereal production per capita in ESAP, 1961-2005. Source: FAO, 2006a*

over the last three decades, which helped Bangladesh avoid a looming hunger crisis, despite high population growth and dwindling amount of arable land (Hossain et al., 2006). The rice-breeding programs at IRRI and its partners in Asian countries demonstrated how AKST requires continuous development of new cultivars to secure the world's food supply.

The rice–wheat system of the Indo-Gangetic Plains. Rice–wheat cropping has been practiced for a thousand years, but it expanded rapidly, particularly in northwest India and Pakistan, only since the mid-1960s, following the Green Revolution. The rice–wheat belt occupied nearly 24 to 27 million ha in South Asia and East Asia. Rice was mostly grown in flooded fields, while the ensuing wheat crop required well-drained soil (Ladha et al., 2004). The system occupied 13.5 million ha in the Indo-Gangetic Plain of South Asia in 2001 (Timsina and Connor, 2001). Rice–wheat systems evolved with the introduction of wheat into traditional rice areas in Bangladesh, eastern India and Nepal and rice into traditional wheat areas in northwest India. The driving force for expansion was the need to intensify cropping to meet an increasing demand for food. It was made possible by the development of short-duration and medium-duration rice and wheat cultivars. Their combined productivity responded to improved nutrient management, pest control and the expansion and improvement of irrigation. The rice–wheat system is complex. Overall optimum management strategies must be established for the alternating and contrasting anaerobic environment required for rice and aerobic environment for wheat. A summary of the sequence of the technology and its effect on productivity is as follows:

- Before the Green Revolution, yield was small and the system operated with few inputs and much human and animal labor.
- As short-season rice and wheat cultivars became available, management focused on expanded irrigation and improved management. During this early expansion there were no environmental issues to counteract the benefits from increasing productivity. Yields increased with further intensification.
- Further intensification included new cultivars, nutrition and mechanization. Early sowing of wheat to avoid heat stress and low yield during flowering and grain filling was a major strategy for yield improvement.
- Subsequently, yield increase slowed and yield declined in some places from a combination of causes. Evidence showed deteriorated soil structure and fertility from excessive cultivation and nutrient extraction from the more intensive system, operating with ever-increasing yields. Problems arose from irrigation with both excessive extraction of groundwater and accumulating salts in regions with low water quality. Decreased solar radiation and increased minimum temperature also contributed to yield decline (Pathak et al., 2003).
- The recent phase was to recuperate yield. Attention to water and labor use and environmental problems led to much new technology across the entire Indo-Gangetic Plain. New techniques and machines for planting enabled more rapid and timely crop establishment. Reduced cultivation and site-specific fertilizer management were reversing soil deterioration. Bed planting was introduced in some places to improve water management and diversify crops away from a strict rice–wheat system. Fertilizer requirements were more precisely defined, and soil and tissue testing enabled more effective and efficient nutrient management. Laser leveling of land, aided by more accurate water requirements, improved irrigation efficiency. Less stubble burning contributed to improved air quality and more soil organic matter. These resource-conserving technologies were to improve farmer income by increasing input efficiency, maintaining crop productivity and enhancing crop diversification (Gupta et al., 2002; Ladha et al., 2003).

Rainfed wheat production in the State of Victoria, Australia. The Australian wheat industry, exemplified by the State of Victoria, had already passed through two phases of development when the rapid development of a lucrative world market for wool following World War II provided an opportunity for significant change (Connor, 2004). It became economical to improve pasture by species composition and fertility. "Sub and super," subterranean clover (*Trifolium subterraneum* L.) and superphosphate fertilizer, became the buzzwords for pasture development. Sheep-carrying capacity of pasture increased markedly and encouraged close integration with wheat production. Pastures were managed for sheep and to build up nitrogen to extract during a cropping phase. Other technology supported the greater economic benefits that flowed to farmers from increased wheat and sheep production. Plant breeding continued, horses were replaced by tractors, and new machines were developed for tilling and harvesting. Herbicides and pesticides became available, and increasingly precise fertilization for pasture legumes, including micronutrients manganese (Mn) and molybdenum (Mo), became possible. Fallowing became less frequent. Wheat yields had risen to around 2 tonnes ha^{-1} by the 1980s. The system was mostly seen as ecologically stable.

With the application of inorganic nutrients, leguminous pastures with increased nitrogen supported profitable sheep and wheat production. The system did not, however, persist into the 1990s because the wool market collapsed. Furthermore, soil acidification and salination were unanticipated environmental effects of the system. Clover growth and nitrogen fixation were reduced, and consequently the overall productivity of pastures and wheat crops declined. The solution lay in liming and changing the water balance to keep the salt at the depth where it had accumulated under native vegetation. As a result, a more diverse system was sought, one that involved less pasture, less fallow, perennials such as lucerne and trees in agroforestry systems, and a wider range of crops including canola, lupine, field pea, fababean, chickpea and lentil. There was also increased use of zero tillage, controlled traffic, yield mapping and precision farming. Nitrogen fertilizer entered the system. Plant breeding continues, now for a wider range of species and by applying new biotechnology techniques such as marker-assisted breeding. But so far genetically modified crops (GMO) have been avoided. Although there has been some local opposition to GMOs, the dominant concern has been to maintain access to overseas markets.

Farmers now require considerable technological and economic skills to manage increasingly complex cropping systems. High yields are possible only in years of greater than average rainfall. In low rainfall years, the focus of management must be to minimize costs, perhaps even not to plant at all. Cropping is no longer a matter of applying established cropping sequences but tactically adjusting crop and management to likely seasonal and economic conditions. Nitrogen fertilization is a good case in point. The objective is to gain yield benefit in years of high potential and avoid low yield in years of low rainfall. This requires a response to weather forecasts in crops, measurement of soil water and nitrogen at sowing, and crop nitrogen content at responsive points during the growing season. Measurements and their analysis are the keys to success in crop choice, crop condition, weed and pest incidence, and the likely success of management.

2.2.3.3 Emerging trends on biofuel production

Biofuels are an important energy source in ESAP, mostly from crop residue, wastelands and wood from forests. The countries with evident pressure on forests are India, Nepal and Thailand. The agricultural and natural resource base of all countries faces a potentially much greater pressure to produce liquid biofuel. The driving forces are worldwide: energy security, climate change, and export to the many industrial countries that set mandatory goals for biofuel use. Within ESAP, India has set a mandatory minimum 5% ethanol in nine states; Thailand has tax incentives to encourage production. Australia, a low-cost sugar producer second only to Brazil, established a cane ethanol market in 2001 to overcome financial hardship among producers. The annual target is to produce 350 million liters by 2010 from a base of 30 million liters in 2001.

Modern technology is best suited to producing ethanol from sugar or starch crops or biodiesel from oilseed crops. Target crops for ethanol in ESAP are sugarcane, cassava, maize, oil palm and coconut, and for biodiesel, jatropha. Ethanol from cellulose, crop residue, biomass crops and trees is possible, but much less efficient for producing energy. Energy efficiency was not high in any case, with output-to-input ratios for ethanol from sugar and starch from mostly < 2; higher, perhaps 4 to 5, only for sugarcane under the production and cultural conditions in Brazil.

At a time when concern is being expressed for the capacity of agriculture to feed an anticipated world population of 9.5 billion, including 6 billion in ESAP, immense additional pressure will be placed on agriculture and forestry. Current technology would require 3.5 tonnes of grain to fuel a motor car with bioethanol for one year—almost seven times the annual grain equivalent needed to provide one person an adequate and balanced diet (Connor and Mínguez, 2006). Put another way, it would take 100 kg of grain to produce the ethanol used to fill a 40-liter tank of a car just once. That caloric content (100 x 4,000 kcal) is equivalent to a survival diet of 2,200 kcal a day for one person for six months.

This simple calculation exposes the enormous increase in agricultural production required if biofuel is to make any significant contribution to private motoring, and the inequality it would engender in developing countries struggling to feed all inhabitants adequately (von Braun and Pachauri, 2006). While farmers will benefit from the additional market, even a small portion of the total liquid fuel requirement produced from agriculture will place enormous strain on the environment. Even at this early stage of the biofuel boom, there have been widespread reports attributing higher food prices to the diversion of agricultural production to biofuel production, such as for maize in Mexico. The notion that there are special fuelcrops that are significantly more efficient fixers of solar energy than current crop species and will not compete with food production or will flourish on land unsuited to agriculture, jatropha is an example, is unproven.

When methods are devised to break down the cellulose for fermentation, stubble and biomass crops will also be targets for biofuel. Stubble, however, is important in maintaining soil structure and fertility. While a portion, perhaps 50%, might be removed with the highest-yielding crops, retention is generally required to sustain productivity. Witness the deleterious result of removing stubble for animal fodder and roofing and root crowns for fuel that is practiced in parts of the rice–wheat system. Energy crops will compete with food crops for land and markets. To the small extent they will be able to contribute, they will require high management and, most importantly, large inputs of water and nutrients to maintain productivity.

2.2.4 Application of AKST to livestock production

Millions of rural households in Asia and the Pacific depend on domesticated animals for food, farm power and income. Livestock is important in the economies of many ESAP countries and has particular cultural significance in India. The livestock sector has been shifting from extensive grazing to more commercial production and changing from rural to urban and periurban production.

This pattern is directly related to increased urbanization throughout the region. The rate of urbanization is highest in East and Southeast Asia and less pronounced in other parts of ESAP (Steinfield, 1998). Between 1950 and 2000 the percentage of people in Asia living in urban areas increased from 16 to 38% (UNFPA, 2001). By 2025 the urban population is anticipated to surpass 54%. In Oceania, which includes the Pacific Islands, Australia and New Zealand, the trend was the same, with a prediction that 84% of the population will reside in urban areas by 2025.

These demographic changes have been accompanied by a shift from large ruminants—buffaloes and cows—to monogastric pigs and poultry. The developing countries have had some of the highest growth rates in producing and consuming livestock and meat products. Asia has had some of the highest growth rates in pork and poultry production, with an estimated 150% increase between 1991 and 2003. ESAP also more than doubled its egg production and accounted for about 50% of world production (Steinfield et al., 2006).

2.2.4.1 Livestock production systems

Animal production systems are of three main types: grazing, mixed farming, industrial. Of these, the mixed farming system dominated in ESAP countries. Grazing systems use native grasslands with little or no integration with crops. Mixed farming involves integrating livestock and crops and provides an opportunity to intensify by using crop residues and manure. Industrial production is capital and labor intensive, detached from land for feed supply and waste disposal (Steinfeld, 1998).

Industrial production systems are increasingly important, particularly for pigs and poultry. Livestock production systems vary from country to country. Sri Lanka has small-scale dairy and poultry producers, with buffalo as the main source of draft power. Nepal also uses buffaloes and bullocks for draft power and ghee production. Sheep and goats are mainly kept for wool and meat. In Southeast Asia swine and poultry dominate livestock production. In Thailand, more than three-quarters of pigs are produced on large industrial farms with more than 500 animals. In Guandong, China, on the other hand, half the pigs are produced on farms with fewer than 100 animals and in Viet Nam, very small operations of three or four pigs dominate.

The growth in the production and demand for poultry and pork has resulted in a growing shift away from pasture systems. As livestock production becomes more intensified, there has been a shift from locally available feed to commercial feed concentrates, particularly in pig and poultry production (Steinfeld et al., 2006).

Only three countries in ESAP—Japan, Mongolia and Singapore—have not increased meat production over the last 20 years. Countries registering more than a 100% increase in production are Brunei, Cambodia, China, Fiji, India, Indonesia, Lao PDR, Malaysia, Pakistan, Philippines, Sri Lanka and Viet Nam. China's share of meat production in the world total increased from about 10 to over 28% in the last 20 years (FAO, 2006a)

2.2.4.2 *Changes in dietary patterns on livestock production*

Though livestock food products are still not a significant part of the diet in developing Asia and the Pacific, consumption is growing rapidly. Developing Asian countries have the world's highest growth rates of production and consumption of food from livestock (FAO, 2006a). The growth in livestock production across ESAP comes from changes in demand and new technology, which have enabled producers to move into more intensive industrial production, thus greatly increasing the supply of pork and poultry meat.

Poultry consumption has shown the fastest growth throughout ESAP. China stood out for its impressive growth in beef consumption, which increased by more than 500% between 1985 and 1995. Growth patterns in South Asia have been more balanced, with poultry showing significant increase in consumption but consumption of other meat products—beef, mutton and goat—increasing only modestly (Steinfield, 1998).

Pork was the most-produced meat. In 1961, it was 30% of total meat production. Beef and veal were next in importance, followed by mutton and lamb. Pork became even more dominant by 2000, when it was over 55% of the meat produced. Poultry meat took second place, beef and veal were third.

Dramatic changes occurred in Asia: protein from livestock in human diets increased more than 130% between 1980 and 2002. The increase in animal products in the human diet was part of a dietary transition that also included higher intake of fat, fish, vegetables and fruits, and a decrease in cereals and tubers (Steinfeld et al., 2006). This transition was directly related to growing urbanization and increasing standards of living throughout ESAP and many of the developing regions of the world. Urbanization, coupled with income growth and increasing globalization, generated a major shift in Asian diets away from staples and toward livestock and dairy products, vegetables, fruits, and fats and oils.

The dynamic Asian livestock sector is growing between 3.5 and 5% each year—more rapidly than crops, such as cereals, vegetables and pulses—driven partly by increasing population, rising income and change in consumer lifestyles. Since animal products are expensive to import, most countries plan to meet the rising demand through increased domestic production. Hence, livestock growers in periurban areas are increasing production and modifying management to respond to the rapid rise in demand. Structural changes are also led by the growth in urban supermarket vendors, intensifying the need to examine opportunities for vertically integrating vulnerable producers. Small-scale producers are not generally a part of the rapid rise in intensive animal production. Yet more than half of the small-scale farmers in Asia rely on livestock as a major source of income and nutrition (FAO, 2006a).

Although most ESAP countries are technically capable of increasing meat, milk and egg production, most face shortages of key feed ingredients, particularly maize and soybean meal. As a result, there is a large and burgeoning trade in feed crops worldwide.

2.2.5 Application of AKST to forest production

Although ESAP contains only about 5% of the world's forests, it had an estimated 25% forest loss over the last decade. The proportional loss (the amount lost relative to the remaining forest) was greatest in Asia (UNFPA, 2001). The cumulative loss of forest cover across Asia and the Pacific between 1990 and 2000 was estimated at about 1.1% (FAO, 2006a). The Philippines had the highest rate of deforestation, followed by Pakistan, Thailand and Malaysia. However, the largest losses occurred in Indonesia and Myanmar (Waggener, 2001). Between 1990 and 2000 the region experienced considerable reduction in forest cover, with the greatest decline in the Southeast Asian islands, followed by continental Southeast Asia and the Pacific Islands. Population pressure and the resultant conversion of land to agriculture was the dominant cause of deforestation across the region. While in percentage the most forest lost was in the smaller Pacific Islands, the forests of insular and mainland Southeast Asia had the greatest population pressure (Brown and Durst, 2003). Because tropical forests contained about half the remaining biodiversity in the world, their destruction was of particular concern (UNFPA, 2001).

2.2.5.1 *Native forest management*

Native forests cover about 25% of the area across Asia and the Pacific. The Pacific Islands, with 65% forest cover, and the Southeast Asian islands with 53% cover, have the highest proportion of land-user forest. Papua New Guinea has the largest rainforest coverage in the Pacific, accounting for the third largest block of tropical rainforest in the world

(Chatterton et al., 2000). South Asia has relatively little forest cover.

Native forests are not limited to terrestrial environments. Asia and the Pacific are also home to the greatest concentration of mangroves. Once thought of as coastal wasteland, mangroves were destroyed at an alarming rate for agriculture, aquaculture and firewood. Almost half the mangrove destruction in recent years has been prompted by the desire to create shrimp farms (UN Atlas of the Ocean, 2002). Over the last 20 to 30 years, with help from the UNESCO Mangrove Programme and other international initiatives, government planners and fisheries experts have become more aware of how mangroves are a nursery for many coastal and aquaculture fish, a buffer that reduces sediment flows to offshore reefs and a barrier against storm surges and tsunamis (Vannucci, 1997). About 90% of all marine organisms spend some portion of their life within mangrove systems (Adeel and Pomeroy, 2002).

Most countries in the region have well-defined policies, laws and programs to regulate the use of forests and the development of forestry activities, although they are rarely consistently enforced. As a result, corruption and illegal logging are significant in many ESAP countries. Indonesia loses $1.4 billion a year as a result of the trade in illegal logging (DFID, 2007). Historically, most ESAP countries have regulated forest management by assigning management responsibilities to government agencies and attempting to enforce strict controls on forest access. Transient upland populations and traditional tenure systems based on common access to forests have often conflicted with government policy initiatives.

Asia and the Pacific had over 552 million ha of forests, of which 477.7 million ha, 86%, were natural forests. (The FAO 2001 figures were 709 million ha, which could be an overestimate due to the forest cover classifications used.) However, only about 249 million ha, 45%, were available and suitable for harvesting (Waggener, 2001). The natural forests throughout ESAP had, until very recently, been seen mostly as a vast source of raw timber to generate export income. However, there is now general agreement on the need to change from a focus on timber exploitation to emphasizing management for sustainable, multiple-use natural forests (Enters, 1997). In the face of increasing deforestation China, New Zealand, Philippines, Sri Lanka, Thailand and Viet Nam have imposed several total, partial, temporary or selective bans on logging in natural and old-growth forests. The results of these restrictions have been mixed. Several case studies indicate that such bans could have unanticipated effects on timber supply, forest harvesting, transport, processing and consumption of forest products. The restrictions affect forest residents and those who depend on forestry for their livelihoods (Waggener, 2001).

The need for sustainable forest management has been clearly recognized throughout ESAP, but few examples of effective management have been implemented on a large scale. In some areas, improved practices such as reduced-impact logging, forest and timber certification and log tracking systems have been introduced as avenues for more sustainable management. Forest certification involves certification by a third party that an area of forest is being managed in accordance with a defined set of standards. Chain of custody certification tracks wood products from a certified forest to the point of sale. However, incorporating social criteria and indicators into forest management and harvest practices was criticized as difficult to assess and interpret in the field (Wollenberg and Colfer, 1996). Although there was no significant effect of timber certification on loss of tropical forests over the last two decades, timber certification is expected to create greater awareness among forest managers of the need to protect the environment and minimize the loss of biological diversity (Thang, 2003).

In addition to the limited use of technology to reduce the effect of logging on the environment and the forest soils, different governments in the region have put in place policies designed to reduce environmental damage and increase the economic returns from forestry. Thailand has a total ban on logging within its borders, the Philippines has banned the export of unprocessed logs and Bhutan has mandated that the country must keep 60% of the land under forest cover (UNEP, 2005). However, the effectiveness of these laws and plans is often greatly reduced because of limited resources, shortage of skilled staff, corruption and weak law enforcement. Rather than the need for new technology, the critical issue has been the lack of political will in most countries to enforce existing policies and regulations (Enters, 1997).

Outside reforestation, natural forest management has focused on forest harvest techniques to minimize the adverse effect on natural regeneration, ground cover and underlying forest soils. In some places traditional cut-and-drag systems have been replaced by less-damaging methods. In Malaysia skyline cables and helicopters are used to minimize degradation from harvesting timber on steep slopes. These and other technologies developed for forestry in temperate areas could be adapted for tropical forests. The real constraint is that most forest harvesting in ESAP is done by private companies, where profit is the driving force in management practice. Portable wood chippers are available, but there are few or no economic incentives for extracting or on-site processing harvest waste. Helicopter logging and cable and skyline yarding represent a large capital investment that might not be justified by the value of the timber, especially if there are no impositions upon the harvester to conserve soil and water or limit the damage to the residual stand. Although cable yarding systems damage soils and understory less than cut-and-drag systems, they may be more difficult to use in the selectively harvested forests common in Asia, rather than clear felling, common in temperate forests.

2.2.5.2 *Nontimber forest products*

Managing native forests to collect and produce nontimber forest products has received little attention, other than as a component of agroforestry and a traditional agroforestry practice. There is not even clear agreement on what nontimber forest products are. The broadest definition would include all biological material harvested from forests for human use. Scale, mode of harvesting and market distinguish nontimber forest products from forest wood products. Nontimber products are usually harvested by individuals, households or small groups and marketed directly by the harvester or through small-scale processing operators. Nontimber forest products generally are forest plants and animals used for food, beverages, forage, medicine and fiber. Although many upland households harvest, process and sell nontimber products to augment household income, few data relate to the number of people employed and the value of the outputs across the region. Nontimber forest products have generally been collected and consumed by those who collect them, others traded and processed before reaching the market. Women and children generally collect and process the nontimber products.

In response to strong market demands, some nontimber products, such as rattan in Malaysia, are domesticated and grown in plantations for commercial use. Such practices meet consumer needs without further depleting natural forest stock (Poh, 1994). Forest products formerly grown in the wild are now grown commercially, including tropical fruits, cocoa, coffee, tea, cardamom, cinnamon, cashew and pepper. While domestication of some species could expand to meet growing market demand, for many rural and upland residents collecting nontimber products remains a significant contribution to subsistence farming. In principle, the harvesting of nontimber products from natural forests should be sustainable. However, in practice this has often not been the case, particularly where changes in land tenure, hydropower projects and logging roads have given outside populations easy access to remote areas (Enters, 1997).

2.2.5.3 *Plantation forestry*

Plantation forestry is another form of management found in the region. Five ESAP countries, China, India, Indonesia, Japan and Thailand, ranked among the world's top ten plantation forestry countries. Together, these countries accounted for 55% of the world's forest plantation resources, 91% within Asia and the Pacific (Brown and Durst, 2003). The average age of Asia's industrial plantations is less than 15 years (FAO, 2001). This has come primarily from the rapid acceleration in plantations in China and the short rotations generally used in that country. Older plantations were mostly in Japan (FAO/RAP, 2003).

With the diminishing availability of large-diameter timber from natural forests in the region, plantation forestry is fast becoming the alternate source for wood in ESAP. The region accounted for more than 80% of forest plantations in the tropics (Enters, 1997). Most of the legally produced industrial wood in the region has come from plantation forests. Most plantation forestry is intensively managed monocultures, mainly pine, teak, poplar, acacia and eucalypts, cultivated for a relatively narrow range of products and species (Enters, 1997).

Plantation forests have considerable diversity in ownership, management, scale and products. Plantation systems have been established to meet the need for fuelwood, poles, wood chips, furniture wood and estate crops, including rubber, oil palm and coconut. Until the last 25 years, forest plantations were largely smallholder or government operated. The growing trend has been increasing private investment and management of forest plantations in response to increasing demand for wood for pulp, furniture and particleboard. Smallholder plantations have sprung up to meet this market in the some ESAP countries, such as the Philippines (Garrity and Mercado, 1994; Pasicolan et al., 1997), Laos (Roder et al., 1995), India and Thailand.

The technological innovations in plantation forestry depend on the production objectives—conservation, fuelwood, fiber, or sawlogs. The technology and management adopted by plantation operators include improved seedling production by using polyethylene bags, centralized nurseries, thinning and pruning for sawlog production. Breeding programs and improved planting material from tissue culture were still relatively minor (Enters, 1997). Improved trees have been limited to large commercial plantations, particularly in China. Virtually no improvement has been done for species commonly used in multipurpose small farm operations. Harvesting technology ranges from manual to completely mechanized, mostly in response to rising labor costs and increasing concern about minimizing soil disturbance.

Although the reduced species diversity and the younger tree age associated with plantation forests provide conditions favorable for pests and diseases, uniform products often compensate for these risks. Plantation forestry is seen by many as a way to address environmental objectives, such as soil conservation, mitigation of erosion, carbon sequestration, and rehabilitation and protection of habitats for important wild flora and fauna.

2.2.5.4 *Wood-processing technology*

Wood-processing capacity in the region has increased significantly over the past 30 years, concentrated in the most industrialized countries. Most of the technological improvements in wood processing have come through adapting technology from industrial countries, such as medium-density fiberboard. Most modern processing machinery has been imported from Europe (Enters, 1997). It has generally been adapted for processing small-diameter trees from forest plantations. Medium-density fiberboard production emerged as a response to shortages in raw material and the newly developed ability to use untapped resources to produce a plywood-like product. Medium-density fiberboard and similar products became price-competitive alternatives to plywood, particleboard and hardboard (Adhar, 1996). Within ESAP, Malaysia is the number one exporter of veneer sheets, Indonesia the number one exporter of plywood, followed by Malaysia, China and New Zealand.

With many governments banning or severely restricting the export of unprocessed logs, a demand has developed for efficiently processing and converting sawn timber into particleboard and other panels. Likewise, some wood previously considered to have little or no value, such as rubber wood, is now being processed for the furniture industry. Largely as the result of the research at the Forest Research Institute in Malaysia, a major market has developed for rubber wood in furniture and panels. More recently, the technology for processing rubber wood has been applied to oil palm stems and research is looking at using oil palm fiber as an ingredient in wood-based boards, pulp and chipboard. Compared with natural wood and plywood products, composite, defect-free fiberboard can be easily produced in large, uniform sizes (Yayah et al., 1995). When the supply of natural fiber begins to dwindle, the panel processing industry will likely introduce nonwood fibers. Most major nonwood fiber processing facilities are in China and India (Enters, 1997).

2.2.5.5 *Agroforestry*

Agroforestry in its simplest form means integrating trees with crops or livestock enterprises in a farming system. Tree farms and nut plantations managed as a monocrop are not considered agroforestry (Beetz, 2002). This approach, used primarily on smallholder systems, has gained widespread attention by government agencies and NGOs to address conservation objectives. Because of its potential for enhanced food security, poverty reduction and environmentally sound land management, an internationally supported research center, the World Agroforestry Centre (ICRAF), is devoted to this research and development.

Technology and management associated with agroforestry include (1) alley cropping, (2) improved fallow systems, (3) silvopasture, (4) windbreaks, (5) mixed agroforests, including breadfruit systems in the Pacific Islands and (6) riparian buffer strips. The technology most widely associated with agroforestry has been alley cropping. This involved incorporating tree hedgerows within crop fields to act as a fallow and improve soil fertility through nitrogen fixation (Craswell et al., 1997). In Asia hedgerows were promoted on sloping fields to reduce soil erosion (Garrity, 1986). While there were some significant success stories on the positive effects of alley cropping, particularly in Alfisols not deficient in phosphorus (Sanchez, 1995), it was not widely adopted because it was designed as a conservation approach rather than meeting the needs of upland dwellers. It was labor intensive and in most cases did not result in sufficiently high economic return to justify the labor. Tree and crop competition for light, water and nutrients led to the failure of many alley-cropping systems to outperform traditional cropping (Sanchez, 1995).

Different views regard the future of agroforestry. One view holds that (1) agroforestry cannot be widely adopted for forest rehabilitation so long as farmers do not have secure land tenure and use rights, (2) resource poor farmers will receive only marginal benefit from expanded agroforestry and (3) future agroforestry effort should focus more on intensively managed small-scale plantations that produce only one or two products for commercial use, including coffee, cacao, and neem (Enters, 1997). In Malaysia, small-scale farmers grew rubber for its wood rather than latex. Other options included fodder crops, living fences and shade trees. Others suggest that the future of agroforestry in much of the developing world lies in intensifying and commercializing the management of traditional homegarden forests (Leakey and Tchoundjeu, 2001). Given the less than optimal adoption of research-designed systems, improving traditional practices that build upon farmer knowledge, skills and resources may be a more successful strategy.

2.2.5.6 Community and social forestry

In the narrowest perspective, community forestry is the governance and management of forest resources by communities for commercial and noncommercial purposes. The core of community forestry is the recognition that communities living adjacent to or in forests have rights to manage them and extract resources to support their livelihoods and traditional knowledge. As such, community forestry has become the focus of policy and training initiatives, rather than technological interventions. In India and Papua New Guinea customary and village ownership of forests has been recognized. Community forestry programs focus on educating villagers to become better stewards of forest lands. As a development strategy, community forestry has become a way many governments in the region involved rural communities in protecting and managing forests (Nurse and Malla, 2005). In practice, community forestry was often interchangeable with social forestry, referring to many activities that involve local people, from managing woodlots, growing trees as a cash crop to household processing of forest products (Casson, 1997).

Over the last 40 years community and social forestry has expanded from a way to meet the fuel and income needs of the rural poor and reducing deforestation and desertification to empowering a community (Poffenberger, 1990; Hidayat, 1998). Examples include community forest management and participatory conservation in the Philippines (Utting, 2000). Other variations include joint forest management in India (Fisher, 2000), village forestry in Lao PDR and collective forest management in China (Gilmour et al., 2004). One model example of community forest management came from Nepal, where over 12,000 recognized forest user groups managed more than a million hectares of forest. One challenge to community management is resolving issues related to forest tenure, ownership, user rights and common access. Another key issue that had not been effectively addressed is how local participatory approaches can be scaled up to affect landscapes (Nurse and Malla, 2005).

Farm forestry has received little if any attention in developing and transferring improved technology. Farm forestry is growing trees on private agricultural land, wasteland and degraded forests. There is an important perceptional difference between social forestry and farm forestry. Planners of social forestry projects emphasize the subsistence return to farmers in fuelwood and fodder, while forest farmers place priorities on trees for cash income. Areas where farm forestry is most successful are where small farms have a long history of producing for the market, where cash returns from trees and agricultural crops can be easily seen (Pasicolan et al., 1997). Examples of this can also be seen in the industrial countries of the region. Australian farm forests often provide ecosystem services similar to agroforestry and community forestry. In Australia, the nature and the size of small-scale forestry differ from county to county but can be classed into two types. The first is based on growing *Eucalyptus globules* for pulp and the second involves producing native hardwoods for saw and veneer logs (Herbohn et al., 2002).

2.2.6 Application of AKST to fisheries production

In ESAP, fisheries have always been vital to food security, supplying animal protein, minerals and vitamins; generating employment, reducing poverty and earning revenue through trade. It is part of the cultural heritage in many parts of India and Bangladesh; fish is important in matrimonial and other social customs and celebrations. Globally, fisheries production from 1950 to 2004 increased steadily with the ESAP contribution. ESAP countries contributed at least 64% to total global production in 2004 (FAO, 2007),

ESAP had about 87% of the 38 million people in global fisheries (FAO-SOFIA, 2006). This possibly represents only those who are full-time fisherfolk and aquafarmers. The number of persons who provide labor for the various stages of fishing and aquafarming and the ancillary industries, including net making, boat and transport carrier construction, fish processing, feed milling, ice making and trading must be immensely more. Bangladesh alone employed about 12 million people in the fishing and aquaculture industry (Department of Fishery of Bangladesh, 2003).

2.2.6.1 *Capture fisheries*

Capture fisheries have either stagnated or dwindled in most of the world. Historically, the vast seas and the inland lakes, rivers and canals were rich sources of fish. With relatively little effort, people could harvest plenty of fish from waters close to the shore and meet their demand. They thought the sea was an inexhaustible source of food. As the human population increased and the demand for fish grew, people gathered more and more knowledge and technology to quickly and safely go farther into the ocean in search of more fish. The modern fishing fleet, with cold storage, processing facilities, fish-scouting airplanes and sophisticated acoustic technology, can detect the size and nature of fish schools in the open sea and at various depths. This technology, coupled with extremely efficient fishing gear, including the purse seine and trawl nets, increased marine production dramatically. But unscrupulous application of technology eventually resulted in overfishing and depletion of the oceans' fishes (FAO-SOFIA, 2006). Despite caution from scientists, many of the rich marine fishing grounds all over the world, including ESAP, were excessively exploited for human food, industrial raw material for fish meal in farm animal feed, vitamin oils, soap, isinglass for wine purification and other uses. As a result, 8% of the marine fisheries have been depleted, 16% overexploited, and 52% fully exploited; 21% moderately exploited and only 3% remain underexploited (FAO-SOFIA, 2006).

The inland lagoons, rivers, canals, floodplains and other open waters were not excepted in many countries (FAO, 2007). Effective enforcement of conservation rules for marine or inland open water fisheries resources is seldom possible. Aquatic habitat change or destruction from massive construction of embankments for flood control, drainage and irrigation, construction of weirs in rivers, excessive surface water withdrawal, aquatic pollution from agricultural pesticides or indiscriminate release of industrial effluents and unplanned construction of rural roads and culverts that obstruct fish movement have all contributed to the destruction of marine fisheries. In at least six ESAP countries, China, India, Japan, New Zealand, South Korea and Thailand, fish catch clearly declined (FAO, 2007).

2.2.6.2 *Aquaculture fisheries*

As opposed to the decline in capture fisheries, aquaculture production since the 1950s increased steadily, with the 1980s described as spectacular, largely from the significant development of aquaculture knowledge, science and technology (FAO, 2007). Significant increase in the global human population, the reduced supply of food fish and the high price of exportable aquatic species from open water because of the increased demand stimulated aquaculture practices to quickly develop and flourish. Farming various aquatic organisms became profitable.

Within global aquaculture, ESAP aquaculture rose from 54% in 1950 to 90% in 2004 (FAO, 2007). The first seven ESAP countries in gross aquaculture production by volume, including aquatic plants, in 2004 were China, India, Philippines, Indonesia, Japan, Viet Nam and Thailand. China alone produced 41,661,660 tonnes, accounting for 78%; the next six countries accounted for 17%; the remaining countries 5%.

The value of ESAP aquaculture products was estimated at nearly US$56 billion, which was about 80% of the global value. Although the rest of the world produced 10% of the global production volume, ESAP contributed 20% of the value, indicating they produced more higher-value items. Within ESAP, China alone accounted for 66% of the total value; six other top countries, Japan, India, Viet Nam, Indonesia, Thailand and Bangladesh, together with China, exceeded 92% of the ESAP value (FAO, 2007).

The estimated numbers of employment in aquaculture of ESAP countries varied greatly, depending on the production and its socioeconomic importance. China had the highest numbers, reflecting its production. In some countries employment could be broken down according to the species involved. For example, shrimp aquaculture in Bangladesh employed about 600,000 people (Karim, 2003).

In many ESAP countries, fish was a major source of animal protein: Cambodia 75%; Bangladesh 63%; Philippines 52%; and China 32% (FAO-SOFIA, 2006). It was not easy to get reliable data on per capita fish consumption, since fisheries products were varied, were not solely for human consumption and often it was hard to separate imported and exported fish products. Fish consumption per capita in selected ESAP countries stayed the same from 1969 to 2002 for some countries, like Japan, while in others, such as Cambodia and China, it increased three- to fivefold. The inference is that the increase in ESAP population increased fish consumption tremendously.

Finfish from freshwater, marine and diadromous species—species that use both marine and freshwater habitats during their life cycle—constituted 46% (24,526,070 tonnes), aquatic weeds 25% (13,453,710 tonnes), mollusks 22% (12,022,658 tonnes), crustaceans 6% (3,324,779 tonnes) and miscellaneous aquatic animals less than 1% (393,037 tonnes) of total production. Value from freshwater, marine and diadromous finfish was 49%, crustaceans 23%, mollusks 14%, aquatic weeds 12% and miscellaneous aquatic animals 2%. Of the total finfish production in 2004 of 24,526,070 tonnes, carp accounted for about 64%; by value, it contributed 48%.

In most ESAP countries, aquaculture started in the freshwater ecosystem with mainly carp and carp-related species. In Indonesia and the Philippines, it started in the brackish water ecosystem, mainly for culturing milkfish in the tidal flats.

When aquaculture began, it was simple and entirely based on stocking wild fry. It used no liming, fertilization, artificial feeding or aeration of the pond. It depended on either rainwater or high tide for its water supply. With time, AKST gradually developed and aquaculture underwent rapid changes. An important milestone was the development of artificial spawning technology. That made it possible to produce quality fish and crustacean fry in an artificial environment on a commercial scale. The technology was first developed and commercially used in the 1950s in China. It soon spread to Bangladesh, India, Thailand and most other Southeast Asian countries. While low levels of aquaculture were the general practice in the beginning, the trend has been toward intensifying pond culture and has been driven by an increasing demand for fish and decreasing amount of land suitable for expansion.

Intensification entailed any combination or all of the following:

- Developing artificial spawning techniques to produce fry of desired species on a commercial scale
- Developing superior brood stock by selective breeding to produce superior genetic quality fry
- Liming and fertilizing the pond to induce the growth of natural food organisms
- Formulating and using balanced artificial feed to promote good growth
- Using pumps to ensure and stabilize the water supply
- Using artificial aeration to ensure oxygen is supplied in all layers of the water to increase carrying capacity
- Using pesticides to control predatory or harmful organisms
- Using probiotics to maintain the quality of pond environment
- Using genetically improved broodstock
- Ensuring freshness of the produce and a good market price with good postharvest care

Aquaculture in Asia has become characterized by its wide diversity in species and culture systems. It includes freshwater, brackish water and marine ecosystems. The species used for aquaculture production includes many finfish, shrimp, crab, oyster, mussels, abalone, sea cucumber and even seaweed. Aquaculture has been practiced in various systems: earthen ponds, tidal flats and paddy fields with peripheral dikes, concrete tanks, raceways, pens, cages and racks. Monoculture, polyculture and integrated aqua-agriculture have been developed to suit the region's diversified aquatic environment. However, pond culture remains the main source of aquaculture production in most countries.

Pen culture appears to be extensive in lakes and lagoons in the Philippines, where it is used mainly for milkfish and tilapia. The pens are enclosures made of synthetic or non-corrosive metallic mesh resistant to salt and sun and to crab cuts. The pen area may be enclosed by mesh on one to all four sides, depending on the topography. The bottoms and the tops of the enclosures are open, with no netting. Aquaculture of tilapia, catfish, sea bass, some species of carp and marine eel in fish cages has been popular in Japan, the Philippines, Thailand and Viet Nam. Like pens, fish cages are also made of synthetic, metallic or plant material, but unlike pens, the cage bottoms are closed. The tops of submerged cages are also closed. The cages are set in large ponds, lakes, rivers and bays. Racks made of synthetic fibers are hung in protected areas in the sea or in backwaters for oyster culture; the practice is common in the Philippines and some other Southeast and East Asian countries.

Aquaculture in seasonal monsoon water and floodlands, either in association with or alternating with paddy rice, has rapidly gained importance in Bangladesh, Cambodia, China and Viet Nam. The favored species are carp, tilapia and prawn. In Bangladesh, about 50,000 ha of low-lying land that allowed only one crop of paddy rice during the dry season from January to May is now under prawn and carp aquaculture during the wet season. This multiple use has created excellent opportunities for rural farmers to enhance their income and nutrition. Some studies have suggested that this culture area should increase to at least 80,000 ha by 2015 (Karim, 2003)

2.2.7 Organic agriculture

Organic agriculture had two faces in ESAP. On one side, a small sector grew certified organic produce for the home market and for export to industrial countries. On the other side, a larger proportion, mostly subsistence farmers, farmed organically because they could not purchase or afford synthetic inputs. Organic farming could still be considered fringe farming and would only benefit a few producers for domestic or export markets. However, in recent years, driven by the rising popularity of organic products and the often higher financial return, many conventional farms are converting to organic farming.

Concerns about the environment and food safety in the use of agrochemicals in agriculture, especially pesticides, have led groups of farmers to form the organic movement and rely upon traditional methods of soil, nutrient and weed management. These methods have been designed to make the best use of natural cycles of nutrient flow, pest and disease control and competition to control weeds. To them, modern organic farmers add new technology not based on synthetic fertilizers or chemicals, and more recently, not on genetically modified organisms. The organic movement spread worldwide and now includes biological agriculture, ecological agriculture, nature farming, permaculture and biodynamics (IFOAM, 1996; FAO-WHO, 1999).

The area under commercial organic cultivation in ESAP is generally less than 1%. It lags behind Europe and Latin America, in part because development and uptake by farmers have been hampered by lack of supportive government policy in many countries (ESCAP, 2002). Bangladesh had the largest proportion of land, 1.9%, devoted to organic agriculture, with Sri Lanka 0.65%, China 0.6%, and Japan 0.56% also relatively large contributors (Willer and Yussefi, 2006). All other countries fell below 0.2% and most below 0.1%. This analysis omitted Australia because it was hard to compare with other data. Australia had, by far, the largest area of certified organic agriculture in the world, 13 million ha and 40% of the world area, but it was not a country with large organic fruit, vegetable or cereal production. The reported area included 11 million to 12 million ha of extensive zero-input grazing land of low productivity, with few products that enter the certified organic market.

2.2.7.1 Crop organic farming

A wide range of cropping techniques was employed to replace external chemical inputs with ecosystem functions (FAO, 2002). Organic management techniques were devised to support an integrated and holistic agroecosystem, which inhibited the growth of weeds, pests and diseases but enhanced favorable biological activity. The holistic and integrated approach fosters beneficial processes and interactions like those occurring in natural ecosystems, encouraging internal stability rather than relying on external control measures. It aims to recycle nutrients, conserve energy, soil and water and to preserve biodiversity.

Developing good soil structure, biological activity and fertility is central to organic farming, because they are crucial to good plant health, which is important in resisting pests and diseases. For example, comparison of soil under organic management and conventional management in kiwi fruit orchards in New Zealand revealed that organic orchard soils had higher pH, higher soil cation exchange, more calcium and magnesium, more potentially mineralizable nitrogen and biomass carbon, greater size and activity of the microbial population and greater earthworm populations, although it had lower phosphate (Pearson et al., 2005). Some of the known organic cropping techniques include:

- Selecting crops and varieties that best suit the climate and agroecological system and have disease resistance or tolerance
- Rotating crops, including fallowing and herbal leys
- Intercropping and using undercrops, including mulching and animal grazing, for controlling weeds and preserving the habitat for beneficial insects
- Using solarization
- Applying animal and green manure, especially legumes, turning in crop residues, composting and using effective microorganisms
- If necessary, using approved mineral-bearing rocks and foliar fertilizers to help return nutrients in organic matter
- Using biopesticides like neem and parasitic insects for managing biological pests
- Using mechanical barriers

FAO warned that comparing yields between organic and conventional systems were meaningful only over time because high yields in conventional farming are often based on "exploitative systems that degrade land, water, biodiversity and ecological services on which food production depends" (FAO, 2002). Conversion to organics from high-yielding conventional systems often results in a drop in gross yield of the marketable commodity; the degree of drop might vary considerably. Conversion from low-input, often traditional systems could raise productivity by optimizing the use of local resources (FAO, 2002; IFAD, 2002). Additionally, conversion to organics in medium-potential areas in the tropics could show good performance (FAO, 2002).

2.2.7.2 Organic livestock

In organic agricultural systems, similar to traditional approaches to agriculture, animals are incorporated into mixed animal agriculture and cropping, often with the addition of agroforestry. At the other end of the spectrum are large single-animal enterprises, such as the dairy industry in New Zealand. To the unpracticed eye, these would look like conventional farms. The difference lies largely in the organic management of pasture, manure disposal, inputs permitted and practices that allow animals to express their innate behavior. Organic animal agriculture practices include:

- managing the soil based on appropriate stocking rates and sympathetic grazing regimes to minimize damage to soil structure and compaction
- providing good-quality drinking water
- providing organically grown feed
- giving all animals conditions that allow them to perform all aspects of their innate behavior, including free access to graze and range on a wide variety of pasture and browsing species
- using natural health remedies as much as possible, with resort to synthetic veterinary medicines as a last option to prevent suffering

Intensive raising of animals on feedlots and battery cage confinement of hens are definitely not organic agricultural practices.

2.2.7.3 Organic aquaculture

Organic aquaculture has lagged behind the development of other organic agriculture. Organic aquaculture can take place in fresh water, brackish water and the sea to produce fish, crustaceans, mollusks and plants. New Zealand has been one of the largest producers outside Europe, with one salmon farm producing 500 to 800 t of organic salmon. Other organic aquaculture in the region includes shrimp in Indonesia, Thailand and Viet Nam; mussels in New Zealand; and salmon in Australia. One constraint has been sourcing acceptable nutrients for the farmed species (FAO, 2002).

Conventional shrimp farming in Southeast Asia has caused a great deal of concern about its negative social and environmental effects. The challenge for organic aquaculture has been to provide much-needed protein-rich food without damaging the environment. Food for the farmed species needs to come from sustainably managed fisheries. It should come from local fishery products not suitable for direct human consumption, free from synthetic additives and contaminants and be fed only to aquatic species with naturally piscivorous feeding habits (FAO, 2002). FAO concluded that with the "introduction of appropriate water and nutrient management techniques, the prospect for the increased production of farmed organic aquatic plants and mollusks is considerable" (FAO, 2002).

2.3 Trends in AKST: Organization and Institutions

2.3.1 Organizations and institutions that helped shape AKST in ESAP

Agricultural development often depends upon the actions of a large number of different actors and organizations, including those involved in agricultural production and marketing, as well as those concerned with research and development, training, extension and public policy.

2.3.1.1 Composition of different AKST organizations in ESAP and their institutional behavior

Knowing the different AKST actors and how they behave is important for understanding how these actors and institutions interact with each other in response to challenges and opportunities. This is especially vital as the nature of farming in this region and elsewhere constantly changes under the backdrop of a fast-paced knowledge economy. Plateauing crop yields, compounded by declining water and land availability, accelerated global trade liberalization, concerns on food safety and demand for standardization of agricultural practices all make the production, marketing and trade of agricultural produce more complex (see chapter 3).

National public research and development institutions within ESAP. Most national agricultural research systems (NARS) in ESAP were established in the 1960s. They are typically organized under a ministry, as an autonomous agency or as a coordinating council (Dar, 1995). Although they differ in operation, they are similar in policy and program formulation. Each has research agencies and stations dedicated to a specific commodity, and they are usually attached to the ministries of Agriculture, Natural Resources, Science and Technology or Higher Education. Most NARS are organized top-down and are government funded but have the autonomy to craft their own research programs.

NARS are organized nationally, regionally and locally. The national research organizations conduct basic and applied research on areas strategic to national interest and importance. Regional centers undertake applied research of regional significance and local research stations perform adaptability verification trials and fine-tuning of technology generated by the national or regional research centers. This system allows for work specialization and complementarity and provides for location-specific technology. Collaborative research is common among members, as well as with the private sector, civil society, Consultative Group on International Agricultural Research (CGIAR) centers and international donors. Collaboration fosters task sharing to provide scientific solutions to common agricultural problems, expands the sources for research and development investment and cultivates long-term partnerships and links.

Private sector participation in research and development with AKST is quite limited and mostly complements, rather than substitutes, for continued public research. The bulk of private research and development has been in developing new crop hybrids, animal breeds, chemical pest and disease controls, veterinary medicines, commercial livestock feeds, food storage, packaging and processing technology. The technology is often most suited to a small subset of the needs of small-scale farmers, is typically capital intensive and is covered by intellectual property rights.

National extension systems within ESAP. Every country in the region has a public department that provides agricultural extension services. Four models of extension systems prevail in most ESAP countries (Sulaiman and Hall, 2005) with approaches that are centralized, decentralized, NGO led or private sector led.

Centralized approach. Under this scheme, extension services are centrally planned, funded and implemented by units attached to the Ministry of Agriculture. Programs are mainly supply driven and use a top-down approach, with little participation from farmers or other stakeholders and with little or no accountability to clients. Technology dissemination is the primary objective. It is unclear if extension has been responsive to the drastically changing information and support needs of farmers in recent decades (Sulaiman and Hall, 2002; van den Ban, 2005).

This is true in India, China and a number of other Asian countries where extension policy is developed centrally in a fairly prescriptive fashion. Although approaches have evolved over the long term, it is not clear how lessons from their experiences are used in developing policy. In fact, development fads and encouragement from international donor agencies seems to be a major source of implementation. While these programs might be conceptually laudable, making them work on the ground is much harder. Furthermore, these major shifts often lock up extension in a particularly operational mode until yet another new idea comes along.

China illustrates quite a different and interesting approach to agricultural extension. The National Agricultural Extension Center under the Ministry of Agriculture formulates national extension policy. The center draws up extension strategies that link agricultural programs with other agencies and provides training and supervision over provincial agents. With the country's move toward a market-oriented economic system, rural extension services have expanded and diversified according to local resource and market development needs (Yonggong, 1998). Arrangements have been restructured to help farmers relate to new market opportunities more effectively. An incentives structure has been developed to allow profit sharing between extension workers and farmers. The policy, while insufficient to provide specific courses of action, allows extension agents and farmers to pursue local, pragmatic program innovations. This has been important in responding to the rapidly changing socioeconomic environment.

Decentralized approach. In response to demand for decentralized governance, this approach promised to improve farmer control and make extension services more demand driven. However, the lack of sufficient preparation by extension management and the institutional inertia in most government bureaucracies has failed to deliver on these promises. Despite this, widespread clamor for decentralization suggests implementation problems might eventually be overcome.

The cases of Indonesia and the Philippines highlight the complications of making broad policy prescriptions. The foreseen benefits of decentralization, primarily the devolution of authority and decisions locally, have not yet been fully realized. The effectiveness of this approach depends on the skills and vision of local government officials. This suggests that policy instruments such as decentralization need to be accompanied with capacity development. Also, local stakeholders need to understand the importance and rationale for strengthening local knowledge networks. Since the performance of extension is dependent on these systems, stakeholders need to have the skills to analyze them, diagnose system failure and design remedial measures. Capacity development is not only necessary to successfully implement decentralized approaches; it is indispensable if local stakeholders are to be more active in the policy process.

For the Philippines, inadequate funding curtailed the effectiveness of devolved extension. Experience suggests that with decentralization came a trade-off between the effectiveness of technology transfer, which seems to have suffered, and the accountability of the system to its clients, which seems to have improved. This has led to the emergence of pluralistic extension systems, such as that provided by private input suppliers, NGOs, farmer associations, agroprocessing companies and private consultants.

NGO-led approach. NGOs have long articulated the needs of small farmers and other socioeconomically vulnerable groups. They have advocated more equitable and sustainable economic development and poverty alleviation programs in Bangladesh, India, Nepal, the Philippines, Thailand and Viet Nam. The range of their activities has varied: agroforestry in Nepal, tea production and vaccine research on cattle disease in India, soil and water conservation techniques in the Philippines. The extent of NGO inputs in research and extension, especially in technology adaptation and dissemination, has been quite large, as has been the magnitude of their organizational network.

In most countries, the relationship between NGOs and the government has often been adversarial rather than cooperative. But, whenever cooperation is possible, results can be extremely fruitful. In India, the government has taken concrete steps to establish close ties with NGOs. The Indian Council for Agricultural Research (ICAR) set up farm science centers to serve as centers for demonstration and training in "scientific farming" to open NGO access to the public research system (Sulaiman and Hall, 2005).

The central role of NGOs and farmer organizations in reducing poverty has chiefly focused on building social capital, catalyzing entrepreneurship and disseminating public information. NGOs rely on the concept of participatory research, where all stakeholders play a role in setting and implementing the research agenda. Initial attempts at conducting participatory research gave greater priority to involving poor people in evaluating new technology, rather than in setting priorities for the research (Hazell and Haddad, 2001). Their involvement in diagnosing problems and field testing technologies has provided national researchers with useful information, resulting in more useful products for farmers (Hazell and Haddad, 2001).

Participatory research promotes organizational and skill-building capacity to communities to help solve collective problems and resolve conflicts. However, it is constrained by the need for multidisciplinary teams willing to work together, respect and value each other's knowledge and appreciate the high initial cost of many personal interactions among team members. In a project to develop pest control measures in Ghana, farmer participation costs increased by 66% and accounted for 80% of researchers' time, although this might lead to higher returns in reducing time needed to identify promising technology (Hazell and Haddad, 2001). Also, developing the farmers' own capabilities in developing improved pest management systems, conducting field trials or breeding could be cost effective in adapting technology to diverse local needs. Thus there are no cost tradeoffs where participation might mean the difference between success and failure in technology development.

Private sector-led approach. Extension services can be completely privatized (Siamwalla, 2001). New Zealand is the only ESAP country with fully privatized extension. Farmers and extension agents sign profit- and risk-sharing contracts. The extension agent serves as a consultant, selling services to farmers for a fee. Thus, consultants are important sources of information and advice to large commercial growers and are valued for customizing advice to individual farms. They also provide expert counsel to international development agencies because they have a collective memory of what worked in formulating new initiatives. There is, however, the risk that employing the same consultants and advisers will lead to adopting old recommendations, some of which have failed in the past.

Technology transfer by the private sector through the system of contract farming is popular in Thailand and the Philippines. The Thai case, depicting the soybean trader sitting astride commodity input and credit markets, is an example of informal contract farming. More formal contract farming exists. A well-known example is in poultry farming pioneered in Thailand by the Charoen Pokphand Company, a firm that later became a large conglomerate, with agribusiness interests in other Asian countries. Charoen Pokphand forged contract growing arrangements with small poultry growers. The arrangements varied from a guaranteed wage contract to a guaranteed price. Charoen Pokphand brought in a hybrid breed from the Arbor Acres Company in the US and set up large automated feed mills, which remained the core of their operations. In Australia, the Grain Growers Association supports the grains industry through direct research and development funding, largely in plant breeding and grain-quality testing. In recent years, the Grain Growers Association has supported research on developing commercially viable biological control agents and development of best management practices.

While these approaches have their strengths, innovations in providing extension services should be viewed not only from an institutional perspective but also from a functional one. Extension can still occur even without organizations, since imparting knowledge has always been between individuals who trust each other, rather than from an external agent. Extension was once understood as "extending the knowledge imparted in class to those who cannot attend the class," suggesting that extension is not just the mechanical transfer of technology or information, but also instructive.

Regional and international research and development institutions. Regional and international research organizations have been set up to meet regional and global demands in agricultural research. In 1960, CGIAR set up 15 international commodity research institutes, a third of which are based in Asia. These international agricultural research centers (IARCs) and many other international programs based elsewhere have done considerable work within ESAP. Aside from making headway in global research on frontier and cutting-edge science, the greatest achievements of these centers have been to encourage open germplasm exchange, support human resource development and training, and create links with the national agricultural research systems.

Traditional, local and indigenous knowledge systems. Traditional knowledge, indigenous knowledge and local knowledge are often used interchangeably to refer to the matured and long-standing traditions and practices of regional, indigenous or local communities, which encompass their wisdom, knowledge and teachings accumulated through generations of experience, careful observation and trial-and-error experiment. In many cases, traditional knowledge has been orally passed on for generations through stories, legends, folklore, rituals, songs and laws. In agriculture, these are built up through generations of farming and managing forest and water ecosystems. While traditional knowledge is entrenched in these communities, it is also considered dynamic because it adapts to and incorporates new knowledge from outside sources to suit gradually changing environments (Grenier, 1998).

Traditional knowledge has helped maintain and improve the livelihood of farming communities. In many rural communities, it governs local decisions in agriculture, health care, food preparation, education and natural resource management (Warren, 1991). Traditional knowledge is being recognized as a base for many sustainable development initiatives, such as sustainable agriculture and natural resource management, enriching global agricultural knowledge. It has produced lessons and insights in addressing rural hunger and poverty and accounted for on site crop genetic conservation, crop diversification, regenerative soil and water management, organic agriculture and ecological pest management. Much of sustainable agriculture has its roots in traditional and indigenous practices that are viable because of generations of innovation and improvement prompted by the sharing of knowledge and resources.

In general, traditional systems are perceived to have great potential because (1) they are inexpensive and may be paid for in goods or services, (2) they are readily available and accessible even to those who do not have cash income, (3) people are more comfortable using them than western technology and (4) when combined with modern practices they provide more options for innovation in dealing with complex agricultural problems. However, traditional agriculture is labor intensive. This may be viewed as either a disadvantage or an advantage, depending on the social circumstances. For example, the additional labor might keep people from other economic activity. On the other hand, it could provide meaningful employment for rural people who would otherwise migrate to urban areas, thus creating adverse social effects such as leaving behind a household without its male head and potentially contributing to urban unemployment and poverty.

In most cases, traditional knowledge is not only socially desirable but also economically affordable and sustainable and poses little risk to rural farmers. Since traditional knowledge evolved gradually within the community, it is appropriate to the needs of the local people (Rouse, 1999). Traditional systems are more directed toward self-reliance and self-sufficiency than some modern technology (Fernandez, 1994). However, traditional agriculture has not been able keep pace with increased population pressure, evidenced by the great famines of the 1950s, 1960s and 1970s in Bangladesh, China and India. Traditional systems appear unable to provide sufficient food for current urban populations. What they can do, however, is provide product diversity equal to, if not greater than the total biomass production of conventional equivalents, while conserving scarce resources and providing food security for the producers (FAO, 2002). In general, the greater the biological diversity of the agricultural system, the greater is its ability to withstand adverse climatic and pest events (FAO, 2002). In addition, there is historical evidence of wetland rice yields in India higher than present yields supported by chemical fertilizers and pesticides. In the 1700s, the yields in 800 villages near Madras were reported to have averaged 3.6 t ha^{-1}, surpassing 10 t ha^{-1} in some areas, whereas the current yield in that region averages 3.1 t ha^{-1}. Genetic diversity was the main weapon against pests and diseases; but, from using about 30,000 traditional rice varieties, India now uses only a few, with 75% of rice produced coming from only 10 varieties (ESCAP, 2002).

Today about 70% of the world's indigenous peoples live in Asia and the Pacific, where they are a major subgroup of the rural poor. Indigenous knowledge and traditional agricultural systems can provide answers to their food security needs. However, resource access is important. Marginalizing many indigenous communities could lead to the eventual loss of traditional knowledge. Many of these communities are being deprived of the ability to lead the lives they value (IFAD, 2002).

Traditional knowledge is increasingly becoming acceptable to the scientific community. In fact, "informal" research is being done in local communities by using traditional knowledge (Stanley and Rice, 2003). In contrast, much past research failed from the lack of knowledge and understanding of local practices. Technology generated by formal research institutions can complement and improve indigenous methods.

Before modern agricultural practices were developed, indigenous communities had already devised methods to ensure the success of their agriculture. A common example of traditional knowledge emanating from communities is use of the neem tree (*Azadirachta indica*) in India as a natural fertilizer, pesticide and medicine. Knowledge of indigenous practices on crop protection and fertilization can be appreciated when developing appropriate programs for pest and soil management within the capability of farmers and that do not cause adverse effects on either the community or the environment (Varisco et al., 1992). It was estimated that in 1985, plant-based medicines, many first discovered by indigenous peoples, valued at US$43 billion were sold in industrial countries (Posey and Dutfield, 1996). As advances in biotechnology broaden the range of life forms containing attributes with commercial applications, the full market value of traditional knowledge will definitely increase.

Traditional knowledge is also important for food security and genetic conservation. In Nepal, a centuries-old seed management system allowed farmers to grow and protect their seeds (Timsina and Upreti, 2002). Modern plant breeding owes much to the landraces bred, conserved and developed by traditional communities over the millennia. These local varieties have been the continuous source of genes used to develop and improve high-yielding varieties.

In India, a study revealed that traditional health control and treatment systems were effective in curing ailments in animals, including dysentery, arthritis, dog bites, coughs and colds, anestrus, wounds, bloat and diarrhea. Although modern veterinary medicines provide quicker cure, traditional treatments are cheaper, locally available, and have fewer side effects (De Amitendu et al., 2004).

Indigenous people have practiced sustainable forest use and management for centuries. Jackson and Moore (1998) found that although forest fire is often destructive, indigenous use and management of fire were significant in forest management and conservation. For instance, in Indonesia and Nepal, fires were intended to maintain grasslands for animal agriculture. In central and northern Australia, aboriginal communities had sophisticated applications of fire that took into account seasons, patterns of burning, specific effects on wildlife and plants, and exclusion of fire from particular areas and vegetation (Jackson and Moore, 1998). Aboriginals also used fire to encourage growth of grasses for target wild animal species, particularly kangaroos and wallabies.

In China, communities in Yunan developed a system of classifying forests and forest systems according to their function and products, such as forests for building materials, cash crops, landscaping and graveyards, and protected rattan (Table 2-2).

2.3.1.2 *Roles of different organizations in generating, disseminating and adopting AKST*

Agricultural entities in ESAP vary in number, composition, capability and performance. These entities include the stakeholders that NARS serves and affects or that can affect NARS. The main roles of stakeholders are varied and their interrelationships evident (Table 2-3). The research agenda crafted nationally reverberates in the activities of research and extension personnel, affecting the decisions of farmers to adopt a technology. Feedback mechanisms allow for the refinement of technology and the accompanying research and extension. Collaboration among stakeholders influences investment, research decisions and information dissemination.

Organizations realize that research, development, training and extension services need to develop and maintain partnerships with farmers, NGOs, producer organizations, agroprocessors, agribusiness houses, traders, retailers and consumers (van Mele et al., 2005; Hall, 2006). Developing wider links is essential for improving the performance of organizations involved. The optimum use of AKST can be best facilitated by addressing the barriers to change caused by some institutional rigidities (Box 2-2).

2.3.1.3 *Transformation of AKST institutions*

As production agriculture became increasingly informed and scientific, new researchable areas have emerged in biotechnology, sustainable agriculture, and information and communications technology (ICT) and climate change. A diversified institutional structure in agricultural research, development and extension has emerged nationally and globally with profound effects on our ability to produce food and manage our natural resources and the environment.

Most research efforts have been done by national public research institutes, state colleges and universities and international research centers. The private sector has played a marginal role, especially in basic research. In recent years, farmer organizations, NGOs and the private sector have emerged as key players. As farmers became more organized, experience gained from participatory research schemes and other rural development projects has been used. This has allowed new approaches to research to emerge that put the farmer at the center of development, not just as a user of the technology. NGOs have complemented the role of the state or filled a gap generated by weakness in public extension agencies. Incentives for research and development have increased private sector biological research. The sector accounts for approximately 80% of plant biotechnology research worldwide (Chaparro, 1999). The private sector has become important in basic and adaptive research, changing members' role from users of the knowledge generated by the public sector to generators of knowledge. Issues of property rights and plant breeder's rights and their effects have also emerged. This evolving institutional environment needs to be considered in strengthening AKST for sustainable agriculture and in developing new approaches of cooperation. Faced with diminishing funds from traditional sources, partnerships among stakeholders should be founded on collaboration and mutual benefit.

In the past 25 years, many ESAP countries have changed how agricultural research and extension is organized and funded. Toward the end of the 1990s, roles of public and international research organizations shifted and support for public agricultural research slowed down (Pardey et al., 2006) (see 2.3.3). Public agricultural research became less understood and more closely scrutinized. Some considered the world's food supply problem solved; some thought that public research was constrained by factors other than research or that the private sector should take over the job (Pardey and Beintema, 2001). Government decisions to continually underinvest in public research exacerbated the global gap in scientific knowledge. For instance, new culti-

Table 2-2. *Examples of known indigenous agricultural practices emanating from traditional knowledge.*

Sector	Indigenous agricultural practice
Crops	Indigenous indicators to determine favorable times to prepare, plant, and harvest gardens; land preparation practices; indigenous ways to propagate plants; seed storage and processing (drying, threshing, cleaning, and grading); seed selection practices; indigenous methods of sowing (seed spacing and intercropping); seedling preparation and care; farming and cropping systems (for example complementary groupings); crop harvesting and storage; food processing and marketing; pest management systems and plant protection methods.
Livestock	Indigenous methods of animal breeding and production; traditional fodder and forage species and their specific uses; animal-disease classification; traditional ethno-veterinary medicine.
Forestry	Management of forest plots and their productivity; knowledge and use of forest plants and animals; understanding of the interrelationships between tree species, improved crop yields, and soil fertility.
Fisheries	Integrated aquaculture production into cropping systems such as the rice-fish systems; use of larva-eating fish.

Source: Grenier, 1998.

Table 2-3. *NARS actors and roles in the generation, promotion, dissemination and adoption of AKST.*

Stakeholder	Main Roles
Policymakers	Authorize the existence of the NARS; Provide mission and resources; Formulate public policies.
NARS	Set research, development and extension (RDE) directions, priorities, policies; Formulate RDE agenda and allocate resources among different priority areas; Plan and conduct research programs/projects; Submit project reports and inform others of research findings; Submit budget requests and undertake resource generation activities; Develop working linkages and establish RDE networks; Provide technical assistance to the public with regards to AKST applications.
Training and Extension agencies including NGOs/PVOs (change agents)	Disseminate knowledge and encourage adoption of new technologies thru training, demonstration trials, field days, distribution of info materials; Provide feedback to researchers on new research agenda and farmers' response to introduced technologies; Improve skills of recipients for adopting new technologies; Provide access and support to specific farmer groups via social mobilization efforts (farmer empowerment).
Farmers	Adopt and adapt research results on a selective basis; Collaborate with researchers and change agents in testing and evaluating the suitability and appropriateness of recommended technologies; Provide feedback on utility of technologies; Combine traditional knowledge and modern technologies based on resource access, appropriateness of technology, and skills; Contribute taxes to government to partially finance public RDE.
Private sector (Agro-industry, lending agencies buyers and distributors)	Adopt and commercialize technologies with business potentials; Provide farmers access to resources, technology and markets; Finance and conduct limited complimentary RDE activities in their area of business; Contribute taxes to government to partially finance public research.
Consumers	Demand farm produce with certain attributes; Contribute taxes to government to partially finance public RDE.
International & regional agricultural R&D organizations	Provide training and technical guidance to local RDE personnel; Provide access to global germplasm and technology; Undertake collaborative programs with the NARS on areas of common concerns; Generate international funding for common global RDE initiatives; Provide leadership and coordination in global information and knowledge exchange.
Donors	Supplement meager RDE funds of recipient NARS; Provide guidance and RDE directions.

Source: Authors' elaboration.

vars carry forward not only the genes of earlier varieties but also the crop breeding and crop selection strategies used by earlier breeders.

Policies and practices that facilitate and encourage accumulating knowledge and adopting technology are equally important. Without them, discoveries and data improperly documented or inaccessible are lost when researchers leave or institutions are unstable. This happens in fund-strapped research agencies in most developing countries; inadequate and irregular funding results in fast staff turnover and limits the functioning of libraries, state-of-the-art laboratories, nurseries, databanks and gene banks.

The limited public funding for agricultural research shifted from the traditional agenda of improving productivity to new concerns. For example, in 2000, NARS began promoting commercially viable technology to accelerate research use (APAARI, 1999).

2.3.1.4 *Interactions and links among AKST organizations*

Since the early 1990s, research managers have recognized the need to work with nontraditional partners and engage in more meaningful research consultations. Donors and policy makers recognized partnerships as a strategy for agricultural development. The advantages of partnerships are obvious: pooling diverse expertise, leveraging scarce resources and enhancing competency. Technology innovations are seldom generated by individual research agencies; they come from transnational knowledge generation, dissemination and application (Chaparro, 1999).

In the same vein, the relationship between public and private sectors in agricultural research and development has changed around the world (James, 1996; Byerlee and Echeverria, 2002; Speilman and von Grebmer, 2004; Hall, 2006). This change arose from the diversity of actors outside the public sector, increasingly complex agricultural development needs, declining financial capability for research investment in developing countries and re-evaluation of the role of the state in research and extension. However, only a few cases of public and private partnerships in agricultural research and extension were successful. Problems included insufficient accounting of the actual and hidden costs of partnership, conflicting goals, lack of transparency, persistent negative perception across sectors, undue competition over financial and intellectual resources, and lack of working models from which to draw lessons and experiences (Spielmen and von Grebmer, 2004). The unresolved issues on intellectual property rights and genetically modified organisms made public and private partnerships increasingly difficult.

In Australia and New Zealand, farmer organizations provided a framework for partnership between researchers and farmers. Farmer organizations were also equal partners

Box 2-2. Barriers to change arising from some institutional rigidities

- Linear approach to technology development and promotion: In this model, extension is a conduit for transferring technologies developed by the research systems with or without participation by the farmers. Though there are limitations of this model, it continues to be the dominant paradigm determining investments in agricultural research and extension. Administration and funding by different departments or ministries further constrain development of relationships.
- Due to the perceived hierarchy between research and extension, the process is top-down with limited feedback and each one blames the other for poor performance of technology a and adoption. Hierarchies also exist between biological and social scientists, preventing interactions among these two groups.
- While farmers, NGOs and the private sector need research and scientific expertise to solve specific problems, what research offers is predetermined technologies. Meetings with the private sector have not moved beyond rhetoric, especially due to mismatch in expectations. While procedures for transferring technologies are in place, arrangements for providing technical expertise to solve problems have not been fully developed
- Narrow focus of extension on technology dissemination: Extension has been limited to the transfer of technologies and does not consider the varied needs of end users (such as market information and support services).
- Evaluation parameters within research organizations favor (1) technology development at the cost of problem solving and (2) reporting only success at the cost of learning from failures and (3) reporting only technical innovations at the cost of process and institutional innovations that facilitated the development and promotion of technologies. Similar is the case with extension, where performance is evaluated in number of farmers adopting a specific technology, inputs distributed and increase in productivity achieved. This restricts extension staff from trying other promising approaches that could potentially increase farmers' incomes.
- Focusing only on farmers as clients has restricted the interaction of research and extension at the cost of interaction and working with a range of other actors like NGOs, agro-processors, traders, private sector and producer associations.
- Interaction among the different actors (public and private; private and NGOs, etc.) are further constrained by high levels of mistrust and lack of mechanisms to develop better understanding. Though some of the public, private and NGO actors have come together as part of specific initiatives promoted by donors, this has been restricted to the duration of the project. Levels of mistrust are still considered too high between NGOs and the private sector. A lack of transparency in research and an inability of many scientists to communicate with different stakeholders have further contributed to the climate of distrust.
- Long chains of command and control constrain the ability of the different organizations, especially the public and private sector to respond quickly to the challenges from the field (or market). This is also constraining development of joint activities even when the policies favor partnerships and linkages.
- The current patterns of funding and governance (mostly public) ensure that the organizations are only accountable to the Ministry/Department funding and governing them with only weak or limited accountability to the clients. Companies in the private sector also behave as if they are only accountable to their shareholders and not to other stakeholders who can potentially influence their operations.

in extension in South Korea and Taiwan. The Bangladesh Rural Advancement Committee (BRAC), an NGO, worked with small-scale farmers on projects in poultry, feeds, diagnostic laboratories, bull stations, fish and prawn hatcheries, planting materials supply and vegetable cultivation extension. Partnership arrangements with farmer organizations for promoting technology were common. Farmer field schools initiated to address pest problems in rice became a platform for joint learning in several Asian countries. Most emerging challenges in agriculture in new marketing arrangements, contract growing, quality management and certification needed community mobilization. Continued learning, problem solving and collectivity supported by the farmer field school, albeit with a changed focus, remained important (van de Fliert, 2006).

Private and private partnerships have also been forged to better serve new markets. A reliable supply of quality produce in supermarkets is of prime importance. Contractual arrangements along the supply chain ensure reliability in volume and quality. Many companies provide seeds, inputs and credit to participating growers and procure the produce at set prices. They have also brought in new technology and provided technical advice to growers. This arrangement appears beneficial, but its success lies in enforcing contracts and maintaining trust. For farmers to gain advantage, they need to understand contracts and negotiate better arrangements.

A different partnership is emerging strongly in ESAP. NGOs formed or strengthened alliances and networks to advocate pressing issues. For instance, PABINI in the Philippines, a network of farmers, academics and researchers, opposes introducing genetic engineering technology. Organizations in research and extension might link with networks to enhance innovation.

The linear model of technology development and promotion—research to extension to farmer—continues to set patterns of interaction and alliance. However, the concept of a national innovation system offers a novel framework in how institutions help innovations feed into economic growth. Partnerships remain important in agricultural development. Forging partnerships, however, requires resources. Nonetheless, new modes of partnerships contribute to institutional change.

Although there is no blueprint for promoting partnerships, supporting stakeholder meetings or holding collaborative activities may help develop them. These partnerships have to be supplemented with effort to evaluate progress and outcomes, and participants must have the vision and willingness to make needed institutional changes (Table 2-4).

2.3.2 Capacity of AKST organizations in generating, accessing, disseminating and adapting knowledge and information

Science and technology drive economic growth. Yet ESAP countries struggle to increase research spending, upgrade their scientific workforce and improve agricultural research facilities. While accomplishing these aims reflects capability, it does not guarantee contribution to knowledge and economic development without support systems that encourage public access, dissemination and application of the knowledge and information gained (Tables 2-5, 2-6).

Agricultural research in ESAP still suffers from lack of political support, insufficient funding, minimal links between researchers and users, and inadequate library and information services (Rao, 1994). Most of the research infrastructure and institutional capacity is also weak (Dembner, 1994). The World Competitiveness Yearbook 2006 placed many developing countries at the lower rung because of inadequacies in science and technology infrastructure and capability. Korea and Singapore had relatively high scientific infrastructure, compared with the lower-ranked Indonesia, Philippines and Thailand. From 2003 to 2006, Indonesia, Malaysia, the Philippines, and Thailand did not improve in the overall world competitiveness ranking. India made a significant upgrade, from 50 to 29. These results stressed the strong need for ESAP countries to develop their own agricultural research capability.

2.3.3 Investment in AKST

Throughout the 1900s, growth in agricultural productivity considerably reduced poverty and hunger and fueled economic progress. Technological advances over the past 50 years have allowed farmers to feed twice as many people

Table 2-4. *Potential ways to facilitate institutional change.*

What to do	How best to encourage adoption to change
Learning from the emerging institutional arrangements in the region	This would necessitate a detailed analysis of cases where the various actors in specific contexts come together and collaborated to solve particular problems or address new challenges. What kind of institutional changes were made? How were these sustained?
Develop a culture of learning within the organization	Institutional learning cultivates new ways of doing things. It specifically asks questions such as what rules, habits and conventions have to be changed to do a new task or to do an old one better? (Hall et al., 2005). Organize "capacity development" programs to address the institutional barriers. Opportunities could be created and if need be specifically funded to bring about change. Bring staff together to reflect on lessons learned and assess the required elements for improvement.
Develop long-term mutually beneficial relationships	Create opportunities to bring different actors together and develop joint activities. Development of collaborative projects and the necessary resources to be mentored over a period of time. Funding could be used to facilitate the development of collaborative projects.
Use better framework of analysis	Similar to the "innovation systems approach": Analyze the patterns of interaction as a framework for planning interventions (World Bank, 2006). This would necessitate detailed exploration of the innovation systems and organization of capacity development programs.

Source: Authors' elaboration.

Table 2-5. *Rank of World Competitiveness (by Factor) of selected countries, 2006.*

Factor	Singapore	China	Malaysia	India	Thailand	Korea	Philippines	Indonesia
Overall Rank	3	19	23	29	32	38	49	60
1. Economic performance	4	3	11	7	21	41	52	61
2. Government efficiency	2	17	20	35	21	47	45	51
3. Business efficiency	7	30	20	19	28	45	44	57
4. Infrastructure	5	37	31	54	48	24	56	61
(a) Basic infrastructure	1	20	35	33	38	29	61	53
(b) Technological infrastructure	3	33	21	43	48	6	37	61
(c) Scientific infrastructure	16	17	38	26	53	12	58	47
(d) Health and environment	15	51	39	57	48	32	53	61
(e) Education	13	51	30	59	48	42	57	61

Source: World Competitiveness Center, 2006.

from less cropland. A large body of evidence closely links improved productivity to investment in agricultural research and development, averaging rates of return of over 40%, particularly for commodities with short production cycles (Byerlee et al., 2006). It is not surprising that in 2000, US$731 billion was invested in sciences worldwide, including public and private research. This represents less than 2% of the world's US$42.4 trillion gross domestic product for that year and an increase of nearly one-third over the inflation adjusted total of just five years earlier (Pardey et al., 2006).

ESAP, excluding Australia and New Zealand, spent about US$142.4 billion or nearly 25% of total global expenditures on research and development, a spending increase of about US$52 billion from 1995 to 2000. This regional trend hid two extremely disturbing developments—a large and growing gap between industrial and developing countries and the miniscule percentage of gross research and development spending for domestic AKST. The overall growth in ESAP masked that this investment was concentrated in only a handful of countries. China, India and Japan accounted for nearly 85% of the region's scientific spending in 1995, climbing to 87% by 2000. In contrast, research spending by most of the other 24 ESAP countries declined about 2%. Agricultural research and development expenditure in 2000 was a mere 5% of global science spending. Funding for AKST within ESAP, with the exception of six industrial countries, Australia, China, India, Japan, New Zealand and South Korea, could be characterized as perennially dismal and declining, if not outright stagnant, with the public sector shouldering the bulk, 92%, of the expenditures. Three typical major funding sources for public research and development were production or export taxes, direct government appropriations and external sources (Dar, 1995).

Table 2-6. *Overall World Competitiveness Ranking of selected countries, 2003-2006.*

Country	2003 (of 61 countries)	2004 (of 61 countries)	2005 (of 61 countries)	2006 (of 61 countries)
Singapore	4	2	3	3
Japan	25	23	21	17
China	29	24	31	19
Malaysia	21	16	28	23
India	50	34	39	29
Thailand	30	29	27	32
Korea, Rep.	37	35	29	38
Philippines	49	52	49	49
Indonesia	57	58	59	60

Note: The rankings are based on four factors: (1) Economic performance; (2) Government efficiency; (3) Business efficiency; and (4) Infrastructure. The technological and scientific infrastructures are under the fourth factor.

Source: World Competitiveness Center, 2006.

Table 2-7. *Total gross domestic expenditures on research and development in ESAP, 1995-2000.*

Region/Country	Total R&D expenditures (million international dollars year 2000)		Share of global total (percent)	
	1995	2000	1995	2000
Global (164 countries)	561,641	730,939	100	100
ESAP (27 countries)	142,380	194,450	25	27
China	19,469	48,247	14*	25*
India	11,678	20,749	8*	11*
Japan	89,964	99,500	63*	51*

*Share of regional total.

Source: Adapted from Pardey et al., 2006.

All national agricultural research in the region received direct government appropriations to finance their activities. In addition, Malaysia and the Philippines had either a production or an export tax on export commodities, which they partially used to augment limited funds for agricultural research and development. In Malaysia, this was done for rubber and palm oil. The Philippines taxed coconut, sugarcane and tobacco for the same purpose. External fund sources consisted of 32 bilateral donors, multilateral organizations and nongovernment foundations who generously supported the establishment of some national research, particularly in Indonesia, Korea, Myanmar, the Philippines and Thailand.

With the globalization of science, the private sector reportedly spent US$663 million on agricultural research in 2000, roughly 8% of the US$8.19 billion total agricultural research investment in ESAP countries. If private firms have limited opportunity to appropriate profits for themselves from providing agricultural technology, they lack the incentive to invest. Hence, the private sector also often relied on knowledge provided by public research. Because of this market failure and long-term risky payoff, the public sector funded most agricultural research and development, especially in developing countries (Tables 2-7 and 2-8).

The International Food Policy Research Institute (IFPRI), using pooled time series and cross-sector data, conducted several studies on the impact of government spending on agricultural growth and poverty reduction in China, India and Thailand. Results showed additional public expenditure on agricultural research and extension improved agricultural productivity the most and was the second most powerful way to reduce rural poverty. Some studies indicated that in low-income countries, a 1% increase in agricultural yield led to a 0.8% reduction in the number of people living below the poverty line (Fan et al., 2002; Byerlee et al., 2006). Over the long term, food prices were especially important because food was a large share of the expense in poor households. Employment and wages in labor-intensive production and value-added processing were also important for poor people, who depended more on wage labor.

Most previous studies on return to investment considered only public research and development expenditure, which made it difficult to compare returns to productivity growth and poverty reduction across investment portfolios. Both China and India have made great strides in reducing poverty dramatically over the last several decades. With more than 500 million people lifted above the poverty line, these two countries contributed a major share of the overall global decline in poverty (Thorat and Fan, 2007). Yet together, they still accounted for more than 40% of the world's poor. Therefore, to reach the millennium goal of halving the global number of poor by 2015 largely depends on their performance in alleviating poverty. Thailand for the past several decades has experienced rapid economic growth that has transformed the country from a predominantly agrarian society to a newly industrialized economy, much like Singapore, South Korea, Taiwan (China) and Hong Kong (China). Since the early 1960s, the Thai economy has achieved one of the highest long-term growth rates among all countries, with gross domestic product growth rates ranging from 5.5 to 11% each year from 1960 to 1995 (Fan et al., 2004). Lessons should be learned from the investment experiences of these three countries. Considering that the estimated returns were fairly recent, the results should be useful in deciding how the public sector can better allocate its limited resources to achieving economic growth, food security and poverty alleviation.

Table 2-8. *Estimated global public and private agricultural R&D, circa 2000.*

Region	Expenditures (million 2000 international dollars)			Share (percent)	
	Public	Private	Total	Public	Private
Global	23,010	12,948	35,958	64	36
ESAP	7,523	663	8,186	92	8

Source: DAR, 1995; adapted from Pardey et al., 2006.

2.4 Effects of AKST on Development and Sustainability Goals

2.4.1 Effect of modern AKST on livelihood, poverty and hunger

2.4.1.1 History of agrarian change and development

Science and technology, especially irrigation and chemical inputs, have been responsible for increased agricultural production and decreased rural poverty in parts of ESAP. However, for resource-poor farmers in drought-prone areas, the benefits have been minimal and have had environmental and social costs. These costs include adverse effects on human and animal health from pesticides, decreased genetic diversity of food crops, intensive use of chemicals, loss of traditional knowledge and practices, loss of local biodiversity, loss of soil fertility and farmer dependency on external inputs. In recent years, this dependency also perpetuated indebtedness, especially among poor farmers, and further inequality of benefits. Much of the ESAP population depends on rice as a staple. There has traditionally been much diversity in rice—50 kinds were cultivated in one part of India, many with cultural importance (Dharampal, 1971). These varieties were lost with the introduction of high-yield varieties and associated farming practices. Some estimates suggest that of the 30,000 strains of paddy rice a few years ago, no more than a dozen are expected to dominate three-quarters of the riceland in Asia (Development Forum, 1989).

2.4.1.2 The Green Revolution, food security and poverty alleviation

The introduction of modern AKST, associated with the Green Revolution, more than doubled cereal production in Asia between 1970 and 1995. Poverty steadily declined and nutrition improved through increased income. However, debates on the effects of modern inputs on the poor include common questions: Do modern varieties help poor farmers absolutely or relatively compared with rich farmers? Do rural workers gain or lose income? Do poor consumers gain or lose nutritionally? Has the economic benefit of using modern varieties been uniformly distributed across the farming families? How has a focus on increasing productivity affected social and ecological systems? No easy general and clear conclusions regarding the consequences of the Green Revolution can be based on the literature. The effects have been hotly debated and different researchers have come to different conclusions (Box 2-3).

Greater production came with a price for social and economic systems and was well substantiated from observations from different countries in Asia. Also, considerable contradictory evidence demonstrates that increasing the productivity of smallholders by providing greater farming inputs did not necessarily alleviate their hunger and poverty (Ladeginsky, 1969a,b; Brown, 1971; Frankel, 1971; Rudra, 1971).

Some studies have shown that in some ESAP countries, the population living in poverty increased despite a rise in the production of cereal per head, the main component of the diet of the poor (Lappe et al., 1982). In the Philippines, rice production increased faster than the growth in population, but it had the most widespread undernutrition in all Asia. Similarly, the government of India, in 1979, was holding 16 million tonnes of surplus food grain in storage, while the per capita consumption of food grain in 1975 to 1977 had fallen below that in 1970 to 1972, even below 1960 to 1962 consumption (Lappe et al., 1982). In India, the total food available to each person actually increased, but greater hunger prevailed because of the unequal access to food and resources.

The remarkable difference in China, where the number of hungry dropped from 406 million to 189 million, begs the question, which has been more effective in reducing hunger, the Green Revolution or the Chinese revolution (Rosset, 2003). It has been suggested that the Chinese revolution's broad changes in access to land paved the way for rising living standards.

Assessments on the effect of the Green Revolution on food security and poverty resulted in mixed conclusions, from using different approaches, methods, locations and periods. Agrarian studies in Asia on the effects of agricultural modernization on poverty and food security varied widely

Box 2-3. The impact of the green revolution in India

The roots of agrarian changes in India go back to the era of colonization, when agri-exports increased under the obligation to pay taxes, which resulted in a diversion of resources away from domestically consumed products, decreased availability of food and increased vulnerability to famine (Arena, 2005). After the second World War, there was a concerted effort to increase food productivity. One such effort was the Green Revolution, whose high yielding varieties (HYV) used larger quantities of nutrients and were more water efficient than earlier varieties, which tended to lodge or fall down if grown in soils with good fertility. HYVs could also be planted more densely and did not require long growing periods. They thus have a more favorable harvest index, i.e., the ratio of the economic yield to the total biological yield (Shiva, 1993). A shorter growing season in some cases allows farmers two to three harvests per year when irrigation is available, often requiring the use of machinery and thus becoming more capital intensive (Bhagavan et al., 1973).

The early benefits of the Green Revolution in India were captured by the big farmers (Lewis, 1970), and technical change had strengthened the political dominance of land owners and accentuated income inequality (Frankel, 1971; Griffin, 1979). In South Asia and particularly India, total food available per person actually increased but hunger prevailed due to unequal access to food and food-producing resources (Rosset and Collins, 1998).

Another impact that has been linked to Green Revolution technologies in India is the increase in indebtedness and the increased rates of suicide of farmers on marginal lands. It has been suggested that this is due to the need for capital, lack of seed saving and thus the dependence on buying seeds and the uniform varietals plantings that expose the farmers to the risk of crop failure during droughts (Shiva, 2004).

in style and temper, but they generally subscribed to three viewpoints (Mohanty, 1999):

- Modernization further exacerbated the inequities.
- Improvement of the economic conditions of the poor and the landless in agriculture reduced existing inequalities.
- Modernization of agriculture had mixed effects.

These views indicate that the measurement of poverty is complex. Different proxy indicators measure economic analysis, welfare and food security. The choice of indicators and their use and interpretation should be considered, along with other factors:

- The loss in diversity and the equivalent monetary value when farmers switch from diverse systems to monoculture
- The extra costs of Green Revolution systems, in chemical inputs and the cost of environmental degradation
- Who will benefit from the surplus

For awhile, the Green Revolution contributed to increased agricultural production (Janvry and Sadoulet, 2002). Since the main objective was to generate more food, little attention was directed to how the benefits would be distributed equitably. The Green Revolution was intimately tied to the purchase of seeds, chemical fertilizers, pesticides and intensive irrigation—all external inputs. Its effect included the high dependency it created on external inputs and the debt that farm families incurred. Alternative knowledge was neglected. The approach seemed to assume that farmers were ignorant; it devalued local and indigenous knowledge (Gadgil et al., 1996). The introduction of pesticides and chemical fertilizers diminished land productivity, creating a need for more and more inputs to reap the same yield, adding an extra financial burden on the farmer (Pereira, 1996). Rosset and Collins (1998) reported that in Central Luzon, Philippines, rice yield increased 13% during the 1980s, but it came at the cost of a 21% increase in fertilizer use. They reported that in the central plains yields went up 65%, while fertilizer use increased 24% and use of pesticides jumped 53%. In West Java, the benefit of a 23% yield increase was virtually cancelled by a 65% increase in use of fertilizers and 19% in pesticides.

The Green Revolution was not neutral. The real wages during 1970/71 to 1973/74 in Uttar Pradesh, when the Green Revolution was making a big impact on yields, showed that wages decreased 18% because large landowners brought in more machinery and migrants to compete with local labor and the landless. In many areas, the Green Revolution failed to raise incomes of the rural poor appreciably or contribute substantially to their effective purchasing power. Also, larger-scale farmers had greater access to subsidies for irrigation and credit from the government (Dogra, 1990).

Credit became a major factor in Green Revolution technology and the consequences of debt repayment took their toll on farmers. Cheap credit in one market may merely have the effect of subsidizing and maintaining expensive credit elsewhere. Some landlords in the Philippines who borrowed cheap credit with land as collateral from the rural banks lent the money to their tenants at interest rates left to their own discretion (Palmer, 1976).

Overall, far less research was done on integrated technology for diversifying the livelihoods of small-scale farmers in developing countries and increasing the sustainability of land use. Little was understood, for instance, about the role of organic matter in soil, reduced tillage systems, use of farm organic resources in combination with inorganic fertilizers and the role of legumes in biological nitrogen fixation. Similarly, research was limited in integrated pest management and in weed and pest control. These were topics of little interest to the private sector and were also in danger of neglect by public research institutions.

India was among the first countries in the world to pass legislation granting farmer rights, protecting them in the Plant Varieties and Farmers' Right Act 2001. Farmers' rights were not just an alternative to breeders' rights (Ramanna, 2006). Their rights should be multidimensional, including rights to conservation of biodiversity and to affordable inputs, rights to equity and justice, and above all, the right to reliable quality seeds. The value of conservation of indigenous diversity was implied. The Plant Variety Protection Act of 2002 did not explicitly include in its definition of breeders, farmers and farming communities who continuously nurture, conserve and improve crop varieties. It subsumed farmers under persons who bred or discovered and developed a new plant variety. To give an example, the Philippines Plant Variety Protection Act of 2002 neither recognized nor protected farmer rights to seeds and to participate in the agriculture of the country. Like most policies and laws that directly affect their lives, farmers in many of the countries were generally unaware of the existence of such a law.

A plant variety protection system is an administrative procedure that an applicant complies with to secure a form of intellectual property rights, called the plant breeder's rights. This right is awarded in recognition of the intellectual creation of innovative citizens, as applied on plant varieties, particularly the transformation of plants through breeding, whether done the classical way or through modern technology, such as genetic engineering.

2.4.1.3 *Effects of biotechnology*

Biotechnology and genetic engineering are increasingly used in a few countries in ESAP, for example, China and India but these two countries together account for only 8% of GM crop production worldwide (FAOSTAT, 2004). Despite the perceived advantages, serious reservations persisted about health and environment implications of large-scale application of biotechnology.

Genetically engineered crops can be sprayed with a herbicide to kill weeds without killing the crop plants (Steinbrecher, 1996). Intensified spraying boosts weed resistance to the herbicide. As weeds become resistant, higher and higher doses of herbicide are needed, leaving larger and larger amounts of chemical residue on the crops and the soil. In addition, the engineered crop may itself become a weed. Alongside the development of herbicide tolerance and pest resistance, some scientists have sought to engineer plants to be resistant to pathogens, such as fungi, bacteria and viruses. The immediate hazard from herbicide-resistant crops is the spread of transgenes to wild relatives by cross-pollination, creating superweeds (Ho, 1998). Although it is true that in certain cases, pesticides have reduced the effect on nontarget organisms, biodiversity, evolution of resistance and genetic contamination are some of the concerns.

In Bt crops, if insects developed resistance to the engineered Bt toxin, conventional farmers would revert to chemical insecticides, while organic farmers would have lost one of their most valuable pest control agents. In addition superbugs could emerge—insects that have adapted their behavior and genetics in unpredictable ways to survive in the constant presence of toxins (Stone, 2002). In certain cases, effects on nontarget organisms have been observed (Hilbeck et al., 1998).

Some studies indicate the presence of transformation-induced mutation in commercial crops poses a potentially large biosafety risk (Wilson et al., 2006). This has led to a call for a transparent manner for testing for each individual product before market introduction (Pryme and Lembcke, 2003).

The difference in approach is wide between farmers acting on their traditional knowledge and the new biotechnologists. The first take a broad and holistic approach to a specific agronomic and socioeconomic situation; the latter tend to look for universal, deep-down, molecular solutions. They offer widely differing solutions for problems dealing with pests, diseases, weeds, water, plant nutrients, soil degradation and yield (Table 2-9).

Genetic modification for disease or pest resistance cannot solve the problem of disease or pest attack because intensive agriculture created the conditions for new pathogens (Ho, 1998). For example, a variety of rice hybrid, IR-36, created to be resistant to eight major diseases and pests including bacterial blight and tungro, was attacked by two new viruses, ragged stunt and wilted stunt.

2.4.1.4 *Agricultural sustainability*

The idea of agricultural sustainability centers on the need to develop technology and practices that do not have adverse effects on the environment and human health and at the same time lead to improvement in food and productivity. Sustainable agriculture approaches come under many names: agroecology, organic farming, low external input farming, ecological agriculture, biodynamic agriculture and permaculture (Ho and Ching, 2003). Sustainability in agriculture has been defined as having two dimensions: natural resource sustainability and socioeconomic sustainability.

Sustainable agriculture requires site-specific technology. For example, organic farms vary in complexity and diversity. Studies show that a particular technology can be successful in one site but not in another (Niggli and Ogorzalek, 2007). Evidence from many grassroots development projects also has shown that increasing agricultural productivity with agroecological practices, including organic agriculture, increases not only food supplies but also incomes, thus reducing poverty, increasing access to food, reducing malnutrition and improving livelihoods of the poor.

The question that arises is whether sustainable agricultural practices such as organic farming can be the solution for the future. The debate on the merits and disadvantages of organic versus conventional agriculture continues to influence decision makers. The benefits of organic agriculture are several:

- There is a thriving demand for organically grown food in urban centers of many Asian countries. The premiums paid for organic food offer an opportunity for poor farmers to increase their income (IFAD, 2002). Organic agriculture has the potential to improve household food security and meet the goals of poverty alleviation and environmental sustainability in ESAP (ESCAP, 2002).
- There may be employment effects: Some organic systems may require more labor, which can be negative or positive. The crop diversification that generally happens on organic farms distributes labor throughout the season. This can contribute to stabilizing employment, re-

Table 2-9. *Sustainable agriculture: Farmers and biotech approaches.*

Problem	Biotech response	Farmers response
Pests & diseases	Single gene resistance; engineered biopesticides	Genetic diversity; indigenous varieties; intercropping; insecticidal plants; crop rotation; integrated pest management
Weeds	Herbicide tolerant genes	Early soil coverage; intercropping; cover crops; allelopathic crops
Water	Drought tolerant genes	Moisture conservation practices; contour ploughing; different varieties for different micro climates; water retaining associated crops (Vetiver grass, etc.)
Plant nutrients	Engineered nitrogen fixing crops and microbes	Soil conservation techniques; multiple cropping with legumes; integrated animal and crop agriculture (manure use); composting; green manure
Soil degradation	Saline and other tolerance genes	Restore degraded soils (composting, green manure, rotation, etc.); avoid destruction of the soil in the first place
Yield	Yield increase for mono-cropping	Polycropping; one crop for multiple functions; use of associated crops and animals (weeds, fish, snails, etc.)

Source: Hobbelink, 1991.

ducing turnover and alleviating many problems relating to seasonal migration (FAO, 2002).

- There are environmental benefits. Contamination of ground and surface waters by synthetic fertilizers, especially nitrate leaching, and pesticides are avoided and sedimentation of waterways from erosion is reduced (FAO, 2002). Calculations on comparative energy use in OECD countries indicate that energy consumption on organic farms is 64% that of conventional farms (FAO, 2002). In a three-year comparative study on organic and conventional strawberry production in China, 98% of the energy inputs in the organic systems were from renewable sources, such as animal manure and biogas, whereas 70% of the energy inputs into the conventional system were nonrenewable, such as electricity, chemical fertilizers and pesticides (FAO, 2002).
- Organic agriculture also makes a positive contribution to dealing with climate change: "Organic agriculture may not only enable ecosystems to better adjust to the effects of climate change but also offer a major potential to reduce emissions of agricultural greenhouse gases. Moreover, mixed farming and the diversity of organic crop rotations are protecting the fragile soil surface and may even counteract climate change by restoring the organic matter content. The carbon sink idea of the Kyoto Protocol may therefore partly be accomplished efficiently by organic agriculture" (FAO, 2002).
- Organic agriculture can be considered more flexible, especially when labor is more readily available and high inorganic inputs or mechanization are limited.

The expansion or benefits of organic agriculture, especially on the need to meet increased food demand, raises major doubts:

- Available technology cannot greatly increase the productivity of organic agriculture because it is constrained by nutrient supply. Agriculture of any type is an extractive activity that cannot retain high fertility and productivity without replacing nutrients exported with the products or lost from the site during production. Although high-yielding crops can be produced organically, this is achieved, once natural fertility has been exploited, only by bringing in nutrients from other areas, as plant remains or animal feces, or by accumulating them in situ in long fallows, as in slash-and-burn farming. The consequence, not evident to most consumers and overlooked by many proponents, is that a much greater land area than is immediately apparent is involved in successful organic production. In contrast, crops can be grown more frequently and often repeatedly with fertilizers on the same land, as in the examples of intensive rice and rice–wheat systems.
- It is the shortage of land that will restrict the contribution that organic agriculture can make to the world food supply. Organic agriculture was the norm at the beginning of the 1900s, when the world population was 1.5 billion. Now there is not enough land or organic matter to support the crop production needed for the present, let alone the anticipated world population.
- Adoption of organic agriculture rates are less than 0.1% of arable and permanent agricultural land in nearly all developing economies in Asia and the Pacific, suggesting that most farmers do not believe organic agriculture can produce food at competitive costs (FAO, 2005). Sometimes production costs per unit of land in organic agriculture are lower than in conventional agriculture. Usually they are higher, which means organic farming is profitable only if the produce can be sold at higher prices. Indeed, prices for organic output are higher, but in developing countries this higher price consigns such produce to niche markets.
- Organic agriculture cannot be the solution to food production for a heavily populated planet. Poor households benefit from greater yields by adopting improved practices. Yield gains from a low base are usually the greatest, but productivity of these systems is probably insufficient to meet future food demand. Nevertheless, the principles of organic agriculture will remain as an important contributor to safe and environmentally friendly food production, since they remain firmly embedded in integrated agriculture.

2.4.2 Improving nutrition and human health

With rapid increase in food production and rise in income, food consumption per capita in ESAP countries has risen significantly during the past 50 years. Since 1990, direct cereal consumption leveled off for the whole region, mainly from the decline in direct cereal consumption in China (Figure 2-8). On the other hand, meat consumption rose in ESAP, led by China's steady increase. The same change was, however, absent in India and Indonesia (Figure 2-9).

In spite of the remarkable growth in agricultural production within ESAP during the last four decades, hundreds of millions of people still live in hunger and poverty. The proportion in developing countries of underfed population—with dietary energy consumption inadequate to sustain more than light activity—was estimated to have fallen substantially in the last 15 years, from around one in three people in 1975, to one in five in 1989. This implies a considerable reduction, from nearly 1,000 million people to just below 800 million. This was considerably influenced by the improving situation in China. South Asia probably improved slowly, according to recent results from India and elsewhere, at around a 0.5% reduction in underweight children each year.

The prevalence of underweight children in South Asia remained the highest in the world, over half the total. Calorie consumption remained low throughout the 1980s, with little change, although this might have improved slightly for some poorer groups, such as the landless. Nutrition in many countries of Southeast Asia improved, reducing underweight prevalence about 1% each year. Food consumption rose during the 1980s, along with marked success in food production. A number of countries changed from net food importing to exporting.

Iron deficiency, a cause of anemia, is the only nutritional problem that increased in many parts of the world. Prevalence is especially high in South Asia, where more than 60% of women are anemic. The worsening anemia is from downward trends in intake of dietary iron and has been caused by reduced production and consumption of legumes with the

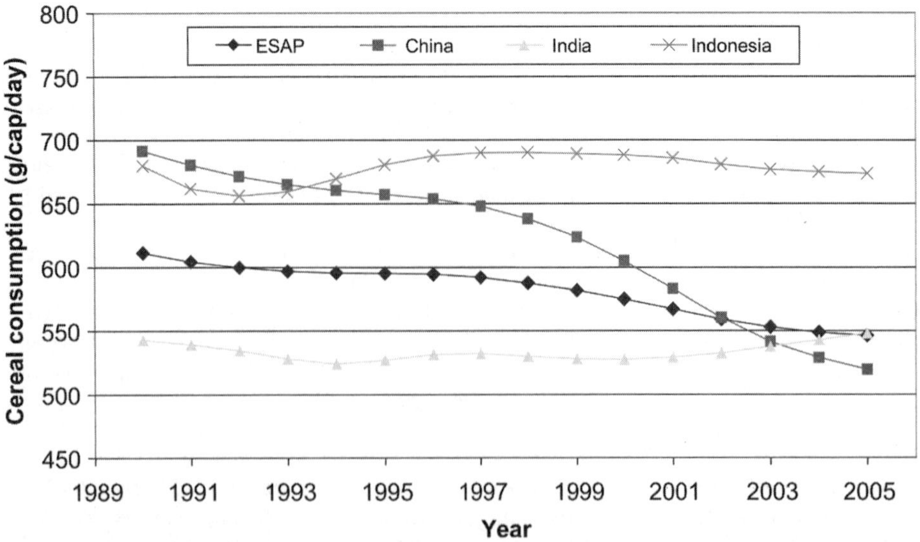

Figure 2-8. *Cereal consumption per country in ESAP, 1995-2005. Source: FAO, 2006a*

Green Revolution. Deficiency of vitamin A affects at least 40 countries. Out of an estimated 14 million with resulting eye damage, vitamin A deficiency blinds up to half a million preschool children each year. Important recent research shows that improving vitamin A status in children in deficient populations reduces mortality among young children by almost one-quarter. Vitamin A supply in some parts of South Asia is so low that deficiency is almost inevitable. The extent of stunting, underweight and wasting in women in developing countries shows that these problems are extensive in developing countries of Asia, particularly low body weight and thinness. Malnutrition in women is generally associated with low birth weight. This has intergenerational effects; malnourished women have small babies, who grow up to be small mothers.

Poverty is clearly a major determinant of nutritional outcome. Rapid economic growth has been a major solution to malnutrition in Southeast Asia. China has far less malnutrition than India. Their average incomes are similar, although allowing for price adjustments puts China considerably ahead. Within India, the relatively low rate of malnutrition in Kerala, one of the poorer states, was parallel to China. The percentages of underweight preschool children were 58.5% in South Asia, 31.3% in Southeast Asia and 21.8% in China in 1990. In 1990 South Asia had 101.2 million underweight preschool children, Southeast Asia had 19.9 million and China 23.6 million (Martorel, 2002). Technology and access to technology and innovation did not benefit many poor people in South Asia. Technology development was geared to market pressure and the needs of the industrial world, not to the needs of countries that had little purchasing power.

Monocropping negatively affected human nutrition. Little-known mammals, birds and snails, which had traditionally served as cheap protein, were killed by pesticides. Traditional plant foods were eliminated because farmers did not prefer them. For example, in South India, sorghum was intercropped, with each acre yielding about 70 kg of different pulses and 10 kg of local oilseeds. The new uniform planting of sorghum varieties reduced the availability of lo-

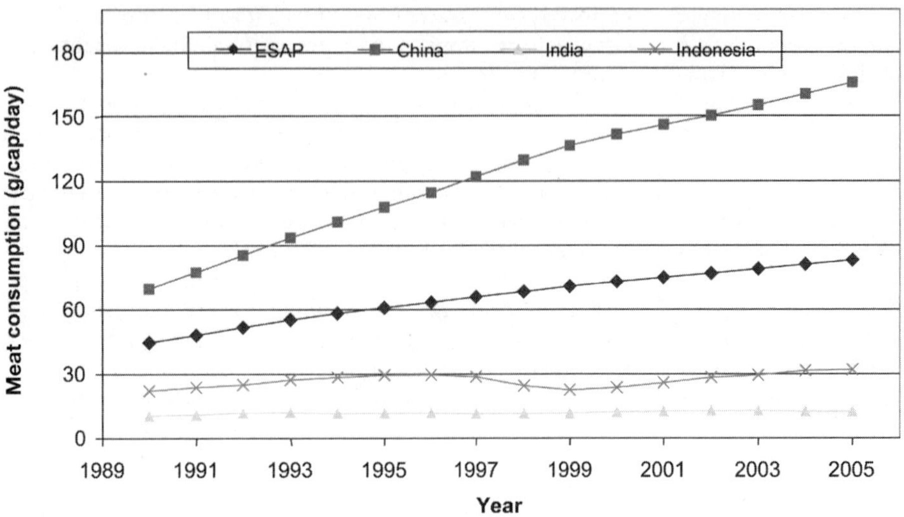

Figure 2-9. *Meat consumption per country in ESAP, 1995-2005. Source: FAO, 2006a*

cal and household protein and fat. The nutritious millets largely grown in semiarid tracts under drought were mostly lost because they were neglected or bypassed.

The chemical pesticides used to protect crops from pests had a direct bearing on human health. Though pesticides may prevent damage by pests and disease and increase production, they are poisons. Pesticide poisoning has always been associated with pesticide use. The developing countries use less than one-quarter of the world's pesticides, but they suffer three-quarters of all pesticide fatalities—about 375,000 people in developing countries are poisoned and 10,000 killed by pesticides each year (Bull, 1982). These figures do not include chronic or long-term effects, such as anemia, leukemia, cancer, birth defects, sterility or suicide. Pesticide use has expanded more rapidly in developing countries than elsewhere. Pesticide imports quadrupled in the Philippines between 1972 and 1978. In 1979, 25% of pesticides the USA exported to developing countries were banned or unregistered in the USA itself.

Some pesticides used were persistent organic pollutants. Despite being present in minute quantities in water and soil, they accumulate in biological systems and, ultimately, in humans, adversely affecting health and reproduction. In addition, pest resistance to pesticides escalates pesticide use, which causes damage to human health, animal health and ecosystems (Nair, 2000; Joshi, 2005).

2.4.3 Effect of AKST on environmental sustainability

Agricultural production and natural resource extraction in forestry and fisheries profoundly intensified throughout ESAP over the past 50 years. Intensified food production has increased food availability but has had trade-offs on sustainability. Often, outside effects of modern agriculture have been masked and their sustainability has been ignored.

2.4.3.1 Effect on soil sustainability

Soil fertility has been declining: soil physical properties have been degraded and nutrients changed adversely, including less availability of major nutrients, deficiency of micronutrients, nutrient imbalances and acidification. The degradation was brought about by incorrect fertilizer use, intensive cropping, depletion of soil organic matter and a decline in soil biological activity. Depletion of primary minerals and organic matter has resulted in micronutrient deficiencies in iron, manganese, zinc, copper, boron, nickel and molybdenum. Over time, heavy crop demand intensifies the severity of the deficiency and exhausts the soil's ability to supply sufficient other micronutrients.

Soil physical degradation has led to accelerated erosion, compaction, crust formation and excessive overland flow. India, Bangladesh, Nepal, Sri Lanka and Bhutan have 140 million ha, or 43% of the total agricultural area of the region suffering from several forms of degraded soil quality (UNEP, 2005). Soil erosion is the most pervasive problem, especially in sloping and unstable agricultural land. Erosion removes the topsoil, where much of the nutrient reserve exists, and consequently causes loss of nitrogen and other nutrients. In China, about one-third of the land, 367 million ha, faces erosion problems (UNEP 2005). In India, 25% of agricultural land has degraded soil, with about 30 million ha of fragile land under cultivation progressively degrading (Dudani and Carr-Harris, 1992).

In intensive agricultural systems in the region, natural soil fertility has declined as a result of crop nutrient removal, nutrient leaching, chemical deficiencies and imbalances. Depletion of plant nutrients (N, P, K, Zn and sulfur) has been the most common chemical degradation. Increasing nutrient imbalances leading to micronutrient deficiency or toxicity of trace elements have been common in continuously irrigated paddy fields.

Soil acidification is enhanced by heavy nitrogen fertilization and adversely affects soil nutrient availability. Many parts of Bangladesh and northern India have acidified and salinized, with a consequent loss of nutrients (Oldeman, 1994). Also, many agricultural lands in Cambodia, Malaysia, Thailand and Viet Nam have experienced chemical soil degradation (Oldeman, 1994). In Australia, Bangladesh, Nepal and Sri Lanka, poor soil nutrient balances were not uncommon. Test plots in IRRI revealed that rice varieties yielding 10 tonnes ha^{-1} in 1966 produced 7 tonnes ha^{-1} in the mid-1990s.

As desertification encroaches, most prone are the arid and semiarid areas. Improper farming techniques of intensive farming and too many animals foraging each unit aggravate the situation. More than half of the 1,977 million ha of dryland in Asia are affected by desertification. Central Asia has more than 60%, South Asia more than 50% and Northeast Asia about 30%. The Gobi Desert in northern and western China expanded by 52,400 km^2 over five years (UNCCD, 1998). Every year, deserts eat up 2,460 km^2 more. Relentless land reclamation, deforestation and overgrazing have led to continued loss of vegetative cover and topsoil. The excessive withdrawal of water upstream in many rivers in arid and semi-arid areas cuts off flows downstream, destroying the riparian ecosystems that rely on the rivers. The denuded land smoothes the way for wind to blow, intensifying sandstorms in areas where the sand originated and in the eastern part of the country and beyond (Yang, 2004).

Soil organisms are important for soil fertility, health and sustainability because they facilitate nutrient cycling and help improve soil structure. Continuous cropping, without considering the capacity of the soil to regenerate, usually results in decline in the amount of soil organisms. Heavy chemical inputs alter the chemical properties of the soil and cause decline in organic matter and humus, the food for microorganisms. Sound soil resource management technologies for efficient and sustainable nutrient cycling such as rotating crops, green manuring and encouraging nitrogen-fixing bacteria and mycorrhizhae were not widely practiced because the dominant production systems focused on short-term productivity.

Soil contaminated by cadmium (in fertilizer), hexavalent chromium, lead, arsenic, trichloroethylene, tetrachloroethylene and dioxin increased, mostly in the northern parts of the region and parts of Australia and New Zealand (UNEP, 2005). Contaminants affecting health from agricultural land in the northwest Pacific and northeast Asia were common in the 1970s (Japan, 2000). Soil contamination from lead and arsenic was prevalent throughout South Asia and Southeast Asia. Irrigation with untreated effluent caused contamination and soil acidification in many areas; in Mongolia, for example, waste disposal and wastewater discharges have been the main causes of soil contamination (UNDP, 2000).

Soil productivity is closely linked with soil organic matter. In some ESAP areas, long-term experiments have shown declining rice and wheat yields (Nambiar, 1994; Cassman et al., 1995; Brar et al., 1998; Yadav et al., 1998, 2000; Duxbury et al., 2000). The major causes observed were a gradual decline in soil nutrients because of inappropriate fertilizer application, a decline in soil organic matter, atmospheric pollution, pest and disease infestation and negative changes in the biochemical and physical composition of soil organic matter (Nambiar, 1994; Yadav et al., 1998, 2000). Observations also have shown that accumulation of nitrogen in soil was better in farms using organic fertilizer than synthetic fertilizer, possibly from a slow release of nitrogen reducing losses (Bhandari et al., 1992; Yadav et al., 2000). Organic fertilizers are known to stimulate nitrogen fixation in soil and may also be responsible for increasing total soil nitrogen (Roper and Ladha, 1995).

2.4.3.2 *Water resource depletion and intensification of water scarcity*

Increasing water withdrawal for irrigation has led to serious environmental consequences, particularly water resource depletion and ecosystem degradation. In an area representing 21% of the world's land, ESAP has 28% of its freshwater resources. However, as the region is home to 53% of the world's population, the water resources for each inhabitant are only slightly above half the world's average.

The hydrology of ESAP is dominated by the monsoon climate, which induces large interseasonal variations in rainfall and river flow. In the absence of flow regulation, most of the water flows during a short season, when it is usually less needed. In Bangladesh, for example, the surface flow of the driest month represents only 18% of the annual average; in Indonesia, 17%. In India, flow distribution of some rivers during the monsoons is 75 to 95% of the annual flow. In north China, about 70 to 80% of the annual rainfall and runoff is concentrated between May and September (FAO, 2006b). This means that irrigation is important for crops produced the rest of the year. Winter wheat, which accounts for over 90% of total wheat sown areas and production in China, is grown between October and the following June. As there is little rainfall during this period, production is heavily reliant on irrigation, which is the largest water user in the water-stressed North China Plain (Yang and Zehnder, 2001).

In many rivers in the region, annual discharge declined from increasing water withdrawal. Some rivers have been completely tapped out during the drier part of the year. The Yellow River, the cradle of China's civilization, stopped flowing in its lower reaches for several months every year during the 1990s. The longest dry-up occurred in 1997—a record of 226 days (Postel, 1999). The consequences of reduced river flows and river dry-ups are serious. The capacity of the river to carry sediment load is reduced, potentially increasing the risk of floods in the lower reach. The dry-ups also adversely affected the aquatic, wetland and estuary ecosystems downstream, in particular the coastal fisheries.

Overextraction of groundwater and consequent groundwater depletion have been widespread problems, especially in semiarid areas. In the North China Plain, the groundwater table has declined over one meter each year (Yang and Zehnder, 2001). In Punjab State in India, the situation has been similar. The rapid decline in groundwater tables reduces availability, on the one hand, and increases the cost of accessing the groundwater, on the other. Poorer farmers have been the most affected. When near the sea or in proximity to saline groundwater, overpumped aquifers are prone to saline intrusion.

Water scarcity has become a major concern in many countries in the region. Increased competition for water between sectors has affected agriculture in China, India, the Republic of Korea, Malaysia and Thailand. The problem is intensifying, mainly from population growth and rapid expansion of the domestic and industrial sectors. Major interbasin transfer programs have been reported in many countries, notably China, India and Thailand.

2.4.3.3 *Water-quality degradation and nonpoint-source water pollution*

Agricultural activities have significantly affected the environment. Water quality is threatened by intensive application of fertilizers, herbicides and pesticides that percolate into aquifers. These nonpoint sources of pollution from agriculture have often taken time to become apparent, but their effects can be long lasting, particularly with persistent organic pollutants. Wetlands are also affected by overextraction of river water and dropping groundwater tables.

Fertilizer runoff from agricultural production, especially nitrogen, contaminates water supplies. For example, Chinese rice farmers often use inappropriate ammonium bicarbonate instead of urea and excessive quantities of nitrogen, 180 kg ha^{-1} or more, leading to low recovery, 35% or less. In addition to reducing farmer profits, the nitrogen lost from crop and livestock production has contributed "dead zones" being created in the East China Sea at the mouth of the Yangtze River (Li and Daler, 2004). Dead zones can devastate fishing grounds and the livelihoods of those who depend on them for sustenance and income. Improving recovery efficiencies will require investment in human capital, for both extension agents and farmers.

In arid and semiarid areas, waterlogging, salinity and alkalinization are serious constraints on agricultural development in irrigated land. The principal effect of salinity is to reduce the amount of water available to the plant by high osmotic concentration of salts in the soil solution. Saline and alkaline cultivated land in China covers about 7 million hectares. In India, waterlogging from irrigation covers about 2.46 million hectares. Salination has also been a serious problem in the Murry-Darling Basin in Australia and a number of other countries in the region. Data for areas actually damaged by salination are sporadic and vary widely among sources.

Animal waste has become a major problem in East Asia and Southeast Asia. With the rapid increases in pig and poultry production in China and Viet Nam, waste runoff from intensive livestock systems has become a major source of nutrient pollution in the South China Sea, one of the most biologically diverse shallow water marine areas. Pig and poultry production has been the primary source of this pollution. Most of this intensive production has been located around periurban centers along or close to the coastline of the East China Sea and the South China Sea. Major hotspots, with the highest concentrations of nitrogen and phosphorus overloads from livestock systems, have been found in the Mekong Delta, the mouth of the Red River and along the whole Chinese coast of the South China Sea. One report has suggested that animal manure accounts for 47% of the P and 16% of the N in these areas. The chemical oxygen demand from untreated piggery effluent accounted for 28% of the current urban industrial chemical oxygen loads in 1996, with the expectation that this will increase to 90% by 2010 (LEAD, 2006).

Increasing concern about management of animal wastes has increased attention by ESAP governments to minimize the effect of nutrient pollution on the landscape and coastal marine systems. While anaerobic digest works on small-scale production systems, such as in the Pacific Islands, large-scale commercial livestock farmers of Southeast Asia will require different technology to adequately treat, dispose of and recycle livestock waste.

Discharge of water excessively laden with organic matter into rivers and canals, especially from intensive aquaculture ponds, was a cause of river water pollution in leading aquaculture countries. However, most destructive aquaculture has been the result of ignorance or nonadherence to "responsible aquaculture practice," also known as good aquaculture practice or best aquaculture practice.

2.4.3.4 *Loss of agrobiodiversity*

ESAP harbors one of the world's richest reservoirs of biodiversity. It is the point of origin of many crop and livestock varieties economically important to humankind. Resource-poor farmers are hugely dependent for their livelihoods on this agrobiodiversity of minor crops, wild plants, wild animals and medicinal plants. They might be insignificant in national statistics but are critically important locally. Biodiversity has been associated with farmer production choices and food security and is a mechanism for coping with environmental uncertainties by spreading and reducing potential risks.

Genetic variability of species comes from genetic resources. Breeders identify desirable genetic traits from genepools and incorporate them into mainstream varieties to produce crops with desirable characteristics, such as improved yield, quality, pest resistance and tolerance to environmental constraints. It is estimated that half the increase in yield of major crops is from genetic improvements through breeding (Chang, 1984).

Agrobiodiversity is being threatened by simplification of ecosystems and species and by varietal replacement. Monocropping has displaced many local and traditional varieties, resulting in genetic erosion. In Indonesia, 1,500 rice varieties disappeared from 1975 to 1990. High-yield rice replaced traditional varieties in about 80 to 82% of fields in Burma, Indonesia, the Philippines and Thailand. The thousands of farmer-developed rice varieties planted in mosaic pattern in agricultural landscapes are no longer planted. Likewise, genetically uniform livestock and poultry breeds have replaced many traditional breeds (Thrupp, 1998). The predominant agricultural approach of monocropping and specialization has often been reinforced by government policies using input subsidies, agricultural extension messages or widespread distribution by governments of modern seeds (Cromwell et al., 1997).

Loss of forest cover, coastal wetlands and other wild, uncultivated areas has further exacerbated the loss of wild relatives and wild foods essential for providing food (Cromwell et al., 1997). Habitat loss has been serious in China, India, the Philippines, Thailand and Viet Nam (ESCAP, 1995).

Cultural diversity, a fourth dimension of biodiversity, has been least appreciated. Traditional and local knowledge is key to using and conserving biodiversity because it embodies the coping mechanisms of local people under the varied and rigorous circumstances that make unique areas productive and sustainable. Local knowledge has been ignored in dominant agricultural systems and much of it is rapidly disappearing (Cox, 2000).

AKST in aquaculture has had both positive and negative effects on the environment. Technology to produce fry of cultivable species obviated using wild fry for farming, saving biodiversity. Many previously fallow water bodies covered with water hyacinth, which used to be the abode of mosquitoes, were now free of the insects. However, coastal shrimp aquaculture was responsible for mangroves being destroyed in many ESAP countries.

2.4.3.5 *Pest and disease incidence and pesticides*

Pests, diseases and weeds have remained significant problems, despite the use of more pesticides. Pesticides may even cause pest problems, when beneficial insects are eliminated or if pest resistance to pesticides evolves. Agronomic practices have their share in causing greater incidence of pests and diseases. For example, outbreaks of the brown plant hopper in rice during the 1980s occurred because pesticides, high nitrogen fertilization, dense planting and continuous irrigation eliminated their natural enemies (Ishii and Hirano, 1959; Heinrichs et al., 1982). Currently, some 500 insect pests, 150 plant pathogens and 133 weed species already have become resistant to insecticides (Brattsten et al., 1986; Altieri and Rosset, 1999).

IPM, developed in the 1980s, was quite successful on selected crops in the ESAP region. Indonesia officially adopted IPM as national policy in 1986 and after five years, it reported a 70% reduction in pesticide use while rice yield increased by 10 percent.

The drive by livestock growers to serve urban markets has led to intensive production, bringing problems of livestock waste, land management and distribution. Greater awareness rose for the potential for transmission of disease from animals to humans. Major diseases that can be transmitted from animals to humans include bovine tuberculosis, Creutzfeldt-Jakob disease and various internal parasitic diseases (Steinfeld et al., 2006). Other examples of the potential dangers of disease transmitted through increased food trade include a 1997 outbreak of foot-and-mouth disease that virtually ruined the pig industry in Taiwan (China). The strain was closely related to strains found in Hong Kong (China) and the Philippines (WHO, 2002). There were also concerns about the rising demand for livestock feed, increased need for veterinary services and training, loss of genetic resources and the need to extend opportunities to small-scale producers to earn cash from livestock (FAO, 2006c).

In ESAP, avian influenza emerged as the most serious threat to animal and human health. The first case of avian flu was reported in a farmed goose in Guangdong, China. The H5N1 avian flu virus spread rapidly across the region, creating transboundary animal disease epidemics. Avian flu outbreaks were reportedly in Cambodia, China, Indonesia, Japan, Laos, South Korea, Thailand and Viet Nam. Countries in the region made massive efforts to cull infected chickens and ducks and to vaccinate healthy birds. In spite of these efforts, incidents of human infection and death occurred among people who worked and lived in close contact with poultry. By 2006-2007, H5N1 had been detected in Bosnia-Herzogovina, Ghana, Hungary, Saudi Arabia, Turkey and the United States.

Effect of agriculture on climate change. Agriculture is a significant contributor to climate change. About 20% of global carbon dioxide emissions, 60% of methane gas emissions and 80% of nitrous oxide come from modern agriculture. By another estimate, livestock accounts for 18% of greenhouse gas emissions, including 9% of anthropogenic carbon dioxide and 37% of anthropogenic methane. Land use, including deforestation, expansion of pastures and land cultivated for feed crops, are the largest contributors to total livestock-related greenhouse gases (Steinfeld et al., 2006). Action should be taken to reduce the overall effect of livestock production on global warming. Both methane and nitrogen emissions can be reduced by better livestock diet and manure management.

2.4.4 Gender, equity and sustainability

2.4.4.1 AKST, workload and time allocation for agricultural production

Women are major stakeholders in agricultural production. This fact is supported through time-use surveys conducted in selected countries, both industrial and developing, in Asia and the Pacific. Women's time contributed to agricultural production is much higher than men's (Table 2-10). In Nepal women work longer than men in all seasons in both rainfed and irrigated agriculture (Sharma, 1995).

With increased migration of male laborers to cities, the agricultural workload of women and children has increased (Balakrishnan, 2005). But the introduction of new agricultural technology decreased the agricultural workload for some women, as in southern Viet Nam, where workload fell about 30% (Ba and Hien, 1996). This saved time was used in other subsistence activities, such as aquaculture on homestead land, home gardening and crafts, sometimes shifting from one activity to another (Felsing and Baticados, 2001). Although additional income might be gained by additional activities, a study in Indonesia and Malaysia showed these additional activities, including aquaculture, added to women's workload, while the profits went to the men (Burgere, 2001).

Though women were the managers and workers, their economic contribution was either not counted or undercounted in the national economy. The agricultural census did not reflect the actual contribution of women in agriculture because of inadequacies in conceptualization, definition of terms and data-gathering methods. After analyzing the gender division of labor, it was found that women contributed much more than men (Joshi, 2000).

Despite women's greater contribution, the predominant image of a farmer in both developing and industrial ESAP countries was male; therefore, policies and programs ignored women's needs and concerns as farmers (ADB and UNIFEM, 1990; Alston, 2004). National statistics, however inaccurate, served as the principal data in framing development policies. These inaccurate data led to undercounting women, both as workers and as those available for work. Women's contributions were either unrecognized or undervalued (Alston, 1998; Siason et al., 2001). Many ignored concerns still need to be understood. Gender-disaggregated data would be necessary for appropriate intervention and policy change. Disaggregated data were lacking or underreported in both developing and industrial countries such as Australia (Alston, 1998; Siason et al., 2001).

2.4.4.2 Gender roles and AKST

Women contributed more time than men in both agricultural production and household activities. The double burden of work reduced the time women had to participate in and benefit from development activities.

The time women and men spent for productive, household, social and religious activities differed significantly by season and environment. It was also significantly influenced by the introduction of technology (Kolli and Bantilan, 1997). Gender division of labor prevails in all social systems. Traditionally, women are allotted most domestic jobs and time-consuming drudgery in the fields. People are slow to perceive what women and men actually do. For instance, both women and men consider fishing a man's job; in fact, women were almost equally involved in fishing in Yunan, China (Yu Xiaogang, 2001). Gender division of labor is not static but changes with time and circumstance (Kusakabe, 2002). Gender division of labor in work outside the home is changing with introduction of agricultural technology, environmental change or economic change. However, it is hard to see drastic change in division of labor in households. Women take on more and more responsibility in production, but their household work remains. This overburdens women with work. It creates physical and psychological problems, there is lack of time for self-development and it enforces gender inequality. Little recognition of women's contribution also prohibits their participation in making decisions.

2.4.4.3 AKST and changes in decision patterns

Women take part in agricultural production, but they make few decisions on technology. Decisions on how to use and manage a technology differ according to the technology used and the activity. According to studies conducted in different countries, women have lagged behind men in making agricultural technology decisions. Examples can be seen in decisions on adopting modern technology in Bangladesh and India (Singh et al., 2000; Rahman and Routray, 2001). Improved technology developed by research and development institutions mainly focused on male workers (Singh et al., 2000). Not involving women in decision making regarding technological production has negative implications for livelihoods and sustainability.

Table 2-10. *Gender, work burden and time allocation in selected Asia and Pacific countries.*

Country	Year	Burden of Work			Time allocation (%)			
		Total work time (min. per day)		Female work time (% of male)	Time spent by women		Time spent by men	
		Women	Men		Market activities	Nonmarket activities	Market activities	Nonmarket activities
Australia	1997	435	418	104	30	70	62	38
Bangladesh	1990	545	496	110	35	65	70	30
Indonesia (urban areas)	1992	398	366	109	35	65	86	14
India	2000	457	391	117	35	65	92	8
Japan	1996	393	363	108	43	57	93	7
Korea Rep.	1999	431	373	116	45	55	88	12
Nepal (rural areas)	1978	641	547	117	46	54	67	33
New Zealand	1999	420	417	101	32	68	60	40
Philippines	1975-77	546	452	121	29	71	84	16

Source: UNDP, 2004.

To build women's decision-making capacity, it is important that women have the same access to information as men. Traditional assignment of market-oriented activities means that introduced technology helps reinforce stereotyped gender roles and reduces the control of women over resources (Kolli and Bantilan, 1997). The rice–fish farming system in Indonesia resulted in increased income (Wardana and Syamsiah, 1990). Although women transplanted, weeded and harvested rice, they made few production decisions and were not involved in farmer meetings and classes.

2.4.4.4 Employment opportunities and income distribution

New AKST in ESAP has created jobs for poor farmers, women and indigenous people and in some cases has helped to reduce poverty. However, the benefits from these new opportunities have varied among gender, class, ethnicity and caste. In most cases, the poorest of the poor did not get equal benefits, compared with richer or middle-income groups. A study conducted in Bangladesh on employment and modern agricultural technology in crop production found the demand for labor increased because of technological changes. However, this demand was mostly met by hiring male laborers; the few women hired were paid significantly lower wages than men. Furthermore, opportunities for women were unequal, and they had less bargaining power both in the conventional hired labor market (Rahman and Routray, 2001) and in contract farming (Singh, 2003). The effect of new technology on women varied with category. In Viet Nam promoting plastic row and drum seeders in rice planting displaced poor women from farming households, who worked as wage laborers in hand weeding and filling gaps. Poor and landless women faced the worst consequences because of lack of alternative jobs and increased debts. Women from better-off families had more time for leisure and other income-generating activities. Progressive men farmers, who had more frequent contact with extension workers, had better-educated wives and used lower seeding rates. This group of women was more likely to benefit from a new technology (Paris and Ngoc Chi, 2005). The farming system also affected the gender decision pattern and income benefits. Female farmers were more involved in farm production and management on vegetable farms and mixed livestock and cash crop farms than in mechanized and capital-intensive production (Hall and Mogyorody, 2007).

Studies of household income distribution revealed that women benefited from small-scale and integrated farming within homesteads, whereas men benefited more from other than subsistence farming (Berman, 2003). Studies in Bangladesh showed that some women involved in growing vegetables had negligible income, but most of their income and vegetables were used for home consumption. Another study showed that women in fish production in Bangladesh got no benefit, because men did the trading and women never knew how much money was earned (Naved, 2000). New groundnut technology in India and intensified aquaculture in Thailand and Viet Nam showed that, while additional income gained was small, women did gain control over it and generally used it for daily expenses (Kolli and Bantilan, 1997; Kusakabe, 2002). If the increased production was more than needed for the home, the extra would be used for trading, and eventually, men benefited from it. Though new technology was likely to change traditional farming into more entrepreneurial systems and add to family income, it was necessary to examine in detail the equity implications of the benefits derived by each member of the farm household. Usually the household was considered as a unit and benefits from certain activities were distributed equally among members. However, case studies showed that to increase weaker groups' choices, it was important that household income have several sources to negotiate priorities. Diversification of sources of income is desirable for addressing risks, increasing household income and controlling economic activities among household members (Kusakabe, 2001).

2.4.4.5 Ownership and control over resources

The effect of AKST also depends upon the ownership and control over the agricultural land, the most basic resource of agricultural production. Land ownership and control is important because it influences the negotiations and decisions of women within the household (Crowley, 2001). It was the single most important contributor to women's economic wellbeing, social status and empowerment (Agarwal, 1994). However, ownership of land may not always give women control over the land, as a study showed in Kerala, India (Arun, 1999). Women's control over key economic resources was more important than economic ownership and was critical to their power within the family. It was important that women had direct access to critical farm inputs to enable them to maximize outputs, challenge ideas of "women's work," gain control over other factors of production and change social norms. Most importantly, there should be a concerted effort to enable women to function as independent farmers who control their own land (Arun, 1999). Enhancing land rights of women requires that those rights become a political priority and a legal possibility; it also requires administrative viability, social acceptability, and moral legitimacy (Crowley, 2001). Complementary policies must address women's limitations in exercising and enjoying their land rights. Control over land is essential because, even with assured land rights, investments in property require access to financial markets, information, extension and other services. Agricultural technology that requires large assets to adopt is more likely to exclude women from the direct benefits. When women earn and control their income, they can use it as a bargaining chip, with the implicit threat of withdrawing it from the household economy (Naved, 2000).

The effect of AKST depends upon the differences in control over assets and technology. The study in Bangladesh showed the choice of new technology and its effect (Meinzen-Dick et al., 2003). The improved vegetables were disseminated to poor women, who could grow them on their homestead, so poor families with only homestead land could also participate. In contrast, one fishpond program focused on those with private fishponds, who were often not poor. Moreover, homestead land was more under women's control; farmland, including fishponds, was more likely to be under men's control. The vegetable program reached women and the very poor, while the output of the private fishpond program went mostly to men.

2.4.4.6 Measures taken for equity and sustainable development

In ESAP, being aware of gender issues and incorporating women's needs and priorities in planning is increasing and some steps have been taken to integrate women's concerns (Kelkar, 2005). Some positive results have been seen. However, there is a long way to go meet the goal and there have been limitations (Rahman, 1999). Some initiatives are microcredit programs in ESAP, such as Grameen Bank in Bangladesh, India and the Philippines (Amin et al., 1998; Milgram, 2005; Holvoet, 2006).

Another effective tool used to empower women is by training poor women in management, trading and marketing, such as managing a small-scale aquaculture enterprise in Vietnamese integrated farming, using a garden, a pond and animal husbandry. In two northern provinces of Viet Nam women gained knowledge from training that helped them make decisions in managing the aquaculture. Once they made such decisions, their position in the household strengthened (Voeten and Ottens, 1997).

In addition to training, group meetings, and saving and credit programs, there was potential for information and communication technology to improve women's and children's access to information and knowledge, enhance their education and accelerate technology transfer. Radio and television were used extensively in several countries to inform and educate rural women about health, nutrition and agriculture. The best-known case studies of information technology's potential benefits for rural women's livelihoods are Bangladesh Grameen Communications' venture of rural women's cell phone enterprises; Pondicherry Village Information Shops; e-Chaupal for market information; SEWA's program on skills development to support women's work in the informal sector; Sri Lanka's Kotmale Project; and information kiosks and telecenters (Balakrishnan, 2005).

Despite the potential, the threat is that an increased digital divide will widen inequality in information, education and knowledge between women and men, rich and poor, urban and rural communities (Kelkar et al., 2005). Therefore, it is necessary to ensure that new agricultural technology is appropriate for the groups of people who most need assistance. Furthermore, it is necessary to assess whether the new technology actually reduces poverty and inequality.

REFERENCES

ADB and Unifem. 1990. Gender issues in agriculture. Proc. Reg. Conf., Manila, 5-6 Dec 1990.

Adeel, Z., and R. Pomeroy, 2002. Assessment and management of mangrove ecosystems in developing countries. Trees 16:235-238.

Adhar, A. 1996. The Indonesian plywood sector—development and challenges. Paper presented at the Indonesian Timber and Forestry Conf., Jakarta 25-26 Mar 1996.

Agarwal, B. 1994. A field of one's own: Gender and land rights in South Asia. Cambridge Univ. Press, UK.

Alston, M. 1998. Farm women and their work: Why is it not recognized? J. Sociol. (Sage, Australia) 34:23-34.

Alston, M. 2004. Who is down on the farm? Social aspects of Australian agriculture in the 21st century. Agric. Human Values 21(4):37-46.

Altieri, M., and P. Rosset. 1999. Strengthening the case for why biotechnology will not help the developing world: A response to McGloughlin, AgBioforum. J. Agribiotech Manage. Econ. 2(3-4):14.

Amin, R., S. Becker and A. Bayes. 1998. NGO-promoted microcredit programs and women's empowerment in rural Bangladesh: Quantitative and qualitative evidence. J. Developing Areas 32:221-236.

APAARI. 1999. National agricultural research systems in the Asia Pacific region—a perspective. FAO Reg. Off. Asia Pacific, Bangkok.

Arena, J.A.N. 2005. Confronting agricultural crisis: Historical food insecurity: The Indian state and the green Revolution. Archive, A journal of undergraduate History, Vol 8. Univ. Wisconsin, Madison.

Arun, S. 1999. Does land ownership make a difference? Women's roles in agriculture in Kerala, India. Gender Tech. Dev. 7:19-27.

Ba, V.T.N., and T.T.T. Hien. 1996. Women in a Vietnamese state farm earn half of family income. Aquaculture Asia 1:42-43.

Balakrishnan, R. 2005. Rural women and food security in Asia and the Pacific: prospects and paradoxes. RAP Publication 2005/30. FAO Reg. Off. Asia Pacific, Bangkok.

Bangladesh Dep. Fisheries. 2003. Brief on Department of Fisheries. Min. Fisheries Livestock, Dhaka.

Barker, R., and F. Molle. 2004. Evolution of irrigation in South and Southeast Asia. Res. Rep. 5. Comprehensive assessment of water management in agriculture. IWMI, Colombo.

Berman, B.K. 2003. Women in small-scale aquaculture in North-West Bangladesh. Gender Tech. Dev. 7:233-263.

Beetz, A. 2002. Agroforestry overview: Horticulture systems guide. Available at www.attra.ncat.org./attra-pub/agroforestry. Nat. Sustainable Agric. Inform. Serv., USA.

Bhagavan, M.R., K. Haraksingh, R. Payne and D. Smith. 1973. The death of the Green Revolution. Halsmere Declaration Group, London.

Bhandari, A.L., A. Sood, K.N. Sharma, and D.S. Rana. 1992. Integrated nutrient management in a rice–wheat system. J. Indian Soc. Soil Sci. 40:742-747.

Bouma, J.A.and C.A. Scott. 2006. The possibilities for dryland crop yield improvement in India's semi-arid regions-observations from the field. CA Disc. Pap. 3. IWMI, Colombo.

Brar, B.S., Y. Singh, N.S. Dhillon, and B. Singh. 1998. Long-term effects of inorganic fertilizers, organic manure and crop residues on the productivity and sustainability of a rice–wheat cropping system in northwest India. p. 335. *In* A. Swarup et al. (ed) Long-term soil fertility management through integrated plant nutrient supply. Indian Inst. Soil Sci., Bhopal.

Brattsten, L.B., C.W. Holyoke, Jr., J.R. Leeper and K.F. Raffa. 1986. Insecticide resistance: challenge to pest management and basic research. Science 231:1255-1260.

Brown, C., and P.B. Durst. 2003. State of forestry in Asia and the Pacific, 2003: status, changes and trends. Asia Pacific Forestry Commission, RAP Publ. 2003/22. FAO Reg. Off. Asia Pacific, Bangkok.

Brown, L. 1994. Who will feed China? W.W. Norton, NY.

Brown, L.R. 1971. The social impact of the Green Revolution. Int. Conciliation No. 581 Carnegie Endowment Int. Peace, NY.

Bull, D. 1982. A growing problem: Pesticides and the Third World poor. Oxfam Publ. Unit, Oxford.

Burgere, C. 2001. Women in aquaculture in Indonesia and Malaysia. p. 65-80. *In* K. Kusakabe and G. Kelkar (ed) Gender concerns in aquaculture in Southeast Asia. Gender Studies Monogr. 12. Asian Inst. Tech., Bangkok.

Byerlee, D., G. Alex, and E. Pehu. 2006. Module 2: Investments in agricultural science and technology. World Bank, Washington DC.

Byerlee, D., and R.G. Echeverria, 2002. Agricultural research policy in an era of privatization: introduction and overview. *In* D. Byerlee and R. Echeverria (ed) Agricultural research policy in an era of privatization. CABI Publ., UK.

Cassman, K., and P. Pingali. 1995. Intensification of irrigated rice systems: Learning from the past to meet future challenges. Geo J. 35:299-305.

Cassman, K.G., S.K. De Datta, D.C. Olk, J. Alcantara, M. Samson, J. Descalsota et al. 1995. Yield decline and the nitrogen economy of long-term experiments on continuous, irrigated rice systems in the tropics. p. 181-222. *In* R. Lal and B.A. Stewart (ed) Soil management: experimental basis for sustainability and environmental quality. Adv. Soil Sci. CRC Press, Boca Raton FL.

Casson, A. 1997. The controversy surrounding eucalpyts in social forestry programs of Asia. Resource Management in Asia-Pacific Program (RMAP) Working Pap. No. 3. The Australian National Univ., Canberra.

Chaboussou, F. 2004. Healthy crops: A new agricultural revolution. Gaia Foundation, London.

Chang, T.T. 1984. Conservation of rice genetic resources: Luxury or necessity? Science 224:251-256.

Chaparro, F. 1999. Towards a global research system: A NARS perspective. Paper presented at the European Forum on Agricultural Research for Development, 7-8 Apr 1999, Wageningen. GFAR, Rome.

Chatterton, P., Y. Bun, C. Hung, K. Whimp, and P. Eddowes. 2000. A future of our forest. Strategies for community forestry and consevation in Papua New Guinea. World Wide Fund for Nature and the World Bank, Suva.

China Ministry of Agriculture. 2005. China agricultural development bulletin. Ministry of Agric., Beijing.

China National Bureau of Statistics. 2006. China statistical yearbook. China Statistics Press, Beijing.

Connor, D. 2004. Designing cropping systems for efficient use of limited water in southern Australia. Eur. J. Agron. 21:419-431.

Connor, D., and M. Minguez. 2006. Letter to the editor. Looking at biofuels and bioenergy. Science 312:1743.

Cox, P.A. 2000. Will tribal knowledge survive the millennium? Science 287(5450): 44-45.

Craswell, E.T., A. Sajjapongse, D.J.B. Howlett, and A.J. Dowling. 1997. Agroforestry in the management of sloping lands in Asia and the Pacific. Agrofor. Syst. 38(1-3):121-137.

Cromwell, E., D. Cooper and P. Mulvany. 1997. Agriculture, biodiversity and livelihoods: Issues and entry points for development agencies. Available at http://www.ukabc.org/odi_agbiod.pdf.

Crowley, E. 2001. Empowering women to achieve food security, land rights. Focus 6. Policy Brief 2. IFPRI, Washington DC.

Dar, W.D. 1995. Support for agricultural research system in Southeast Asia—impacts on growth and development. APAARI Publ. 1995/6. FAO Reg. Off. Asia Pacific, Bangkok.

Dawe, D. 2005. Economic growth and small farms in Suphan Buri, Thailand. Presentation at the FAO Symp. Agric. Commercialization and Small Farms, Rome, 4-5 May 2005.

De Amitendu, P., H.P.S. Arya, B. Tudu, and A. Goswami. 2004. Indigenous technical knowledge in animal husbandry. Available at http://www.cipav.org.co/lrrd16/8/arun16059.htm. Livestock Res. Rural Dev.16(8).

Dembner, S.A. 1994. Editorial. Forestry research. Unasylva 45(177):4-5.

Development Forum. 1989. Seeds breeding extinction. UN Rural Advancement Fund Int. Dev. Forum 37(3):12-13.

DFID. 2007. Tackling illegal logging in Indonesia. Available at http://www.dfid.gov.uk/casestudies/files/asia/indonesia-illegal-logging.asp. DFID, UK.

Dharampal, S. 1971. Indian science and technology in the 18th century: Some contemporary European accounts. Impex, New Delhi.

Dobermann, A., C. Witt, and D. Dawe (ed) 2004. Increasing the productivity of intensive rice systems through site-specific nutrient management. IRRI, Los Banos.

Dogra, B. 1994. Green Revolution: no joy for the poor. p. 242-244. *In* Return to the good earth—damaging effects of modern agriculture and the case for ecological farming. 2nd ed. Third World Network, Penang.

Dudani, A.T., and J. Carr-Harris. 1992. Agriculture and people. South–South Solidarity, New Delhi.

Duxbury, J.M., I.P. Abrol, R.K. Gupta, and K.F. Bronson. 2000. Analysis of long-term soil fertility experiments with rice–wheat rotations in South Asia. p. 7-22. *In* I.P. Abrol et al. (ed) Long-term soil fertility experiments with rice–wheat rotations in South Asia. Rice–Wheat Consortium Pap. Ser. 6. Rice–Wheat Consortium for the Indo-Gangetic Plains, New Delhi.

Enters, T. 1997. Asia Pacific Forestry Sector Outlook Study: Technology Scenarios in the Asia-Pacific Forestry Sector. Asia-Pacific forestry towards 2010. Working Pap. APFSOS/WP25. FAO, Rome.

ESCAP. 1995. Preparatory meeting for the Ministerial Conference on Environment and Development in Asia-Pacific. Econ. Social Comm. Asia and the Pacific (ESCAP), Bangkok.

ESCAP. 2002. Organic agriculture and rural poverty alleviation: potential and best practices in Asia. Econ. Social Comm. Asia and the Pacific (ESCAP), Bangkok.

Fan, S., L. Zhang and X. Zhang. 2002. Growth, inequality and poverty in rural China: The role of public investments. Res. Rep. 125. IFPRI, Washington DC.

Fan S., S. Jitsuchon, and N. Methakunnavut. 2004. The importance of public investment for reducing rural poverty in middle-income countries: The case of Thailand. DSGD Disc. Pap. 7. IFPRI, Washington DC.

FAO. 1999. Irrigation in Asia in figures. Water Rep. 18. FAO, Rome.

FAO. 2001. State of the world's forests 2001. Available at http://www.fao.org/DOCREP/003/Y0900E. FAO, Rome.

FAO. 2002. Organic agriculture, environment and food security. FAO, Rome.

FAO. 2005. Selected indicators of food and agriculture development in Asia-Pacific Region 1994-2004. RAP Publ. 2005/20. FAO, Bangkok.

FAO. 2004. FAOSTAT, Database of FAO. www.fao.org. FAO, Rome.

FAO. 2006a. FAOSTAT, Database of FAO. www.fao.org. FAO, Rome.

FAO. 2006b. AQUASTAT, Database of FAO. www.fao.org. FAO, Rome.

FAO. 2006c. The state of food and agriculture in Asia and the Pacific. FAO-RAP Publ. 2006/03. FAO, Rome.

FAO, 2007. FAO fishery information, data and statistics unit (FIDI). Available at http://www.fao.org/figis/servlet/static?dom=collection&xml=global-aquaculture-production.xml. FIGIS Data Collection, FAO, Rome.

FAO–RAP, 2003. Report of the workshop on strengthening food and agricultural statistics in the Pacific and support of food security and poverty reduction strategies and programes. Nadi Fiji, 10-13 Nov 2003. Available at ftp://ftp.fao.org/docrep/fao/007/ad509e/ad509e00.pdf. FAO, Rome.

FAO–SOFIA. 2006. The state of world fisheries and aquaculture. FAO Fisheries Dep., Rome.

FAO–WHO. 1999. Guidelines for the production, processing, labeling and marketing of organically produced foods. Codex Alimentarius Commission. FAO, Rome.

Felsing, M., and D. Baticados. 2001. The role of women in aquaculture in the Philippines: obstacles and future options. p. 85-92. *In* K. Kusakabe and G. Kelkar (ed) Gender concerns in aquaculture in Southeast Asia. Gender Studies Monograph 12. Asian Inst. Tech., Bangkok.

Fernandez, P. 1994. Indigenous knowledge and development monitor. Available at http://www.nuffic.nl/ciran/ikdm/2-2/articles/fernandez.html. Indigenous Knowledge and Dev. Network.

Fisher, R.J. 2000. Decentralization and devolution in forest management: A conceptual overview. *In* T. Enters et al. (ed) Decentralization and devolution of forest management in Asia and the Pacific. RECOFTC Rep. 18 and RAP Publ. 2000/1. FAO Reg. Off. Asia Pacific, Bangkok.

Frankel, F.R. 1971. India's green revolution: Economic gains and political costs. Princeton Univ. Press, NJ.

Friedman, H. 1990. The origins of Third World food dependence. p. 26. *In* H. Bernstein et al. (ed) The food question. Earthscan, London.

Gadgil, M., N.S. Heman, and D.M. Reddy. 1996. People, refugia and resilience linking soil and ecological systems. *In* F. Berkes and C. Folke (ed) Management practices and social mechanisms for building resilience. Cambridge Univ. Press, UK.

Garrity, D.P. 1986. Tree–soil crop interactions on slopes. p. 299-318. *In* C.K. Ong and P. Huxley (ed) Tree–crop interactions: a physiological approach. ICRAF, Nairobi and CAB Int., UK.

Garrity, D., and A. Mercado. 1994. Reforestation through agroforestry: Market-driven, small-holder timber production on the frontier. p. 239-263. *In* J.B. Raintree and H.A. Francisco (ed) Marketing of multipurpose tree products in Asia. Proc. Int. Workshop held in Baguio City, Philippines, 6-9 Dec. 1993. Winrock Int., Bangkok.

Grenier, L. 1998. Working with indigenous knowledge: A guide for researchers. Available at http://www.idrc.ca/openebooks/847-3/. IDRC, Ottawa.

Griffin, K. 1974. The political economy of agrarian change: An essay on Green Revolution. Harvard Univ. Press, Cambridge, MA.

Groosman, T., A. Linnermann, and H. Wierema, 1991. Seed industry development: A North–South perspective. Pudoc, Wageningen.

Gupta R.K., P.R Hobbs, J.K. Ladha, and S.V.R.K. Prabhakar. 2002. Resource-conserving technologies: Transforming the rice–wheat systems of the Indo-Gangetic Plains. Asia-Pacific Assoc. Agric. Res. Inst. and FAO Reg. Off. Asia Pacific, Bangkok.

Hall, A.J. 2006. Public–private sector partnerships in a system of agricultural innovation: concepts and challenges. Int. J. Tech. Manage. Sustain. Dev. 5(1).

Hall, A., and V. Mogyorody. 2007. Organic faming, gender, and the labor process. Rural Soc. 72:289-316.

Hatibu, N., and J. Rockstrom. 2005. Green-blue water system innovations for upgrading of smallholder farming systems—a policy framework for development. Water Sci. Tech. 51(8):121-131.

Hazell, P., and L. Haddad. 2001. CGIAR research and poverty reduction. Paper prepared for the Tech. Advisory Comm. CGIAR. IFPRI, Washington DC.

Heinrichs, E.A., W.H. Reissig, S.L. Valencia, and S. Chelliah. 1982. Rates and effect of resurgence—inducing insecticides on populations of *Nilaparvata lugens* (Homoptera: Delphacidae) and its predators. Environ. Entomol. 11(6):1269-1273.

Heong, K.L., and M.M. Escalada (ed). 1997. Pest management of rice farmers in Asia. IRRI, Manila.

Heong, K.L., P.S. Teng, and K. Moody. 1995. Managing rice pests with less chemicals. Geograph. J. 35:337-349.

Herbohn, J.L, S.R. Harrison, N. Emtage, and P. Norman, 2002. The current status of farm forestry in Australia. A report for the RIRDC/Land & Water Australia/FWPRDC Joint Venture Agroforestry Program. Available at http://epring.uq.edu.au/archive/00004228.

Hidayat, H. 1998. Empowering communities through social forestry. p 160-165. *In* Proc. IGES Int. Workshop Conservation Strategies for Asia and the Pacific. Inst. Global Environmental Strategies, Hayama, Kanagawa, Japan, 21-23 July 1998.

Hilbeck A., M. Baumgartner, P.M. Fried and F.Bigler. 1998. Effects of transgenic *Bacillus thuringiensis* corn-fed prey on mortality and development time of immature *Chrysoperla camea*. Environ. Entomol. 27:460-467.

Ho, M.W. 1998. Genetic engineering: Dream or nightmare? Third World Network, Penang.

Ho, M.W., and L.Li. Ching. 2003. The case for a GM-free sustainable world. Independent Sci. Panel, London.

Holvoet, N. 2006. The differential impact on gender relations of "transformatory" and "instrumentalist" women's group intermediation in microfinance schemes: A case study for rural South India. J. Int. Women's Studies 7:36-50.

Hossain, M., M.L. Bose, and B.A.A. Mustafi. 2006. Adoption and productivity impact of modern rice cultivars in Bangladesh. Developing Econ. 44(2):149-166.

IFAD (Int. Fund Agric. Dev). 2002. IFAD strategy for rural poverty reduction in Asia and the Pacific. Available at http://www.ifad.org/operations/regional/2002/pi/pi.htm. IFAD, Rome.

IFOAM (Int. Fed. Organic Agric. Movements). 1996. Basic standards for organic agriculture and processing and guidelines for coffee, cocoa, and tea: Valuation of inputs. IFOAM, Bonn.

IRRI. 2006. Bringing hope, improving lives. Strategic plan 2007-2015. IRRI, Los Banos.

Ishii, S., and C. Hirano. 1959. Effects of

fertilizers on the growth of the larvae of the rice stemborer, *Chilo suppressalis* Walker. Growth response of the larva to rice plants cultured in different nitrogen level soils. Japan J. Appl. Entomol. Zool. 2(3):198-202.

Isvilanonda, S., A. Ahmad, and M. Hossain. 2000. Recent changes in Thailand's rural economy: evidence from six villages. Economic and Political Weekly 35 (52-53):464449.

IWMI. 2002. World irrigation and water statistics 2002. IWMI, Colombo.

James, C. 1996. Agricultural research and development: the need for public–private sector partnerships. Issues in Agric. 9. CGIAR Sec., World Bank, Washington DC.

Janvry, A., and E. Sadoulet. 2002. World poverty and the role of agricultural technology: direct and indirect effects. J. Dev. Studies 38(4):1-26.

Japan Ministry of the Environment. 2000. Policies and programmes. Available at http://www.env.go.jp/en/pol/leaflet1.html [Geo-2-159]. Japan Min. Environ, Tokyo.

Joshi, M. 2005. Perils of pesticides. Centre for Environ. Education, Ahmedabad, India.

Joshi, S. 2000. Technological change and women's participation in crop production in Bangladesh. Gender Tech. Dev. 4:255-270.

Karim, M.R. 2003. Present status and strategies for future development of shrimp farming in Bangladesh. p. 1-8. *In* M.A. Wahab (ed) Environmental and socioeconomic impact of shrimp farming in Bangladesh. Tech. Proc. BAU-NORAD Workshop, 5 Mar 2002, BRAC Center, Dhaka.

Kelkar, G. 2005. An overview of progress in achieving gender equality and poverty reduction in South Asia. Proc. Reg. Conf. Dev. Effectiveness through Gender Mainstreaming: Lessons learnt from South Asia. Delhi, 10-12 May 2005.

Khush, G.S., W.R. Coffman, and H.M. Beachell. 2001. The history of rice breeding: IRRI's contribution. p. 117-135. *In* W.G. Rockwood (ed) Rice research and production in the 21st century. Symposium honoring Robert F. Chandler Jr. IRRI, Los Banos.

Kolli, R.D., and C. Bantilan, 1997.Gender-related impacts of improved agricultural technologies: identification of indicators from a case study. Gender Tech. Dev. 1:3:371-393.

Kusakabe, K. 2001. Intensification of aquaculture: does it empower women? p. 11-26. *In* K. Kusakabe and G. Kelkar (ed) Gender concerns in aquaculture in Southeast Asia. Gender Studies Mono. 12. Asian Inst. Tech., Bangkok.

Kusakabe, K. 2002. Gender issues in small-scale inland fisheries in Asia. p. 49-58. New approaches for the improvement of inland capture fishery statistics in the Mekong Basin. Ad hoc expert consultation, Udon Thani, Thailand, 2-5 Sept 2002, FAO, Mekong River Comm., Erewan Press, Bangkok.

Ladeginsky, W. 1969a. The Green Revolution in Bihar—the Kosi area: A field trip. Econ. Polit. Weekly 4:147-162.

Ladeginsky, W. 1969b. The Green Revolution in Punjab: A field trip. Econ. Polit. Weekly 4:773-782.

Ladha, J.K., H.U Neue, R. Wassmann, and M.S. Aulakh. 2004. Mitigating greenhouse gas emissions from rice–wheat cropping systems in Asia. Environ. Dev. Sustainability 6(1/2):65-90.

Ladha, J.K., H. Pathak, A. Tirol-Padre, D. Dawe, and R.K. Gupta. 2003. Productivity trends in intensive rice–wheat cropping systems in Asia. *In* J.K. Ladha et al. (ed) Improving the productivity and sustainability of rice–wheat systems: issues and impacts. ASA Spec. Publ. 65, ASA-CSSA-SSSA, Madison.

Lappe, F.M., J. Collins, and D. Kinley. 1982. United States-sponsored development: is it possible? p. 483. *In* R. Anderson et al. (ed) Science, politics and the agriculture revolution in Asia. AAAS, Washington DC and Westview Press, Boulder.

LEAD. 2006. Livestock's long shadow: Environmental issues and options. Available at http://www.virtualcentre.org. LEAD, London.

Leakey, R.R.B., and Z. Tchoundjeu, 2001. Diversification of tree crops: domestication of companion crops for poverty reduction and environmental services. Exp. Agric. 37:274-296.

Lewis, J.P. 1969. Wanted in India, a relevant Radialism. Policy memoirs No.36. Woodrow Wilson School of Public and International Affairs, Princeton Univ., NJ.

Li, D., and D. Daler. 2004. Ocean pollution from land-based sources: East China Sea, China. Ambio 33(1-2):107-113.

Lipton, M., and J. Litchfield. 2003. Preliminary review of the impact of irrigation on poverty with special emphasis on Asia. AGL/MISC/34. Land and Water Dev. Div., FAO, Rome.

Liu F.M., Y.Q. Wu, H.K. Xiao, and Q.Z. Gao. 2005. Rainwater-harvesting agriculture and water-use efficiency in semi-arid regions in Gansu Province, China. Outlook Agric. 34(3):159-165.

Martorel, R. 2002. Long-term effects of improved childhood nutrition enhancing human potential in guatemalan adults through improved nutrition in early childhood. Available at http://www.unsystem.org/SCN/archives/rwns02vol1/ch04.htm. Cornell Univ., Ithaca.

Meinzen-Dick, R., M. Adato, L. Haddad, and P. Hazell. 2003. Impacts of agricultural research on poverty: findings of an integrated economic and social analysis. FCND Disc. Pap. 164. IFPRI, Washington DC.

Milgram, B.L. 2005. From margin to mainstream: Microfinance, women's work and social change in the Philippines. Urban Anthropol. 34:341-384.

Mohanty, B.B. 1999. Agricultural modernization and social inequality—Case study of Satara District. Econ. Polit. Weekly 34(26). June-July.

Molden, D., and C. de Fraiture. 2004. Investing in water for food, ecosystems and livelihoods. Comprehensive assessment of water management in agriculture. Available at www.iwmi.org/assessment. IWMI, Colombo.

Nair, A. 2000. Pesticides in India. Environment and health source book. TOXICSLink, New Delhi.

Nambiar, K.K.M. 1994. Soil fertility and crop productivity under long-term fertilizer use in India. Indian Council Agric. Res., New Delhi.

Naved, R.T. 2000. Intra-household impact of the transfer of modern agricultural technology, a gender perspective. FCND Disc. Pap. 85. IFPRI, Washington DC.

Niggli, U., J. Early, and K. Ogorzalek. 2007. Organic agriculture and environmental stability of the food supply. Available at www.fibl.org/index.php and www.wwwf.org.

Nurse, M., and Y. Malla. 2005. Advances in community forestry in Asia. Regional Community Forestry Training Center for Asia and the Pacific (RECOFTC), Bangkok. Workshop on capitalism and sharing of experiences on the interaction between forest policies and land-use patterns in Asia, 24-28 Jan 2005, Kathmandu.

Oldeman, L.R. 1994. The global extent of soil degradation. *In* D.J. Greenland and T. Szaboles (ed) Soil resilience and sustainable land use. Available at http://www.isric.nl/GLASOD.htm [Geo-2-161]. CAB Int., Wallingford.

Oweis, T.Y., and A.Y. Hachum. 2003. Improving water productivity in the dry areas of West Asia and North Africa. *In* J.W. Kijne et al. (ed) Water productivity in agriculture: limits and opportunities for improvement. CABI, Wallingford and IWMI, Colombo.

Palmer, I. 1976. The new rice in Asia: Conclusions from four country studies. UN Res. Inst. Social Dev. (UNRISD), Geneva.

Pardey, P.G., and N.M. Beintema. 2001. Slow magic: Agricultural research and development a century after Mendel. Food Policy Rep. IFPRI, Washington DC.

Pardey, P.G., N.M. Beintema, S. Dehmer, and S. Wood. 2006. Agricultural research: A growing global divide? Food Policy Rep. IFPRI, Washington DC.

Paris, T., and T.T. Ngoc Chi. 2005. The impact of row seeder technology on women's labor: a case study in the Mekong Delta, Vietnam. Gender Tech. Dev. 9:2:157-84.

Pasicolan, P.N.A., Udo de Haes, and P.E. Sajise, 1997. Farm forestry: an alternative to government-driven forestry in the Philippines. Forest Ecol. Manage. 99(102):261-74.

Pathak, H., J.K. Ladha, P.K. Aggarwal, S. Peng, S. Das, Y. Singh et al. 2003. Climatic potential and on-farm yield trends of rice and wheat in the Indo-Gangetic plains. Field Crops Res. 80(3):223-34.

Pearce, A. 1980. Seeds of plenty, seeds of want. Clarendon Press, Oxford.

Pearson, A., J. Reid, J. Benge, and H. Moller. 2005. Soil quality on ARGOS kiwifruit orchards, 2004-2005. Res. Rep. 05/02. Available at http://www.argos.org.nz. ARGOS Agric. Res. Group Sustainability.

Peng, S., and G.S. Khush. 2003. Four decades of breeding for varietal improvement of irrigated lowland rice in the International Rice Research Institute. Plant Prod. Science 6:157-64.

Peng, S., K.G. Cassman, S.S. Virmani, J.E. Sheehy, and G.S. Khush. 1999. Yield potential trends of tropical rice since the release of IR8 and the challenge of increasing rice yield potential. Crop Sci. 39:1552-1559.

Pereira, W. 1996. The inefficiency of western agriculture. Organic farming source book. The Other India Press, Delhi and Third World Network, Penang.

Poffenberger, M. 1990. Keepers of the forest. Ateneo de Manila Univ. Press, Manila.

Poh, L.Y. 1994. Malaysia. p. 55-71 *In* P.B. Durst et al. (ed) Non-wood forest products in Asia. RAPA Publ. 1994/28. FAO, Bangkok.

Postel, S. 1996. Dividing the waters: food security, ecosystem health, and the new politics of scarcity. WorldWatch Pap. 132. WorldWatch Inst., Washington DC.

Postel, S. 1999. Pillar of sand: Can the irrigation miracle last? W.W. Norton, NY.

Pray, C., J. Huang, R. Hu, and S. Rozelle. 2002. Five years of Bt cotton in China: The benefits continue. Plant J. 31(4):423-430.

Pryme, I.F., and R. Lambcke. 2003. In vivo studies on possible health consequences of genetically modified food and feed with particular regard to ingredients consisting of genetically modified plant materials. Nutr. Health 17(99):1-8.

Rahman, S., and J.K. Routray. 2001. Technological change and women's participation in crop production in Bangladesh. Gender Tech. Dev. 5:267-287.

Rahman, A. 1999. Micro-credit initiatives for equitable and sustainable development: Who pays? World Dev. 27:67-82.

Ramanna, A. 2006. Farmer's rights in India: A case study. FNI Rep. 6/2006 The Fridtjof Nansen Institute, Lysaker, Norway.

Rao, Y.S. 1994. Forestry research in Asia and the Pacific: Focus on FORSPA. Unasylva 45(177):6-14.

Robinson, R.A. 1996. Return to resistance. Breeding crops to reduce pesticide dependence. AgAccess, Davis CA.

Roder, W., B. Keoboualapha, and V. Manivanh. 1995. Teak (*Tectona grandis*), fruit trees and other perennials used by hill farmers of northern Laos. Agrofor. Syst. 29:47-60.

Roper, M.M., and J.K. Ladha. 1995. Biological N2 fixation by heterotrophic and phototrophic bacteria in association with straw. Plant Soil 174:211-224.

Rosset, P. 2003. Lessons from the Green Revolution. Do we need new technology to end hunger? Available at http://www.foodfirst.org/media/opeds/2000/4-greenrev.html. Food First, San Francisco.

Rosset, P., and J. Collins, 1998. Lessons from the green revolution. Food First Publ., San Francisco.

Rouse, J. 1999. Global dissemination of indigenous knowledge: contradiction of the way forward?. Indigenous Knowledge Initiative. Available at http://www.worldbank.org/afr/ik/global ik990615.htm. World Bank Washington DC.

Rudra, A. 1971. The green and greedy revolution. South Asian Rev. 4:291.

Sanchez, P. 1995. Science in agroforestry. Agrofor. Syst. 30:5-55.

Shah, T., D. Molden, R. Sakthivadivel, D. Seckler. 2001. The global groundwater situation: opportunities and challenges. Econ. Polit. Weekly 36(43):4142-4150.

Sharma, M. 1995. Gender implications of changes in technology and cropping patterns for labor use in rice-based farming systems in Nepal. HS 95-7, Asian Inst. Tech., Bangkok.

Shiva, V. 1993. Monocultures of the mind. Nataraj Publ., India.

Shiva, V. 2004. The suicide economy of corporate globalization. Available at http://www.countercurrents.org/glo-shiva050404.htm.

Siamwalla, A. 2001. The changing character of technology and its impact on research and extension. *In* The evolving roles of the state, private and local actors in rural Asia. Available at http://www.adb.org/document/books/rural asia. Asian Dev. Bank and Oxford Univ. Press, UK.

Siason, I.M., E. Tech, K.I. Matics, P.S. Choo, M. Shariff, E.S. Heruwati et al. 2001. Women in fisheries in Asia. Global Symp. Women in Fisheries. 6th Asian Fisheries Forum, 29 Nov. 2001, Kaohsiung, Taiwan.

Singh, S. 2003. Contract farming in India: Impacts on women and child workers. Gatekeepers Ser. No. 111. IIED, London.

Singh, G., G. Singh, and N. Kotwaliwale. 2000. A report on agricultural production and processing technologies for women in India. Gender Tech. Dev. 3:259-278.

Spielman, D., and K. von Grebmer, 2004. Public–private partnership in agricultural research: an analysis of challenges facing industry and the CGIAR. EPTD Disc. Pap. 113. IFPRI, Washington DC.

Stanley, R., and J. Rice, 2003. Fisher knowledge? Why not add their scientific skills to the mix while you're at it? Proc. Putting Fishers' Knowledge to Work, 27-30 Aug 2001. Available at http://www.pac.dfo-mpo.gc.ca/sci/sa-mfpd/downloads/ResDoc ShelfRF5.pdf. Fisheries Centre, Univ. British Columbia.

Steinbrecher, R.A. 1996. From Green to Gene Revolution: the environmental risks of genetically engineered crops. Ecologist 26:6.

Steinfeld, H. 1998. Livestock production in the Asia and Pacific region: current status, issues and trends. World Animal Rev. 90(1):14-21.

Steinfeld, H., T. Wassenaar, and S. Jutzi. 2006. Livestock productions systems in developing countries: status, drivers, trends. Rev. Sci. Tech. 25:505-514.

Stone, G.D. 2002. Fallacies in the genetic modification wars: implications for developing countries and anthropological perspectives. Curr. Anthrop. 43:4.

Sulaiman, V.R., and Hall, A.J. 2002. Beyond technology dissemination: reinventing agricultural extension. Outlook Agric. 31(4):225-233.

Sulaiman, V.R., and Hall, A.J. 2005. Extension policy at the national level in Asia. Plant Prod. Sci. 8(3):308-319.

Thang, H.C. 2003. Current perspectives of sustainable forest management and timber certification. Paper presented at the 12th World Forestry Congr., 2003, Quebec City, Canada.

Thorat, S., and S. Fan. 2007. Public investment and poverty reduction: Lessons from China and India. Econ. Polit. Weekly 42(8):704-710.

Thrupp, L.A. 1998. Linking biodiversity and agriculture: challenges and opportunities for sustainable food security. Available at http://www.wri.org/wri/sustag/lba-home.html. World Resources Inst., Washington DC.

Timsina, J., and D.J. Connor. 2001. Productivity and management of rice–wheat cropping systems: Issues and challenges. Field Crops Res. 69:93-132.

Timsina, N. and B. Upreti, 2002. Loosing traditional seed management systems: A threat to small farmers' food security in Nepal. Available at http://conference.ifas.ufl.edu/ifsa/papers/e2/EP3/doc.

UN Atlas of the Oceans, 2002. Mangroves. UN Atlas of the Oceans. http://www.oceansatlas.org

UNCCD. 1998. The social and economic impact of desertification in several Asian countries: inventory study. Interim Sec. UNCCD, Geneva.

UNDP. 2000. Human development report 2000. Oxford Univ. Press, UK.

UNDP. 2004. Human development report 2004. Oxford Univ. Press, UK.

UNEP. 2005. Global environmental outlook 3. Fact Sheet. Asia and the Pacific. www.recoftc.org/site/index. UNEP, Nairobi.

UNFPA (UN Population Fund). 2001. State of the world population 2001. Available at http://www.unfpa.org/swp/2001. UNFPA, NY.

Utting, P. 2000. Forest policy and politics in the Philippines. UN Res. Inst. Social Dev., Geneva, and Ateneo de Manila Univ. Press, Manila.

Van de Fliert, E. 2006. The role of the farmer field school in the transition to sustainable agriculture. *In* A.W. van den Ban and R.K. Samanta (ed) Changing roles of agricultural extension in Asian nations. B.R. Publ., New Delhi.

Van den Ban, A.W. 2005. Changes in society require changes in agricultural extension: some observations. Asian J. Extension Educ. 23:1-16.

Van Mele, P., A. Salahuddin, and N.P. Magor (ed) 2005. Innovations in rural extension: Case studies from Bangladesh. CABI Publ., Wallingford.

Varisco, D., A. Dubaie, M. Jazm, M. Mahyub, and A. al-Wali al-Khuleidi. 1992. Indigenous plant protection in Yemen. Study commissioned by the Yemeni-German Plant Protection Project (GTZ) for the Min. Agric. Water Resources.

Voeten, J., and B.J. Ottens. 1997. Gender training in aquaculture in northern Vietnam: A report. Gender Tech. Dev. 1:413-432.

Von Braun, J., and R.K. Pachauri. 2006. The promises and challenges of biofuels for the poor in developing countries. IFPRI, Washington DC.

Waggener, T.R. 2001. Logging bans in Asia and the Pacific: An overview. *In* Durst et al. (ed) Forests out of bounds: Impacts and effectiveness of logging bans in natural forests in Asia-Pacifc. Asia-Pacific Forestry Comm., RAP Publ. 2001/08. FAO Reg. Off. Asia Pacific, Bangkok.

Wardana, I.P., and L. Syamsiah. 1990. Gender analysis in rice–fish farming system: a case study at Binong, West Java, Indonesia. Asian Faming Systems Research and Extension Symp.: Sustainable farming system 21st century Asia, 19-22 Nov 1990, AIT, Bangkok.

Warren, D. 1991. Using indigenous knowledge in agricultural development. World Bank Disc. Pap. 127. World Bank, Washington DC.

WHO. 2002. Future trends in veterinary public health. Tech. Rep. Ser. 907. WHO, Geneva.

Willer, H., and M. Yussefi. 2006. The world of organic agriculture: Statistics and emerging trends 2006. FiBL and IFOAM, Bonn.

Wilson, A.K., T.L. Jonathan, and R.A. Steinbrecher. 2006. Transformation-induced mutations in transgenic plants: analysis and biosafety implications. Biotech. Genetic Trade Rev. 23.

World Competitiveness Center. 2006. World Competitiveness Yearbook 2006. Available at http://www.worldcompetitiveness.com/OnLine/App/Index.htm. Int. Inst. Management Dev., Lausanne, Switzerland.

Yadav, R.L., B.S. Dwivedi, and P.S. Pandey. 2000. Rice–wheat cropping system: assessment of sustainability under green manuring and chemical fertilizer inputs. Field Crops Res. 65:15-30.

Yadav, R.L., D.S. Yadav, R.M. Singh, and A. Kumar. 1998. Long-term effects of inorganic fertilizer inputs on crop productivity in a rice–wheat cropping system. Nutr. Cycl. Agroecosyst. 51:193-200.

Yang, H. 2004. Land conservation campaign in China: integrated management, local participation and food supply option. Geoforum 35:507-518.

Yang, H., and A.J.B. Zehnder. 2001. China's regional water scarcity and implications for grain supply and trade. Environ. Planning A 33:79-95.

Yonggong, I. 1998. Institutional and policy reforms of rural extension in China during the transition towards a market economy. Training for agricultural and rural dev., 1997-98. FAO, Rome.

Yu Xiaogang. 2001. Indigenous women's knowledge of sustainable fishery. p. 27-38. *In* K. Kusakabe and G. Kelkar (ed) Gender concerns in aquaculture in Southeast Asia. Gender Studies Mono. 12. Asian Inst. Tech., Bangkok.

Zhu, Z.L., and D.L. Chen. 2002. Nitrogen fertilizer use in China—contributions to food production, impacts on the environment and best management strategies. Nutr. Cycl. Agroecosyst. 63: 117-127.

Zhu, Z.L., M. Giordano, X. Cai, and D. Molden. 2003. Yellow River comprehensive assessment: basin features and issues. Working Pap. 57.IWMI, Colombo

3

Influence of Trade Regimes and Agreements on AKST

Coordinating Lead Authors:
Dale Wen Jiajun (China/USA), Dev Nathan (India)

Lead Authors:
Lim Li Ching (Malaysia), Sukhpal Singh (India)

Contributing Author:
Marcia Ishii-Eiteman (USA)

Review Editors:
Pascal Bergeret (France), Hira Jhamtani (Indonesia)

Key Messages

Key Messages

1. While greater openness to international trade in the region has been associated with higher rates of growth of per capita income (and a reduction of inequality between developing Asia and the developed countries), there has also been an increase in income inequality within the Asian countries. With the stress on trade, investment in agriculture and rural development has been neglected. Conversion of agricultural land to industrial or urban use has resulted in the displacement of many rural inhabitants (with indigenous or tribal peoples, small farmers and the landless disproportionately represented) who are not the beneficiaries of the resulting industrial or other non-agricultural employment, increasing the rural-urban disparity. Development policies could be more balanced by increased investment to agriculture and rural development and more emphasis on non-tradable sectors.

2. Greater equity (through, for example, land rights for women) and a reduction of social exclusion can also increase productivity and therefore provide gains from international trade. In many countries, women have been drawn in large numbers into export-led manufacturing, with resulting improvement in incomes, but there are also problems of wage differentials, short-term industrial working life, occupational hazards and health risks. Trade policy can be made more inclusive to take into account the specific needs of marginalized segments of the community such as women, indigenous peoples and local small producers, as well as to recognize them as holders of AKST.

3. There are vulnerabilities in international trade caused by rising energy costs, volatility of international markets and overdependence on imports for food and agricultural technology. The implications include worsening terms of trade, endangering of food security and changes in the nature and location of AKST generation and dissemination in the region. Alternative and domestic markets are also options for better and sustainable development, as these are the more easily attainable and locally relevant ways of dealing with issues of sustainability and food security. Comprehensive safety net measures and social welfare systems could help to protect well-being in a situation of growing risk and uncertainty. While South Asia, in particular, has yet to work out ways to effectively deal with the opening up of agricultural markets and reducing subsidies of various types, this is less of a problem in East and Southeast Asia, where there has been more stress on achieving transitions to more productive methods of cultivation and higher value crops. Measures to increase productivity are superior to providing subsidies, which distort factor use and can be a fiscal burden.

4. OECD agricultural subsidies do not allow Asian small-scale producers to compete in external markets or domestic markets without tariff support and have detrimental impacts on their livelihoods and food security. Furthermore, many of the developing countries in Asia, in particular the least developed countries, have limited fiscal capacity and are unable to provide the support allowed under WTO regulations. Along with eliminating agricultural subsidies in the OECD countries, there may be a case for providing support to producers in least developed countries. In addition, countries could consider the need for safeguard measures based on food security, livelihoods and rural development criteria. National trade policies could also balance the interest of net food buyers with the requirement for rural development. Overall, there is scope to promote increases in productivity, through enhanced provision of public goods, such as infrastructure, research, irrigation, etc., to enable agricultural producers to be competitive.

5. Anti-dumping measures have been used by developed countries to protect domestic producers against competition from developing country exports. Competition considerations in anti-dumping law and practice could be introduced, as well as restricting the use of anti-dumping measures only to situations where there is an evidence of predatory intent.

6. The secular decline and wide fluctuations in prices of primary commodities have severe negative impacts on the livelihoods of millions of small producers. Price stabilization measures can be combined with orderly changes in these prices, along with promotion of alternate uses of commodities and movements up the value chain.

7. Least developed countries, including the small Pacific islands, are unable to match the competitiveness of larger and more complex economies. Special, differential market access, for given time periods, both within ESAP and with industrialized countries, can help LDCs benefit from international trade. OECD countries, such as Japan and South Korea, face challenges in reducing protection of their agriculture. It needs to be considered whether agricultural subsidies are better used in promoting environmental improvements and enabling transitions to alternate livelihoods. Major agricultural exporters (Australia and New Zealand) face the challenge of integrating environmental concerns into price competitive agricultural systems. This involves developing new technologies (such as feeds that reduce methane emissions) and methods of pricing, in taking account of what are now externalities.

8. With developing Asia's characteristic of labor abundance relative to land and capital, there is a comparative advantage in crops, such as vegetables, fruits and flowers, which use more labor per unit of land and capital, as against, e.g., cereals. In order to utilize this comparative advantage both AKST and extension would require more attention to high value crops, such as fruits, vegetables, flowers.

9. Vertical integration/coordination of food systems has marginalized primary producers and the dominance of retail chains may further this trend. This does not mean that there can be no countervailing power, e.g., of organized primary producers, that would improve the share of primary producers. The most important requirements of small farmers in this changing environment are better access to knowledge, technology and capital, along with facilitation to overcome collective action problems in dealing with scale requirements. Management capacity, crucially centered around knowledge, is as important as physical capital but is the most difficult thing to produce.

10. Trade agreements do not sufficiently address environmental, social, labor and health dimensions. While sanitary and phytosanitary (SPS) measures may promote better human and animal health and environmental standards, they have been used as trade barriers, which have led to trade loss, diversion and higher costs for developing countries. Bilateral and regional free trade agreements also restrict policy space and make it more difficult for governments to implement and enforce environmental, social and health protective measures. Governments need the policy space to be able to take these measures. Other considerations, such as multilateral environmental agreements, labor standards and social development instruments, could be given at least equal weight.

11. Pesticide use has increased rapidly in the ESAP region with consequent health, environmental and social impacts. Pesticide residues and the use of banned chemicals lead to problems in meeting SPS standards for agriculture-based export protects. In order to utilize the potential for exports, more attention to AKST and extension is necessary to reduce pesticide use and eliminate banned chemical use. Various alternatives, such as Integrated Pest Management, organic agriculture and agroecology exist. The challenge is to mainstream and promote their adoption with necessary policy and investment support.

12. There is good opportunity in organic and fair trade markets and their social, sustainability and ethical objectives often overlap. If the overlap is encouraged, it could increase the volume of trade and improve working conditions and livelihoods of producers. Developing country producers' ability to meet organic and fair trade standards can be facilitated through better access to locally-developed AKST. There could be benefits to small producers through mainstreaming organic and fair trade markets, provided the ill effects of conventional supply chains are avoided.

13. Increasing international trade in agricultural commodities has often led to over-exploitation of natural resources in ESAP countries. There are positive examples of learning and technology development and systems of culture that have reduced pressures on natural stocks. The challenge is to address new problems such as environmental change and erosion of biodiversity. One option is to provide systems of compensation for the provision of environmental services, in order to increase the supply of environmental public goods that are often linked to particular forms of land use and cultivation.

14. In any intellectual property rights (IPR) regime there is a trade-off between rewarding development of knowledge and inhibiting the spread of knowledge and the capacity for reverse engineering, which are both crucial for development. IPR standards under trade agreements have contributed to a shift in AKST, by facilitating private sector dominated research and consequently privately-generated and owned AKST. IPRs may restrict access to research materials, tools and technologies, as well as to plant material for farmers, with consequences for food security. While some national level action has been taken to break monopolies and encourage competition, there is no international mechanism to deal with such issues. Increasing funding and support for public sector research that delivers publicly available outputs is an option to address the growing private sector dominance. Implementation of farmers' rights on seeds is critical to ensure conservation of agricultural biodiversity and associated AKST and can provide an important counterbalance to formal plant breeders' rights and patents. Recognition and protection of traditional/indigenous knowledge remains a challenge. There are questions about whether patentability and ownership of such knowledge are appropriate and what processes are needed to protect them and further to share the benefits of protection.

15. There, however, has been a concentration in agricultural research and development and extension, on a few major crops and tradeables, varieties and traits, to the neglect of locally relevant crops and technologies, which have been marginalized both in the private as well as the public sector. More investment in research on agroecosystems and locally adapted technologies could be used in order to develop approaches that promote food security and environmental sustainability.

16. The current restrictions in various countries, including those of the EU, on imports of transgenic crops, means that the export potential of transgenic crops is limited. At the same time, many developing countries in Asia lack regulatory and monitoring capacity to import transgenic crops. The precautionary principle and the principle of prior informed consent are the key elements in the Cartagena Protocol on Biosafety, but have not been implemented. Given all these considerations, more investment and research prioritization can be considered for independent biosafety and long-term risk-related research and for non-genetic engineering AKST.

17. Though per capita carbon emissions in ESAP developing countries are lower than those in the developed countries, it is likely that there will be pressures on these countries to reduce emissions and shift to low carbon economies. While biofuels may provide prospects for the development of new sources of energy from agriculture, there is the threat of converting natural forests and agricultural lands into monoculture plantations. Furthermore, there is the issue of corporate or community ownership of such initiatives. These developments may have implications for food security, biodiversity, sustainability and livelihoods. Establishing decentralized, locally-based, highly-efficient energy systems is one option to improve livelihoods and reduce carbon emissions.

18. While carbon and other GHG emissions use global public space (absorption space), a price on carbon emissions, along with necessary changes in con-

sumption patterns, could help induce a technological shift to a low-carbon economy. A system of tradable emissions can be devised on an equitable basis, based on the Rio principle of "common but differentiated responsibility." In the event of some countries refusing to participate in a globally-mandated GHG protocol, such "free-riding" can be discouraged by allowing all participating countries to use WTO rules-sanctioned import duties, based on direct and indirect carbon content of products, on export from non-participating countries. Since the opportunity costs of not using forests in an extractive manner are very high, in terms of the foregone livelihoods of some of the poorest peoples, a system of international payments for "avoided deforestation" would combine justice with achieving a necessary measure for reducing global carbon emissions.

19. Hazardous waste is often exported for disposal in countries with lax or poor enforcement of environmental regulations. Without leading to a loss of jobs in developing countries, the disposal of hazardous wastes could be regulated by international coordination of these regulations, supported by civil society and other actions to secure their implementation.

3.1 Context

The influence of national, regional and international trade regimes, agreements, intellectual property rights and the regions' response to them and the role of AKST in addressing these is assessed in this chapter. After a broader context setting on trade agreements and regimes, the assessment on WTO and AKST elaborates on impact of biotechnology along with issues of intellectual property rights. The combination of the changing composition of demand for agricultural commodities in favor of higher quality foods, like fish and meat products and the comparative advantage of labor-abundant Asian developing countries in the production of labor-intensive agricultural commodities, have together brought a change in the composition of agricultural output. Globally as well as in this region, there has been concern about the effects of trade agreement on environment, health and other social dimensions (see 3.6).

The structure of world trade is changing. From the early trade of manufactured goods for raw materials, in the post-Second World War period there was a growth of inter-firm trade, as firms became transnational and set up vertically integrated production bases in different countries. More recently, however, there has been a globalization of production and supply chains, in general a globalization of value chains. With this, rather than vertical integration within a country or corporation, there is a splitting up of parts of a value chain across countries. Trade figures don't capture the change in trade within value chains, since, other than in transport equipment and machinery, a distinction is not made in trade between components and whole products. But there are many analyses of the growing importance of intra-industry trade, referred to as "outsourcing" (Feenstra, 1998) or "vertical specialization" (Yeats, 1998).

With this change in the structure of trade, in which Asia has participated perhaps more than any other region, there has been a double shift, one in the composition of trade and two, in the poles of world trade. In the composition of commodity trade there has been a shift from agricultural products (food and agricultural raw materials) which used to account for nearly 50% of exports in 1960 to just 7% in 2001 and a corresponding increase in exports of manufactured goods from les than 20% in 1960 to almost 70% in 2001 (Table 3-1).

The growth of the Asian economies and the greater importance of trade in their economies have together made Asia an important pole of world trade. The triad of world trade (US, EU and Japan) has turned into a quad, with "Asia other than Japan" joining in as a new pole of world trade (Gibbon and Ponte, 2005).

Within this pattern of world trade there is also a growth of South-South trade. In 2001 in developing Asia 41.5% of exports went to developing Asia itself (UNCTAD, 2004). But this trade is concentrated in the economies of East Asia. It is mainly of a production-sharing type, resulting in a "triangular trade" pattern, i.e., the more advanced economies within East Asia, e.g., Republic of Korea, export intermediate products to China, where they are inputs for production to be re-exported to developed countries (UNCTAD, 2005).

Table 3-1. *Distribution of exports (%) by commodity group.*

Year	All food items	Agricultural raw materials	Ores and metals	Fuels	Manufactured goods
1960	18.9	30.0	1.9	30.0	18.8
1970	14.0	18.3	2.0	36.6	28.6
1980	6.8	4.1	2.2	62.4	23.6
1990	7.7	2.9	1.8	22.7	63.8
2001	5.3	1.7	2.9	14.5	69.3
1960	17.4	16.7	3.6	9.9	51.3
1970	13.2	10.6	4.0	9.2	60.9
1980	11.1	3.7	4.7	24.0	54.2
1990	9.3	3.0	3.6	11.0	70.5
2001	7.4	1.8	3.0	9.1	74.1

Source: UNCTAD, 2004.

In the case of agricultural products, South-South trade is not of a triangular nature. It represents final export to meet growing demand, based on the growth of incomes in developing countries. In the middle- and low-income countries growth of income leads to a growth in demand for agricultural commodities, more than in developed countries, benefiting those economies that mainly export agricultural commodities (UNCTAD, 2005). Within Asia, for instance, Vietnam has increased its exports of rice, coffee and fish, both to markets within the region and to developed countries. But as in other developing countries of Asia, there has not been a one-sided reliance on exports of agricultural commodities, but also a push in exports of manufactures, labor-intensive manufactures, in particular.

The pattern of consumption of food differs from one country to another. But what is common is a falling share of grain and a switch to higher quality foods, like meat, fish and milk products. Such a switch, however, may be the result of growing inequalities in food consumption. The lower sections may have gross deficits even in basic calories while the upper sections diversify their food consumption into higher value foods.

3.1.1 Free trade agreements in ESAP

Of the 33 countries in the ESAP region, 22 are currently members of the World Trade Organization (WTO), with about 6 more countries in the process of accession negotiations. Thus, the rights and obligations under the multilateral trade regime, via the WTO, play an important role in ESAP countries.

Of particular concern are the free trade agreements (FTAs) between developing countries and developed countries like the United States. These North-South FTAs are very comprehensive in scope and extend into the realm of domestic policies (Gibbs and Wagle, 2005), covering areas beyond trade in goods, to include the opening up of services, government procurement, protection of intellectual property rights (IPRs) and creation of new investment privileges and protection (such as binding dispute settlement mechanisms that allow investor-state disputes). Bilateral and regional FTAs can be "WTO-plus", with provisions that go beyond WTO obligations (Gibbs and Wagle, 2005). Thus, the "policy space" for developing countries to pursue national development and socioeconomic goals may be significantly reduced.

The U.S. FTAs in particular seem to be used to influence partners in larger or multilateral negotiations and "to establish precedents that consolidate the U.S. position on issues where it has serious differences with its trading partners (such as on GMOs, geographical indications or audio-visual services)" (Gibbs and Wagle, 2005). Foreign policy and security issues also play a part. Of relevance to agriculture, FTAs do not establish disciplines on agriculture subsidies in the major developed countries and this exposes farmers in the developing partner country to unfair competition (Gibbs and Wagle, 2005). The U.S. FTAs, for example, do not have commitments on anti-dumping or agricultural subsidies and cover all products (i.e., in terms of obtaining market access), with the exception of "sensitive" ones like sugar. This creates the potential for imbalances in the agreement.

As U.S. FTAs generally ask for agricultural tariffs to be lowered to zero, although with varying time periods of implementation, many developing country farmers would be unable to compete with the influx of subsidized U.S. agricultural products and may be adversely affected. For example, under the North American Free Trade Agreement (NAFTA), from 1993 to 2003, exports of U.S. agricultural produce to Mexico more than doubled, climbing from $3.6 billion to $7.9 billion. Over a similar period, Mexico lost nearly 2 million agricultural jobs, according to Mexico's National Employment Survey (The Washington Post, 2007). Ratification or accession to UPOV 1991 is a requirement in U.S. FTAs with Bahrain, the Central American countries—Costa Rica, El Salvador, Guatemala, Honduras and Nicaragua—under CAFTA, Chile, Colombia, Morocco, Oman, Peru and Singapore.

Such obligations remove the flexibility afforded under the TRIPS Agreement that allows countries to choose the option of a sui generis system of plant variety protection, which could be tailored to protect farmers' rights (TWN, 2005). The UPOV 1991 system currently restricts farmers' rights to use and save seed and prohibits them from exchanging or selling seeds of the varieties it protects, thereby subjecting poor farmers who depend on farm-saved seed to dependence on commercial breeders.

There are also numerous South-South FTAs, which may be able to promote South-South trade and allow countries to export goods for which they face market access barriers in the North. South-South FTAs may be more equitable in that there is less of an imbalance between the negotiating partners and they are less likely to be as comprehensive in scope as the North-South FTAs, since they tend to focus mainly on trade in goods.

Nonetheless, the South-South FTAs have to be also assessed carefully to ensure that the overall benefits outweigh the costs for the countries concerned and that any sectoral implications as a result of the liberalization of tariffs on goods are properly addressed to help with adjustment costs. For example, under the Thai-China FTA, agricultural tariffs have been lowered on 116 types of fruits and vegetables, including garlic and onions, from 1 October 2003. Since then, Chinese garlic has entered the Thai market in large quantities and at lower prices, with the result that garlic growers and small traders in Thailand have lost their livelihoods (Narintarakul and Silarak, 2005). A major issue in "asymmetric" North-South FTAs would be to ensure that policy autonomy is retained for national development and the particular needs of each society.

3.1.2 Major players: their roles and interactions

The roles of different countries and block of countries in influencing trade policy depend on their positions as importers or exporters or both, driven by the size of population and their food production. More recently, large corporations have become major players and have influenced trade policies.

China and India, with their large and growing markets will have a substantial influence on the pattern of trade within Asia. Because of the size of their populations, even at a lower per capita income level these economies are larger and their influence is likely to be greater than that of Japan or Republic of Korea during a corresponding period of their ascent as manufacturing powers (UNCTAD, 2005).

With China already achieving basic calorie requirement levels (accepting the regional and socioeconomic-based inequalities in such an aggregate measure), there is likely to be a further rise in demand for livestock products, oil crops, vegetable oils, fruit and vegetables (FAO, 2002b) and with the boom in home construction for wood. In India there is a lag of a decade or so, as compared to China, in the changes in the pattern of consumption. There is still scope for increase in consumption of basic calories and an increase in per capita food consumption at the bottom of the economic scale, though above that consumption patterns have changed. Some of the boom in demand is negative, as in the case of demand for wildlife products in East and Southeast Asia for food and medicine, which has driven the illegal wildlife trade in the region (World Bank, 2005a).

Both Japan and Republic of Korea have substantially protected rice markets. But for other agricultural commodities, including fishery and meat products and vegetables, they present large and growing markets. Recently, however, Japan has reduced levels of protection. There, however, are still issues of further reducing tariffs, enabling existing agricultural producers to shift to other livelihoods and also enhancing supply of environmental services.

Overall in Asia, the dietary pattern is changing not only towards higher value foods but also to the use of semi-processed and processed food stuffs, which are more convenient even for home cooking and even towards more consumption of food outside the home (Popkin, 1993). The last trend is particularly influenced by the large-scale entry of women into the non-home-based labor force. Within the ESAP region, Australia and New Zealand are both large agricultural exporters. They are among the Cairns group of countries that press for free trade in agricultural commodities. They are joined by Thailand and Vietnam, both of which are efficient agricultural producers.

The other major players are the developed economies of North America, Europe and Japan. Their policies have affected the trade prospects of Asian countries in many agricultural commodities. They were all substantial importers of sugar, but subsidized production of corn-based and beet-based sugar have changed the picture. Since the early 70s, US sugar imports have declined from more than 5 million tonnes per year to just more than 1 million tonnes per year. While Japan's sugar imports have fallen from 2.5 million to 1.5 million tonnes per year, the EU has changed from a net importer of 2.5 million tonnes in the early 1970s to a net exporter of 5 million tonnes (Mitchell, 2005) at the lower price created by its subsidies, thus further undermining the price received by farmers and producers in developing countries (Robbins, 2003).

Countries, however, are not the only actors in the region. Trade policies themselves are influenced by the various lobbies in the countries. Even WTO policies are influenced by corporate interests, which maintain substantial lobbying presences at WTO headquarters in Geneva (ActionAid, 2006).

Corporations, particularly the big multi-national corporations, influence economic decisions in the region not only through their lobbying with governments and international bodies, but also through their economic practices. Production is more and more being organized in global value chains, as mentioned in the introduction to this chapter. Agricultural commodities, in particular, are organized in what are called buyer-driven or retailer-driven value chains. The big food retailers and producers have substantial influence over prices that are paid to producers of primary agricultural commodities at the end of the value chain. In their product markets there is oligopolistic competition. The price pressures of this competition are passed on to the producers through lower prices. It is competition among producers with no better alternative, which enables buyers to gain lower prices from small producers (Singer, 1950).

Concentration is not only a factor among buyers in agricultural product markets. It is also increasing in agricultural input markets (UNCTAD, 2006b). But the entry of China's national agricultural research organizations and small seed producers in India, both in Bt cotton, has increased competition in some input markets (Fukuda-Parr, 2007).

Consumers have recently entered as policy makers in international markets. Particularly in developed countries, consumers and consumer lobbies have become more active in demanding certain standards. This is sometimes reflected in improved SPS standards, some of which have become contentious issues between developing and developed countries. There are also other standards, environmental standards, for which consumers have shown a willingness to pay a premium on standard prices and include organic food, shade coffee and certification of sustainably harvested wood products.

A study by the International Trade Centre (ITC) of UNCTAD and the WTO surveyed the European market for organic foods and beverages. A major conclusion of the study was that demand for these products is growing rapidly (see 3.4.5) and that insufficient supply rather than demand is the problem for these markets (Kortbech-Olsen, 2001). There is also a growing demand for organic foods in the urban centers of many Asian countries, though it is still quite limited. China's production of organics under the Green Label was close to $12 billion, which almost matches the size of the US market, the largest organic market in the world (IFAD, 2005) India's exports of organics was just about $15 million, though a lot of organic production is consumed locally and not marketed outside the locality.

There are also other forms of consumer standards, as the Forest Stewardship Council (FSC) certification. The certification is expected to be based on environmental and community-role criteria. The "Bird Free" symbol of "Shade grown coffee", again, certifies the non-destruction of forests to cultivate coffee. These certification systems provide for some premiums on price.

Unlike consumers and consumer lobbies, producers' associations have not had the same impact. In the first few post-WWII decades, marketing boards for many agricultural commodities, such as coffee and cocoa, tried to control prices and reduce outputs. But the weaknesses of these opposition from buyers' lobbies and the spread of production outside export quotas (e.g., Vietnam's entry into world coffee markets) undermined the boards. The post-WWII experience would lend itself to the hypothesis that changes in the location of production cannot be managed through export quotas, as countries keen to expand their own export opportunities are likely to undercut established producers.

3.1.3 National policy trends

Within the ESAP region there are a number of differences in national policy. In OECD members, Japan and the Republic of Korea, there is a strong protection to their rice producers, often justified on the basis of national culture, or tastes. For Australia and New Zealand exports of agricultural commodities (including livestock) are an important source of national income. They, along with other members of the Cairns group press for removal of restrictions on trade.

In developing Asia, there is a difference between East and Southeast Asia and South Asia. In most of East and Southeast Asia the proportion of the population dependent on agriculture has come down substantially over the last few decades. Among them, for Thailand and Vietnam, agricultural exports are important but account for a declining share of total exports, as manufactured exports have increased. But in Indonesia and the Philippines there is a large proportion of population still dependent on agriculture, more like the picture in South Asia, where there is still something like 50% of the population dependent on agriculture as the mainstay of their livelihoods. The high numbers of people dependent on agriculture and the relative stagnation in agricultural technology and yields, along with insufficient growth of labor-intensive manufactures have made it difficult to move more people out of agriculture.

The above differences within Southeast and South Asia are reflected in different national policies. In countries like Thailand and Vietnam there is a stress on increasing productivity, so as to retain or improve their competitive positions in world agricultural trade. In China too there is a similar stress on improving productivity and moving into high value agriculture. In both cases the attempt is to improve infrastructure and provide research and technology development and marketing support.

But in South Asia (as also in Indonesia and the Philippines) there is a much greater stress on protecting domestic producers, with subsidies, from international competition. While there are moves to diversify into high value agriculture, these are not as consistent. In the "Green Revolution belt" of Punjab-Haryana, continued minimum price support to wheat and rice continues to stall attempts at diversification, as the rates of return from assured grain prices inhibit a shift towards more risky, if higher return, crops (Joshi et al., 2004; Rao et al., 2006).

Along with the above, there are also trends to opening up sections of the agricultural markets, for instance in cotton. In India this has led to a fall in cotton prices, affected as they are by competing imports from subsidized producers, like those in the US (Philip and Jenniah, 2006). "Between the period 1990 and 2005 the import of cotton lint increased at a compound growth rate of over 75%, growing in geometric multiples to domestic production. The price witnessed a decline of more than 55% between the years 1996 and 2003. In terms of individual years, the prices dipped as low as US$1000 per tonnes in the year 2002," (Philip and Jenniah, 2006). The plight of cotton farmers was compounded by the many instances of sale of spurious Bt cotton seeds. The destitution of many farmers has resulted in numerous suicides. This has become a frequently recurring political issue.

Overall, South Asia, in particular, has yet to work out ways to effectively deal with the opening up of agricultural markets. It is less of a problem in East and Southeast Asia, affecting a much smaller proportion of the population and where there has been more of a stress on achieving transitions to more productive methods of cultivation and higher value crops.

One problem that has recently come to prominence is that of displacement of agriculturists from their lands, taken over for industrial use. In this displacement without compensatory jobs, small farmers and the landless, along with indigenous and tribal peoples are disproportionately represented. This has exacerbated the problem of the rural-urban divide, which is part of the growing inequality in developing Asia. The numerous rural protests in both China and India are witness to the social tensions caused by the growing rural-urban divide in these major countries of ESAP. On the one hand, greater openness to international trade in the region and the resulting growth of Asia as a manufacturing centre of the world, have created many jobs, lifted tens of millions out of poverty and reduced the global inequality between developing Asia and the developed countries. On the other hand, there has been a neglect of the agricultural sector (after the initial first period of Green Revolution and agricultural export growth), increasing the rural-urban divide in the contemporary situation. The emphasis on trade has led to neglect of rural development and of non-tradable sectors of the economy.

Policies to reduce gender equality (e.g., land rights for women, recognizing women as holders and developers of AKST); to reduce social exclusion (of indigenous and tribal peoples, low castes, minorities); to increase public investment in the supply of rural and agricultural public goods, such as infrastructure (roads, communication, health and education) and policies that encourage investment in research for neglected and non-internationally traded crops are options to reduce the rural-urban disparity (see Chapter 5). Education and skill building can help rural poor people to benefit from the new jobs in manufacturing and services that are being created in ESAP.

3.1.4 Trends in private and public institutional roles

Some of the studies referred to in this chapter point to the weakness of "world governance on questions of corporate conduct and competition" (UNCTAD, 2006b). Whether it is competition policy or corporate governance, there has been a globalization of economic processes, but not a globalization of the regulatory framework. This is an important public issue that affects agriculture and trade.

As an example, when China, as a concession to the growing trade surplus with the USA, agreed to buy soya, prices of soya immediately went up from $7.70 in December 2003 to $9.82 per bushel in March-April 2004 and when China completed its purchases, the price promptly fell to $5.93 in August 2004. It was estimated by the Chinese Academy of Science (http:www.chinafeed.org.cn/cms/_code/business/include/php/218139.htm) that China overpaid $1.5 billion on this purchase; what is of greater interest is the next part of the story. Because of the high import prices of soya, many processing plants in China went into the red and as many 64 out of 90 soya mills are now partly or wholly owned by the same soya trading companies, ADM, Cargill, Bunge and Luis Dreyfus. International trade still does not have the regulations and organizations to deal with such cartel behavior.

3.1.5 Trade and food security

Regional Experience of Trade and Food Security

ESAP and developing Asia in particular, has seen rapid reduction of poverty and improvement in household and individual access to food. At the international trade level, this is largely due to its rise in exports of manufactures. At the same time, there are substantial gains from international, particularly, regional trade in food grains. Carrying and transport costs can be lowered as regional trade in food grains becomes part of the national food management system of countries, particularly smaller countries. The ability to utilize regional trade to supplement domestic production depends on the country concerned having adequate foreign exchange reserves, otherwise it can be subject to unwanted external pressures. Along with this there is the negative effect of cartel arrangements between exporters and importers being more likely in a regional than in a global context (World Bank, 2006a).

Farmer households do react to market prices in deciding between production alternatives. But market prices are lowered by subsidized exports, something done not only by developed countries but also by developing countries, as are rice exports by Thailand, India and Vietnam.

Thus, given the twin realities of power relations and subsidized exports, countries cannot depend entirely on market-based individual household production decisions to set domestic food production levels. Subsidized exports can justify import duty to the extent of the subsidy.

In the absence of improvements in public service delivery like irrigation and other agricultural infrastructure, adequate research and extension and adequate institutional credit and marketing channels, poor producers remain trapped in low productivity states. Poor and food insecure households can benefit from expanded opportunities of trade provided that those constraints are addressed. Giving voice to poor producers' interests by placing these issues on the policy agenda is crucial for fostering reforms that unleash the productivity potential of poor people and increase their bargaining power.

Importantly, however, food security issues are related not only to poor producers but also to poor consumers. Low prices of food, brought about through imports of cheap food, when combined with increased productivity, can lead to both higher real wages and increased farm incomes. Internal political economy considerations, i.e., the strength of different lobbies, determine the level at which food prices are set. Setting import duties higher than the extent of subsidies provided by exporting countries would further erode the food security gains of higher real wages. The possibility of substantial unofficial trade to take advantage of price differences in neighboring countries, in fact, sets a limit to the extent to which import duties can be greater than transport costs.

As detailed below, small producers' livelihood are often threatened by imports. In the manner of providing domestic support, however, measures to increase productivity are superior to providing subsidies to continue high-cost production. They would not only increase national productivity but also can strengthen the fiscal position as compared to subsidies.

Household food security would be improved by allowing farm households to choose their own mix of crops and livelihoods, reacting to market prices and their own aspirations, rather than have the mix of crops dictated by administrative decision. Farmers in many areas of Nepal, Bangladesh, Indonesia, etc., are themselves moving into areas of comparative advantage, like vegetables and other such crops, which require more labor than cereals. But developing competitiveness in new areas of production requires substantial support, especially in improving quality and building capabilities, for instance, in meeting Sanitary and PhytoSanitary standards.

Thus, uniform rules on the nature and measures of support cannot be applied to industrial and developing countries alike, in particular, to LDCs. Least developed countries, including the small Pacific islands, are unable to match the competitiveness of larger and more complex economies. Special, differential market access, for given time periods, both within ESAP and with industrialized countries, can help these and other LDCs benefit from international trade.

Developing countries in general and LDCs in particular (including the small island nations of the Pacific), with narrow markets and not-so-developed capabilities, need to be provided specific support to build on areas of comparative advantage. If they are bound by the restrictions of the WTO, disallowing the benefit or support that is crop specific, they may well be unable to undertake the necessary diversification of production that can increase household incomes and thus food security.

Prices of primary commodities, like coffee, however, are subject to substantial fluctuations, threatening the food security position of producing households. But measures of price stabilization can be combined with steps to encourage diversification of product use, as is the case with palm oil. Further, as lower cost producers, or producers willing to accept lower returns enter the market, higher cost producers need support to move into other areas production, with disincentives for not doing so.

In all of the above measures of changes in production structures responding to comparative advantage, gradual change would reduce the social costs of the transformation compared to "big bang" type of change and would thus be more desirable.

Where women have participated in the commercial process fostered by trade, they have gained in household and social position, though often at the cost of an increased workload. But the frequent exclusion of women from long-distance trade may be tackled by access to capital, training and facilitation measures.

The nonmarket access rights of tribal or indigenous peoples to land and forests, which are important for their food security, may be eroded through trade agreements which open up land to the market. At the same time, the increased scale of production fostered by commercialization cannot be sustained without a transformation of indigenous property systems in the direction of individualization or regulated commons, so as to link investment and returns.

Trade and the Rural Poor

With underdeveloped infrastructure, the upland and mountainous areas of Asia suffer from social deprivation due to political neglect and remoteness. The current process of international trade increases the risk of further marginalization, disempowerment and desperation, unless it is specially adapted for these areas (IFAD, 2001).

The limited accessibility, fragility, marginality and diversity of the mountain areas generally require diversification of resource use and production. But international trade, guided by short-term profitability and external demand, promotes narrow specialization in few specific products. It encourages indiscriminate resource-use intensification and over-extraction of niche opportunities, with little concern for their environmental and socioeconomic consequences. The process of globalization is so rapid that mountain communities do not have sufficient lead-time and capacity to adapt (IFAD, 2001). There are other agriculture-based poor, concentrated in the large arid and semi-arid regions, dependent on rainfed agriculture. They share some characteristics of poverty with the upland poor.

In many Asian countries, small farmers have been affected by competition from imports that are cheaper than their products. Their organizations have been raising the alarm and requesting assistance from their governments, e.g., Asian Farmers Group for Cooperation request that WTO continue to allow Asian countries to protect their agricultural products (Antara News Agency, 2000).

The Sri Lankan agricultural sector has come under heavy pressure from increasing competition arising from cheap imports resulting from import liberalization. The comparative and competitive advantage of Sri Lanka to produce particular commodities will need to be considered in selecting IFAD's interventions in future projects (IFAD, 2002). There have been protests of Sri Lankan farmers who were adversely affected by cheap imports. Protests were held by potato farmers, chili and onion producers and chicken farmers against cheap and ruinous imports (Samath, 1999).

In 2000 the U.S Agriculture Department accused the Philippine government of violating WTO rules when the import of US chicken was limited to curtail dumping. According to the Minimum Access Volume (MAV), only 19,000 tonnes could be imported to safeguard the local chicken industry. (The Philippine Daily Inquirer, 21 July 2000). About 330,000 workers or a third of a million in the chicken industry were affected.

The economic reforms in China, especially on the occasion of China's entry into the WTO, have led to concerns by some senior officials as well as experts that there may be adverse effects on the competitiveness and livelihoods of local farmers. "China's leaders worry that economic reforms could be placing more burdens on farmers than they can bear. Farmers are on the receiving end of the earliest and sharpest changes from the new policies that China agreed to implement to gain entry to the WTO. . . . According to a report by China's State Council, the country's WTO commitments are likely to wipe out the livelihoods of 13 million farmers who grow wheat, rice and cotton, while creating new ones in non-grain crops for only about 1.5 million. Some economists reckon that China will eventually need to find jobs for about 200 million farmers as its market reforms continue. "The Chinese farmer is in a very unenviable position. The impact of reforms on agriculture is profound" (Ke Bing-sheng, director general of the Research Centre for Rural Economy in China's Ministry of Agriculture as quoted in Goodman, 2002). These concerns have materialized, as manifested in soybean (Box 3-1).

Indian farmers have in recent years faced competition from imported skimmed milk. "The import of 17,000 tonnes of skimmed milk powder from Denmark at zero duty a couple of years ago resulted in a political uproar in Punjab. New Zealand has dumped a large quantity of butter oil into India. Even after paying an import duty of 35.2%, the butter oil imports have been at less than US$1,000 per tonne against the prevailing global price of US$1,300 per tonne. Domestic prices crashed, coming down by 10-15 percent. Highly subsidized imports of milk flowing into India will only further marginalize millions of milk producers. Thousands of dairy cooperatives which pulled the poverty-stricken masses into a path of economic emancipation will collapse faced with cheap and highly subsidized imports" (Sharma, 2002).

Indonesian farmers in several sectors, including poultry, rice and corn have been affected by cheap imports on different occasions in recent years. As Indonesia has attempted to adjust its import policies with WTO agreements through lowering import duties and lifting bans on various commodities local producers say the flood of imports is forcing them out of business.

Rice is the staple food for most Indonesians and is a strategic commodity for the country, grown by 40 million farmers. Before to 1998, i.e., before the reforms in the country following the Asian financial crisis, the price of rice was kept at low levels by the government's food agency, BULOG, by implementing a buffer stock policy. Farmers were given production input subsidies (Suparmoko, 2000). Although the 1997 crisis was rooted in the banking sector and exchange rate policy, the IMF demanded trade liberalization measures in both the agricultural and manufacturing sectors. This included ending the monopoly of the BULOG on food imports and marketing and cutting the import tariff on rice to zero (Oxfam, 2005). From 1996 to 1999, rice imports more than doubled, reaching 4.7 million tonnes. Since BULOG was unable to defend the floor price promised to producers, farmers were left to sell their crops at low prices. In late 1999, the government stepped in to restrict the flood in imports and in 2000 re-introduced a levy equivalent to an import tariff of 30%.

3.2 Developing Countries' Issues in Trade Agreements

3.2.1 Subsidies and market access

Developed countries (or industrial countries) share of world agricultural exports remains about 63% (Aksoy, 2005); Asia and the Pacific together have a share of 13.9% of world agricultural exports in 2000-01, which is almost the same as in 1980-81. This is in contrast to the change in the shares in

Box 3-1. Case Study: The future of soybean in danger?

Soybean was first domesticated in China almost five thousand years ago. The legendary Emperor Shennong (which literally means the Emperor of Magic Agriculture) made soybeans the only legume of his five life-sustaining grains. Millennia of cultivation have produced an enormous range of varieties, as well as the vast body of indigenous knowledge associated with it.

Prior to 1995 China enjoyed a long history of exporting soy. During WTO negotiations, China made huge concessions in the agricultural sector cutting the tariff for soy imports to 3%. Since then, soy imports have soared. In 2003, soy imports reached 20.74 million tonnes (doubling within three years), and China became the world's biggest soy importer. In 2005, soy imports equaled 26.5 million tonnes, 1.6 times of domestic production. Most of the imports are GM soy from the US, Brazil and Argentina. Conversely, in the mid-1990s, China exported more than 1 million tonnes of non-GM soybeans per year to South Korea and Japan. In recent years, the export has steadily dropped to 200-300 thousand tonnes per year, partly because buyers are concerned about genetic contamination by the GM imports.

Needless to say, the massive import surge and declining export has a huge negative impact on domestic producers. Heilongjiang, the Northeast province of China, produces about 40% of the country's soybeans. About 20 million small farmers used to grow soybeans. According to a news report in September 2006, the soybean price in Heilongjiang dropped to 28 cents per kilogram in 2005. This was below the production cost even if the labor cost was not counted. Consequently, in 2006, the province saw the area of cultivated soybean shrink by 25%. Millions of soybean farmers scrambled to switch to other crops, or simply abandoned the land to join the huge population of migrants in a desperate search for jobs.

Another big loser is the future of the soybean: with the rapid and massive bankruptcy of huge number of small growers, the incredible biodiversity of soy varieties accumulated over millennia and the indigenous knowledge associated with it is dying out as well.

manufacturing exports, where developing countries, particularly those of Asia, have substantially increased their share of world manufacturing exports.

What accounts for the high share of developed/industrial countries in agricultural exports and the relatively low share of developing Asia? A much commented upon factor is that of high subsidies and tariffs for agricultural products. The combination of tariffs (border protection) and direct subsidies were 44.9% of farm gate prices in 2000-02 (Aksoy, 2005). This support was down from 62.5% in 1986-88, but still very high. Among OECD countries, only Australia and New Zealand, had low levels of total support, which went down from 10.6% in 1986-88 to just 3.6% in 2000-02. In contrast to the OECD countries, developing countries as a whole reduced average agricultural tariff rate from 30% in 1990 to 18% in 2000 (Aksoy, 2005).

Developing countries, in particular LDCs, are exempt from reducing the so-called *de minimis* support. The important problem here is that developing countries' budgetary positions do not allow them to reach even the allowed *de minimis* support.

It is necessary to first consider the nature of the world food market. Here we take the example of rice, since rice is of critical importance to food security in most of the countries studied. The world rice market is neither deep nor very competitive (Tabor et al., 2002). The rice market is less dominated by import demand from Asia than it was two decades ago—Asia accounted for two-thirds of global rice demand in the 1970s, but this figure has come down to a third in the late-1990s. The number of traders in the rice market has increased and there are now numerous small traders, involved in what is called smuggling, but is better regarded as unofficial trade. But world rice prices at below $150 per tonne are dominated by the major exporters. All of which use various forms of support to subsidize rice exports. The USA provides the largest subsidy to rice export, $143 per tonne of paddy (Wailes and Durand-Morat, 2005) or about $530 per tonne of exports, if all of the subsidy were attributed to exports (Tabor et al., 2002).

The major Asian exporting countries also subsidize rice exports. Thailand provides loans at subsidized rates; Vietnam provides credit subsides, while India allowed exporters to buy rice at subsidized prices supposed to be for below the poverty line (BPL) households. Consequently, although the exporters are also lower cost producers than the importers, competition between exporters is "less on productivity gains and more on the degree to which domestic markets are protected and exports subsidized" (Tabor et al., 2002).

Vietnam is said to have the lowest rice production costs in the world (UNEP, 2005). This has allowed it to enter the market for rice exports in medium to low qualities of rice. Over the 1990s Vietnam's rice exports have grown at 13% in quantity and at least 12% in value terms (UNEP, 2005). In response to the low export prices of rice, some of the major rice exporters, like Thailand and Vietnam, have proposed the formation of a cartel. This has been rejected by India, which has continued to undercut its rivals in the low end of the market (mainly Pakistan and Vietnam) by selling highly subsidized rice.

Subsidies to exports mean that global rice prices are not a good guide to marginal costs in supplying world rice requirements. This is the first reason why domestic food production cannot be determined by pure global price-based decisions. International rice prices would have to be revised upwards and domestic rice production would then also be higher than that which would be dictated at existing international rice prices.

In China, for instance, sugar prices were higher than world market prices. With the news that China would join the WTO, sugar and sugarcane prices began to fall. Sugarcane prices fell from Y 230 per tonne in 2003 to just Y 170 per tonne in 2004, bankrupting small producers (Oxfam, 2003). An option is to allow import duties, equal to the extent of subsidy paid by OECD countries and for as long as these subsidies, in whatever form they are given, continue to be in place.

3.2.2 Agreement on agriculture and fiscal support

The Agreement on Agriculture (AoA) limits the extent support that governments can provide to their agricultural producers. The aggregate measure of support (AMS) that developing country governments can provide is quite high, at 20% of the value of agricultural production. This does not compel developing countries to reduce their support, which is much less than that allowed. But countries dependent on exports of agricultural commodities, like the West African countries that export raw cotton, are pressing for the elimination of developed country (OECD) support to agriculture, as this support depresses world prices and enables, say, USA to export its cotton at prices that eliminate or reduce the presence of West African producers from the market. Middle-level developing countries like India are trying to get agreements that will maintain their own existing levels of support while reducing the levels allowed to developed countries. There are complex bargaining positions in the negotiations that are currently underway.

The issue we need to consider is: Is the sovereign right of governments to decide on the AMS curtailed by the AoA? Or, are developing country governments and LDCs in particular, being forced to reduce their levels of support because of WTO agreements?

The AMS ranges from less than 2% in the case of Bangladesh to about 8 to 10% in the case of India and Vietnam. In both cases the AMS is below the permissible WTO limit. What keeps the AMS at the present levels is not the limit set by the WTO, but the fiscal weaknesses of the governments concerned.

Given that all developing country governments face considerable resource constraints, which in fact restrict the AMS, one needs to ask what is the right balance between price-support or input-subsidy measures and productivity-enhancing investments? Price support measures in food grains have negative effects on food buyers, who include not only laborers but also small-scale farmers. This is a negative effect of whatever positive merit there might be in price support for farmers.

On the other hand, investments in infrastructure, including irrigation and public research and extension will have productivity enhancing effects. Given the admittedly low productivity of many sectors of food production in developing Asia, it is necessary to concentrate on productivity increasing measures. Such productivity increases will pay for the costs of the support.

Conversely, price support measures can lead to various distortions, both in product and input markets. For instance, subsidies for use of electricity in India have led to overuse of electricity. There is the well-known case of overuse of urea. Further, many of these input supports programs though targeted at protecting farmers, mainly benefit the input-producing enterprises.

Besides various types of domestic support, there are also explicit export subsidies. They can take various forms, such as low interest loans or longer-term loans, both financed out of public subsidies and other related promotional measures. Export subsidies can also take the form of food aid. Food aid, unlike other export subsidies, is not subject to the Uruguay Round AoA schedule of reductions. Food aid is often used by developed countries (now even some developing countries like India) to dispose of surpluses. The effects of food aid on the market are similar to that of export subsidies—they depress prices locally and reduce incentives to local producers, where the aid is being distributed. Contemporary experience (Sharma, 2005) shows that distribution of food aid can reduce local prices and thus serve as a disincentive to local producers to increase production.

3.2.3 Tariff escalation

Tariff escalation refers to the practice of increasing tariffs as commodities progress along the value, moving from raw materials to processed products. Moving up the value chain also means that the country and its producers are less affected by price fluctuations, as both intermediate and final product prices tend to fluctuate less than raw material prices. But such movement up the value chain is inhibited by the practice of increasing tariffs with stages of processing. For instance, the tariff on oranges is less than the tariff on orange juice. This makes it difficult, if not impossible, to use the developed country markets to make the shift from selling raw materials to selling processed products.

Tariffs on fresh, i.e., unprocessed fruit and vegetables in developed countries range from 0.9% for fresh fruits in Canada to 9.2% in the EU. For processed fruits the EU tariff rates are above 20%, with many facing tariffs of 50% (Diop and Jaffee, 2006).

Trade restricting measures can be classified as:

- Economic: Measures which affect pricing, competition and market entry or exit. For example, Quotas and domestic content requirements;
- Social: Measures that protect public interest like health, safety and environment. For example, quality standards, food safety measures and environmental regulations; and
- Administrative: Measures that are administrative formalities. For example, customs valuation, classifications and clearance procedures.

The technical barriers to trade (TBT) (Table 3-2) are regulations and standards governing the sale of products into national markets which have, as their primary objective, the correction of market inefficiencies stemming from externalities associated with production, distribution and consumption of these products. These externalities may be regional, national, transnational or global. These barriers include measures that protect public interest such as health, safety, environment and social cohesion. These could be food safety measures, environmental measures or quality standards. Depending on the policy instrument, TBT could be in terms of import bans—total or partial, technical specifications like process, product or packaging standards, or information remedies like labeling requirements. They could apply either to domestic as well as import products, or only imports or some imports. The compliance with these measures could mean either loss of markets or higher costs to the importers (Roberts, 1999). A study of technical barriers to US agricultural exports for 1996 showed that they were more concerned with reducing risks in the area of food safety and commercial animal and plant health protection. They

Table 3-2. *Classification of Technical Barriers to Trade (TBT).*

Criteria of Classification	Types of Barriers
Policy instrument	• Import bans—total and partial • Technical standards—process, product, and packaging • Information remedies—labeling and voluntary claims
Scope	• Uniform—both for domestic production and imports • Border (universal)—only for imports • Border (specific)—only for some imports
Regulatory goal	• Producer/processor interest—commercial animal and plant health protection, compatibility • Consumer interest—food safety, quality attribute • Natural environment—protection and conservation of environment

Source: Roberts, 1999.

were implemented through process and product standards mainly in the case of food safety and total and partial bans and process and product standards in the case of animal and plant health protection. On the other hand, non-risk reducing measures were few and mainly with respect to quality attributes. Many countries use very blunt instruments such as import bans that excessively restrict imports well beyond what is necessary for protecting the health of their people, plants or animals. The level of protection involved in some cases is equivalent to tariffs of more than 10% (Hoekman and Anderson, 1999).

An analysis of technical barriers to US agricultural exports in 1996 showed that:

- 80% were risk-reducing measures;
- 60% were about commercial animal and plant health protection (CAPHP);
- 25% were about food safety;
- More than 50% in were in CAPHP and 75% in food safety category in terms of process and product standards;
- Non-risk reducing (quality attribute) were also mainly in terms of process and product standards;
- 85% of barriers were under SPS agreement with an average trade impact per barrier being US$17 million;
- Major restriction by barriers was in market access or market expansion;
- Most of the barriers were in East Asia, the Americas and Europe;
- Major products facing barriers were fruits, vegetables, grains and feed grains, animal products (beef and pork) and seed (Hoekman and Anderson, 1999).
- On the other hand, from the USA alone there were numerous technical barriers to developing country exports, amounting to up to 56 detentions per million dollars of imports.

The WTO agreement on TBT sets standards for labeling and packaging of agricultural products as recommended by the Codex Alimentarius Commission (CAC). The CAC, on which both the TBT and the SPS measures agreements of WTO are based, was established by FAO and WHO in 1962 which recommends food safety and labeling standards. In the 1980s, the CAC came out with general labeling standards and nutritional labeling standards. After this in the Tokyo round of GATT, an agreement on technical barriers to trade was negotiated. The TBT agreement which has been now signed by all the WTO members is applicable to all products including agricultural goods and food but its provisions do not apply to SPS measures (Swinbank, 1999).

The TBT agreement covers labeling of food, quality requirement for fresh food products, packaging requirements and labeling of textiles in the agrofood sector (Chawla and Kumar, 1997). Although the public debate on the use of technical barriers to trade has focused on use of these measures to protect consumer and the environment interest, a large number of these measures actually protect the commercial interest of producers by reducing the probability of biological risks to crops and livestock (Roberts, 1999). There is no doubt that TBT will remain an important issue in international regulatory and trade policy forums for the foreseeable future.

3.2.4 Sanitary and Phytosanitary (SPS) measures and AKST

The SPS measures agreement of WTO, reaffirms the right of countries to set their own health and safety standards, provided that they are justifiable on scientific grounds and do not result in unjustified barriers to trade. SPS measures include all relevant laws, decrees, regulations, requirements and procedures including, inter alia, end product criteria; processes and production methods; testing, inspection, certification and approval procedures; quarantine treatments including relevant requirements associated with the transport of animals or plants, or with the materials necessary for their survival during transport; provisions on relevant statistical methods, sampling procedures and methods of risk assessment; and packaging and labeling requirements directly related to food safety (Swinbank, 1999). The SPS measures, thus, encompass food additives, contaminants, toxins, drug or pesticide residues in food, certificate of food, animal or plant health safety, processing methods, food labeling, plant or animal quarantine, requirements for prevention, control or establishment of pest or disease and sanitary requirements for imports. Whereas the sanitary provisions relate to food and animal health, the phyto-sanitary provisions cover plant health aspects of products (Chawla and Kumar, 1997).

For the purpose of the definitions, "animals" includes fish and wild fauna; "plant" includes forests and wild flora; "pests" includes weeds; and "contaminants" include pesticide and veterinary drug residues and extraneous matter (Adopted from Swinbank, 1999: Original source GATT, 1994). The SPS standards comprise articles on basic rights and obligations, non-discrimination, harmonization, transparency, equivalence, regionalization, risk assessment and control, inspection and approval procedures; and are based on Codex Alimentarius Commission (CAC) guidelines of FAO/WHO which is nothing but application of Hazard Analysis and Critical Control Points (HACCP). This method is about improving and controlling processes as variability in processes can cause quality problems; and is product-specific in nature.

The basic rights and obligations clause means that members have the right to take SPS measures necessary for the protection of human, animal or plant life or health provided such measures are consistent with the provisions of the agreement, are based on scientific principles and do not arbitrarily or unjustifiably discriminate between members where identical or similar conditions prevail. The harmonization provision calls for members to base their SPS measures on international standards where they exist though members can adopt more stringent SPS measures if there is a scientific justification as per the agreement. Under the agreement, members are also to recognize the SPS measures of other members as equivalent to their own if the exporting member objectively demonstrates to the importing member that its measures achieve the importing member's appropriate level of SPS protection (principles of equivalence). Further, if members wish to apply more stringent measures than the international standards, then they are obliged to base their risk assessment and level of SPS protection on scientific evidence and their levels should not be more trade restrictive. Members are also required to consider objective geographical and ecological conditions rather than national boundaries to apply SPS measures (regionalization clause). Under the transparency clause of the agreement, members are to ensure that all SPS measures and changes in them are notified in a transparent manner through a single national enquiry point. Finally, the control, inspection and approval procedures are to be applied in no less favorable manner for imported products than for like domestic products (Swinbank, 1999).

Critique of SPS measures. Since both the agreements (TBT and SPS Measures) are relatively new and technical, there is a certain amount of confusion and a lack of differentiation between the two measures. For example, shelf life regulations can be adopted as an SPS measure or a TBT measure depending on the exact purpose. Therefore, knowing the objective of a measure is critical to determine whether a measure is subject to the discipline of TBT or SPS agreement. Similarly, the range of measures given in the SPS agreement is not totally inclusive. For example, measures introduced to control the spread of weeds would generally be covered by the SPS agreement. But, the agreement is not clear enough about the concerns of those who believe that use of genetically modified organisms (GMOs) could lead to cross-pollination and GMO genes into the natural flora. In this context, the USA challenged the EU's labeling requirement for certain products produced from GMOs under the TBT rather than under the SPS agreement arguing that it is not aware of any information that GM foods differ as a class in any way from products produced by other methods (Swinbank, 1999).

Secondly, the differences in standards across countries are very difficult to resolve even with the best scientific advice. The examples of disputes under WTO umbrella in this field include that of beef hormones, irradiated food, cheese made from unpasteurized milk and genetically modified foods (Hoekman and Anderson, 1999). Though the SPS agreement does not impose international standards on members, it does enhance the importance of international standard setting agencies as it encourages members to base their SPS measures on international standards and that national provisions have to be justified on scientific grounds if they are more stringent than international standards. Over time, it tends to impose a de facto set of international standards worldwide.

From the developing countries' and the Indian prospective, the SPS measures set very high standards which are not suitable for these countries either because they have higher cost of compliance or are not required in their contexts. Further, no lead time has been given to these countries for implementing these provisions. It is also argued that what was designed in the Western contexts (CAC guidelines) has been imposed on the developing world. There is also hypocrisy in the practice of these provisions as there is lack of transparency and prevalence of discrimination against the developing world. For example, under Codex standards, the raw material for some types of cheese like mozzarella, cheddar has been restricted only to cow milk in the Codex standards on the basis of the argument that these cheeses were traditionally made from cow milk. This means that there may be difficulties in exporting cheese made from buffalo milk (Chawla and Kumar, 1997).

An SPS measure becomes a barrier:

1. When domestic standards are lower than those for imports
2. When standard conformity assessment is different/not recognized by two countries as it duplicates costs of product testing (Table 3-3).

There is also no doubt that the SPS barriers can lead to import bans which means higher cost of compliance (15-40% of FOB value) for the developing country exporters which, in turn, could lead to reduced trade or diversion of trade between exporters due to high cost. The developing countries are also likely to find it difficult to implement these standards as there is lack of SPS control systems, lack of awareness and understanding of standards, lack of technical abilities to implement standards and organizational structures are not geared for such standard setting (Henson and Loader, 1999). There are also problems of multiplicity of standards organizations which leads to duplication and lack of coordination and small size of firms/farms.

Due to the TBT and SPS provisions of WTO, India has faced non-tariff barriers for its products. In 1997 Indian fishery products were banned by EU and were put on automatic detention by the US (Scheuplein, 1999). There were numerous detentions in 2000-2001under the SPS provisions.

Cases of SPS restrictions on Indian food exports subsequently have included:

- UAE ban on Indian meat imports (for 10 companies) due to health and hygiene reasons;
- EU ban on Indian fish imports due to lack of SPS standards especially in canning (only 90 out of 404 plants approved for fishery exports to the EU);
- Fruit flies in fresh fruits and vegetables necessitating treatment before exportation: e.g., mango (stone weevil) to Australia; mango, citrus and flowers to Japan; and grapes to China;
- Groundnut and spices (EU, Italy and Germany) and Chilies (Spain) due to aflatoxin and chemical residues;
- India was delisted from the list of approved countries in EU for import of egg powders, two years ago, for non-submission of Residue Monitoring Plan (RMP);
- Dairy products export problems include mastitis in bovines and foot and mouth disease in cattle and buffalo; Somatic Cell Count (SCC) based pricing in first world; and input sector related problems like quality of fodder which affect milk quality.
- "Karnal bunt" in wheat and Iran's rejection of Indian wheat sent by two private exporters due to quality problem; and
- Forty Indian basmati rice consignments detained in 1999-2000 by the USFDA because of filth and pesticide residue.
- Under the WTO agreement, India had obligated itself to comply with the SPS provisions by the end of 1997. In the food sector, this includes strengthening of the national food export control system.

A study of quality control and monitoring practices in two of the commodity sectors in India (fisheries and spices) found that there were serious problems of maintenance of hygiene and quality standards and processes at the primary production or procurement level. For example, the fishing boats did not have ice on their streams when they arrived at the pier. The appearance of the boats was dirty and it did not seem possible under those conditions that they complied with hygiene standards. When fish and shrimp were unloaded from the boats, they were dumped into piles sometimes very carelessly and in an unorganized manner. There was no separation of fish from the general walking areas and every one appeared to have free access to any place on the pier or any pile of fish. On the other hand, the processing centers were excellent at maintaining quality and hygiene standards and they had HACCP in place and in operation. But this may not be the case with all the 400 processing facilities in India. Most of the quality and hygiene problems at the primary produce level were due to lack of awareness and lack of infrastructure like potable water and landing facilities. Similarly, in spice production and processing, the major problems were in production which is carried out by small-scale farmers who lack knowledge of quality and hygiene and do not have an incentive to maintain them. Here too, the processing plants had all the quality systems in place, but the contamination takes place at the farmer and the trader level (Scheuplein, 1999).

Table 3-3. *Maximum Residue Limit (MRL) under Codex and PFA for milk/milk products.*

Residue and Product	Codex MRL	PFA MRL
Lead in butter	0.05 ppm	2.5 ppm
Lead in milk	0.02 ppm (suggested)	No
Aflatoxin in milk	0.05 ppb (600 times higher than under PFA)	0.03 ppm

Source: Chawla and Kumar, 1997.

Options. At the international level, there is a need to make the WTO system more transparent. The farmers' organizations should be allowed to participate, either through their governments or directly, into the standard setting bodies like the CAC so that farmer concerns could be brought into the body and its rules and recommendations.

Further, since domestic markets do not value quality, the farmer is not encouraged to maintain high quality standards of the produce. Therefore, what is required is not end-product testing for exports but monitoring of the entire commodity chain to maintain quality and hygiene standards. It is here that the application of HACCP comes in as a process control concept which places the burden of ensuring safety on the members of the food chain which include farmers, traders, processors and distributors. There is a serious need to link farmers with processing and exporting agencies and firms so that quality can be ensured right from the raw material production stage. This can be achieved through appropriately designed arrangements like contract farming or the procurement cooperative alignment with processing and marketing companies.

3.2.5 Anti-dumping measures and AKST

After the removal of all other non-tariff barriers under the WTO regime, the anti-dumping measures are the most important non-tariff barriers as they are being used as a protectionist measure with little connection with dumping or fair trade. The anti-dumping disputes (15.4% of total) were next only to import restrictions (on goods) related disputes (38.4% of total) brought to the WTO during 1995-2003 (Rameshan, 2004). It is not the use of the anti-dumping measures but their very existence that can have significant trade effects like collusive behavior among domestic and foreign firms (Zanardi, 2004). This is already evident in the fact that in the recent past, there has been a steady increase in the number of anti-dumping actions by both the developed and the developing countries. The exporters in many developing countries find that, as their exports rise, there are increasing pressures from developed country industries for the levy of anti-dumping duties on the ground that goods are being dumped. Thus, anti-dumping measures might counter balance the tariff reductions accomplished by various GATT rounds.

Anti-dumping measures are intended to prevent the import of products at prices lower than those at which they are sold within the exporting (home) country markets. It is a type of penalty against imports to protect the domestic industry. All members of the WTO are obliged to set up their own anti-dumping authorities to prevent injury to domestic industry.

The WTO agreement on anti-dumping measures stipulates a rigorous framework for dealing with the problem of dumping. The anti-dumping measures, as per the agreement, can be initiated only when (1) an existence of dumping is identified; (2) injury to industry is measured; and (3) causal link between dumping and injury to industry is established. All these steps require strong technical and analytical support (Panchmukhi, 2001). Dumping is defined as the introduction of a product of one country into the commerce of another country at less than the normal value of the commodity (Gupta, 1996). The principal criterion for determining dumping is whether the price of the product exported from one country to another is less than the comparable price in the ordinary course of trade for the product, when destined for consumption in the exporting country. In the absence of the domestic price, the highest comparable price for the like product for export to any third country in the ordinary course of trade or the cost of production of the product in the country of origin plus a reasonable addition of selling cost and profit are relied on. No matter which standard is used, in each case, it is enjoined that due allowance shall be made for differences in conditions and terms of sale, difference in taxation and other differences affecting price comparability (Kaul, 1997).

The conditions for imposition of anti-dumping duties to offset or prevent dumping are that:

1. The anti-dumping duty shall not be greater than the margin of dumping.
2. No anti-dumping duty shall be levied by reason of exemption from or refund of duties for taxes borne by a product when destined for domestic consumption in the exporting country.
3. No anti-dumping duty shall be levied unless it is determined that the effect of dumping is such as to cause material injury to an established industry (Kaul, 1997).

Anti-dumping duties can be of several types i.e., ad valorem duty, specific duty and dumping margin duty. Besides anti-dumping duty, the other measures against dumping can be provisional measures or duties, price undertakings and voluntary export restraints. Provisional measures are used to prevent injury being caused during the anti-dumping investigation and can be in the form of provisional duty, security deposit or withholding of appraisement. These measures are normally limited to four months and expire with the conclusion of the proceedings. Provisional duties are refunded if no evidence of dumping and injury is found and the difference is reimbursed if the final duty is less than the provisional duty. Price and voluntary export restraint undertakings are voluntary undertakings given by any exporter to the effect that the exporter agrees to increase the prices or to cease/ reduce exports to the area in question at dumped prices in order to satisfy the authorities that the injurious effect of dumping has been eliminated (Gupta, 1996). When petitions result in voluntary export restraints, exporters are allocated with export licenses based on firms' foreign market shares in the past. Thus, forward looking exporters have an incentive to enlarge their market shares by dumping more at present and thus securing larger profits under the export restraint (Zanardi, 2004).

Until recently, most intensive use of anti-dumping actions has been made by the US, Canada, the EU and Australia in that order. Canada was the first country to adopt an anti-dumping legislation in 1904 followed by Australia in 1906 and several others by 1920. After the passing of the anti-dumping code during the Tokyo round of GATT in the 1970s, many developing countries also started passing anti-dumping legislation with India doing it in 1985 (Zanardi, 2004). By the end of June 1997, 76 members (with EU countries counted as one) had submitted notification of their anti-dumping legislation or regulations to the WTO's committee on anti-dumping practices and by the end of 2001, 94 countries (with EU countries counted individually) had their anti-dumping laws in place. By the end of 1996, the WTO member countries reported 900 anti-dumping measures, including price undertakings, being in force which rose to 1119 by the end of 2000. The major sectors affected by these measures were base metals, mostly steel, chemicals, plastics, textiles, machinery and equipment and agriculture and food in that order (Ghate, 1998; Zanardi, 2004)). The "Big Four" i.e., the US, the EU, Canada and Australia still account for more than 40% of all anti-dumping investigations (Bhattacharyya and Gupta, 2001).

By 2001, more than 90% of worldwide imports were potentially subject to anti-dumping actions compared with only 71% in 1990 (Zanardi, 2004). And, the developing countries are the major targets of anti-dumping actions. They faced 38% all cases during 1990-94 which rose to 42% during 1995-99 (Bhattacharyya and Gupta, 2001). On the other hand, Argentina, Brazil, Mexico, India and South Africa emerged as major users of anti-dumping actions accounting for 1/4th of all anti-dumping investigations since 1995 (Bhattacharyya and Gupta, 2001). The WTO Anti-Dumping Measures agreement excludes the use of AD in a retaliatory fashion in line with the non-discriminatory principle of the WTO (Zanardi, 2004).

During 1980-2001, 4597 anti-dumping investigations were initiated and the largest four users (Australia, Canada, EU and the USA) each had a double digit share and altogether filed 64% of all anti-dumping petitions. But, in more recent times (1995-2001), only the seven largest uses together reach a share of more than 64% with new ones being Argentina, India and South Africa who have even larger shares than Australia and Canada. India initiated a total of 192 anti-dumping investigations during 1980-2001 with most being after 1996 (Zanardi, 2004). India has been one of the major users as well as victims of the anti-dumping measures. India initiated 140 anti-dumping cases during 1995-1999 compared with only 15 during 1991-94 and 45 during 1993-1997 with definitive duties in 11 cases (Panagariya, 1999) and it was the highest among the developing countries, accounting for 15% of all cases in the developing world. India imposed its first ever provisional anti-dumping duty in January, 1993. The index of such anti-dumping initiations was 1875 per dollar of imports for India compared with only 100 for the USA.

India also faced very costly anti-dumping actions for its exports: 779 per dollar of exports in terms of index, compared with only 100 for the USA (Mattoo and Subramanian, 2000). In 1998 alone, India faced one case of anti-dumping for every $2.74 billion of exports as against only 15 such cases faced by the US for every $45.46 billion of exports. India was next only to Ukraine in this regard. In fact, more than 15% of all final measures imposed under anti-dumping investigations were aimed at India (Bhattacharyya and Gupta, 2001).

Over the period 1980-2001, 113 countries were targets of anti-dumping investigations and during the recent period of 1995-2001 alone, 93 countries faced anti-dumping investigations with prominent ones being from Asia i.e., China, South Korea, Japan and Thailand which together accounted for 30% of all cases. In fact, China has faced about 15% of all (2416) anti-dumping cases filed by the WTO members up to the end of 2003. Due to this, China has recently set up an early warning system on 189 goods of export importance mainly including textiles, home appliances, steel and furniture which account for 60% of China's exports to the USA (Joseph, 2004). India's share in all anti-dumping actions suffered went up from 0.9% in the 1980s (1981-87) to 3.72% by the late 1990s (1995-2001) (Zanardi, 2004). Also, it is increasingly the developing world countries which are targeting more of other developing world countries (50% cases) besides the developed countries targeting developing countries. But, most of the cases in Japan, South Korea and the EU have been settled with price undertakings as the Japanese avoid courts and litigation by tradition. On the other hand, India had all its anti-dumping investigations settled through anti-dumping duties only (Zanardi, 2004).

The USA imposed anti-dumping duty on Indian preserved mushrooms along with those from China and Indonesia in 1999. The dumping margin calculated for India was the highest (243%), followed by China (198%) and Indonesia (22%). The USA imposed company specific anti-dumping duties on Indian firms which ranged from 7-243% though the effective rates were ranging from 7% to 15% as other firms were not exporting any more (The Economic Times, March 1, 1999). The EU investigated 28 exporters from India, the highest number followed by China (24) and South Korea (20) during 1998-2002 mainly in iron and steel, chemicals and textiles. On the other hand, the EU suffered most from USA and India in 2002 with 25% of the cases each by the two countries (Silberston, 2003).

There is also significant evidence of retaliation in anti-dumping actions. Twelve countries simultaneously targeted to protect the same industry group wherein same product was subject to anti-dumping duty both at home and abroad. It is difficult to accept the fact that an industry that is injured by imports from a country can be causing injury to the very same industry in another country (Bhatt, 2003).

Decisions of the WTO panels on anti-dumping measures. The working of the WTO panels on anti-dumping so far has shown that it is able to build confidence in the dispute settlement mechanism of the body. This is evident in the case of US Anti-Dumping Act of 1916 where the WTO panel and the Appellate Body have unequivocally held that the US Act, which provides for specific action against dumping in the form of civil and criminal proceedings and penalties, is inconsistent with the WTO agreement on anti-dumping (Satapathy, 2000a). Similarly, the WTO panel ruling on India's complaint against anti-dumping measures by the EC on imports of bed linen from India, in favor of India, suggests that WTO panels cannot be manipulated. In particular, the measures of anti-dumping by the EU were rejected. The EU is one of the four major traditional users of these measures along with the US, Canada and Australia and has a long experience and administrative and legal setup. Secondly, the panel has ruled against the EU practice of zeroing negative price differences in the calculation of dumping margins. This finding of the panel against the zeroing practice would now force the prevailing practice in some of the developed countries to change. This will mean that in many cases, the dumping margins may disappear or come down below the *de minimis* level for the developing country exporters, requiring no anti-dumping duties (Satapathy, 2000b).

Further, the EU did not even collect data for examining the effect of all economic factors on an industry which led the WTO panel to reject the EU's claim on injury to the industry because of dumping of imports. This means that in all the countries, much more economic analysis to determine injury to industry and to attribute it to dumping will be required. The panel even questioned the sample used for determining injury for the domestic producers as the EU found domestic industry to consist of 35 producers but used data on other and lesser number (17) of producers. The panel also argued that before imposing anti-dumping duties, possibilities of constructive remedies should be explored by the developed countries. The EU had rejected India's request to offer price undertakings and by doing so, EU had failed in its obligation to explore constructive remedies to the problem of dumping as provided in the Agreement on Anti-Dumping Measures (Satapathy, 2000b).

Anti-dumping system has been able to sustain and grow in practice due to public perception of "dumping" which is different from the rules and regulations and its relevance as a safety valve, political expediency due to impact of liberalization and globalization, lobbying by pressure groups and differences in competition standards among nations (Tharakan, 1999). Anti-dumping actions have implications for foreign investment flows. There seems to be a coincidence between anti-dumping cases and inward investment. The evidence from the EU and the US shows that anti-dumping actions have substantially increased the incidence of manufacturing investment by Japanese firms in these regions. What it means is that imports are being replaced by local production by foreign firms which can still practice price discrimination or sales below full production cost. But, at the same time, anti-dumping actions lead to large welfare losses. Anti-dumping duties can also have negative impact on export competitiveness of an industry if duties are imposed on products that go as inputs into that industry (Bhat, 2003).

There are many problematic aspects of the Agreement. The definition of dumping favors the party imposing anti-dumping duties. Dumping is considered to exist if the export price of a product is less than the comparable price of the product or like product in the domestic market in the ordinary course of trade. However, when the average export and domestic prices of a product are calculated, domestic sales prices below total cost are considered beyond the ordinary course of trade and therefore, excluded. But, all export prices are included. This, artificially, raises the domestic price. Also, if no home market prices can be found, the sales price in a third country—the so-called surrogate country—can be used for comparison. Since, different countries have different levels of economic development and comparative advantage in different sectors, the arbitrary choice of a surrogate country may easily lead to finding of dumping. For example, while investigating dumping by the Chinese firms, the US authorities often use, as "surrogate" country, market economies with higher cost of labor and raw material or countries where economic reform is proceeding more slowly and production in many sectors is less efficient than in China. This will naturally lead to the non-market economy being considered to be dumping. This practice has been now done away with by the EU in case of Russia but still prevails for other so called non-market economies and even Russia in non-EU markets (Silberston, 2003). Even use of constructed value price in the absence of availability of home market or third country prices is prone to inherent subjectivity as the costs which go into constructed value price vary greatly among countries and companies. The concept of injury is also problematic if domestic and import markets are simultaneously expanding, but the domestic market is expanding at a slower rate and the imported products are garnering the lion's share of the market. Further, the presence of dumping may not indicate injury but rather may be related to other local and international factors. Therefore, it is very difficult to establish a strong link between dumping and injury (Silberston, 2003).

Even selling below total cost is a normal business practice in some situations. For example, a firm may have to sell below total cost in order to attract skeptical customers or to meet existing competition in a foreign market, without any intention to dominate the market, especially if the product is new and un-established. It is unreasonable to subject such practices to anti-dumping investigations. Further, the anti-dumping laws are also country-specific instead of being firm-specific as the country does not really represent costs of particular firm and all firms from a country should not be targeted. Another problem with the practice of these laws is that though the agreement recommends "lesser duty" than the margin of dumping if that suffices to prevent injury, many developed countries do not follow it and impose duty equal to margin of dumping as there is no obligation under the agreement which only refers to the desirability of the practice (Reich, 2003). Many firms and countries resort to back-to-back anti-dumping petitions in order to benefit from trade effects of anti-dumping litigation which discourages imports in their markets (Zanardi, 2004).

Besides, the use of anti-dumping duties to protect domestic industry from imports may be misplaced if the difficulties of domestic producers result from their own inefficiency. In this situation, the anti-dumping duties tend to penalize the more efficient foreign producers. Also, because of the difficulties in finding out the origin of a product due to global sourcing, it is problematic to identify the agency responsible for dumping. The anti-dumping agreement also does not define the concept of export price and the globalization of production further leads to difficulties in determining export price as products are the result of global souring. There are even problems with defining domestic industry (Didier, 2001).

Then, there is also an overlap and a contradiction between anti-dumping laws and the competition policy. Since anti-dumping actions aim at reducing anti-competition practices, they are a part of the competition policy. But, sometimes actions like price undertakings are anti-competition and in conflict with the competition policy of WTO (Bhattacharyya and Gupta, 2001). Some firms may also resort to anti-dumping in order to foster collusive agreements between/among domestic or foreign firms as this action will give relief from foreign competition or a domestic firm will use this threat to negotiate a collusive agreement with a foreign firm. This kind of practice was found in the USA (Bhattacharyya and Gupta, 2001).

3.2.6 Options

The above discussion shows that despite the WTO agreement on anti-dumping measures, there will be widespread use of these measures against developing country exports as well as dumping into these countries. Anti-dumping is also seen as a necessary valve in the presence of trade liberalization and globalization to protect domestic firms from foreign competition. There is a need to introduce competition considerations, end practices of cumulation of market shares in injury determination (except in cases where there is evidence of collusion) and introduce some form of counterfactual analysis in measuring injury margins (Tharakan, 1999). Anti-dumping duty should be imposed only if it is established that there was a predatory intent on the part of the exporting country. If the market is already rapidly declining, dumping by any exporter can be ignored (Silberston, 2003).

3.2.7 Commodity Prices

Prices of primary commodities or commodities like coffee, tea, sugar, cotton, etc., in international trade are subject to two kinds of problems. The first is that there are often substantial, even violent short-term price fluctuations. Just a 10 to 15% swing in production can lead to major price changes. The second problem is that there is a tendency for a long-term or secular decline in the terms of trade for commodities.

These price fluctuations and secular decline in terms of trade have substantial effects on the producers of these commodities. The producers in developing countries are often small-scale producers, dependent on income from the commodity for their livelihoods. A sharp fall in price means a substantial fall in income. Income may fall below not only what the producers are normally accustomed to but also below the minimum required to pay debts acquired in the course of production. This is the micro-economic effect of price falls in commodities.

The market for commodities is again most often dominated by a few large buyers. It is a classic monopsonistic position with millions of small sellers and a handful of big buyers. The market for coffee, for instance, is dominated by a few big buyers like Nescafe and Volcafe, while there are millions of producers and sellers of coffee.

The problem of many small producers is compounded by the entry of new producers into the market. For instance, in the market-based reform in Vietnam, large numbers of farmers took to coffee production as a cash crop. These new producers may be willing to accept a lower net income, one which is an improvement for them but lower than what traditional producers are used to. In the manner that sections of manufacturing have relocated to areas of lower wage costs, sections of commodity production can also relocate to areas where small producers are willing to settle for less than what traditional producers earned. This type of competition among small producers in developing countries is a feature of current commodity production.

While the above analysis puts competition among producers as a key factor in low and fluctuating prices for commodities, another analysis points to the non-cognizance of environmental costs as leading to low prices for commodities (Dasgupta and Maler, 1990). When the social costs of production are higher than private costs, there is a subsidy on the basis of non-valuation of environmental resources, which are production resources for the small producers.

More recently, yet another factor, that of competition between developed and developing countries, has entered the picture in leading to low prices of some commodities. This is the case of those commodities, like cotton and sugar, which can be produced in both tropical and in temperate or semi-temperate conditions. Take the case of sugar, which can be produced both from tropical cane and temperate beet. In the EU sugar producers are paid twice what they would get in the international market. At the same time, EU also export about 5 million tonnes of sugar at the lower world price, thus "undermining further the price received by farmers in developing countries" (Robbins, 2003). But there are still many commodities, like coffee, tea or bananas, which are not grown in OECD countries and are thus not affected by subsidies and protection by OECD countries.

While commodities are largely produced in a competitive environment, the markets for manufactures are much more monopolistic, leading to a secular decline in terms of trade for commodities (Singer, 1950). Productivity increases are passed on to consumes and buyers as prices fall. But in the case of manufactures, the monopolistic market position of producers enables them to keep the benefits of technological advances that lower costs of production. This analysis leads to the conclusion that the only development path for developing countries is to diversify away from commodity production into manufacture; something that many Asian countries have successfully accomplished.

Along with providing some adequate return to labor in commodity production, there is also the problem of enabling transitions in commodity production. If incomes are assured then will there be a shift from high-cost to low-cost producers? This is the positive function of the competitive market mechanism. The market by itself, however, brings about this transition in an entirely ruthless manner, leading to the destitution of the displaced producers. To bring about an effective transition commodity interventions have to meet two objectives: providing a reasonable income to the producers and enabling an orderly transition from high cost to lower cost producers.

There are two types of interventions in commodity markets: (1) price stabilization, with buffer stocks; and (2) output regulation, with quotas and restrictions. The second, however, requires a few producers or a few organized groups of producers, to organize a cartel. This is what OPEC has done successfully to control crude oil production. Prices are then allowed to take their own levels.

In the Bretton Woods Conference, where the IMF and World Bank were born, Keynes proposed a commodity board which would operate buffer stocks. At each point of time, a base price would be set and fluctuations allowed of 10% on either side of the base price. If, however, stocks increased at the end of the year, then the base price would be marked down by 5% and vice versa for decreases in stocks. The annual reduction of prices would have the effect of enabling lower cost producers to increase their share of production, while higher cost producers would leave the sector. But unlike the violent adjustments of the market system this adjustment would be brought about in a gradual and thus less painful manner. There could be stability, but not stagnation, as producers enter or leave the market.

The US was opposed to such a scheme that would stabilize commodity prices and it was not taken up after Bretton Woods. But some commodity boards did come up in the post-Second War period. Among these commodity boards only the coffee board had what is known as an economic clause, meaning it would undertake market stabilization activities. The others confined themselves to trying to set export quotas.

The major problem with export quotas is that they tend to freeze production among existing producing. What about new countries that wish to enter? A big factor in the collapse of the International Coffee Agreement (ICA) was the spread of production outside the countries with quotas. Indigenous peoples in upland areas (e.g., Vietnam or India) have often taken to coffee production as a substitute for forms of swidden, or introduced coffee into the swidden mix. All this developed production centers outside of the traditional coffee growing areas. It is not possible to manage these changes in the location of production through export quotas.

In the 1970s there was a series of UNCTAD-inspired commodity agreements. But the experience of these commodity boards has not been encouraging. In price determination they played a positive role, in that coffee buyers purchased coffee from a handful of boards rather than large numbers of small operators. But the quotas were used, particularly in the African countries, to favor those ethnic groups from which the ruling sections came (Gibbon and Ponte, 2005). In years where high prices were paid for products, the amount paid to farmers was kept low in order to increase the amount retained by the boards. In years of low prices the price stabilization efforts were swamped by exchange rate fluctuations. Large resource transfers were needed in those years, well beyond the capacity of the governments concerned.

The commodity agreements were discontinued because the main consuming countries withdrew financial support.

In 1989 the economic clause in the coffee agreement was suspended and it was not renewed in 1994. What this meant was the removal of the national coffee boards from the bargaining and, therefore, from the price equation and its substitution by a buyer-driven commodity chain (Gibbon and Ponte, 2005).

With the rise of the Washington Consensus there has been a shift of emphasis. In the first place there is an emphasis on respecting "market fundamentals", rather than dealing with market failure. Second and perhaps following from the first, the search is not for schemes to stabilize prices, not to reduce volatility, but to reduce uncertainty with regard to volatility. Third, the attempt is to develop market-based instruments, futures and insurance systems to help producers deal with price risk. It is expected that the development of such market-based systems will allow for announcement of harvest prices, which would enable producers to plan their investments.

The proposed system is an elaborate market-based system of futures swaps, options and derivatives. These have worked in the US and Europe, where, for instance, livestock raising is done in large units. But even then it has been found that less than 10% of OECD producers use these instruments. Why is the proportion so low? It is possibly due to the fact that the producers actually depend on the high level of government subsidies, amounting to $19,000 per producer per annum (ITF, 1999). On top of that, they rely on tariff and non-tariff barriers to shield them from foreign competition. Further, the incomes of farm families are now mainly from non-farm sources. Finally, up to 40% of commodity production is done on contract basis with the large buyers, rather than with any price hedging systems (Economic Times, Delhi, 6 Feb 2006).

It is these measures, rather than the elaborate systems of market-based instruments that provide income security for farm households. These futures instruments are actually used by processing and marketing companies, which buy from producers and in turn pass on some of the benefits to producers. Finally all these market-based schemes link up to commodity markets in Chicago, New York, London and other financial centers. New large-scale commodity markets are coming up in China and India, but they are still at an early stage. In any case, they cannot substitute for price stabilization measures.

Among developing countries, Mexico has tried to implement market-based insurance and futures schemes. India is in the process of developing the secondary markets for commodities, which already has a larger turn-over than the Mumbai stock market. The Mexican Agricultural Products Options Program (APOP) has large lots in which operations can be conducted. Corn farmers, who have been devastated by NAFTA-induced competition from subsidized US corn and are typically small farmers, operating less than one hectare of land, have a very low participation in the APOP price insurance schemes. There is more participation in wheat and cotton, where production units are larger (ITF, 1999). The Mexican experience is certainly not very encouraging about the possibility of using such market-based insurance and futures schemes.

In trying to extend price insurance schemes to developing there is thus one critical problem of small size of lots in which production is undertaken. Unless farmers join together in groups it will be impossible to participate in such insurance schemes. Collective action problems come in the way of such participation. But, it is important to overcome such collective action problems—in the increasingly buyer-driven commodity chains, small producers need to combine in order to strengthen their bargaining position.

Countries like China or India, with large volumes of commodities traded on the market, could possibly set up futures markets and price insurance schemes. But in smaller countries there would not be scale for such institutions. In this there is a possible role for regional institutions. The commodity exchanges in India could serve all of South Asia. But then profits from the trade would also accrue to Indian institutions.

Finally there is the need to compare price insurance mechanisms with price stabilization measures. The costs and benefits of the two would need to be compared. A crucial factor in the comparison is that while price stabilization would benefit all small producers, price insurance schemes would tend to exclude small producers from their ambit. If it is important not to exclude the poorest and smallest producers from the likely benefits, then price stabilization would certainly be superior to price insurance.

In insurance schemes the cost would be directly borne by participating producers. This would be direct deduction from their incomes, to be passed on to those, generally far richer then themselves, who would profit from their ability to take risks. On the other hand in price stabilization measures the costs would be borne, if done nationally, by taxpayers. If done internationally, the costs could even be borne, as Keynes had proposed, by the countries in surplus or rich countries. Given that the countries most dependent on commodities' export are among the poorest, there is a strong case for internationally-funded action. Should this burden be borne by taxpayers in the contributing countries or by consumers of the commodities? A tax on, for instance, consumption of coffee could be used to fund a buffer stock scheme for coffee. If taxes are to fund a buffer scheme for a commodity, it would certainly be appropriate that the costs be borne by those who consume that commodity. This would further directly relate consumers with producers, strengthening the moral connection between the two.

What a buffer stock system should not do is to allocate quotas, whether between countries or within countries. There would inevitably be disputes about market shares. And, how would new countries be able to enter the market? The former Director-General of UNCTAD proposed export quotas, in which case, " . . . within each country, means would have to be found for distributing that country's quota among its domestic producers . . . " (Corea, 1992). Experience has shown that this increases the power of those who take these decisions, power which can be used to garner a portion of the income from the producers and also to favor one group or community against another. It is better to leave such decisions to individual produces operating in the market. As Keynes's scheme proposed if, when demand is less than supply and stocks pile up, there is a step-by-step reduction in prices, then the higher cost producers are likely to exit from the field. Such a process would involve the market mechanism in fostering competition and thus efficient, low-cost production, something administrative allocations are unlikely to achieve. While reducing the excesses of competition, it is also necessary to avoid stagnation.

Allied to a buffer stock operation, there is need for what the Committee of Eminent Persons (UNCTAD, 2003) proposed—action to develop other uses of the commodities and to support producers to move up the value chain. Developing new uses of traditional commodities is one way to expand the market for that commodity. Cassava producers are working to spread the use of cassava not just for animal feed but as the raw material for a food additive, monosodium glutamate, which is very popular in East and Southeast Asia, though there are doubts on whether this is a good substance or not. Lac which is traditionally used as a lubricant is being replaced by artificial substances for this use. But a new food use has developed for lac—coating fruits. This coating both protects the fruit from insect attacks, increasing its shelf life and also makes it attractive to look at.

For instance some vegetable oils can also be used as diesel substitutes. The use of palm oil has been developed for this purpose. The Malaysian government taxes palm oil exports in years of good prices. In years of low prices, then the money so collected is used to subsidize the use of palm oil as diesel substitute, thus increasing demand for palm oil when prices are low. Finding new uses for commodities would help to increase demand for them. Other vegetable oils could also have similar uses.

The other measure recommended by the committee is for producing countries to move up the value chain. But the movement up the value chain is hindered by the strong barriers the developed countries have adopted in the form of escalating tariffs. This tariff issue needs to be resolved first. But after that there will still remain the currently inadequate capacity of many LDCs to undertake these movements into processing on their own.

Another form of movement up the value chain is not into processing, but into specialized products; in other words to switch from generalized commodities to specialized products. "Shade grown coffee" which commands a price premium as being environmentally friendly is one such high value product. Better qualities of coffee also command higher prices and have a more stable market. Some processors are working with Central American coffee growers to enable them to switch to higher-quality and higher-value products. To what extent the producers get the benefit of higher quality depends on the manner of their integration into the supply chain, something which we will discuss later on. But the very fact of producing a higher value product, something that cannot be easily done by other producers, is likely to increase the bargaining power of the producers.

Existing WTO rules do not rule out the possibility of international supply side management, with coordinated action by producers and producing countries. The EU's Agreement on Dairy Products was part of the WTO system; it set minimum export prices for milk powder, milk fat and cheese. When it was terminated, the reason was not that it contravened WTO rules but that its members saw no further need for it.

In the recent steel crisis in the USA and EU, for instance, the WTO Director General, Supachi Panitchpakdi, proposed just such a scheme to reduce production, "The long-term solution to the problem that has arisen can be found only through the adoption by producing countries of an agreement providing staged reduction in production. Such an agreement could be negotiated under the umbrella of the WTO and supported by the establishment of a World Trust Fund to provide adjustment assistance to industries which would be required to reduce production and compensate workers who lose their jobs" (Robbins, 2003).

Supply side management with the objective of obtaining remunerative prices is explicitly allowed by GATT (Chapter on Trade and Development, Part IV, Article XXXVI), "Given the continued dependence of many developed countries on the export of a limited rage of primary products, there is need wherever appropriate, to devise measures designed to stabilize and improve conditions of world markets in these products including, in particular, measures designed to attain stable, equitable and remunerative prices thus permitting an expansion of world trade and steady growth of real export earnings of these countries" (Robbins, 2003).

It is a seemingly inevitable feature of such commercial crop intensification that it leads to a specialization in production and thus reduces the range of local production. In the Himalaya-Hindukush region it is reported that families that used to produce and consume over 20 varieties of food items; consequent upon commercialization they now consume only 5 (Nagpal, 1999).

Such commercialization of products has often, even usually, been accompanied by monoculture of the products. Tea, rubber, potatoes and a host of other upland crops are often grown in plantation monocultures. But the traditional upland cultivation system, both of swidden agriculture and the home garden, is based on multi-species, multi-storey cultivation. Dedicated monocultures would destroy an important part of the value of the uplands, both to the mountain communities themselves and to the world at large, as biodiversity is an important global public good produced in the uplands.

Work done at a number of upland research institutes, such as the Institute of Botany and the Institute of Ecology both at Yunnan (Xie, 1993), has developed models of human-made communities of trees and vegetation that could mimic the diversity of the home gardens. Choosing the combination of trees and crops, with an eye both to their commercial possibilities and to their use value for the farmers, could yield an overall value that is higher than that of single stand plantations.

New developments in the market also promote such diverse stands even with commercialization. For instance, there is now a growing market for "shade grown coffee" as against the traditional "sun coffee", which involved the cutting of huge areas of forests to turn them into coffee plantations. Similarly, in the Himalayan uplands too different tree and annual crops and grass are being simultaneously cultivated in farmers' plots. In Meghalaya, farmers plant bay leaf trees and broom grass in the same plots. In other areas large cardamom is grown in the forest. In Kunming there are experiments to grow vanilla, a high value aromatic crop, in the natural shade of forests, rather than in greenhouses, as is the currently done in the Caribbean islands. Coffee plantations now contain pepper vines, while cashews are combined with pineapple, other fruits and turmeric.

What this shows is that commercialization and intensification of production need not necessarily lead to monoculture plantations. Under what conditions will one or the other occur? This needs further investigation and analysis. But a few preliminary points can be made. Where there is a known synergy between different components of the agro ecosystem, for instance bay leaf trees and hill broom grass and both or all components have commercial value, then farmers are likely to take up the simultaneous cultivation of more than one plant/tree.

Further, where the farmers undertaking the commercial production are locally resident farmers and not distant corporations, then the farmers are also likely to respond to the use values of other components of the agro ecosystem that do not have commercial value, but can be of various uses to the farmers. On the other hand, distant corporations, concerned with their commercial profits will see these other plants or trees as weeds and seek the single-minded maximization of production of the commercial crop in which they are interested. While farmers would have a multi-valued function, including even use values in their assessment, corporations have a single-valued function, based on the maximization of the commercial income from what they sell.

The introduction of diversity into the shade serves a very important economic function—that of protecting against the risk inherent in commercial systems of production. In the mid-1990s prices of tobacco collapsed and in the early 2000s it is coffee prices that have collapsed. What this shows is that monoculture commercialization carries serious risks. Particularly where the crops require large external inputs, as is the case with tobacco compared to coffee, there is a danger of falling into serious debt when prices collapse. A mix of commercial crops needs to be promoted so that farmers are protected against excessive risks in any one market.

At the same time the limits of such self-insurance, as it were, should be noted: it is costly. It is estimated that the loss of income due to choosing a mixed cropping pattern may be as much as 10 to 15% in India (Kabeer, 2005). This probably holds for crops between which there is no synergy. But for insurance, a better method is to pool risks, something that requires well developed insurance and financial systems. This is something that countries like China or India, with large commodity markets might be and are able to carry out. But for smaller countries, a regional approach might be needed. Of course, there will be issues of power and domination within such regional arrangements, issues that are probably better negotiated in regional forums, rather than on a one-to-one basis. And, as the volume of production in a country grows, it might set up its own forward trading and insurance arrangements.

Even these market-based instruments for dealing with fluctuations in commodity prices have their limits. They need to be well-funded. More important there is an important contradiction—if they work, the resultant subsidies may not bring about the needed changes in the structure of production, in particular the reduction in output of those products that are over-supplied, or, if they do bring about such a change, there is an enormous social cost that accompanies the change. The market method of bringing about a change in production is through the destruction of livelihoods of those in such sectors. The numerous suicides of farmers in different parts of India are testimony to the social violence of pure market-based transformations. The challenge is to fashion ways of bringing about change in structures of production that do not carry such social violence. The Group of Eminent Persons (UNCTAD, 2003) proposed the formation of a fund to promote diversification of production by commodity producers and also the search for and promotion of new uses of products.

3.3 Trade Agreements, Intellectual Property Rights and AKST

IP (intellectual property) is driven by technology and business tactics. Intellectual property rights (IPRs) are not natural rights but rather privileges granted to inventors to reward them for inventions. There are many types of IPRs like patents, trademarks, plant breeders' rights and copyrights. Patents in agriculture are important for promoting agricultural research and development (Alam, 2004). This conferment of the privilege of monopoly is supposed to be an incentive for innovation and to enable recovery of cost. Any IPRs system has to balance the privilege given to inventors and corporations owning the IPRs with the public interest. The public interest includes consumer welfare, the right of other producers to use technology, the right to develop, sustainability and environmental protection.

3.3.1 The TRIPS agreement and other IPR regimes

WTO Agreement on Trade-Related Aspects of Intellectual Property Rights (TRIPS). The Trade-Related Aspects of Intellectual Property Rights (TRIPS) was established as part of the WTO in 1995. The TRIPS agreement has resulted in a very significant shift in the balance in the IPRs regime away from the public interest towards the monopolistic privileges of IPRs holders. Since TRIPS is a legally binding international framework enforceable in the WTO through the threat of trade sanctions, it has been able to effectively disseminate a model of IPRs regime throughout the world to its over 130 member states. TRIPS has therefore instituted a basically "one-size-fits-all" system of IPRs, where similar standards are set for countries of differing levels of development. It is in the developing countries where the unsuitability and effects of the inappropriate provisions are most adversely and acutely felt.

Before TRIPS, where patent law existed, most countries provided for "process" patents but not product patents. So different people could use different processes to produce the same product and that allows many products to enter the market and consumers can have competitively priced products. Research and innovation was also encouraged and a good example was pharmaceutical products.

Most developing countries, before TRIPS, did not allow patents on food and medicines even if they had patent laws in operation. Patents on biological resources were also not allowed in almost all countries. Countries were free to choose the scope of patents, the term of patent protection (usually from 5 to 15 years depending on the national laws) and other safeguards to meet their socioeconomic objectives.

Developed countries in their developing stages did not allow patenting and other IPRs or had very narrow scope of IP protection. Many of them also discriminated between nationals and foreigners, favoring the former. This was to promote domestic research, innovation and creativity. For example, Switzerland only allowed patents on pharmaceuticals and agricultural chemicals in the 1970s. Having reached industrial status, these countries then sought to have high IPR standards around the world to protect the technological advantage and market dominance of their major industries especially those in the pharmaceutical, agriculture, biotechnology and information technology sectors.

TRIPS sets mandatory "minimum standards" but these are based on standards of developed countries in the late 1980s to early 1990s when TRIPS was negotiated. Therefore the standards are actually very high and have serious adverse impacts on the development prospects of developing countries. Article 27.2 provides that an "invention" can be excluded from patentability, if it is necessary to protect *ordre* public or morality and the grounds include to protect human, animal or plant life or health; and to avoid serious prejudice to the environment.

Thus inventions can be excluded from patentability on grounds contained in national patent laws. The grounds for excluding patents are not exhaustive in TRIPS, so countries can decide what those grounds are, that are in line with the protection of *ordre* public and morality. There are also other provisions that give a WTO member flexibilities and safeguards at the national implementation level. It is therefore important to understand and interpret TRIPS in a proper way.

Under TRIPS Article 27.3(b), a WTO Member has to allow for the patenting of the following: non-biological and microbiological processes for production of plants and animals; and "microorganisms". With TRIPS, for the first time there is an international obligation to patent microorganisms. But many countries interpret this to exclude "naturally-occurring microorganisms" as these are discoveries. Gene sequences and other parts of microorganisms are not specifically mentioned and many countries exclude these in their national laws, too.

A WTO Member may exclude the following from patentability: essentially biological processes for production of plants or animals; and diagnostic, therapeutic and surgical methods for treatment of humans or animals. IP experts and scientists have observed that it is illogical to exclude patents on biological processes but mandate patents on microbiological processes. This was a concession to the biotechnology industry that was already bioprospecting and commercializing microorganisms and TRIPS is openly acknowledged today as the result of successful industry lobby.

The criteria for patentability should also be carefully understood and applied. Patent principles and law were designed for mechanical inventions. Applying patent law to biological resources raises ethical, religious and socioeconomic issues. The patenting of gene sequences and microbiological processes also raises scientific questions on the legitimacy of patents in this area.

TRIPS Article 27.3(b) also requires new plant varieties to be patented or protected by a sui generis system or a combination of both. Many countries reject patents and are trying to develop or have developed national laws on plant variety protection that can protect plant breeders' rights as well as farmers' rights (see also 3.3.5). But they are under pressure to adopt the 1991 International Convention on the Protection of New Plant Varieties (UPOV) as the "sui generis" system, but this is more like a patent and favors plant breeders at the costs of small farmers.

IPR provisions in Convention on Biological Diversity (CBD). The entry into force of the United Nations Convention on Biological Diversity (CBD) in 1994 (before WTO agreements came into force) raised important issues on access to biological resources and the fair and equitable sharing of benefits arising from the use of such resources, between countries of origin or source and user countries. There are provisions in the CBD that directly deal with IPRs. The provisions are in Article 16 and appear to be finely balanced. Article 16.5 states: "Contracting parties, recognizing that patents and other intellectual property rights may have an influence on the implementation of this Convention, shall cooperate in this regard subject to national legislation ad international law in order to ensure that such rights are supportive of and do not run counter to its objectives."

This clause seems to recognize the IPRs can have a negative effect on implementing the CBD and that contracting parties have to cooperate to ensure that IPRs are supportive of and do not run counter to the CBD's objectives. However, the clause itself has a conditioning term, namely, that the cooperation is subject to national and international law. It is also balanced by Article 16.2.

Article 16.2 states that access to and transfer of technology to developing countries shall be provided and/or facilitated under "fair and most favorable terms, including on concessional and preferential terms where mutually agreed." In the case of technology subject to patents and IPRs, "such access and transfer shall be provided on terms which recognize and are consistent with the adequate and effective protection of intellectual property rights. The application of this paragraph shall be consistent with paragraph 3, 4 and 5 below."

Article 16.3 states that each contracting party shall take measures with the aim that parties (especially developing countries) that provide genetic resources are provided access to and transfer of technology which makes use of those resources, on mutually agreed terms, including technology protects by patents and IPRs, in accordance with international law and consistent with paragraphs 4 and 5.

Tensions between TRIPS and CBD. There are several areas of tension between critical aspects of TRIPS and the CBD and of relevance to many countries as they are signatory to CBD and TRIPS. Following are some examples:

- *Differences in rationale, origins and overall framework.* TRIPS is an international agreement drawn up with the encouragement and active support of large corporations to promote their technological dominance and gain additional margins of profit through obtaining private monopolies. The IPRs model contained in TRIPS is tilted heavily in favor of the rights and benefits of IPRs holders. Because WTO members are obliged to fulfill TRIPS obligations, TRIPS has facilitated the extension of its particular model of IPRs to the wide membership of the WTO. TRIPS is basically a commercial treaty with commercial objectives that largely benefit strong private corporations. The principles of environmental protection or human development are not central to TRIPS and are in fact marginalized by it, although there are references to or exemptions made on behalf of the environment, human and animal health and public order.

The establishment of the CBD was prompted mainly by the growing concern about the rapid worldwide loss of biodiversity, recognition of the important role of traditional knowledge and the rights of local communities that developed and hold the knowledge and the need to regulate access to and the sharing of benefits deriving from the conservation and sustainable use of biodiversity, including genetic diversity. One of the CBD's central aspects is to the recognition of the need to regulate the behavior and effects of private corporations and researchers and constrain their rights of access and benefits within a larger framework that stresses the goals of environmental protection and the rights of sovereign states to their resources and the rights of local communities within them. Many of the tensions betweens TRIPS and CBD stem from these differences in the overall rationale and framework of the two regimes.

- *National sovereignty versus rights of foreign IPR holders.* Based on the principle of national sovereignty enshrined in the CBD, countries have the right to regulate access of foreigners to biological resources and knowledge and to determine benefit sharing arrangements. TRIPS enables persons or institutions to patent a country's biological resources (or knowledge relating to such resources) in countries outside the country of origin of the resources or knowledge. In this manner, TRIPS facilitates the conditions for misappropriation of ownership or rights over living organisms, knowledge and processes on the use of biodiversity takes place. The sovereignty of developing countries over their resources and over their right to exploit or use their resources, as well as to determine access and benefit sharing arrangements, is compromised.
- *Conflict between private rights of IPRs holders and community rights of traditional knowledge holders.* In the preamble of TRIPS, it is recognized that "intellectual property rights are private rights". In TRIPS, the award of IPRs over products or processes confers private ownership over the rights to make, sell or use the product or to use the process (or sell the products of that process). This system of exclusive and private rights is at odds with the traditional social and economic system in which local communities make use of and develop and nurture, biodiversity. For example, seeds and knowledge on crop varieties and medicinal plants are usually freely exchanged within the community. Knowledge is not confined or exclusive to individuals but shared and held collectively and passed on and added to from generation to generation and also from locality to locality. The CBD has several provisions that acknowledge this and also that aim at protecting community rights, the key provision being Article 8(j).
- *Differing treatment of innovators using modern knowledge and traditional knowledge.* Related to the different ways in which the CBD and TRIPS treat private and community rights is the difference in their treatment of knowledge holders or innovators using modern and traditional technology. Whilst the CBD adequately recognizes the nature and crucial role of traditional knowledge and practices in biodiversity conservation and use (for example, see article 8(j) of the CBD), TRIPS is constructed in ways that effectively deny this and instead rewards additions to knowledge (even if very slight and minor) made through modern technology. This different treatment for modern technology and traditional knowledge is also associated with discrimination against local community rights.
- *System of prior informed consent of states and communities (under CBD) versus unilateral patent actions by private companies and researchers (under TRIPS).* Article 15.4 of the CBD states that "access to genetic resources shall be subject to prior informed consent of the Contracting Party providing such resources, unless otherwise determined by that Party." Thus, intending collectors of biological resources or of knowledge relating to these have to provide sufficient information of their work and how it is intended to be used and obtain consent, before starting the work. The PIC requirement is thus a measure to prevent misappropriation of resources and knowledge and to facilitate fair benefit sharing.

  In TRIPS, there is no provision that applicants for patents or other IPRs over biological resources have to obtain prior informed consent. There is thus no recognition in TRIPS of the rights of the country in which the biological resource or knowledge of its use is located. Thus, patent applicants can submit claims on biological resources or knowledge to patent offices in any country (that recognizes such patentability) and the patent offices can approve the claims without going through a process even of checking with the authorities of the country or countries of origin. Thus, whilst the CBD has set up a PIC system as a check against misappropriation or biopiracy, TRIPS on the other hand facilitates the possibility of such misappropriation by not recognizing the need for and thus omitting a mechanism of PIC.
- *Differences in benefit-sharing arrangement.* A key aspect of the CBD is that it recognizes the sovereign rights of states over their biodiversity and knowledge and thus gives the state rights to regulate access and this in turn enables the state to enforce its rights on arrangements for sharing benefits. Access, where granted, shall be on mutually agreed terms (Article 15.4), shall be subject to prior informed consent (Article 15.5), countries providing the resources should fully participate in the scientific research (Article 15.6) and, most importantly, each country shall take legislative, administrative or policy measures with the aim of "sharing in a fair and equitable way the results of research and development and the benefits arising from the commercial and other utilization of genetic resources with the contracting party providing such resources. Such sharing shall be upon mutually agreed terms".

  Under TRIPS, there is no provision for the patent holder on claims involving biological resources or related knowledge to share benefits with the state or communities in countries of origin. In fact, there is little that a country of origin can do to enforce its benefit-sharing rights (recognized in CBD) if a person or corporation were to obtain a patent in another country based on the biological resource or related knowledge of the country of origin. If the patent laws, the administration of approvals, or the courts of a particular country operate in a context that is favorable to granting such patents, there is little that can be done by a country of origin to ensure that biopiracy does not take place, or that if it takes place that it can get a remedy.

- *Treatment of the environment.* Protection of the environment is at the heart of the rationale and provisions of the CBD. The objectives of the Convention are "the conservation of biological diversity, the sustainability use of its components and the fair and equitable sharing of the benefits arising out of the utilization of genetic resources" (Article 1). Countries are obliged to develop strategies and plans to conserve and sustainably use of biodiversity in sectoral and cross-sectoral plans and policies (Article 6); to carry out in situ and ex situ conservation (Article 8, 9); to minimize adverse impact on biodiversity whilst also carrying out remedial action in degraded areas (Article 10); and to conduct environmental impact assessment on and minimize adverse effect of projects (Article 14). In particular, Article 19 asks parties to consider the need for an international biosafety protocol, which has been established—The 2000 Cartagena Protocol on Biosafety (CBD, 2000). This Protocol is meant to deal with the safety aspects of biotechnology and international transfer of genetically-modified organisms.

TRIPS does not have environmental protection as part of its objectives. It does, however, have provisions that enable members to exclude patents on environmental grounds as stated in Article 27.2 (see above). This provision provides some scope for members to take the environment into account in their IPR policies. Article 27.3(b) of TRIPS also allows for exclusion from patentability of plants and animals other than microorganisms and essentially biological processes other than microbiological processes. Whilst the article at first reading enables the exclusion of patentability for plants and animals, in fact it has opened the door to worldwide patenting of genes and microorganisms and patenting of genetically-modified organisms, including modified plants and animals. Many environmental groups and scientists are concerned that patents granted on life forms would hinder the process of scientific research by researchers that do not own the patents; and also that the incentive of providing monopoly rights to companies to produce GMOs would contribute to the proliferation of genetic-engineering application that have adverse effects on biodiversity.

3.3.2 Farmers' access to AKST vs. breeders' rights

The importance of the conservation and sustainable utilization of plant genetic resources (PGRs) for food and agriculture is broadly recognized today. One of the areas for global action relates to farm conservation. Farmers not only use seeds and related AKST; they are key players in the process of conservation and improvement of plant varieties. Their activities ensure crop evolution whereby new varieties arise through genetic recombination, mutation and hybridization within and between cultivated and wild plant populations (Brush, 1994).

With the importance of farmer protection and public interest protection from the patent regime in agriculture, many developing countries like Thailand, Zambia, Bangladesh and Costa Rica provide farmer rights in their legislations. The Indian Protection of Plant Varieties and Farmer Rights Act, 2003 also provides for farmer rights to use, reuse, exchange and even sell (unbranded) seed, has researcher exemption, creates a national gene fund and provides for compulsory licensing in case of public interest. Farmers' rights are valuable as they promote equity, conservation and preservation which are so crucial for sustainable agriculture. But so far as protection of farmer varieties is concerned, there are problems of identifying one from another, duration of protection and passing on the benefits to community (Alam, 2004).

A recent comparative analysis of the protection to plant varieties and farmer rights in the patent laws of the various Asian countries shows that only India and Malaysia recognize the protection of farmers' interests as one of the objectives of the law and almost all the countries have based their definition of plant variety and essential derived variety on the UPOV with only Bangladesh, India, Malaysia and Thailand excluding microorganisms expressly and only China and South Korea not defining essentially derived varieties (EDVs).

On definition of breeders again, except India and Thailand, other countries specifically recognize "discovery" as a ground which could hurt farmer interest as any breeder could discover a variety which rightfully might have been invented by farmers. Only India and Malaysia recognize "evolution" and "genetic manipulation" as one of the criteria for breeders respectively. Surprisingly, most of the countries, except India, do not define farmers as they are not given any rights. This is due to the fact that UPOV has been followed which only provides breeder rights. Indian definition of farmer is broad enough. Except India, Malaysia and Thailand which accommodate farm varieties to some extent, mostly UPOV laws have been followed for criteria for granting protection to plant variety which is NDUS—new, distinct, uniform and stable.

TRIPS requires protection of plant varieties as against new plant varieties under UPOV. Breeders have exclusive rights over agricultural and horticultural varieties and even export and import is in the hands of breeders. Most countries provide Plant Breeder Rights (PBRs) for 20-25 years for trees and 15-20 years for other plants except India which has initially shorted protection but extendable and Malaysia which is biased against farmers' varieties. In all cases, the PBR can be forfeited if variety does not fulfill the claims made or if it is detrimental to the environment or the public order with Bangladesh even going further by making provision for invoking food security, monopolies or rights of the communities.

Most of the countries also provide exemption to the rights granted to plant breeders but not as wide as in case of India. The exhaustion of breeder rights is provided by Pakistan, Sri Lanka, South Korea and Philippines (UPOV-style) whereas others are silent on this. The provision may have implications for sale for seed from harvested crops or subsequent sale of variety after it has been put into the market by the right holder. Most importantly, so far as farmer rights are concerned, India provides very comprehensive rights which encompass saving, sowing, resowing, exchanging, sharing or selling his/her farm produce including seed of a protected variety provided that farmer is not entitled to sell branded seeds of a protected variety. A farmer in India is also entitled to registration of his newly developed variety like a breeder and for reward under the Gene Fund for conservation of genetic resources of landraces and wild relatives of economic plants. But, most other countries have not granted rights to farmers. Further, India and Pakistan also safeguard farmers against sold variety failing to perform but no other country has such provision. Most countries have compulsory licensing of a protected variety provision in public interest. Indian law also prevents terminator technology. Further, only Bangladesh, India and Thailand provide for community rights and benefit sharing and common gene fund (Kumar and Sahai, 2003).

Also, implementation of TRIPS can have a negative impact of farmers' access to AKST. Article 27.3(b) of TRIPS is a major driving force of the biotechnology industry and provides the legal protection for the development of GMOs, which are patented. Furthermore, countries like the United States allow patents on plants and animals and there is enormous pressure on developing countries to adopt similar standards for IPRs. All these have implications for farmers around the world. Patented seeds cost more and threaten farmers' rights to save, reuse, exchange and sell seeds, or even access to the seeds. This is already evident in the case of Bt cotton seed in India as discussed below.

Monsanto-Mahyco Biotech Limited is charging rupees (Rs.) 1250 per 450 gram packet of Bt cotton seed from its licensees as trait value of seed which is nothing but royalty for transfer of Bt technology to about 20 Indian seed companies, for which it has a patent under the TRIPS regime. It also collected Rs. 50 lakh from each of its sub-licensees as non-refundable fee which is illegal as per MRTP commission that monitors trade practices in India. MRTP has already initiated investigations against the company for overcharging for Bt cotton seed, which is considered an unfair trade practice by a monopoly as the company is the only Bt cotton seed seller in India. In US, it charges a royalty of only Rs. 573 per acre (Janaiah, 2006). The real cost of seed is said to be Rs. 500. The company on its own reduced the trait value fee to Rs. 900 per packet after the initiation of the case, but in US, the company charges only Rs. 108 per packet which is much lower than its rate for India (Times of India, New Delhi, April 21, 2006). The Bt cotton seed costs only Rs. 550 per packet per acre in China, Rs. 250 in South Africa and Rs. 1000 in Mexico. Only in India and Argentina, it is priced very high, i.e., Rs. 1800 and Rs. 1900 per acre respectively (Janaiah, 2006). The company has its patented technology based Bt seed being sold in India with the help of many licensees. Thus, TRIPS has already become a barrier due to high price of the Bt cotton seed so far as poor farmers are concerned.

There are conflicting reports on the performance of the new seed in India and it has been banned in Andhra Pradesh for three years due to poor performance. It is due to this prohibitive high price of Bt seeds that some farmers in India have resorted to illegal and spurious Bt cotton seeds being sold by local traders and farmers, especially in Gujarat where the so called Bt seed is available for Rs. 300-800 per packet. Thus, a large proportion (50%) of total Bt cotton area in India is under illegal and spurious varieties. The Supreme Court of India has recently asked the company to bring its trait value to levels which it charges in China within a month. Thus, Monsanto may have to slash its trait value fee to Rs. 40 per packet from Rs. 900 per packet of 450 grams of Bt cotton seed. But, the company is likely to appeal against the order.

In several developed countries, patenting of plants, plant varieties and traditional knowledge associated with their use is already taking place and has been accelerating since TRIPS. In that process, "biopiracy" or the misappropriation of biological resources and traditional knowledge is taking place, as plants and seeds originating in developing countries are being patented, usually without the knowledge or consent of these countries of origin.

Between 1985 to 1999, about 11,000 patents on plants had been registered in the US (ActionAid, 1999). In the European Union, patent law has been extended to microorganisms and genes of plants, animals and humans. The biotechnology industry is racing to map the genomes of the world's staple food crops with a view to patenting the vital and most interesting genes. The farmers of developing countries that developed the world's food crops would have no effective rights over the varieties, due to the patenting being carried out by the transnational companies. Only 10% of seed in the developing world is purchased commercially with some poor farmers buying seed only once in five years; hence patents pose a threat to farmer livelihoods and global food security (ActionAid, 1999).

There is no system of informed consent to notify communities involved of the intentions of genetic collectors even if the "invention" relies upon the knowledge and insight of local people. In some of those countries where there are patents on plant varieties, farmers are being prosecuted for alleged violation of IPRs. These developments could be reproduced in developing countries in the future.

3.3.3 Public and private sector research and development

How have IPRs, especially the availability of IPRs for living organisms affected public and private research and development in AKST? Plant breeding has shifted from the public to the private sector since the early 1980s for soybean, wheat, and cotton; 75% of the plant biotechnology patents originate in the private sector (Atkinson et al., 2003; Gepts, 2004). There is evidence the shift occurred with the introduction of TRIPS. The shift coincided with the consolidation of agribusinesses controlling agrochemicals, seeds and biotech traits (Lesser and Mutschler, 2002; UNCTAD, 2005). For agrochemicals, the three leading corporate groups alone are estimated to represent approximately half of the market. For seeds, four corporations have about 30% of the market share, but the figures may mask much stronger market concentration for major crops in specific regional markets. The figures obscure the outstanding degree of consolidation in some of the major seed country markets (UNCTAD, 2005). There is a strong potential for demand complementarity between agrochemical and seed businesses.

Another structural change has been increased coordination with a trend towards heightened strategic cooperation amongst large competitors in the agricultural biotechnology sector and vertical coordination upward and downward along the food chain described in the introduction (UNCTAD, 2005)

The incentives for extensive mergers ". . . along with the breadth . . ." of protection accorded to patent holders (in many cases seed or biotech companies), and the increased market concentration of companies engaged in agricultural biotechnology is giving large corporations unprecedented power vis-à-vis growers and other stakeholders. In particular, the privatization and patenting of agricultural innovation (gene traits, transformation technologies and seed germplasm) have supplanted the traditional agricultural understandings on seed and farmers' rights, such as the right to save and replant seeds harvested from the former crop (UNCTAD, 2005). In some cases, this has resulted in a drastic erosion of traditional farmers' rights and changed farmers from "seed owners" to mere "licensees" of a patented product (UNCTAD, 2005). The synergy and vertical integration offered by the alliance of traditional seed industry and biotech have facilitated a race to buy seed companies by the biotech and agrochemical giants.

The combination of biotech and seed companies has been crucial to the market penetration of GM varieties. Some of the largest agricultural biotechnology companies in Europe and the United States have emerged as significant players in the rapidly growing Brazilian seed market (UNCTAD, 2005). By these acquisitions the largest biotech companies have established global corn and oil-seed business through which to commercialize crop enhancement products in Brazil, a country that had for long resisted GM crops. ESAP countries are a major market for the global biotech and agrochemical giants, thus it is conceivable they would employ similar strategy in ESAP regions. While the synergy and vertical integration can be good thing for business, it raises serious concerns for AKST development. The companies' overriding profit-seeking motives may not always be compatible with the goals of poverty/hunger reduction and sustainable development.

The recent development in IPR regime is also a driving force of market consolidation and concentration in the sector. Led by changes in US patent system, growing proprietary rights have been granted to agricultural innovation. This leads to increasing number of patents, patents being increasingly issued on fundamental technologies, multiple claims over various aspects of a technology. Due to these reasons, even giant companies often find it difficult to avoid infringing patents when conducting product development research. "Monsanto and DuPont, DuPont and Syngenta, Monsanto and Syngenta, Syngenta and Dow have all filed suits against one another involving claims of patent infringement . . . Besides litigation, 'defensive patenting' (companies tend to patent as much as they can to deter litigation though the threat of reciprocal suits) has become common practice within the industry" (UNCTAD, 2005). This thus creates a need to consolidate patent portfolios, thus acts as an incentive for the extensive mergers and acquisitions in the agricultural biotechnology and seed businesses.

The asymmetries between the developed and the developing world in aspects like agricultural systems, market institutions and research and regulatory capacity, which raise transaction costs for the latter, increase doubts whether poor people can benefit from the biotechnology development in terms of spillover or trickle-down effects. China is the only country to have developed GM technology in the public sector with other developing countries depending on imports or local adaptations of imported varieties. Further, GM crops are not targeted at poor farmers and marginal environments as they are not attractive to the private sector agencies involved in this technology (Pingali, 2005). In India, the policy towards GM crops was more of preventive nature in terms of IPRs and trade, precautionary in terms of biosafety and permissive on food safety and consumer choice while being promotional on public research investment (Paarlberg, 2000).

The crowding of IPRs and the increasing concentration of them in corporations is also jeopardizing research. According to the UNCTAD study, "Academic scientists engaged in agricultural research report problems of access to important technologies due to an overlapping set of intellectual property (IP) rights on research tools and genetic contents. The reasons would lie in the increasing number of patents being issued, increasing patent breadth and uncertain ownership of rights, all resulting in IP congestion and uncertainty. The accumulated transaction costs involved (tracking down owners, conducting negotiations and multiple royalty payments to administer) have created a major access obstruction that is hampering agricultural research, according to some commentators."

Gepts (2004) used the case of golden rice to explain the problem, "The development of the pro-vitamin A-rich, "golden" rice (Ye et al., 2000) provides a stark example of how quickly an invention can get lost in a "thicket" of IP rights. Seventy IP or tangible property rights belonging to 32 companies and universities had been used in the development of this rice line (Kryder et al., 2000). In addition, MTAs (material transfer agreements) further complicated the situation. "Freedom to operate" (FTO) was achieved by providing a license to a large biotechnology company, Zeneca, covering not only the pro-vitamin A pathway in rice but also in any other crops, in exchange for a humanitarian use (defined as a maximum of U.S. $10,000 revenue from golden rice) in developing countries (Potrykus, 2001). Clearly, such a solution was made possible in part because of public relations concerns on the part of the major holders of IPRs, mainly large, multinational biotechnology companies. However, this "segmentation" of the potential market did not solve fundamentally the issue for researchers, farmers and consumers in developed countries."

Gepts also points out the negative impacts and challenges by the IPRs regime on public research: "Public institutions are faced with similar 'thickets of IPRs,' despite the fact that they have been responsible for much of the basic research leading to the initiation and continued development of biotechnology in the first place (Atkinson et al., 2003). The fragmentation of IPRs covering technologies (so-called "enabling technologies") and plant materials among many companies and institutions also created FTO problems. Biotechnology companies have dealt with these problems by developing their home-grown technology, licensing technology from other companies and by acquiring or merging with other companies, and thus assembling a complete IP portfolio allowing them to commercialize new technologies, including transgenic cultivars of major field crops such as maize, soybean and cotton. Left out of this equation are many horticultural crops or specialty crops with smaller markets in developed countries and subsistence crops in developing countries.

A recent initiative from some leading public universities and private foundations promises to address the FTO issue. The Public-Sector Intellectual Property Resource for Agriculture (PIPRA; www.pipra.org) intends to establish "best practices" encouraging the greatest commercial application of publicly funded research, while also retaining rights to allow public institutions to fulfill their responsibilities toward the public at large. It will also establish a database providing an overview of IPR currently held by public institutions, with up-to-date information on the licensing status of these IPRs. In addition, it will also attempt to pool patents or other IPRs to develop "technology packages" of complementary patents, which would provide FTO to public sector researchers and reduce transaction costs associated with obtaining licenses to develop transgenic cultivars (Atkinson et al., 2003). While actions such as those proposed by PIPRA attempt to address the FTO issues, they do not fundamentally alter the framework in which current public research has come to operate. The public-sector research "culture" has a long tradition of open sharing of genetic resources, germplasm and research findings. This has led, among other things, to extensive genetic resources collections with broad availability. This tradition of open sharing and exchange is now severely challenged and raises several concerns with regard to the availability of biodiversity for research and cultivar development.

In response to bioprospecting by corporations, the gene banks of the centers belonging to the Consultative Group for International Agricultural Research (CGIAR), such as the Centro Internacional de Mejoramiento de Maíz y Trigo (CIMMYT; Mexico), the International Rice Research Institute (IRRI; The Philippines) and the Centro Internacional de Agricultura Tropical (CIAT; Colombia), which hold more than 500,000 germ plasm accessions, have instituted an MTA (http://www.sgrp.cgiar.org/MTA_E.pdf). This MTA seeks to protect the germ plasm or breeding lines and associated information distributed by the CGIAR center from ownership or IP claims by the recipients of this material. Obviously, this MTA does not cover further breeding uses leading to improved materials. It is noteworthy that most of the germ plasma in these gene banks were donated by Southern countries and has been and continues to be accessible on an open access basis. Yet the genes and improved varieties derived from such material (usually developed by Northern corporations or agents) often enjoy proprietary protection under the current IPR regime. In the growing enclosure of genes and biodiversity, the developing countries are getting the raw deal.

3.3.4 Technology dissemination and transfer

A strong IPR system is normally advocated to stimulate innovation. However, for most developing countries, the extra innovations generated by stronger PIPRs (private IPRs) would be meager, as agents in these countries possess poor innovative capabilities according to IPR criteria. As even Primo Braga (1996), who is quite sympathetic to TRIPS, admits, there is little evidence that stronger PIPRs encourage greater R&D in developing countries. Thus, one of the main concerns of developing countries with the adoption of the TRIPS agreement has been the extent to which the new rules will affect the transfer of technology, a vital element to foster economic development. As 97% of world patents are held by developed countries (UNDP, 1999), the cost from paying royalties may significantly outweigh the benefits from (the insignificant) additional knowledge that the system extracts from nationals of developing countries.

It has been argued that higher standards of IP can lead to transfer of technology, as foreign corporations would be encouraged to invest in developing countries and make use of their technologies. However, there is also a counter-argument that foreign firms that have obtained patents in developing countries are able to make inroads and profits in these countries without having to produce the patented products there, as they can import the products and sell them at monopoly prices.

There are several ways in which a strong IPR regime can hinder access of developing countries to technology (see Khor, 2002). Obstacles to technology transfer make it difficult for developing countries and their corporations to upgrade productivity which is necessary for them to compete successfully. They thus impede competition. Firstly, a strict IPR regime can discourage research and innovation by locals in a developing country. Where most patents in the country are held by foreign inventors or corporations, local R&D can be stifled since the monopoly rights conferred by patents could restrict the research by local researchers. Strict IPR protection, by its apparent bias, may actually slow the pace of innovation in developing countries and increase the knowledge gap between industrial and developing countries. In such situations, the IPR system favors those who are producers of proprietary knowledge, vesting them with greater bargaining powers over the users (Oh, 2000). The UK Commission on Intellectual Property Rights also provides analysis and examples of how the patent system might inhibit research and innovation (CIPR, 2002). Secondly, a strict IPR regime makes it difficult for local firms or individual researchers from developing or making use of patented technology. Thirdly should a local firm wish to "legally" make use of patented technology, it would usually have to pay significant amounts in royalty or license fees. As pointed out earlier, TRIPS increases the leverage of technology-suppliers to charge a higher price for their technology. Many firms in developing countries may not afford the cost. Even if they could, the additional high cost could make their products unviable. Moreover, there could be a large drain on a developing country's foreign exchange from having to pay foreign IPR holders for the use of their technology. Many developing countries with serious debt problems will be unable to afford to pay the cost of using the technologies.

Fourthly, even if a local firm is willing to pay the commercial rate for the use of patented technology, the patent holder can withhold permission to the firm, or impose onerous conditions, thus making it impossible or extremely difficult for the technology to be used by the firm. Patent holders can refuse to grant permission to companies in the South to use the technologies, even if they are willing to pay market prices; or else the technologies may be made available at high prices (due to the monopoly enjoyed by the patent holders). Companies in the South may not afford to pay at such prices and if they do their competitiveness could be affected.

3.3.5 Indigenous, traditional and institutional knowledge

Local or traditional knowledge (TK) refers to information held by local or indigenous people with regard to biodiversity in this case (Brush and Stabinsky, 1996). Indigenous people are defined as descendants of preconquest, traditional people of a certain geographic area, with a common history, culture, language and customary law. TK encompasses information about, for example, crop landraces and their agronomic or culinary characteristics or the medicinal qualities of native species. TK is an essential aspect of an indigenous group's cultural survival; it has been developed through generations of intimate contact with the biological materials (Mauro and Hardison, 2000). It is transmitted in many ways, including apprenticeship with elders and specialists and oral tradition (including poems, songs and music; Posey, 2002). Although indigenous people comprise only some 5% of total world population, they have a disproportionately large role in the maintenance of and knowledge about biodiversity because they are located primarily, although not exclusively, in biodiversity centers. Furthermore, with regard to crop biodiversity, indigenous or local farmers play an important role in in situ (on farm) conservation of landrace varieties (Brookfield et al., 2002). TK is not, however, limited to the knowledge of indigenous people but encompasses knowledge (and associated heirloom varieties) of local, nonindigenous communities in modern societies as well (e.g., Bérard and Marchenay, 1996).

Traditional knowledge is now widely recognized as having played and as still playing crucial roles in economic, social and cultural life and development, not only in traditional societies but also in modern societies. Even today, the majority of the world's population depend on traditional knowledge and practices for food and medicines. Eighty percent of the world's people rely on indigenous knowledge for their medical needs and half to two-thirds of the world's people depend on foods provided through indigenous knowledge of plants, animals, insects, microbes and farming systems (RAFI, 1997). This recognition has heightened in recent years as a result of the increased awareness of the environmental crisis; the role of some modern technologies, production methods and products in contributing to this crisis; and a growing appreciation that local communities (especially in developing countries) have a wide range of traditional knowledge, practices and technologies that are environmentally sound or "friendly" and that have been making use of the manifold and diverse biological and genetic resources for food, medicines and other uses. The knowledge of local communities, farmers and indigenous peoples on how to use the many forms and types of biological resources and for many functions, as well as on how to conserve these resources, is now recognized as being a precious resource that is critical to the future development or even survival of humankind. At the same time, this precious knowledge is maintained and thrives in the context of the traditional ways of social and economic life and customary practices of the traditional communities. Their rights to their knowledge, to the use of their knowledge and to the products arising from such use must be recognized. The misappropriation of their resources, their knowledge or the products of their knowledge would not only violate their rights, but also adversely affect the conservation and use of the knowledge and of biodiversity (as the IPRs obtained by corporations and other institutions may erode the communities' rights to continue using their resources or to continue with their traditional practices).

The position of traditional knowledge and the rights of local communities is now widely accepted with acknowledgment: (1) of the role and importance of traditional knowledge; (2) that for traditional knowledge to be maintained, the social and economic context in which it developed and is applied has to be maintained; (3) that for this context to be maintained, the rights of local communities to their resources and knowledge have to be recognized and respected; and (4) that misappropriation of these rights can erode the basis of traditional knowledge and thus adversely affect the prospects of sustainable development.

There are proposals to encourage countries to use their options under TRIPS and the CBD in favor of sustainable development. Each country should interpret the agreements in ways that are most appropriate for itself, maximizing the creative use of provisions of each agreement to suit the country's chosen policies.

A major drawback of this approach is that developing countries in general have limited capacity (in terms of policy-making, legal and administrative expertise) to analyze the international agreements and to formulate national policies and draft legislation with the sophistication required. Thus, they may not be able to make full use of the flexibilities in TRIPS and the CBD. Also, for this approach to work, developed countries would have to allow the developing countries to make use of the flexibilities in the agreements and not unduly put pressure on them when they do so.

This approach is an attempt to harmonize the traditional knowledge system and western IPR system. There are already some existing cases that are noteworthy. For example, India has already seen its practice in Kerala state where Jeevani—a drug with anti-fatigue properties—has been patented by TBGRI under a benefit sharing formula with Kani Tribe. The drug was extracted from a plant called *arogyapacha* in local language and was developed based on lead knowledge given by the tribe. For this kind of development, it is important that the system of protection takes into account the ethical norms of the community involved, intention of protection (trade or health), GI protection and benefit sharing mechanisms for cumulative innovations (Harilal, 2006).

Meanwhile, many representatives of indigenous communities are advocating rejection of the application of an IPRs system based on their worldviews. In June 1999, a group of 114 indigenous peoples' organizations from many countries around the world, as well as another 68 indigenous peoples' support groups, issued a joint indigenous peoples' statement on the TRIPS agreement (Tebtebba Foundation, 1999). Some of the key points of the statement are as follows:

I. Nobody can own what exists in nature except nature herself . . . Humankind is part of Mother Nature, we have created nothing and so we can in no way claim to be owners of what does not belong to us . . . [W]estern legal property regimes have been imposed on us, contradicting our own cosmologies and values.
II. We view with regret and anxiety how Article 27.3b of the Trade-Related Aspects of Intellectual Property Rights (TRIPS) of the World Trade Organization (WTO) Agreements will further denigrate and undermine our rights to our cultural and intellectual heritage, our plant, animal and even human genetic resources and discriminate against our indigenous ways of thinking and behaving.

The indigenous peoples' representatives are of the view that the IPRs regime threatens the rights, way of life and knowledge of indigenous peoples. They also reject the application of an IPRs system of indigenous peoples which is based on collective innovation and collective rights. Thus they are advocating that the international agreements need to modify to include diverse worldviews. This was also presented in a statement on behalf of indigenous peoples at a roundtable on Intellectual Property and Traditional Knowledge at the World Intellectual Property Organization (WIPO) in November 1999. According to the statement: "We believe that the challenge for WIPO and governments, as well as other international multilateral organizations, is to maintain an open mind and be more daring in exploring ways and means to protect and promote indigenous and traditional knowledge outside of the dominant IPR regimes. WIPO should not insist in imposing that the IPR regime it is implementing, particularly patents, is what should be used to protect traditional knowledge. Other forms of protection should be explored and developed in partnership with indigenous peoples and other traditional knowledge holders. Any effort to negotiate a multilateral framework to protect indigenous and traditional knowledge should consider indigenous practices and customary laws used to protect and nurture indigenous knowledge in the local, national and regional levels." (Tauli-Corpuz, 1999)

3.3.6 National and regional responses, impact on developing countries

There is a trend for bilateral free trade agreements (FTAs) between developing and developed countries (especially the United States) to oblige the countries concerned to allow for the patenting of plants and animals and this is often under pressure from the developed countries.

In the case of new plant varieties, there are pressures for developing countries to adopt the 1991 International Convention on the Protection of New Plant Varieties (UPOV) as the "sui generis" system, but this is more like a patent and favors commercial plant breeders at the cost of small farmers and even public researchers. Malaysia and Thailand have adopted sui generis plant variety protection laws that strike a better balance for small farmers, but in ongoing negotiations of bilateral FTAs with the United States, they are pressured to take on UPOV 1991. China became a Member of UPOV 1991 on 23 April, 1999. As a WTO Member, China also has TRIPS obligations and the challenge is to ensure that the flexibilities and safeguards are maximized so that the public interest and long-term sustainable development of the country are assured.

The shortcomings and inherent inequities in existing intellectual property systems, especially patents, are increasingly acknowledged. A comprehensive assessment and the net adverse impact of IPRs on developing countries can be found in the report of the International Commission on Intellectual Property Rights, entitled "Integrating Intellectual Property Rights and Development Policy" (2002). This Commission was initiated by the UK government and chaired by a leading US lawyer, Professor John Barton. Literature survey, commissioned papers, consultations and country visits were undertaken to "incorporate voices from both developed and developing countries: from science, law, ethics and economics and from industry, government and academia" [for a full report, see www.iprcommission.org].

The obligations on developing countries to implement TRIPS are estimated to result in increased payments by them of US$60 billion a year (Finger, 2002). The net annual increase in patent rents resulting from TRIPS for the top six developed countries in this field are estimated to be US$41 billion—with the top beneficiaries being the US with $19 billion, Germany $6.8 billion, Japan $5.7 billion, France $3.3 billion, UK $3 billion and Switzerland $2 billion (World Bank, 2002). Developing countries that will incur major annual net losses include South Korea ($15.3 billion), China ($5.1 billion), Mexico ($2.6 billion), India ($903 million) and Brazil ($530 million).

The World Bank's patents rents estimates, already high enough, significantly understate the actual costs to developing countries, as these only measure the direct outflow of patent rents from these countries (Weisbrot and Baker, 2002). In addition there are economic distortions as the IP protection causes goods to sell at prices far above their marginal costs, thus given rise to "dead-weight cost". Citing other studies, they estimate the deadweight costs to be twice the size of the estimated patent rents.

In addition, there are costs for administering and enforcing IP laws and policies, requiring law reform, enforcement agencies and legal expertise. World Bank project experience indicates that it will cost a developing country $150 million to get up to speed on three new WTO areas (IPRs, SPS and customs valuation) (Finger, 2002); this amount is more than a full year's development budget in many LDCs.

Compared with the outcome of the market access negotiations, the TRIPS amounts (i.e., net rents) are big money (Finger, 2002). The US obtained 13 times more benefit from annual patent rents arising from TRIPS than from liberalization of industrial tariffs with Germany, France and UK gaining 3.6 times more. Conversely, the loss from TRIPS obligation is 18 times greater for Korea than gains from Uruguay Round tariff liberalization and the costs outweigh benefits 7 times for Mexico and 4.7 times for China.

Well-known trade economists who advocate free trade have also written harshly on the imbalances of TRIPS and the adverse effects on competition caused by the upward harmonization of IP standards induced by TRIPS. For example some have argued that the TRIPS Agreement be removed from the WTO because the WTO is meant to be about mutual gains in trade and IP protection is a tax on poor countries' use of knowledge, hence constitutes a wealth transfer to the rich countries (Bhagwati, 2001). Others argue that if it is not removed at least some of its provisions should be renegotiated (Srinivasan, 2000). The arguments put forward that high IP standards benefit developing countries center around the encouragement of local innovation and the likelihood that foreign enterprises would be more willing to transfer technology and to invest.

"These a priori arguments are based on the premises that first IPR protection of the type imposed by TRIPS is needed to encourage innovation and second that foreign enterprises place a significant weight on the strength of IPR protection regime. The theoretical justification for and even more importantly the empirical evidence in support of both these premises is not at all strong. . . . It would appear that patent protection as a spur to innovation does not appear to be powerful in the real world. And the cost to the general public of restricting access to new technology through patenting may be high."

3.3.7 Ways forward

The shortcomings and inherent inequities in existing intellectual property systems, especially patents, are increasingly acknowledged, with concerns over the net adverse impact of intellectual property rights (IPRs) on developing countries, who remain net IPR importers. The WTO Agreement on Trade-Related Aspects of Intellectual Property Rights (TRIPS) sets global minimum standards on IPRs. There is debate over the role of IPRs in development, with some claiming that high IPR protection is necessary to ensure returns to research investment and innovation. Yet, evidence shows that the monopoly of knowledge afforded by IPRs can be detrimental to development goals. Historically, IPRs were applied mainly to mechanical inventions, or to artistic creations, however the assignment of IPRs to living things is of relatively recent origin in developed countries. While these issues are still being debated and not fully addressed yet, regrettably, higher protection of IPRs, even going beyond that required under the TRIPS Agreement, is increasingly advocated in free trade agreements, particularly with developed countries like the United States.

IPR standards under trade agreements have contributed to a shift in AKST, by facilitating private sector dominated research and consequently privately-generated and owned AKST. Patents and to some extent plant variety protection (PVP), have played a part in the major consolidation of the global seed and agricultural input corporations, many of which are also developing transgenic crops. The need to consolidate patent portfolios and hence ensure freedom to operate appears to have created incentives for this consolidation. In this private sector dominated context, market forces rather than food security needs have dictated the direction of research in general. At the same time, public sector research is either stagnating or declining and also faces barriers in terms of IPRs preventing access to research materials, tools and technologies. Public sector research needs to be strengthened and better funded. The objective should be to ensure that research is oriented to address the needs of poor and small farmers. There is a need for governments to consider the use of competition law (e.g., antitrust) to respond to the high level of concentration in the private sector. While some national level action has been taken to break monopolies and encourage competition, there is no international mechanism to deal with such issues.

The international trade regime raises issues of relevance, adequacy, affordability and access to AKST; in particular, IPRs may restrict access to plant material for farmers and threaten farmers' rights. For farmers and rural producers, knowledge is increasingly becoming an economic good for which they are willing to pay and are paying significant costs. However, IPRs may restrict access to plant material for farmers. Patented seeds cost more as patent owners have a monopoly and can charge high prices. There are considerable dangers to food security if seeds are overpriced to the exclusion of poor and small farmers. The consolidation of the global seed and agricultural input corporations and their subsequent monopoly over the agricultural chain also results in high prices for agricultural inputs.

The spread of private IPRs is also considered to be a threat to the rights of farmers to save, use, exchange and sell seeds that have been subject to proprietary claims, even though it is farmers who have played a crucial role in conserving, developing and making available plant genetic resources that are the basis of food and agriculture and these are the very practices that have formed the basis of their traditional role in conservation and development. IPRs can thus stifle local innovation and research. Furthermore, Genetic Use Restriction Technologies (GURTs) or "terminator" technologies can be used to biologically prevent seeds from germinating in order to protect proprietary claims of IPR-holders. This has tremendous impact on small farmers and indigenous communities and has been heatedly debated under the Convention on Biological Diversity, under which a de facto moratorium on field-testing and commercialization of GURTs exists.

Currently, farmers' rights are not yet adequately protected through effective means, both domestically and internationally. The International Treaty on Plant Genetic Resource for Food and Agriculture is a start, as it acknowledges the role and contribution that farmers have played in conserving and developing plant genetic resources. Parties have an obligation to protect and promote farmers' rights, including the right to save, use, exchange and sell farm-saved seed/propagating material. However, these rights are subject to national law. Implementation of farmers' rights at the national and international level is critical to ensure continued conservation and maintenance of agricultural biodiversity and associated AKST and provide an important counterbalance to the rights accorded to formal plant breeders under PVP and patents.

3.4 Trade and Technology Options

Pesticides and genetic engineering provide examples of technology options in agriculture that largely exemplify a flow of trade from developed to developing countries. On the other hand, fisheries, aquaculture and forest products are examples of technology options being implemented in developing countries and of the products being traded from developing to developed countries.

3.4.1 Composition of output and relationship to technology development

The rice market in Asia is less dominated by imports than it was two decades ago. Asia accounted for two-thirds of the global rice demand in 1970s, but this has come down to a third in the late-1990s (Tabor et al, 2002). This is due to the regional spread of HYV-rice, which has increased domestic production in most Asian countries, but has substantially reduced rice diversity.

The growth of the sugar industry in the developed countries, due to the development of technology to extract sugar from corn and beet, propped up by substantial subsidies, has almost eliminated Asia developing countries' possibilities of exporting sugar. With Asia as a whole being a labor-abundant region, it could be expected that comparative advantage in international trade would lie in the production of labor-using products, like vegetables, fruits and flowers, as against the less labor-using products, like cereals. Calculations for Bangladesh showed that the Domestic Resource Cost (DRC, i.e., the cost of all inputs, including land, labor, capital, used in production) in vegetables is only about 10% of the export rice, as against 60% for aromatic rice and more than unity for other rice. At the same time, in import price terms the DRC of other rice is also around 60 to 70%. Thus, while development of rice is beneficial in import substitution terms, it is not beneficial in export terms. Thus Bangladesh and most other Asian economies with similarly abundant labor have turned to export of vegetables, fruits and flowers. The production of these "new export crops" has grown across most countries of Asia.

It, however, is not only the more abundant and cheaper labor in ESAP developing countries that is the factor enabling Asia to undertake export production of fruits, flowers and vegetables. It is also depends on the advances in transport (containerization), packaging and communication technology (ICTs). The extent to which it is profitable to shift perishable agricultural commodities long distances depends on transport costs. As fuel prices rise, which they will by all indications, small differences in production costs might be neutralized by higher transport costs. Thus, while making use of the international trade possibilities currently available, countries may also find it necessary to consider alternatives in the event that fuel prices and transport costs rise substantially. The growth of demand in some agricultural commodities, however, has triggered some changes in technology or the widespread adoption of some technologies. This has been the case, for instance, in both fish and forest products. In fish there has been a shift from capture fisheries to culture fisheries. In 2002, Asia accounted for above 90% of the quantity and 70% of the value of aquaculture, both freshwater and marine (FAO, 2004b). This is a technology whose widespread adoption was induced by the shortages resulting from over-harvesting of wild fish.

Similarly, in the case of wood products and Non-Timber Forest Products (NTFP) there has been an initial depletion of natural stocks and then a shift to plantation of valuable species. Asia in 2005 accounted for more than 50% of plantation forests in the world (FAO, 2006). In a number of NTFP too collection from the wild has been replaced by culture or plantation as wild stocks have been depleted. A well-known example is that of orchids. Initially collected from the wild and with the growth of demand, subject to depletion, tissue culture has now replaced such collection in most countries and regions. Regions like Northeast India, however, still continue collection rather than tissue culture.

High prices of timber have stimulated the development of substitutes for wood in different uses, some using artificial substances, like plastic, other fast-growing species, like bamboo and still others, former waste material, like the trunks of aged rubber trees.

A broad conclusion can be drawn from these experiences. Initially increasing trade (both international and national) in agricultural commodities that are collected from the wild, led to over-exploitation of natural resources. But this has been followed by changes in both technology (aquaculture, plantation) and management systems (community-managed, or individual household-based in the place of open access systems) and the development of substitutes. There are positive examples of learning and technology development and systems of culture that have reduced pressure on natural stocks. They have also created new problems of waste management, environmental change, biodiversity conservation and increasing social inequality (Table 3-4).

3.4.2 Pesticides

At the most basic level, pesticides are intended to kill organisms; they include herbicides, insecticides and fungicides, as well as algicides, insect and animal repellents, antimicrobial and cleaning products, wood and material preservatives and insect and rodent traps. Besides harming target insects, weeds and fungi, pesticides also affect wildlife and human health. Some have immediate lethal effects including death, some cause acute illness at even minute levels of exposure and others have been found to cause chronic (long-term) health and environmental harm.

Today, organochlorine pesticides, organophosphorus pesticides, pyrethroids, herbicides such as 2,4-D, glyphosate and paraquat, and fungicides are commonly used. With increasing evidence of negative effects, efforts have been undertaken to ban or restrict some pesticides, but in general, their use in developing countries is still widespread.

National and global concerns over food security drove the intensification of agricultural production in the South, epitomized by the Green Revolution and the adoption of synthetic chemical pesticides. Pesticide reliance became widespread across much of Asia and Latin America, where the Green Revolution had been widely embraced (Rosset et al., 2000).

Table 3-4. *Asia's export share by product, 1980-91 to 2000-01.*

Item	1980-81	1990-91	2000-01
Tropical Products			
East Asia and Pacific	2.5	1.6	1.3
South Asia	0.9	0.8	0.5
Temperate Products			
East Asia and Pacific	3.7	3.2	3.0
South Asia	0.5	0.4	0.6
Seafood, fruits and vegetables			
East Asia and Pacific	3.0	5.1	5.7
South Asia	0.6	0.6	0.8
Other processed products			
East Asia and Pacific	2.5	1.7	1.8
South Asia	0.2	0.2	0.1
Total			
East Asia and Pacific	7.1	11.7	11.9
South Asia	2.1	2.0	2.0

Source: Aksoy and Beghin, 2006.

Thus the spread of Green Revolution-type agriculture throughout most developing countries was accompanied by a rapid rise in pesticide use (Rosset et al., 2000). Along with the CGIAR, the agricultural research and development agencies and universities of many countries focused on breeding seeds to increase plant uptake of nitrogen, so as to boost yields, which frequently required increasing pesticide use to control pest outbreaks.

However, promising increases of yield were offset by rising costs associated with increased use of chemical inputs. In the Central Plains of Thailand, yields went up only 6.5%, while fertilizer use rose 24% and pesticides jumped by 53%. In West Java, profits associated with a 23% yield increase were virtually cancelled by 65% and 69% increases in fertilizers and pesticides respectively (Rosset et al., 2000).

While multinational chemical companies based in the US or Europe account for the bulk of worldwide production and sales, local pesticide industries have also expanded, growing rapidly in countries favoring high input agriculture. For example, the pesticide industry in India is now the fourth largest in the world and second largest in the Asia-Pacific region after China. Estimates of its total market value vary between US$850 million and US$911 million. According to the Pesticides Manufacturers and Formulators Association of India, there are around 55 basic producers and 300 pesticide formulators, as well as numerous small-scale manufacturers. Around 200-odd generic pesticide products are made in India (CSE, 2001).

Pesticide manufacturers are the most direct drivers of pesticide use, acting on their own as well as through public agencies. They have increased pesticide sales through extensive marketing, advertising, supply to extension agencies or workers and local or district leaders and through partnerships.

Policy drivers include decisions by many developing countries to focus on export-led agricultural growth, which is typically accompanied by high pesticide use. Many governments also focused on increasing yield through adoption of Green Revolution technologies. Extension workers and government media channels like television and radio with high penetration into rural areas have been used to disseminate pesticide application related information. States shifted to a more "science-led" rather than farmer-led agriculture and also linked farmers' access to credit and capital to their acceptance of Green Revolution packages of seeds, fertilizers and pesticides. National quotas, priorities and directives for farmers were established in many regions (e.g., wheat and sugarcane in India, rice in Indonesia). National government research and extension systems removed farmers' decision-making power through direct state intervention in pest management via calendar spraying regimes and enforced control methods (Meir and Williamson, 2005).

Technological drivers include both public and private research and development of new technologies in seeds, machinery, fertilizers and pesticides. Institutional arrangements that contributed to the development of Green Revolution technologies included the international research community (e.g., CGIAR), the national agricultural research systems (NARs), academic institutions, research stations and the private sector. International donor agencies and bilateral agencies have also indirectly supported the spread of pesticides by supporting shifts towards Green Revolution technologies and/or have supplied pesticides directly in agricultural aid packages (Shiva, 1991; USAID, 2004).

International financial institutions such as the World Bank have contributed directly to increased pesticide dependence, traditionally providing them in fixed packages of inputs that farmers are required to use by the terms of their contract (Ishii-Eiteman and Ardhianie, 2002), or indirectly, by imposing structural adjustment conditions on borrower countries that require shifts towards high value export crops that result in increased pesticide dependence (Hammond and McGowan, 1992; Korten, 1995; Oxfam America, 1995; McGowan, 1997); by promoting intensified production without offering training in Integrated Pest Management (IPM) and leaving pest control advice up to pesticide companies (Hamburger and Ishii-Eiteman, 2003) or by providing emergency rehabilitation or reconstruction loans that encourage or promote increased pesticide use (Karel, 2004).

Recent external reviews of World Bank lending have found that a majority of projects likely to affect pesticide use failed to provide plans for introducing or implementing IPM in a meaningful way and were considered more likely to increase farmers' dependence on pesticides (Tozun, 2001; Karel, 2004). Past reviews also acknowledge the Bank's difficulty in implementing its IPM policy, but suggest that compliance is likely to improve in future (Liebenthal, 2002; Sorby, et al., 2003). The World Bank's "poor record of compliance" with its pest management policy has been linked to its practice of "actively open(ing) the door" to pesticide companies through programs geared towards modernizations of agriculture, liberalizations and privatizations (FAO, 2001). Nonetheless, other UN agencies like the FAO have helped the move towards IPM, providing examples of how developing countries have been able to adopt AKST beneficial to farmers in the face of powerful trade interests (see 3.4.3).

Partnerships and linkages between the pesticide industry and public agencies have also encouraged the opening of new markets for industry products. The French pesticide company, Rhône-Poulenc Agro, for example, joined a World Bank program in West and Central Africa in the late 1990s, in order to "break into the cocoa, coffee, rice and vegetable [pesticide] markets which account for around 40% of the crop protection market in [West Africa]" (Rhône-Poulenc, 1998).

Social drivers include perceived inefficiencies in low external input farming as compared to Green Revolution agriculture. Changing food consumption preferences and patterns, with a shift towards more meat and grain in many regions, have led to increased production of specific crops such as wheat and rice. As newer generations of farmers lost much of traditional AKST in countries that embraced the Green Revolution, they naturally resorted to the Green Revolution technologies that surrounded them (Shiva, 1991; Rosset et al., 2000; Meir and Williamson, 2005).

In China, the situation is slightly different. Self-sufficiency in food formed a central component of national policy. The agricultural systems focused on the use of external inputs and mechanization of agriculture to increase yields (Xiaoyun et al., 1997). Agriculture was characterized by extensive monoculture and use of HYVs, chemical fertilizers, pesticides and biotechnological products. The collectivization model of agricultural production was followed until the mid 1980s, after which the Household Production Responsibility System emerged (Xiaoyun et al., 1997; Wen, 2005), within which technological change has become the primary engine of agricultural growth.

China relied heavily on chemical fertilizers and pesticides to achieve short term yield gains. Central planning offices compelled the planting of Green Revolution crops, thus increasing the demand for pesticides to control the associated pest outbreaks (Xiaoyun et. al., 1997). Widespread loss of traditional AKST, including non-chemical approaches to pest management, occurred among peasant communities, who were required to adopt the collectivization model and expert advice of agricultural scientists (Hamburger, 2002).

Since 1975, the value of pesticide imports into China has grown from US$76 million to $293 million in 1994 (Pretty, 1995). A more recent spur for the growth of the Chinese pesticide industry has been the growth of pesticide exports and collaboration with multinational pesticide companies since the opening up of the Chinese economy. During this period, the Chinese Ministry of Chemical Industry signed cooperation agreements with Dupont, Ciba-Geigy, Bayer, BASF and Rhône-Poulenc and established joint ventures with Dupont, Ciba-Geigy, Zeneca and Agrevo to produce herbicides and insecticides. The Chinese government has also supported the pesticide industry by subsidizing importation of raw materials (although this type of assistance is decreasing quickly), tax exemption, lower costs for raw materials allocated through the central planning mechanism and preferential electricity rates and bank loans (US Embassy Beijing, 1996).

According to data from Nanshen Pesticide Company, China produced 250,000 tonnes of pesticide active ingredients in 1995, equivalent to 1.5 million tonnes of formulated product (PAN-UK, 1996). Data from 2000 indicate that China is the second largest producer in the world of agrochemicals by volume, of which 35% is exported (Dinham, 2005). In 2004, China's pesticide industry experienced high production and growth in exports (China CCM, 2005). Domestic use of pesticides in Chinese agriculture has continued to grow and China has become one of the primary exporters of cheap pesticides to Asian markets.

3.4.3 Technology choice for sustainable agriculture: integrated pest management

As the Green Revolution model of agriculture began to break down in ESAP, with increasingly evident health and environmental impacts, farmers, scientists and governments began to look for alternatives, including Integrated Pest Management (IPM). IPM is generally understood to focus on maintaining pest populations at economically acceptable levels through a systems approach that can include cultural practices, soil, field and habitat management, use of resistant varieties, biological and sometimes chemical control strategies (Shennan et al., 2005). Organic farmers have taken IPM a step further and have eliminated synthetic pesticides from farming practices. This is also known as non-pesticide management (NPM). Pest management among organic farmers can range from simple input substitution (e.g., use of biopesticides) to more comprehensive ecological approaches.

Advances in ecological understanding of pest population and community dynamics in rice fuelled the development of a more nuanced and comprehensive approach to pest management (Kenmore et al., 1984; Settle et al., 1996). FAO's paradigm-shifting work in Asia in the late 1980s provided (1) the scientific demonstration that pesticide-induced pest outbreaks were, at times, responsible for crop failures in rice; (2) the ecological evidence that removing pesticides would restore yields and system stability; and (3) the policy insight that a number of directives (e.g., ban on pesticides, removal of pesticide subsidies and national support for IPM) could transform the situation.

Participatory field-based educational processes in pest management replaced conventional "transfer of technology" methods (Röling and Wagemakers, 1998). IPM programs that utilize non-formal education methodologies and build on—rather than replace—farmers' traditional knowledge, have longer lasting success in farmers' adoption of and innovation in AKST, than training methods that disseminate fixed instructions for input use and pest control (Mangan and Mangan, 1998).

The IPM Farmer Field School (FFS) methodology pioneered in Southeast Asia typified this knowledge process and was subsequently adapted by governments, NGOs and farmers' associations. As such, IPM has evolved from a classical and technological insect management approach towards one in which the focus is on education and social change, whereby farmers develop the scientific research skills to test hypotheses and manage pest populations (Matteson, et al., 1994; Ooi, 1998). This, of course, is expensive; but it builds the knowledge of the farmers, which is the base of improvements in production.

Meanwhile, demand for pesticide-free, organic and fair-trade produce in export markets is growing and has created new markets for Southern producers (IFOAM, 2003), although farmers must negotiate complex and costly certification processes. Burgeoning consumer interest in "green" and "pesticide-free" products, particularly in countries with growing middle class populations (e.g., Thailand, China, India), has supported the emergence of new domestic markets that encourage transition towards IPM.

IPM has met with significant success in rice producing Asian nations like Indonesia, Vietnam, China, India and Sri Lanka (Pretty, 1995, 2001). Millions of farmers have reduced pesticide use through IPM, without experiencing reduced yields (Heong and Escalada, 1998; Mangan and Mangan, 1998; Barzman and Desilles, 2002). Yield advantages of IPM have been particularly strong in the South and thus have significant policy implications for food security in developing countries.

Some actors have questioned the ability of pesticide-free IPM methods—and sustainable and organic agriculture more generally—to produce adequate quantities of food. However, a growing body of literature demonstrates the high productivity of both organic and low-external input systems, particularly when the production of multiple outputs is calculated (Pretty, 2000; Pretty and Hine, 2000; FAO, 2002a; Parrot and Marsden, 2002).

The community-wide economic, health and environmental benefits of IPM have been widely documented. IPM Farmer Field Schools, in particular, have led to improved farm profitability and yields; significant reductions in pesticide use; improved occupational health, reductions in medical costs and lost working time caused by pesticide poisonings; reduced environmental harm; positive social impacts at the individual farmer and community level; better returns on government investments in extension and longer-term advances in food security (ter Weel and van der Wulp, 1999; Mancini, 2006; van den Berg and Jiggins, 2007).

It is clear that IPM is an example of AKST that not only provides an alternative to harmful pesticides, but that also brings benefits in its own right. The challenge is to mainstream its adoption, while providing the necessary policy support. A growing number of bilateral donor agencies are investing in ecological IPM strategies. The Global IPM Facility, FAO and EU have provided considerable technical and policy assistance to countries seeking to develop national IPM programs and to establish favorable policy environments.

3.4.4 Genetic engineering

Genetic engineering, also called modern biotechnology or genetic modification, is a departure from conventional breeding, involving the transfer of genetic material from one organism to another, often unrelated, species. This results in a transgenic organism containing new genes or novel combinations of genes.

The introduction of genetically engineered (GE) crops (biotech crops, genetically modified crops or transgenic crops) has been accompanied by controversy over the role of genetic engineering in addressing agricultural problems in both developing and developed countries. Advocates cite potential yield increases and reductions in pesticide applications, among other factors. Critics point to environmental and health risks and widening socioeconomic disparities as significant drawbacks.

Although GE technologies have the potential to affect both traded and non-traded products, most applications to date have involved highly traded agricultural commodities (Diaz-Bonilla and Robinson, 2001). Agricultural commodities such as soybean, maize and canola, for the purposes of food, feed or processing use, are the major genetically modified organisms (GMOs) that are currently traded internationally.

In addition, two GE traits, herbicide tolerance and insect resistance, have thus far dominated the market. In 2006, herbicide tolerant crops accounted for 68% of the global GE crop area, insect resistant (Bt) crops, 19% and stacked genes for the two traits, 13% (James, 2007). Almost four out of every five hectares of GE crops are engineered to withstand the application of proprietary herbicides sold by the same company that markets the GE seed and thus have little, if any, relevance to farmers in developing countries who often cannot afford to buy these chemicals (FOEI, 2007).

The major exporters of GE crops and their products are the US, Argentina and Canada, with Brazil recently joining the ranks. Analyses show that in 1961/1963 developing countries as a whole had an overall agricultural trade surplus of US$6.7 billion, but that this has gradually disappeared so that by the end of the 1990s trade was broadly in balance. The outlook to 2030 suggests that the agricultural trade deficit of developing countries will widen markedly, reaching an overall net import level of US$31 billion (Bruinsma, 2003). Given the current limited distribution and traits of GE crops, it is likely that the major GE crops importers will continue to be developing countries, with the exception of a few large agricultural developing country exporters. Furthermore, the cautious stance of the European Union towards GMOs and the overwhelming public opposition there, has led to the domestic market in the EU being largely GE-free, or at the very least, only allowing GMO products that are clearly labeled for consumer choice. This restricts the export market for GE crops.

The changing focus to trade in agricultural commodities and export-oriented agriculture may have serious ramifications for developing countries. As farmers and peasants directly link to the international market, economic forces increasingly influence the mode of production characterized by genetically uniform crops and mechanized and/or agrochemical packages (Altieri, 2003). This situation is expected to be aggravated by genetic engineering, whose development and commercialization is increasingly concentrated in a few corporations, accompanied by the increased withdrawal of the public sector as the major provider of research and extension services to rural communities.

Even if the rural poor benefit from GE crops, because GE crops are mainly traded cash crops, this benefit would be likely reduced. Technological crop improvements tend to lower the market price and therefore the value of the farmer's marketable surplus (Santaniello, 2003). Moreover, the great majority of GE crops cultivated today are used as high-priced animal feed to supply rich nations with meat.

GE crops have therefore not addressed the main agricultural problems and challenges facing farmers in most countries, neither have they proven to be superior to conventional crops (FOEI, 2007). It remains to be seen if large-scale production and trade of commodity GE crops has positively affected overall food security, although the opposite has been argued for some countries.

For example, in Argentina, one of the main exporters of GE soybean, adverse impacts have been observed, including the loss of food diversity and food sovereignty (Pengue, 2005) The export-oriented, commodity-production system is most likely to drive smaller farmers that are not able to face uneven competition out of business. Thousands of small- and medium-scale farmers in Argentina have been forced out of the production system, due to the expansion of GE soybean (Pengue, 2005). This phenomenon is not new or unique to Argentina. In many developing countries, due to historical and colonial inequalities, rural food-producing societies have been pushed off the best land most suitable for farming, into marginal areas (Rosset, 2005). The best lands were converted to production for export and this trend has continued post-independence. Land is increasingly concentrated in the hands of the wealthy, leaving the rural areas in many developing countries today characterized by extreme inequities in access to land, security of tenure and quality of land farmed.

The marginalization of the majority then leads to narrow and shallow domestic markets, leading land-owning elites to orient their production to export markets where consumers have purchasing power. In an ever-vicious cycle, elites become less interested in the well-being or purchasing power of the poor at home. By keeping wages and living standards low, this preempts the emergence of healthy domestic markets and thereby reinforces export orientation (Rosset, 2005).

The increased focus on agricultural export commodities, particularly GE crops, influences the type of AKST that is generated. The potential implications of technologies for agro-ecological stability and for sustainability and equitability have fundamental consequences for the planning of future agricultural research strategies (Bruinsma, 2003). Reluctance to challenge the belief that GE crops can benefit the small farmer and relieve world hunger has led to massive investments in GE technology to the neglect of other more promising but less glamorous approaches (Jordan, 2002). This has led to a disproportionate focus on GE research and investment into those technologies.

Already, in the last decade, national government and international donor support for agricultural research has declined significantly. While more and more funds go into biotechnology research, including GE, other key areas into agricultural alternatives, such as organic research, attracts only a fraction of investment compared to conventional and biotechnological approaches (Parrott and Marsden, 2002). Research in ecology and natural resource management, as well as socioeconomics, are trailing behind (Bruinsma, 2003).

Furthermore, a number of recent World Bank loans are facilitating the introduction of GE crops in Southern borrower countries (Ishii-Eiteman, 2002; Karel, 2004). Through these loans, the Bank is financing the research, development, field-testing and mass release of newly created transgenic crops (World Bank, 2002). Other Bank loans with implications for developing country uptake of GE technology have focused on introducing or revising IPR laws around genetic resources and/or have included research contracts or grants in support of biotechnology (World Bank, 1999ab; Karel, 2004).

While some analysts argue that all this means that more efforts should be made to redirect research focus towards public sector agricultural biotechnology research, including on genetic engineering (e.g., FAO, 2004a), others call for a reassessment of research priorities, so that more resources and research are directed towards alternative and proven approaches, that could better meet the needs of the poor, such as sustainable or organic agriculture, or agroecology (e.g., Jordan, 2002; Parrott and Marsden, 2002; Rosset, 2005).

In addition, a particular situation has developed with respect to research on GE crops. While there has been a large research focus on GE technology advances such as developing GE crops that may bring benefit, there has been rather less focus on biosafety research, i.e., looking at the health, environmental and socioeconomic risks. This is important, as in determining research priorities, it is critical to understand how new technologies, including GE, affect and influence the lives and livelihoods of the poor (Bruinsma, 2003). While the potential benefits need to be considered, so do the potential risks.

It is clear that any introduction of GE crops must assess not just potential health, environmental and socioeconomic impacts, particularly in the longer-term, but must also take into account structural, regulatory and economic evaluations that relate economic, political, social and scientific context of GE crops to their region of adoption.

3.4.5 Technology choice for sustainable agriculture: a pro-peasant research agenda

The increasing shift to private sector-driven, GE technology research and knowledge generation privileges farmers that can take advantage of GE crops and these are unlikely to be small or poor farmers in developing countries. Would GE crops be able to increase crop production and, at the same time repel pests, resist herbicides and confer adaptation to stressful factors commonly faced by small farmers? Thermodynamic considerations suggest that they cannot (Jordon, 2002).

Traits important to indigenous and small farmers (such as resistance to drought, suitable quality for food or fodder, competitive ability, performance on intercrops, compatibility with household labor conditions and more advantageous maturity, storage quality, taste or cooking properties, etc.) could be traded for transgenic qualities that may not be important to farmers (Altieri, 2003). Under this scenario, risk will increase and farmers may lose their ability to adapt to changing biophysical environments and to produce relatively stable yields with a minimum of external inputs, while supporting food security.

A pro-peasant research agenda comprises the following elements: creation of safeguards against homogenization and in situ conservation and rural development in GMO-free centers of origin (Altieri, 2003). The maintenance of pools of genetically diverse material, geographically isolated from any possibility of cross-fertilization or genetic contamination by uniform GE crops, is necessary as genetic uniformity or changes in the genetic integrity of local varieties could have considerable impacts. Moreover, biological and cultural diversity and the associated local skills and resources, are needed for rural populations to maintain or recover production processes.

Furthermore, the maintenance of traditional agroecosystems is a sensible strategy to preserve in situ repositories of crop germplasm (Altieri, 2003). However, this cannot be done in isolation from the maintenance of sociocultural organization, including of the need to organize small farmers into groups to strengthen their collective bargaining positions, particularly in facing corporate players (see 3.4.5). The process must be linked to rural development efforts that give equal importance to local resource conservation, food self-sufficiency and some level of market participation. In order for peasants to have a competitive edge, they need to be able to produce "unique" agricultural crops (i.e., GE-free) for niche markets. Such "uniqueness" is crucial for maintaining the stability of local farming systems in times of uncertainty.

AKST for sustainable agriculture should thus fully involve farmers and develop technologies that are low-cost, readily available and responsive to diverse local conditions, without posing risks, particularly to the diversity base of poor farmers. It is difficult to see how traded GE commodity crops can meet these criteria.

3.4.6 Fisheries and aquaculture

The liberalization of trade has led to a big increase in exports of fish and fish products from developing countries, as a whole and Asia in particular. Fisheries now generate more foreign exchange than any other traded food commodity, such as rice, coffee, tea or cocoa (FAO, 2004b) (see 2.2.4).

However, there are changes within this trade, though the geographical pattern remains the same. First, there is the shift from export of raw material to be processed in developed countries to export of processed fish. The development of fish processing capacity and knowledge in developing countries of Asia has enabled them to bring about a shift in the location of processing. The lower wages in Asian countries compared to the former processing countries (EU, Japan and US) has facilitated this shift in location. Moreover, the highly perishable nature of fish also favors the shift of processing to the source of raw materials. There is also a learning or capability-building process, whereby labor and management in Asia have learnt and invested in the technology of processing.

Second, there is also a shift to exports of live fish. Most of it is for ethnic markets in the developed countries. The migration of large numbers of Asians has led to the growth of a market for live fish from their countries of origin. Some of the live fish is also of the ornamental variety for aquariums. In both cases, the development of transport and logistics technology have enabled a growth in this sector of trade, which now accounts for about 10% of fish trade.

With the growing world fish trade and the possibility of reasonably elastic export earnings, there were initial trends towards over-exploitation of fish resources. At least 25% of fish varieties in the world are reported to be substantially over-exploited. There was an increase in the proportion of overexploited and depleted stocks from around 10% in mid-1970s to about 25% in early 2000s (FAO, 2004b). Besides bans on fishing, often brought about by the collapse of certain sectors, such as cod in the North Atlantic, there have also been technological shifts towards aquaculture, both of the freshwater and marine varieties. This is a very major technology change in response to the growing demand for fish along with the relatively fixed fish resource available (see 2.2.4).

While the conduct of aquaculture has its own problems, which will be dealt with later, it has certainly enabled a growth of production without endangering available stocks of wild fish, as trade based on capture fisheries tends to do. So far, in Asia, aquaculture has developed substantially for freshwater fish. Hatchery-based marine aquaculture is not as developed. Most marine aquaculture, as for prawn and seaweed, still depends on collection of seed from the wild. The jump to true aquaculture, with hatchery rearing of fry, has yet to be developed for many marine species. The type of marine aquaculture developed for salmon and trout has yet to be developed for the fishes of Asia. Recently, Japan has developed technology for sustaining bluefin tuna broodstock in offshore cages, leading to the first closed-cycle breeding of tuna. In Indonesia, fishers are replacing cyanide harvesting of reef fish with hatchery-raised juveniles of aquarium fish.

As pointed out earlier, most of the global fish trade is from developing to developed economies. There are some technology and production concerns arising from this specific nature of trade. There are also other concerns arising from other aspects of global trade.

Two concerns that arise from the developing to developed nature of the trade relate to the meeting of quality standards, particularly those of Sanitary and Phytosanitary Standards (SPS) (see 3.2). A not entirely unrelated matter is that of traceability, something insisted on by the developed country fish retail chains that have to contend with supplier responsibilities.

SPS problems have led to many temporary bans on imports of fish from Asian countries, particularly shrimp from various countries. Fish and fish products represented the largest category, above 25%, of food safety and quality alerts in the EU. Frequently there have been bans on imports of fish from various Asian countries. Initially they were met with cries of trade barriers. But after some time, the various Asian countries have begun to take measures to comply with these SPS standards. While they add to cost, the reduction of pesticide or veterinary drug residues or elimination of growth hormones are certainly desirable in themselves.

With a large part of Asian aquaculture being carried out in small farms, traceability is certainly a problem. But as an example from Bangladesh, the Noakhali Gold Project, shows, this can be tackled along with that of meeting SPS standards by linking groups of small producers with the larger processing and packaging units. The meeting of SPS and traceability, however, is more a matter of management methods than one of technology. This intervention was promoted by a donor-funded project.

Consumers in many parts of the world are concerned about the ecological impacts of different types of fishing and aquaculture. Endangered and charismatic species, like the sea turtle, often a by-catch (or collateral damage) of tuna fishing have aroused concern. This has led to the attempts to develop technologies that are more specific to the species to be harvested and eliminate or substantially reduce by-catch. The Convention on International Trade in Endangered Species of Wild Fauna and Flora (CITES) requires certificates of origin for cultured species on the endangered list, before they can be traded.

A trade issue that has come up recently and is likely to play a more important role in the future is that of "dumping" (see 3.2). With not only lower wages, but also more efficient and large-scale processing and production techniques, Asian countries are low cost producers in a range of fishery products. Their exports at sustained lower prices threaten livelihoods of producers in developed economies.

With growing demand for fish and fish products (fish is a superior good, in that its consumption increases as world per capita income rises), capture fisheries obviously cannot meet market demand. The attempts to increase the quantity of capture fish has frequently led to the collapse of various fisheries, most spectacularly North Atlantic cod fishery. Subsidies to fishing boats have contributed to increases in the over the limited catch capacity over which countries are competing. This is a classic case of the tragedy of the unmanaged commons, compounded by subsidies. Aquaculture has developed as a technology that can increase production beyond the natural limits of capture fisheries. It can help reduce pressure on wild stocks and thus help ecosystem rehabilitation. However, aquaculture too has its own ecological problems.

The best-known ecological problem is related to marine shrimp culture. For one, it has been based on collection of fry and juveniles from the wild, leading to an over-collection of such stocks. With numerous individuals collecting wild fry in an open access manner, there is no way in which sustainable harvesting limits can be maintained.

More important, however, has been the degradation of coastal environments by shrimp farming. Mangrove forests, important as the spawning ground of numerous species of fish, have been destroyed in the course of shrimp farming. Saline water intrusion has further degraded coastal lands. The inability to sustain productivity in shrimp farming has further led to the financial collapse of shrimp farming in many areas and a locational shift to other areas. More recently, attempts have been made to mitigate these negative impacts, through zoning and replanting of mangrove areas in Thailand.

There are other impacts of aquaculture, including marine aquaculture. There are concerns about escape of cultured into wild stocks, spread of pathogens from the former to the latter, discharge of effluents and solid wastes and so on. Some of these concerns have been met by improved site selection and improved management practices.

There is also the matter of dependence on fish meal prepared from "trash" fish, with the likely depletion of these stocks. This is the case for carnivorous fish, like salmon and shrimp. But, with the exception of marine shrimp, the bulk of aquaculture production in Asia comprised omnivorous/herbivorous fish, while 74% of aquaculture production in developed countries was of carnivorous species. There is a need to develop feed feeds that reduce dependence on fish meal. This would reduce dependence on capture fisheries.

With pressure to increase production per hectare of the earth's surface and to increase the incomes of small cultivators, there has been an extension of aquaculture into systems of rice monoculture. This extension tries to utilize the synergies between rice and fish, either in simultaneous or in alternating systems of cultivation. While this is a new system of production, it is, however, a new management practice rather than a new technology.

Fisheries has seen little of the genetic improvement of stocks to increase yields. These have yielded spectacular results in agriculture, with wheat yields going up by 50% and rice yields by 25%. In terrestrial animal management, there have been higher yields of milk or meat with genetically improved stocks.

In fisheries the attempt to genetically improve stocks, through selection and breeding, was first undertaken in the North Atlantic for salmon and trout. Such an attempt was then made for tilapia, a fish of African origin, but now widely cultured across Asia. The genetic improvement of tilapia was undertaken in the public sector, with the World Fish Center playing the leading role. This attempt resulted in what is known as GIFT (Genetically Improved Fish Tilapia), which was then distributed to various countries in Asia. The improvement in the rate of growing in GIFT as compared to other tilapia, however, was just 10%. Possibly this rate of increase in yield is not enough to result in its widespread adoption by small farmers, as the increase in yield could easily be negated by poor management or insufficient inputs.

The success of the GIFT project, however, illustrates that it is not necessary, as some argue, that research and development of new technology be undertaken in the private sector and that the incentives of high rents from patents or licenses are essential to provide the incentives for investment in research and development. Scientists and officials working in public sector institutions can as well develop new technologies.

While GIFT itself is in the public domain, it is intriguing that its further development has been handed over to a Norwegian private sector company. Having made the necessary initial breakthrough there seems no reason why the further development of GIFT could not have been left also in the public domain.

The question of the relation between public and private domains is also raised by a new marine based medical development. Australian firms are testing the use of brown seaweed (*Undaria pinnatifida*) as an anti-viral agent, including its use in treating HIV. The medical use of *Undaria* is an established practice in the Korean peninsula. This possibly is the original knowledge on the basis of which an innovation is being developed. What should be the relation between the original knowledge and the likely subsequent patent?

3.4.7 Forestry

In the ESAP region, a net loss of forests of about 792,000 ha per year in the 1990s was reversed into an annual gain of 1 million ha, largely due to increased plantation activity in the region, particularly in China. However, in South and Southeast Asia there continued to be an annual net decline in forest area of about 2.7 million ha per year (see 2.2.3).

With the growth of trade in timber, there was an initial

(up to the 1990s) increase in wood extraction from forests. The revenue from timber was an important source of central government revenue in a number of countries, like the Philippines and Indonesia. This revenue was largely used for accumulation and infrastructure development outside of the forest areas themselves. The indigenous peoples who lived in and around these forests were marginal in the political equations of these countries. Consequently, forest revenue was not reinvested within the forest-based communities themselves, but outside for national interests. Extraction did not take account of sustainability considerations. Clear-felling resulted in the destruction of trees that were not the object of extraction and thus increased the extent of deforestation.

The extraction of timber was later followed by the transformation of forests into plantations, the best example being palm oil plantations in Malaysia and more recently in Indonesia. This has, however, had implications for biodiversity and the ecological services that forests supply. Along with this, there was the development of tissue culture in order to propagate trees on a large scale.

The one country in which wood extraction was linked to local ownership of the forests is China. After the late-1970s reform, large areas of forest were handed over to either local collective ownership (i.e., village-based) or household ownership under the household responsibility system. Of course, under neither system of collective or household ownership did the owners have the right to sell or mortgage the forests, i.e., they had the right to manage the forest and the right to the income from it, but could not dispose of the forest. This truncated property right meant that forests could not become real estate, subject to speculation as, for example, urban land is.

The result of this property reform was that the immediate owners had an interest in sustainably harvesting the timber from the forests. This has led to China playing a leading role in the development of plantation forests, i.e., a subset of planted forests consisting primarily of introduced species. Thus the technology change, from extraction from natural forests to the planting of forests for extraction, came about not just through the increased demand made by economic growth and trade but also by the change in property or institutional system. Had there been merely a change in availability due to depletion of natural forests, as has occurred in most other countries, there would have been just a shift in the origin of supply of timber to other countries that still had available forests. There need not have been a shift to plantation forests, as has not happened in India.

The shift of supply, however, is also one of the responses of the Chinese and Indian markets. Both countries have instituted some forms of "logging bans" in the aftermath of devastating floods in the late 1990s and have shifted to importing timber to meet local demand. The difference between the two is that, as pointed out above, in China there has been a growth of plantation forests and in India there has not been such a change.

Further, with the ownership of the forests in the hands of the state, the external diseconomies that are borne by forest-dwellers are not part of the relevant cost economies. This problem can be taken care of by the above-mentioned institutional change whereby property rights are allotted to the forest-dwellers.

There is yet another problem of costs that is related to technology change. As mentioned, the costs of extraction are the relevant costs so long as extraction is the method of production, if it can be called that. But such extraction, in the face of growing external demand and the attempts of the producers to maximize their own incomes, leads to the depletion of the resource, certainly where the resources are extracted in an open access property situation. This has been frequently observed in the case of medicinal plants and herbs. There is often a sudden increase in demand, as the modern pharmaceutical industry discovers ways of utilizing indigenous knowledge to develop new kinds of drugs and medicines. This then leads to an increase in demand for the raw material, which the forest dwellers also collect in the largest possible quantities in order to maximize their own incomes and a consequent rapid depletion of the resource.

The way out of this situation has often been the domestication and thus planting of the required plant material. This then means a shift in production from collection to growth or culture of the plant material. This technology shift is necessary to be able to provide supply in a sustainable manner. This has occurred in numerous Non-Timber Forest Products (NTFP). A change in technology enables a shift to sustainable production for the market. Deforestation continues at an alarmingly high rate, but that the net loss of forest area is slowing down thanks to forest planting, landscape restoration and natural expansion of forests on abandoned land (Steinfeld, 2006).

Forests are increasingly being conserved and managed for multiple uses and values and they play a crucial role in climate change mitigation and in the conservation of biodiversity and of soil and water resources. If managed sustainably, forests also contribute significantly to local and national economies and to the well-being of current and future generations.

While Asia reported a decrease in wood removals in recent years, Africa reported a steady increase. It is estimated that nearly half of the removed wood was fuelwood. The question of fuel use, specifically that of use of wood for fuel, is usually seen in energy models as a function of household income. This is incorrect. Sample studies have shown that in many areas the proportion of wood as fuel does not go down with income in rural areas (Nathan and Kelkar, 1997). This is so when fuel is not purchased on the market, but is collected with household labor. This household labor is basically that of women and it has a low opportunity cost, i.e., its possible alternate uses yields very low income. It is this low-cost or even non-costed women's labor that is the cost to the household. As a result, unless income earning opportunities for women increase and their opportunity cost goes up, there is not much saving of women's labor in such activities.

In order to bring about the desired energy transition, i.e., away from wood to other commercial fuels, what is crucial is the increase in the income earning opportunities for women. This, for instance, has happened on a large-scale in China where women are part of what is called the labor force, participate in income earning activities. As against this, in India, the participation of women in the labor force is much lower in many parts, even higher income areas, but there is still a high reliance on collected fuel (Nathan and Kelkar, 1997).

Along with the entry of women into the labor force, the household shift from collected to purchased fuel seems to play a role in the energy transition. Where a household begins to buy fuel, then the economies of inferior goods comes into play and as income rises, the proportion of wood as fuel goes down.

This analysis points to the importance of entry into the commercial world of trade, both as income earning producers and as buyers of fuel, for bringing about a change in technology adoption. A technology change, from wood as fuel to gas or other commercial fuels, depends on the gendered economic factors of women's income earning and household purchase of fuels.

3.4.8 Organic agriculture and fair trade

There are increasing opportunities in organic and fair trade products, which are emerging as important niche markets that are growing at a high rate around the globe. Asia alone has 20 countries growing organic produce with 60,000 enterprises and 0.6 million ha in organic cultivation, which is 15% of all farms and 2.6% of total area under organic farming worldwide (Raynolds, 2004).

Organic and fair trade movements contribute not only to environmental and economic sustainability, but also help rural livelihoods in a sustainable manner. Organic farming is one form of sustainable agriculture with maximum reliance on self-regulating agroecosystems (Browne et al., 2000).

In globalized markets, whether or not local producers can gain access to global value chains and at which point, is likely to be an important factor in determining whether they will benefit from trade liberalization (Eapen et al., 2003). This has meant that the access of developing countries to enter developed world markets is dependent on their ability to enter the global value chains or production networks of lead firms. The newly emergent organic produce supply chains tend to exclude small producers due to reasons of high certification costs, smaller volumes produced and tighter control by the chain leaders in the absence of any local market outlets for the organic producers (Raynolds, 2004; Singh, 2006a).

There is therefore a need to mainstream organic and fair trade movements to ensure the participation of large number of producers in developing countries in these markets, without bringing in the ills of conventional chains. Thus, there is need for policy thrust and support for such market-oriented sustainability and livelihood initiatives.

It is argued that organic production is suited for small farmer participation as it is labor intensive and compatible with traditional peasant practices. However, export of organic products involves certification, documentation, record keeping and auditing which makes it industrial in nature and counters the traditional norms and practices of peasant producers. Also, price premiums are likely to decline as economies of scale are attained in marketing and the supply base expands at a rate unmatched by market expansion (Krissoff, 1998).

The organic value chains are very complex due to the process importance in being organic. But, the farmers and the laborers are the weakest links in the chains driven by importers, exporters and retail chains (IDS, 2003; Kabeer and Mahmud, 2004). It is only the fair trade and alternative trade networks which still provide some scope for participation of the small and marginal organic producers (Yussefi and Willer, 2003; Raynolds, 2004).

Further, in international markets, increasingly, organic trade and ethical and fair trade concerns are beginning to overlap (Raynolds, 2004). An increasing number of fairly traded goods are also organic (70%) and the organic movement is moving towards including social rights and ethical trade in its standards. If there is consumer pressure for this overlap, then there would be considerable implications for the volume of trade, the developing country producers' ability to meet the requirements and for the working conditions and livelihoods of producers (Browne et. al., 2000). Whereas ethical trade is people centered, environment focused and animal centered, the fair trade approach emphasizes partnerships with producers for improving the status of disempowered groups through alternative trading organizations. It works though Self Help Groups (SHGs) for provision of fair price to primary producers, with focus on gender equality, market access and long term relationship (Tallontire, 2001).

The exclusion of small farmers from participating in global food chains does not appear to be, in any way, automatic. There have been cases of success when public or private assistance to the growers in terms of technical assistance and supply of input credit was made available. In some places in Brazil, small dairy farmers have gone for collective tanks to meet the scale requirement, though the large farmers will still have an advantage, as they do not face the transaction cost involved in collective use of physical assets. The dairy companies and cooperatives encourage the use of collective tanks, even by financing or facilitating credit for milk producers in some cases (Farina, 2002). Similarly, National Dairy Development Board (NDDB) in India is implementing a clean milk production program with price incentives, in a small dairy cattle holder context.

Market access for small producers depends on (1) understanding the markets; (2) organization of the firm or operations; (3) communication and transport links and (4) an appropriate policy environment (Page and Slater, 2003). Insofar as the role of the government in the commodity chain is concerned, it can proactively help the stakeholders in the chain to identify the opportunities and threats in the global commodity chains. It can also assist producers to enter the chains (Kaplinsky, 2000). If, in a given country, a few chains command majority of the organic sector, then development policies and programs need to learn how to deal with this handful of big companies.

However, it is equally important to promote good business practices that optimize retailer-supplier relations, protecting both sides. This can be initiated by establishing or improving contract regulations and business rules of practice some of which are already available in the form of legal provisions in the US and Argentina. These practices can also be forced by private sector codes of practice. These changes and the basic requirements they impose on growers are conditions that will have to be met if the growers are to be able to tap the powerful market of the supermarkets. Therefore, it is crucial that government and donor agencies help small farmers and entrepreneurs to make the investments in equipment, management, technology, commercial practice and the development of strong and efficient organizations to meet the market requirements.

Global buyers can have a role to play in assisting suppliers to improve practices and become compliant. The standards need to be flexible and interwoven with local conditions if they are to benefit poor workers. They must also involve local stakeholders who reflect the interests of workers in the process of standards setting and monitoring. The policy challenges on standards include standard setting, monitoring compliance, providing assistance to achieve compliance and sanctions on non-compliance. Much depends on how standards are implemented, monitored and verified (IDS, 2003).

Thus, major conditions for successful interlocking between agribusiness firms and small producers include increased competition for procurement instead of monopsony, guaranteed market for farmer produce, effective repayment mechanism, market information for farmers to effectively bargain with companies, large volumes of transactions through groups of farmers, for lowering transaction costs and no alternative source of raw material for firms (Kristen and Sartorius, 2002).

The main requirements of small farmers in this changing environment are better access to capital and education. Management capacity is as important as physical capital but is the most difficult thing to provide. Further, collective action to deal with scale requirements needs to be designed to satisfy new product and process standards or to avoid exclusion from the supply chain. Collective action through cooperatives or associations is important not only to be able to buy and sell at a better price but also to help small farmers adapt to new patterns and much greater levels of competition (Schwentesius and Gomez, 2002).

Small farmers also require professional training in marketing and in technical aspects of production. There is also a need to strengthen small farmer organizations and provide technical assistance to increase productivity for the cost competitive market, provide help in improving quality of produce and to encourage them to participate more actively in the marketing of their produce in order to capture value added in the supply chain.

On the other hand, regulation of supermarket chains to control or mitigate their market power can be an effective tool to ensure the presence of small growers in value chains as seen in the case of the banana trade regime in the pre-WTO period in the EU policy, single channel (monopoly) exports by producer bodies in some exporting countries like South Africa and regulation of domestic import markets in France (Gibbon, 2003).

Though there are concerns about the ability of the small farms and firms to survive in the changing environment of agribusiness, there are still opportunities for them to exploit product differentiation with origin of product or organic products and other niche markets. However, the major route has to be through exploitation of other factors such as external economies of scale through networking or clustering and such other alliances like contract farming (Kirsten and Sartorius, 2002). The experience of contract farming across the globe suggests that it is not the contract per se which is harmful as a system but how it is practiced in a given context. If there are enough mechanisms to monitor and use the contract for developmental purposes, it can certainly lead to a betterment of all the parties involved, especially small and marginal farmers (Singh, 2006b).

3.4.9 Livestock

International trade accounts for only 8-13% of total production of livestock products. It is high in bovine and poultry meat and milk and low in pig meat. Livestock and Livestock Products (LLPs) account for about 1/6th of value of all agricultural trade. Meat exports make up about half of this total value with bovine, pig and poultry meat as three major types. The subsectors of pig meat and poultry meat have grown by 6 and 14% during the last decade. Dairy products account for 1/3rd of value of LLP exports and have grown at the rate of 3% during the last decade (FAO, 2001).

Major global players in exports of LLPs are Australia, USA, Canada and EU in beef and pork and Brazil, EU, China and Thailand in poultry (Perry et al., 2005). Developing countries are net importers of LLPs with dairy products and poultry dominating the scene. Though least developed countries have more pastureland per head of rural population than in the developing countries, but stocking density and meat production per animal are lower. Still LLPs account for 4% of their GNP. Thus they may have comparative advantage in small ruminant production but productivity is lower (FAO, 2001).

Further, a large part of the global trade in livestock products especially dairy is intra-industry trade which is under the control of large global players. In many countries in Asia, export trade is with a few large players who are able to meet new quality standards like SPS measures. Still, even these large players have suffered from SPS restrictions in some markets e.g., UAE banning Indian meat imports from 10 firms by name for not meeting the hygiene standards. There is also growing vertical coordination of the sector especially in chicken and pigs where large processors and retailers work directly with primary growers of such products who essentially provide all inputs and the grower gets wages for his/her labor and supervision costs. In some countries like Thailand, there is almost 100% contract production in poultry and piggery sectors. In this process, small growers are getting marginalized due to small scale and lack of bargaining power due to lack of effective producer organizations. In some places like India, producers' organizations have been able to bring large number of milk producers under a common platform and are significant players in domestic markets and even are foraying into global markets.

Like crop sector, subsidies to producers in developed countries, especially EU, remain an obstacle to fair trade in livestock products. Even in ESAP, there are large exporters i.e., Australia and New Zealand which are globally competitive and do not subsidize their livestock producers. Once subsidies in the developed world go, developing world can benefit from freer trade in livestock products. Dairy exports are likely to increase from Oceania, South Asia, South

America, Eastern Europe and Southern Africa. But, due to SPS measures, developing countries including those from ESAP face compliance costs related to export standards especially in meat and meat products (FAO, 2001; Perry et al., 2005). Within the ESAP developing world, net exporting countries like Thailand, press for reduction of trade barriers in developed countries and low income net importers like Bangladesh or Indonesia promote import substitution, wherever possible.

Though extensive grazing is still dominant, intensification and industrialization (known as "factory farming," especially in poultry and piggery) is growing even in the developing world to achieve higher efficiency in production as there is growing problem of direct competition for raw materials between food for humans and feed for animals/birds. There is also the threat to wild species due to competition for feed from commercialized livestock systems. There has been loss of genetic diversity in livestock which is a threat to the sustainability of the livestock sector. It is estimated that one livestock breed a month has become extinct over the last seven years. In Vietnam, the percentage of indigenous sows declined from 72 in 1994 to 26 in 2002. Of its 14 breeds, five are vulnerable, two in critical state and three face extinction. Similarly, in Kenya, introduction of Dorper sheep has eliminated the pure bred Masaai sheep. This is important for developing world as though less productive, many breeds at risk of extinction are unique in their characteristics which may be useful to deal with challenges like climate change, animal diseases and rising demand for specific products.

There have been many technological breakthroughs like cross breeding, Artificial Insemination (AI) and Embryo Transfer (ET) in the sector and the potential of biotechnology is immense not only in primary production but also in livestock product processing and value addition. Indigenous knowledge and its practice in livestock rearing, animal husbandry and hygiene maintenance come in as alternatives to the growing "factory farming" system. But, the dominance of trade by large players may not attend to these concerns unless there is market pressure on them. The environmental, animal welfare and intellectual property rights issues in LLPs are also becoming crucial to deal with for better benefits from trade as they present both threats and opportunities for developing countries.

Due to many problems in conventional supply chains like mad cow disease and foot and mouth disease, there is emerging organic or natural livestock products market which combines principles of ethical trade as well by focusing on ethical treatment of animals (Steinfeld et al., 2006). Organic dairy products have emerged as an important component of the livestock products market wherein dairy and poultry have shown greater growth rates than beef and pork. In USA, there is even certified organic livestock production in many states with eggs and dairy being the fastest growing sectors. In Europe, it is EU, Austria, France and Denmark which have large production of organic dairy and other livestock products. In Latin America, it is Brazil and Argentina which had significant organic livestock activity. On the other hand, in ESAP region, organic activity is not that widespread in livestock sector.

The success stories in LLP exports were the result of strong private sector contributions of capital and state-funded support such as the NDDB in India. This support led to improved management and entrepreneurial drive, export of value-added products instead of live animal exports and vertically integrated or coordinated systems, which included small-scale producers and a strong focus on marketing. Thailand is a good example of moving from frozen poultry meat to cooked products after the avian influenza outbreak (Perry et al., 2005). The vertical coordination was achieved through contract production or corporate production.

But recent market pressure to require that livestock products come from environments free of animal infectious diseases rather than simply testing for product safety might tip the balance away from contract farming by smallholders and negatively impact rural livelihoods (Perry et al., 2005). There is also need to carry out assessments of different models of vertical coordination to assess their impacts on livelihoods of primary producers.

The newer issues of animal identification and traceability, differential safety infectivity of live animals versus products and animal health status at product source, besides product certification and animal welfare are challenges that have to be met in order to benefit from trade in LLPs. Capacity building would be important to meet global and other quality norms and more participation in standards setting bodies is required for developing countries to benefit from global LLP trade.

In ESAP, Australia and New Zealand, with their high volumes of livestock production and high per capita incomes, could take the lead in capacity building for the region. They could also pioneer research to reduce the GHG emissions related to current methods of livestock production.

3.5 Environmental, Health and Social Dimensions in Trade Agreements

3.5.1 Trade, environment and sustainable development

The relationship between trade and environmental, health and social dimensions, as well as with sustainable development, is complex. Actions in one area affect the other areas, directly or indirectly. Any impact assessment of trade in agricultural products would depend on which perspective is used as the starting point, whether it is environmental protection, or resource management and biodiversity conservation, or health concerns, or trade. Another issue to take into account would be whether short-term or long-term considerations are being examined.

While environmental, health and social dimensions are acknowledged to be important, they are often perceived as potentially conflicting with trade objectives (see Koester, 2001). In this regard, there is a need to move from a simplistic and selective "balance and trade-offs" approach, which cannot deal with complex realities, towards a more holistic approach, which implies a complex integration of the various perspectives mentioned above, with recognition that there will be conflicts of interests requiring policy decisions that are in favor of long-term ecological and economic sustainability, human/animal health and safety, social justice, cultural rights and ethics.

The WTO's legally-binding rules impact on the economic and social well-being of a WTO Member and its dispute settlement system and enforcement mechanism (including trade sanctions) make the WTO a powerful body when compared to the United Nations which also has legally binding treaties on environment and natural resources management and on social issues such as the ILO Conventions.

Therefore it is not surprising that "WTO-inconsistent" allegations are often made against environmental negotiators or WTO Members seeking to take strong national environmental or health or social measures at the international level. For example, in recent multilateral environmental agreements (MEAs) such as the Cartagena Protocol on Biosafety, there were intensive negotiations over the hierarchy of agreements (Mackenzie et al., 2003). Major developed countries that are producers and exporters of genetically modified organisms (GMOs) wanted trade agreements to prevail over MEAs. Developing countries and some developed countries, such as Norway and the European Union, wanted to ensure the supremacy of MEAs. The result is the approach of "mutual supportiveness" between trade agreements and MEAs, with a stated preambular paragraph affirming the equal status of all the agreements.

In practice and because of the WTO's formal and enforceable dispute settlement system, this could have the effect of creating a legal hierarchy through its decisions with respect to United Nations agreements, which was actually not the intention of countries that negotiated the trade agreements and the establishment of the WTO. Thus, the struggle between trade on the one hand and environmental, health and social dimensions on the other hand, continues.

3.5.2 Trade at any cost?

However, the WTO is not about "trade at any cost" even though the policy freedom of Members has been reduced. WTO agreements have a context for trade. For example, the preamble of the Marrakesh Agreement Establishing the World Trade Organization (1994) affirms " . . . the objective of sustainable development, seeking both to protect and preserve the environment. . . ." A number of WTO agreements also provide for various types of review and amendments, such as the Agreement on Trade-Related Aspects of Intellectual Property Rights (see 3.3.1).

Article XX of GATT (1994), which provides general exceptions to trade liberalization, is of crucial importance. This is because the body of WTO-related rules does not contain general exemptions of an environmental nature, nor does it provide a special status for MEAs. Article XX of GATT contains several general exceptions, among them for trade-restricting measures (a) "necessary to protect human, animal and plant life and health"; and (b) "relating to the conservation of exhaustible natural resources if such measures are made effective in conjunction with restrictions on domestic production or consumption".

This means that WTO Members may adopt or enforce measures for these purposes, even though they restrict trade. There are, however, conditions for measures (including import bans) taken under Article XX. First, there must be no "arbitrary or unjustified discrimination between countries where the same conditions prevail". Thus a Member cannot put restrictions (on health or environmental grounds) on an imported product, without having the same restrictions on similar domestic products. Secondly, the restrictive measures must not be "a disguised restriction on international trade". Thus, there is scope for WTO Members to take protective measures and to restrict trade of certain products, including agricultural products, for environmental and health purposes.

Despite the exceptions and special provisions in the WTO agreements, these are not enough and have limited scope in ensuring environmental protection, sustainable resource management and the safeguarding of human, animal and plant health. The current international trading system is also not able to ensure social equity; while inequalities between the developed world and Asia have gone down, there has been an increase in inequality within countries (Macgillivray, 2006).

Where there are possible conflicts between the WTO and other agreements, the situation raises even more concerns, as it could mean that the WTO could be effectively adjudicating on those other agreements. The WTO Dispute Settlement Mechanism cannot be the judge of non-WTO Agreements and may not be the best way to resolve disputes in these important areas of policy-making (Shaw and Schwartz, 2005). The difficulties were evident in the recent dispute led by the United States against the European Communities on the European approval procedures for GMOs. Although the WTO Dispute Panel did not rule on the legality of the procedures or on the right of national governments to ban GMOs or to take restrictive measures, the case illustrated the inappropriateness and even discomfort of the trading system in dealing with biosafety (and hence, environmental, health and socioeconomic) issues (Bernasconi-Osterwalder and Oliva, 2006; FOEI, 2006; Lim and Lim, 2006; Palmer, 2006).

Therefore, MEAs and other social development instruments with their own compliance mechanisms are necessary (e.g., the Cartagena Protocol on Biosafety has a Compliance Committee) to ensure that these agreements are implemented fully. Trade forums are not appropriate to be the judge and arbiter of sustainability.

3.5.3 Standards for environmental, health and social dimensions

It is important to recognize the validity of other standard setting bodies such as MEAs. For example, during the negotiations of the Cartagena Protocol on Biosafety, many countries wanted a provision on the setting of international biosafety standards under the Protocol. Major developed countries such as the United States, Canada, Australia and Japan rejected this, arguing that standard setting bodies such as the Codex Alimentarius Commission, the International Office of Epizootics and the bodies of the International Plant Protection Convention would be sufficient.

The compromise was Article 2(5) of the Cartagena Protocol: "The Parties are encouraged to take into account, as appropriate, available expertise, instruments and work undertaken in international forums with competence in the area of risks to human health."

Therefore, the standards set in UN MEAs such as the

Convention on Biological Diversity and the Cartagena Protocol on Biosafety are arguably legitimate and thus actions taken under these MEAs are WTO-consistent. This is indeed what the European Communities argued in the WTO dispute on biotech products; it implied that MEAs such as the Cartagena Protocol are setting international standards and that its regulatory processes are consistent, with both WTO rules and the Protocol (Shaw and Schwartz, 2005).

The issue of trade and labor standards is highly controversial. The WTO Agreements do not deal with any core labor standards. But some industrialized countries believe that the issue should be studied by the WTO as a first step toward bringing the matter of core labor standards within its ambit. WTO rules and disciplines, they argue, would provide a powerful incentive for Member nations to improve workplace conditions.

On the other hand, many developing and some developed countries believe the issue has no place in the WTO framework. These countries argue that efforts to bring labor standards into the arena of multilateral trade negotiations are little more than a smokescreen for protectionism. Many developing countries believe that the campaign to bring labor issues into the WTO is actually a bid by industrialized nations to undermine the comparative advantage of lower wage trading partners.

In 1996, after heated discussions, WTO Members identified the ILO as the competent body to deal with labor standards. WTO members said they were committed to recognized core labor standards and that these standards should not be used for protectionism. The economic advantage of low-wage countries should not be questioned. The WTO and ILO secretariats were asked to continue their existing collaboration. There is currently no work on the subject in the WTO.

It is apparent then that when dealing with the interface of trade and social dimensions such as labor standards, that the WTO is not the appropriate forum. Nonetheless countries must have the adequate policy space to implement labor standards and the ILO Conventions, in order to promote opportunities for women and men to obtain decent and productive work, in conditions of freedom, equity, security and dignity.

On the other hand, the effort to include socioeconomic considerations in the Biosafety Protocol was strongly resisted by developed countries. The result is a general provision in Article 26, as follows: (a) A decision on import under the Protocol or under its domestic measures implementing the Protocol, may take into account socioeconomic considerations arising from the impact of LMOs (living modified organisms, the term used for GMOs in the Protocol) on the conservation and sustainable use of biological diversity, especially with regard to the value of biological diversity to indigenous and local communities; and (b) The Parties are encouraged to cooperate on research and information exchange on any socioeconomic impacts of living modified organisms, especially on indigenous and local communities.

It would be important for developing countries to conduct research and studies to contribute to this international process of research and exchange of information among governments, international and NGOs on the socioeconomic aspects of GMOs (Second Conference of the Parties serving as the Meeting of the Parties to the Protocol, 2005). At the national level, decision-making on GMO policy and specific GMOs would also greatly benefit from such studies. Many countries allow for socioeconomic considerations to be taken into account when taking a decision on whether or not to allow the import of a GMO into the country.

The proliferation of bilateral and regional free trade agreements (FTAs) in ESAP countries may have implications for national policy space, making it more difficult for governments to implement and enforce environmental, social and health protective measures. Of particular concern are the FTAs between developing countries and developed countries like the United States. These North-South FTAs are very comprehensive in scope and extend into the realm of domestic policies (Gibbs and Wagle, 2005).

The investment chapter of US FTAs, for example, includes provisions on expropriation and mechanisms for investor-state dispute settlement. These have proved to be problematic in the NAFTA (North American FTA, which has been in force for more than 10 years) context, as foreign investors have successfully challenged government activities and public policies, such as those aimed at environmental protection (Gibbs and Wagle, 2005). It is not inconceivable that health or social measures may also be affected. Furthermore, FTAs that include compensation provisions for expropriation of investment by direct or indirect means could lead to claims against government regulations aimed at enhancing public welfare or protecting the environment, if they are perceived to affect an investor's profitability.

3.5.4 Pollution havens

Different countries have different environmental standards. These differences could be used in international trade systems to export products meant for disposal, for instance, to countries where environmental standards are particularly lax, so-called pollution havens. There have been instances of ships sent to Bangladesh or India for breaking up, not having the hazardous substances removed and dealt with in the originating country.

Ship-breaking, to take this example, creates many jobs in Bangladesh, India, etc. This is based on the lower labor costs involved in these developing countries. But the ship-breaking activities can still be carried out in developing countries with lower labor costs, with prior removal and proper disposal of hazardous substances in the originating countries. Obviously this would be more expensive that the export of these substances to "pollution havens". But examples show (e.g., that of the French aircraft carrier Clemenceau, that had to taken back to France for removal of asbestos and other hazardous substances, before being re-sent to India for breaking up) and general economic analysis would bear out, stricter responsibilities for disposal of hazardous substances, which is also likely to be a more capital-intensive activity, can be combined with jobs for labor-intensive activities, like breaking up ships. Along with action, often initiated by press and civil society organizations, to have more stringent environmental standards in existing "pollution havens", there could also be a role for international coordination of environmental standards to deal with disposal of hazardous substances.

3.5.5 Technology choices

When we look at the range of AKST and associated technologies, on what then should we base our decisions as to whether a particular technology is appropriate? It cannot be just on the basis of trade considerations. A holistic assessment of technology requires the careful and comprehensive examination of environmental, health, safety, legal, socioeconomic and ethical dimensions. It also requires an understanding of the short, medium and long-term effects of a technology.

Concurrently, there is the possibility of a reform of international and national trade laws and policies where necessary, with courage and political will among decision-makers and implementation, with political will and commitment, of international environmental agreements and social development instruments. Finally, ensuring effective public participation and monitoring to ensure compliance with sustainable development principles, laws and programs can help guide policy-makers.

3.6 Climate Change and Trade

3.6.1 Asia in the global climate change equation

Developing Asia's economic growth has largely been based on carbon-based technologies, developed in an era of cheap carbon. Though the per capita emissions of developing Asia are still much below the levels of the USA or Europe, yet the large size of the economies means that total emissions from developing Asia are very large.

In the pre-Kyoto discussions it was argued that the industrialized North were responsible for carbon emissions and hence it was these countries that should take action to reduce carbon emissions (Agarwal and Narain, 1991). Along with this it was proposed that the developing countries should be given incentives to adopt carbon-efficient technologies, through trading based on per capita rights.

The carbon-intensive growth of developing Asia has changed the global equation with regard to actions for reducing carbon emissions. The developing world as a whole now accounts for almost 50% of annual carbon emissions. China is the second largest emitter, after the USA; while India is the world's fifth-largest emitter. Further, land use change resulting in deforestation itself accounts for 20 to 25% of global emissions, with Brazil and Indonesia being the two largest emitters.

In designing policies for mitigating climate change or reducing carbon emissions, three factors now stand out. First, the developed countries bear historical responsibility for the magnitude of the problem; there is question of global justice in distributing burdens for reduction of carbon emissions. Two, without the involvement of the developing countries, particularly the large economies of China and India, not much of a dent can currently be made on the scale of emissions. Third, sectors of agriculture, such as forests (through conversion of land for agricultural use) and livestock also contribute substantially to global greenhouse gas emissions.

3.6.2 International trade instruments for environmental objectives

There is often, even usually, more than one way of producing a commodity, the negative environmental effects of which are different. But the price of the commodity would be the same, irrespective of the method used in its production. For instance, coffee grown in the shade of existing forests would sell for the same price as coffee grown in plantations. If the output of the latter process were higher, then the net income from the environment-friendly coffee process would be lower than from the environment-unfriendly plantation process. From the side of the producers there would be a disincentive to carry on the environment-friendly process.

One approach to this problem would be that of using the "polluter pays" principle in international trade. A tax or import duty could be imposed on each commodity, depending on the amount of carbon emitted in its production, the extent of forest clearance carried out, the loss of biodiversity through the production process and so on. The more negative externalities involved in a production process, the higher would be the import duties on its product. This requires a recognition that processes to produce a product can have different effects and that a product's effects are not restricted to its quality in use.

The "shrimp-turtle" case (in which the WTO panel ruled that the US had a right to take action to conserve exhaustible resources and could require the use of turtle-extruder devices in harvesting shrimp) provides a precedent for extending trade measures, import duties or even prohibitions to cover various environmental externalities in production processes (Stiglitz, 2006). There could import duties for carbon emissions, loss of biodiversity, forest clearing and so on. Such an import duty, based on direct and indirect carbon use in production, could also be used as a manner of dealing with "free riders" who do not subscribe to international agreements on GHG emissions. The result would be to favor commodities produced with environment-friendly instead of unfriendly processes and higher costs for unfriendly processes. The tax would be paid by those who use environment-unfriendly processes and those who consume the resulting products. Such a tax on the production of negative externalities could make international trade somewhat more environment-friendly than it currently is. Low carbon-using processes, e.g., that of the Chinese village of Liuminying, which has developed an integrated gas-energy-fertilizer system, based on animal and field waste, would then have a price advantage over similar products of more carbon-using technologies; or bird-friendly coffee in managed agro-forests would be cheaper than sun-coffee in plantations (ICRAF, 2006).

Adding carbon taxes is also likely to make certain commodities les amenable to international trade. Transport to more carbon-using destinations, such as those covered by jet transport, is likely to become less profitable than transport to less carbon-using destinations. This will promote low food-mile destinations over high food-mile destinations, affecting the existing pattern of international trade. Further, high-value commodities will be less affected than low-value commodities. Prices will go up of, say, cut flowers, which is likely to reduce demand for the same. This would affect developing Asia's export of cut flowers.

Utilizing import duties and other trade instruments in order to bring various negative (and positive) environmental externalities into the picture, would require building an accounting framework for environmental factors, something in which some progress has been made; but a lot still remains to be done (Daly and Cobb, 1989; McDonough and Braungart, 2002; ISAR, 2004; Bainbridge, 2007).

As agreed at Rio, there is a "common but differentiated" responsibility for reducing carbon or GHG emissions. The historical responsibility of the developed countries in having used up most available global carbon space means that they should bear the major burden of reducing emissions. At the same time, the large developing economies, such as China and India, also need to undertake measures to reduce emissions in order to make an impact on global emissions. A system of a global carbon tax (akin to the proposed Tobin tax on hot capital movements) could be instituted, along with a system to redistribute the revenues from the carbon tax. There is no reason why the country that pays the carbon tax should get the revenue. In fact, it should be the other way around and counties that emit the least should benefit the most from the revenue. The carbon revenue could then be distributed on the basis of both population and per capita incomes,

A carbon tax would have the disadvantage in that it does not, by itself, set a limit to the total emissions. It would induce technological change, by making more carbon-intensive products and processes more expensive than less carbon-intensive products and processes. But there is no necessary limit to total emissions. On the other hand, a system of tradable emissions, with total emissions and its distribution, set through international decisions could have the same technology effect, but also set a cap on total emissions. In this case too, the tradable emission quotas could be distributed positively with population and negatively with per capita income. A negative relation of tradable emission quotas would mean that countries with higher per capita incomes, which have already more than used up their proportion of global emission space, would get less than developing countries, which have much lower per capita income and also lower per capita emissions.

There are various proposals in discussion on linking trade with climate change. But to be acceptable and workable, it would seem that a proposal needs to be based on an equitable distribution of burdens, based on both historical and present positions.

3.6.3 Carbon markets

In the Kyoto Accord, targets were set for the developed countries to cut emissions, along with provision for carbon trading through the so-called Clean Development Mechanism (CDM). The carbon market, as it has since developed, has three components: (1) project-based transactions in the CDM, where the buyers purchase additionality; (2) trading of greenhouse gas emission allowances under the cap-and-trade regimes as in the EU; and (3) voluntary carbon market, as in the US and Australia (World Bank, 2006b). The carbon market was a $325 billion market in 2005.

The CDM has shifted the emphasis on making the transition to a low carbon economy from polluting industries in the developed countries to industries in the developing countries, where the costs of such transformations are supposedly lower. In ESAP, China and India have been the main beneficiaries of CDM payments. This, however, does not result in any change in emissions from the developed countries, for whom it is a "business as usual" situation. Further, doubts have been raised about whether any real additionality has been achieved through CDM projects (see UNCTAD, 2006a; Carbon Trade Watch, 2007).

With regard to the EU emissions trading system, two points of criticism have emerged (World Bank, 2006b). First, the allowable emissions for each country have been set very high and therefore there has been little need to trade in or reduce emissions. In fact, the high level of carbon allowed resulted in a crash in the European carbon market, where the price of a tonne of carbon fell from $30 in 2000 to just $2 in 2006. Second, emission rights have been given free to industries, in what has been called a "grandfather" approach, i.e., as a patrimony. Instead of paying for emissions, polluters are given polluting rights as property (Carbon Trade Watch, 2007). This does not put any pressure on them to reduce emissions.

The carbon trade approach has not worked to stimulate investment in renewable-energy technologies. Again, as prices of carbon-using commodities are not affected, there is pressure to switch to a low carbon economy. As discussed below, another and probably more effective approach would be that of imposing a tax on carbon emissions.

3.6.4 "Avoided deforestation" in carbon trades

In the current carbon trading system, carbon offsets are granted for additional growth of forests. Under the Clean Development Mechanism (CDM) of the Kyoto Protocol, payments can be made for reclaiming land to forests. But this does not take into account the incentive to clear existing forests—for the timber they provide or to convert the land to other uses, such as oil palm plantations, or, as is likely given the current emphasis on bio-diesel, to plantations for sugarcane or corn to produce ethanol or *jatropha* plantations.

A 15-country coalition of rainforest nations, led by Papua New Guinea (see www.rainforest.coalition.org) has proposed a change in the method of carbon credits for forests to include payment for "avoided deforestation." Such avoided deforestation has an opportunity cost, in terms of livelihoods foregone. This opportunity cost needs to be compensated in order to provide an incentive to maintain existing forests intact. Taxes on carbon emissions can be used to pay small landowners, local communities and indigenous peoples to keep their forests intact, as is done in Costa Rica.

The introduction of the notion of opportunity costs in terms of livelihoods foregone is a shift from the Kyoto concern with simple costs of technologies. In the Kyoto-system, the costs of reducing a ton of carbon could be lower in the developing countries, when compared to developed countries. Consequently, a large part of CDM trade involved purchasing offsets from developing countries. But besides the cost of utilizing there is another notion of cost that comes into the picture, that is, of opportunity costs or the livelihoods foregone.

The method of financing such an "avoided deforestation" initiative could be of a number of different types, including payments out of a carbon tax, or even from a new environmental financing facility, based on Special Drawing Rights (SDRs). These SDRs could be distributed not, as now, on the basis of existing credits with the IMF but on a combination of per capita income, population and the country's existing emissions (or non-emissions). The notion of the opportunity cost of livelihoods foregone in computing social costs (Coase, 1960) can be combined with that of the declining marginal utility of income as income increases, to argue (see Chichilnisky and Heal, 2000; Nathan, 2003) that the distribution of rights can be proportionately higher for low income countries or peoples, such as indigenous peoples.

3.6.5 Market for biofuels

The market for biofuels, while growing is still quite small when compared with the market for fossil fuel. Trade in ethanol, the major biofuel, was 3 billion liters in 2004, as against crude oil trade of 920 billion liters. But with various governments taking measures to increase use of biofuels (both China and India have policies for biofuels to account for at least 5% of total fuel consumption by 2015), the market for biofuel can only grow. The imposition of a carbon tax will, of course, give a strong boost to the market for biofuels.

Brazil is the main exporter of biofuel, ethanol. Its main export markets are the USA and India. The other internationally traded biofuel is palm oil. The palm oil consortium, headed by Malaysia, has a policy of subsidizing the use of palm oil as biofuel, whenever the price of palm oil falls in the market (Roundtable on Sustainable Palm Oil, www.rspo.org) In the early years of this decade there has been a surge of palm oil exports for biodiesel to the EU (UNCTAD, 2006a).

There are a number of issues that come up in this emerging biofuels market. First, is that of the conversion of forest lands into biofuel plantations. Such conversion would reduce the carbon-reducing impact of biofuels and needs to be taken into account. It could also lead to an increase in the prices of food and thus reduce well-being of buyers of food. The second is that of the role of communities and small farmers or corporations. Forms of technical and financial assistance may be required to enable local communities, including forest-dwelling indigenous peoples and small farmers to benefit from the growing biofuels market. Without such safeguards the benefits of this new market could end up being monopolized by the large corporations and thus reducing its likely contribution to poverty-reduction in developing Asia.

3.6.6 Options

The options discussed above (carbon trade, biofuels, compensation for avoided deforestation through a global fund, taxes on carbon and other environmental factors, tradable emissions and the required environmental accounting) together amount to a substantial shift (even a paradigm shift) in thinking on the interaction of trade and environmental issues. The big question mark is over whether the existing sets of institutions of international trade and finance can formulate and implement the required policies, or whether a new set of institutions (supranational, national and local) will be required to manage the new economic-ecological paradigm, which brings together economic and ecological issues, rather than separate them, as has so far been the basis of international trade. With a business as usual approach, there is the very real likelihood of a worldwide depression, greater in intensity than that of the 1930s (Stern, 2007). The challenge before the global economy is whether the necessary measures and the likely institutional changes will be brought about only after such a crisis strikes, or whether these steps can be taken in advance of and thus mitigate or lessen the likely effects.

References

ActionAid. 1999. Crops and robbers: Biopiracy and the patenting of staple food crops. Actionaid, London.

ActionAid International. 2006. Undue influence: Exposing undue corporate influence over policymaking at the World Trade Organization. ActionAid, London.

Agarwal, A., and S. Narain. 1991. Global warming in an unequal world: A case of environmental colonialism. Centre Sci. Environ., New Delhi.

Aksoy, M. A. and J.C. Beghin (ed) 2005. Global agricultural trade and developing countries. Manas Publ., New Delhi and World Bank, Washington DC.

Alam, G. 2004. State of the Indian farmer: Technology generation and IPR Issues. A millennium study. Vol. 5. Academic Foundation, New Delhi.

Altieri, M.A. 2003. Socio-cultural aspects of native maize diversity. Paper prepared for the Secretariat of the Commission for Environmental Cooperation of North America. As part of the Article 13 initiative on maize and biodiversity: The effects of transgenic maize in Mexico. Available at http://www.cec.org/files/PDF//Altieri-e.pdf.

Antara News Agency. 2000. Asian farmers ask WTO to maintain protection of agri products/ Jakarta, Indonesia, 19 April 2000.

Atkinson, R.C., R.N. Beachy, G. Conway, F.A. Cordova, M.A. Fox, K.A. Holbrook et al. 2003. Public sector collaboration for agricultural IP management. Science 301:174-175.

Bainbridge, D.A. 2007. True cost environmental accounting for a post-autistic economy. Post-autistic Econ. Rev. 41:23-28. Available at http://www.paecon.net/PAEReview/issue41/Bainbridge41.htm.

Barzman, M. and S. Desilles. 2002. Diversifying rice-based farming systems and empowering farmers in Bangladesh using the farmer field school approach. *In* N. Uphoff (ed) Agroecological innovations. Earthscan, London.

Bérard, L., and P. Marchenay. 1996. Tradition, regulation and intellectual property: local agricultural products and foodstuffs in France. p. 230-243. *In* S.B. Brush, and D. Stabinsky (ed) Valuing local knowledge: Indigenous people and intellectual property rights. Island Press, Washington DC.

Bernasconi-Osterwalder, N., and M.J. Oliva. 2006. EC-Biotech: Overview and analysis of the Panel's interim report. Centre Int. Environ. Law (CIEL), Washington DC.

Bhagwati, J. 2001. Intellectual property protection. 14 February 2001. The Financial Times. Available at http://www.columbia.edu/~jb38/FT%20Letter%20on%20IPP.pdf.

Bhatt, T.P. 2003. Globalization of anti-dumping and its impact. Foreign Trade Rev. 38(1-2):54-95.

Bhattacharyya, B., and M.D. Gupta. 2001. Anti-dumping—road to reforms. Focus WTO 3(1):1-9.

Brookfield, H., C. Padoch, H. Parsons, M. Stocking (ed) 2002. Cultivating biodiversity: Understanding, analysing and using agricultural diversity. ITDG, London.

Browne, A.W., P.J.C. Harris, A.H. Hofny-Collins, N. Pasiecznik, and R.R. Wallace. 2000. Organic production and ethical trade: Definition, practice and links. Food Policy 25(1):69-89.

Bruinsma, J. (ed) 2003. World agriculture: Towards 2015/2030. An FAO perspective. Earthscan, London.

Brush, S. 1994. Providing farmers' rights through in situ conservation of crop genetic resources. A report to the Commission on Plant Genetic Resources, Univ. California, USA.

Carbon Trade Watch. 2007. The carbon neutral myth. Available at www.carbontradewatch .org

CBD. 2000. Cartagena Protocol on Biosafety to the Convention on Biological Diversity: Text and annexes. Sec. Convention on Biological Diversity, Montreal.

Chawla, N.K., and R Kumar. 1997. International food laws and their implications for the Indian dairy industry. Indian Dairyman 49(5):107-114.

Chichilnisky, G., and G. Heal. 2000. Who should abate carbon emissions? An international viewpoint. *In* G. Chichilnisky and G. Heal (ed) Environmental markets: Equity and efficiency. Columbia Univ. Press, NY.

China CCM. 2005. 2004-2005 Annual report on China's pesticide market. Available at http://www.chinaccm.com/4S/4S04/4S0401/news/20050509/171928.asp.

CIPR. 2002. Integrating intellectual property rights and development policy. UK CIPR, London. Available at http://www .iprcommission.org/papers/pdfs/final_report/reportpdf.pdf.

Coase, R. 1960. The problem of social cost. J. Law Econ. 3 Oct 1960.

Corea, G. 1992. Taming commodity markets: The integrated programme and the common fund in UNCTAD. Manchester Univ. Press, UK.

CSE. 2001. Endosulfan conspiracy. Available at http://www.cseindia.org/html/endosulfan/endosulfan_page1.htm. Centre for Science and Environment, India.

Daly, H.E., and J. Cobb. 1989. For the common good: Redirecting the economy toward community, the environment and a sustainable future. Beacon Press, Boston MA.

Dasgupta, P., and K.-G. Maler. 1990. The environment and emerging development issues. Proc. World Bank Ann. Conf. Dev. Econ. World Bank, Washington D *In* C.

Diaz-Bonilla, E., and S. Robinson. 2001. Biotechnology, trade and hunger. p. 7-10. *In* IFPRI 2000-2001 Ann. Rep. Biotechnology: Two perspectives. Available at http://www .ifpri.org/pubs/books/ar2000.htm. IFPRI, Washington DC.

Didier, P. 2001. The WTO anti-dumping code and EC practice—issues for review in trade negotiations. J. World Trade 35(1):33-54.

Dinham, B. 2005. Corporations and pesticides. *In* J. Pretty (ed) Pesticide detox: Towards a more sustainable agriculture. London, Earthscan.

Diop, N., and S.M. Jaffe. 2006. Fruits and vegetables: Global trade and competition in fresh and processed product markets. *In* M.A. Aksoy, and J.C. Beghin (ed) Global agricultural trade and developing countries. Manas Publ., New Delhi and World Bank, Washington DC.

Eapen, M., J. Jeyaranjan, K.N. Harilal, P. Swaminathan, and N. Kanji. 2003. Liberalisation, gender and livelihoods: The cashew nut case. Working Pap. 3—Phase 1: Revisiting the cashew industry. IIED, London.

FAO. 2001. Mid-term review of the global IPM facility, April-June. FAO, Rome.

FAO. 2002a. Organic agriculture, environment and food security. FAO, Rome.

FAO. 2002b. World agriculture: Towards 2015/2030. Summary report. FAO, Rome.

FAO. 2004a. The state of food and agriculture, 2003-04. Agricultural biotechnology: Meeting the needs of the poor? Agric. Ser. 35. FAO, Rome.

FAO. 2004b. The state of world fisheries and aquaculture 2004. Available at http://www .fao.org//docrep/007/y5600e/y5600e00.htm. FAO, Rome.

FAO. 2006. Global forest resources assessment 2005: Progress towards sustainable forest management. FAO, Rome.

Farina, E.M.M.Q. 2002. Consolidation, multinationalisation, and competition in Brazil: Impacts on horticulture and dairy products systems. Dev. Policy Rev. 20(4):441-457.

Feenstra, R. 1998. Integration of trade and disintegration of production in the global economy. J. Econ. Perspect. 12(4):31-50.

Finger, J.M. 2002. The Doha agenda and development: A view from the Uruguay Round. ADB, Manila.

FOEI. 2006. Looking behind the US spin: WTO ruling does not prevent countries from restricting or banning GMOs. Friends of the Earth Int., Amsterdam.

FOEI. 2007. Who benefits from GM crops: An analysis of the global performance of genetically modified (GM) crops 1996-2006. Available at http://www.foei .org/publications/pdfs/gmcrops2007full.pdf. Friends of the Earth Int., Amsterdam.

Fukuda-Parr, S. 2007. Human rights and national poverty rights reduction strategies. Economic rights working paper 4. Human Rights Inst., Univ. Connecticut.

Gepts, P. 2004. Who owns biodiversity, and how should the owners be compensated? Plant Physiol. 134:1295-1307.

Ghate, J.P. 1998. Anti-dumping and safeguards actions in the perspective of World Trade Organisation. Manage. Account. 33(7):530-533.

Gibbon, P. 2003. Value chain governance, public regulation and entry barriers in the global fresh fruit and vegetable chain into the EU. Dev. Policy Rev. 21(5-6):615-625.

Gibbon, P., and S. Ponte. 2005. Trading down: Africa, value chains and the global economy. Temple Univ. Press, Philadelphia.

Gibbs, M., and S. Wagle. 2005. The great maze—regional and bilateral free trade agreements in Asia: Trends, characteristics, and implications for human development. UNDP Asia-Pacific trade and investment initiative, Colombo.

Goodman, P. 2002. Article on China's agriculture. International Herald Tribune, 26 Sept. 2002.

Gupta, R.K. 1996. Anti-dumping and countervailing measures—The complete reference. Sage, New Delhi.

Hamburger, J. 2002. Pesticides in China: A growing threat to food safety, public health, and the environment. China Environ. Ser. 5:29-44.

Hamburger, J., and M. Ishii-Eiteman. 2003. The struggle to reduce reliance on pesticides in World Bank projects: Can community-based monitoring lead to improved policy compliance? Pesticide Action Network North America, San Francisco.

Hammond, R., and L. McGowan. 1992. The other side of the story: The real impacts of World Bank structural adjustment programs. The Development Gap, Washington DC.

Harilal, M.S. 2006. Sculpting for a global market: Initiatives and hurdles for indian ayurvedic sector to participate in global gains. Pap. presented at the 5th Development Convention of the ICSSSR institutes in South India held at the CMDR Dharwad, 27-28 April.

Henson, S., and R. Loader. 1999. Impact of sanitary and phyto-sanitary standards on developing countries and the role of the SPS agreement. Agribusiness 15(3):355-369.

Heong, K.L., and M.M. Escalada. 1998. Changing rice farmers' pest management practices through participation in a small-scale experiment. Int. J. Pest Manage. 44:191-197.

Hoekman, B., and K. Anderson. 1999. Developing country agriculture and the new trade agenda. Policy Res. Pap. 2125. Dev. Res. Group, World Bank, Washington DC.

ICRAF. 2006. RUPES report. ICRAF, Nairobi.

IDS. 2003. The Cost of Compliance—Global Standards for Small Scale Firms and Workers. IDS Policy Briefing 18. Inst. Dev. Studies, Brighton.

IFAD. 2001. Assessment of rural poverty, Asia and the Pacific. IFAD, Rome.

IFAD, 2002. Country program evaluation report for Sri Lanka. IFAD, Rome.

IFAD. 2005. Organic agriculture and poverty reduction in Asia: China and India focus. Rep. 1664. IFAD, Rome.

IFOAM. 2003. Organic trade a growing reality. An overview and facts on worldwide organic agriculture. IFOAM, Bonn.

ISAR. 2004. A manual for the preparers and users of eco-efficiency indicators. Intergovernmental working group of experts on international standards of accounting and reporting. Vol 1. UNCTAD, Geneva.

Ishii-Eiteman, M. 2002. Pesticides and pest management. *In* Marketing the earth: The World Bank and sustainable development. Report available at www.halifaxinitiative.org. Friends of the Earth/Halifax Initiative, Canada.

Ishii-Eiteman, M., and N. Ardhianie. 2002. Community monitoring of integrated pest management versus conventional pesticide use in a World Bank project in Indonesia. Int. J. Occup. Environ. Health 8(3):220-231.

ITF. 1999. Dealing with commodity price volatility in developing countries: A proposal for a market-based approach. Int. Task Force Commodity Risk Manage. Developing Countries (IITF). The World Bank, Washington DC.

James, C. 2007. Global status of commercialised biotech/GM crops: 2006. ISAAA Brief 35-2006. Executive Summary. Available at http://www.isaaa.org/Resources/Publications/briefs/35/executivesummary/default.html.

Janaiah, A. 2006. Why a rethink on GM crops is needed. The Hindu Business Line, Mumbai, April 12.

Jordan, C.F. 2002. Genetic engineering, the farm crisis and world hunger. BioScience 52:523-529.

Joshi, P.K., A. Gulati, P.S. Birthal, and L. Tewari. 2004. Agriculture diversification in South Asia. Econ. Polit. Weekly 39:2457-2467.

Kabeer, N., and S. Mahmud. 2004. Globalisation, gender and poverty: Bangladeshi women workers in export and local markets. J. Int. Dev. 16:93-109.

Kabeer, N. 2005. Gender equality and women's empowerment: a critical analysis of the third Millennium Development Goal. Gender Dev. 13:13-24.

Kaplinsky, R. 2000. Globalization and unequalisation: What can be learned from value chain analysis? J. Dev. Studies 37(2):117-146.

Karel, B. 2004. The persistence of pesticide dependence: A review of World Bank projects and their compliance with the World Bank's pest management policy, 1999-2003. Pesticide Action Network North America, San Francisco.

Kaul, H.N. 1997. Anti-dumping protectionist measures. UNI Econ. Service 19(15):133-138.

Khor, M. 2002. Intellectual property, biodiversity and sustainable development. Zed and Third World Network, Penang.

Kirsten, J., and K. Sartorius. 2002. Linking agribusiness and small-scale farmers in developing countries: Is there a new role for contract farming? Dev. Southern Africa 19(4):503-529.

Koester, V. 2001. Cartagena Protocol: A new hot-spot in the trade-environment conflict. Environ. Policy Law 31(2):82-94.

Kortbech-Olsen, R. 2001. Organic agriculture: Definition and overview of trends in world markets. ITCUNCTAD-WTO, Geneva.

Korten, A. 1995. A bitter pill: Structural adjustment in Costa Rica. Inst. Food Dev. Policy, San Francisco.

Krissoff, B. 1998. Emergence of U.S. organic agriculture—can we compete? Discussion. Am. J. Agric. Econ. 80(4):1130-1133.

Kryder, R.D., S.P. Kowalski, and A.F. Krattiger. 2000. The intellectual and technical property components of pro-vitamin A rice (GoldenRiceTM): A preliminary freedom-to-operate review. ISAAA, Ithaca NY.

Kumar, U., and S. Sahai (ed) 2003. Status of the rights of farmers and plant breeders in Asia/ Gene Campaign, New Delhi.

Lesser, W., and M. Mutschler. 2002. Lessons from the patenting of plants. p. 103-118. M. Rotschild, and S. Newman (ed) Intellectual property rights in animal breeding and genetics. CABI, UK.

Liebenthal, A. 2002. Promoting environmental sustainability in development: An evaluation of the World Bank's performance. Oper. Eval. Dep. World Bank, Washington DC.

Lim, L.C., and L.L. Lim. 2006. Summary of the conclusions and recommendations of the WTO Dispute Panel interim report on GMOs. TWN Briefings (#5) for the Third Meeting of the Parties to the Cartagena Protocol on Biosafety, Curitiba, Brazil, 13-17 Mar 2006. Third World Network, Penang.

Macgillivray, M. (ed) 2006. Inequality, poverty, and well-being. UNU-WIDER, Palgrave Macmillan.

Mackenzie, R., F. Burhenne-Guilmin, A.G.M. La Vina, J.D. Werksman et al. 2003. An explanatory guide to the Cartagena Protocol on Biosafety. IUCN, Gland, Switzerland.

Mancini. F. 2006 An evaluation of farmer field schools in the cotton areas of southern India: Impacts on farmers' health, the environment, labor use, farming systems, and livelihoods. PhD. Thesis. Wageningen Univ., The Netherlands.

Mangan, J., and M. Mangan. 1998. A comparison of two IPM training strategies in China: The importance of concepts of the rice ecosystem for sustainable insect pest management. Agric. Human Values 15:209-221.

Matteson, P.C., K.D. Gallagher, and P.E. Kenmore. 1994. Extension of integrated pest management for planthoppers in Asian irrigated rice: Empowering the user. p. 656-668. *In* R.F. Denno and T.J. Perfect (ed) Ecology and management of planthoppers. Chapman and Hall, London.

Mattoo, A., and A. Subramanian. 2000. India and the multilateral trading system post-Seattle: Defensive or proactive? Paper presented at the World Bank-NCAER Workshop on South Asia and the WTO, New Delhi, 20-21 Dec 1999.

Mauro, F., P.D. Hardison. 2000. Traditional knowledge of indigenous and local communities: international debate and policy initiatives. Ecol. Applic. 10:1263-1269.

McDonough, W., and M. Braungart. 2002. Cradle to cradle: Remaking the way we make things. North Point Press, NY.

McGowan, L. 1997. Democracy undermined, economic justice denied: Structural adjustment and the aid juggernaut in Haiti. The Development Gap, Washington DC.

Meir, C., and S. Williamson. 2005. Farmer decision-making for ecological pest management. *In* J. Pretty (ed) The pesticide detox: Towards a more sustainable agriculture. Earthscan, London.

Mitchell, D.O. 2005. Sugar policies: An opportunity for change. p. 141-160. *In* M.A. Aksoy and J.C. Beghin (ed) Global agricultural trade and developing countries. Manas Publ. and World Bank, New Delhi.

Nagpal, S. 1999. Food security in the Hindu Kush Himalayas, Economic and Political Weekly 34:2717-2720.

Narintarakul, K., and B. Silarak. 2005. Thai-China free trade agreement: Who benefits? *In* C.C. Bamford and A. Bamford (ed) Free trade agreements and their impact on developing countries: The Thai experience. p. 79-118. FTA Watch, Nonthaburi.

Nathan, D. 2003. Environmental services and the case for local forest management. *In* D. Nathan et al. (ed) Globalization and indigenous peoples in Asia: Changing the local-global interface. Sage, New Delhi.

Nathan, D. and G. Kelkar. 1997. Wood energy: The effect of women's unvalued labour. Gender Tech. Dev. 1(2):205-224.

Oh, C. 2000. IPRs and biological resources: Implications for developing countries. Briefing Pap. Third World Network, Penang.

Ooi, P. 1998. Beyond the farmer field school: IPM and empowerment in Indonesia. Gatekeeper Ser. 78. IIED, London.

Oxfam America. 1995. The impact of structural adjustment on community life: Undoing development. Oxfam, Boston.

Oxfam International. 2003. Rigged rules and double standards. Trade, globalisationa dn the fight against poverty. Oxfam, UK.

Oxfam International. 2005. Kicking down the door—How upcoming WTO talks threaten farmers in poor countries. Briefing Pap. 72 Oxfam, London.

Paarlberg, R.L. 2000. Governing the GM crop revolution: Policy choices for developing countries. FAE Disc. Pap. 33. IFPRI, Washington DC.

Page, S., and R. Slater. 2003. Small producer participation in global food systems: Policy opportunities and constraints. Dev. Policy Rev. 21(5-6):641-654.

Palmer, A. 2006. The WTO GMO dispute: Implications for developing countries and the need for an appeal. Gene Watch UK, Roy. Soc. Protect. Birds (RSPB), Forum Biotech. Food Security, and GM Freeze, London.

Panagariya, A. 1999. The WTO trade policy review of India, 1998. The World Econ. 22(6):799-824.

Panchmukhi, V.R. 2001. World Trade Organisation and India—Challenges and perspectives. *In* G.K. Chadha (ed) WTO and the Indian Economy. Deep & Deep, New Delhi.

Parrott, N., and T. Marsden. 2002. The real Green Revolution: Organic and agroecological farming in the South. London: Greenpeace Environ. Trust, London.

PAN UK. 1996. Pesticide use in China: A health and safety concern. Pesticide News 34.

Pengue, W.A. 2005. Transgenic crops in Argentina: The ecological and social debt. Bull. Sci. Tech. Society 25:314-322.

Perry, B., A.N. Pratt, K. Sones, and C. Stevens. 2005. An appropriate level of risk: Balancing the need for safe livestock products with fair market access for the poor. PPLPI WP No. 23. ILRI, Nairobi.

Philip, L.M., and A. Jenniah. 2006. The Indian cotton dilemma: Caught in the international cotton mesh. Oxfam, New Delhi.

Pingali, P. 2005. From the green to the gene revolution: How will the poor fare? Indian J. Agric. Market. 19(1):1-16.

Popkin, B.M. 1993. Nutritional patterns and transitions. Popul. Dev. Rev. 19(1):138-157.

Posey, D.A. 2002. Commodification of the sacred through intellectual property rights. J Ethnopharmacol 83:3-12.

Potrykus, I. 2001. Golden rice and beyond. Plant Physiol. 125:1157-1161.

Pretty, J. 1995. Regenerating agriculture: Policies and practice for sustainability and self-reliance. Earthscan, London.

Pretty, J. 2000. Can sustainable agriculture feed Africa? Environ. Dev. Sust. 1:253-274.

Pretty, J. 2001. The rapid emergence of genetic modification in world agriculture: Contested risks and benefits. Environ. Conserv. 28(3):248-262.

Pretty, J., and R. Hine. 2000. The promising spread of sustainable agriculture in Asia. Nat. Resour. Forum (UN) 24:107-121.

RAFI. 1997. Conserving indigenous knowledge: Integrating two systems of innovation. RAFI and UNDP, NY.

Raman, M. 2004. Effects of agricultural liberalization: Experiences of rural producers in developing countries. TWN Trade & Dev. Ser. 23. Third World Network, Penang.

Rameshan, P. 2004. Trade disputes under WTO: Trends and features. Foreign Trade Rev. 38(3&4):19-37.

Rao, P.P., P.S. Birthal, and P.K. Joshi. 2006. Diversification towards high value agriculture: Role of urbanization and infrastructure. Econ. Polit. Weekly 41(26):2747-2753.

Raynolds, L.T. 2004. The globalisation of organic agro-food networks. World Dev. 32(5):725-743.

Reich, A. 2003. Institutional and substantive reform of the anti-dumping and subsidy agreements—Lessons from the Israeli experience. J. World Trade 37(6):1037-1061.

Rhône-Poulenc. 1998. Contribution of international private sector companies to poverty alleviation in low income countries. Rhone Poulenc 1998 press release. Available at http://www.rhone-poulenc.com. (After Rhône-Poulenc Agro merged with Aventis, the Web site was removed.)

Robbins, P. 2003. Stolen fruit: The tropical commodities disaster. Zed Books, London.

Roberts, D. 1999. Analyzing technical trade barriers in agricultural markets: Challenges and priorities. Agribusiness 15(3):335-354.

Röling, N., and M. Wagemakers (ed) 1998. Facilitating sustainable agriculture: Participatory learning and adaptive management in times of environmental uncertainty. Cambridge Univ. Press, UK.

Rosset, P. 2005. Transgenic crops to address Third World hunger? A critical analysis. Bull. Sci. Tech. Society 25:306-313.

Rosset, P., J. Collins, and F.M. Lappe. 2000. Lessons from the green revolution: Do we need new technology to end hunger? Tikkun Mag. 15(2).

Samath, F. 1999. Sri Lanka: Cheap imports ruin chicken farmers. South-North Dev. Monitor, 1 Sept. 1999.

Santaniello, V. 2003. Agricultural biotechnology: Implications for food security. Proc. 25th Int. Conf. Agric. Econ. 18-22 Aug. IAAE, Durban.

Sathapathy, C. 2000a. WTO panel rulings on anti-dumping.I. Econ. Polit. Weekly 35(48):4164-4166.

Sathapathy, C. 2000b. WTO panel rulings on anti-dumping: II. Econ. Polit. Weekly 35(49):4298-4300.

Scheuplein, R.J. 1999. An analysis of the safety of Indian food exports—the quest for international safety equivalence. Indian Food Indust. 18(6):364-381.

Schwentesium, R., and M.A. Gomez. 2002. Supermarkets in Mexico: Impacts on horticulture systems. Dev. Policy Rev. 20(4):487-502.

Settle, W.H., H. Ariawan, E. Tri Astuti, W. Cahyana, A.L. Hakim, D. Hindayana et al. 1996. Managing tropical rice pests through conservation of generalist natural enemies and alternative prey. Ecology 77(7):1975-1988.

Sharma, D. 2002. Destroying India's white revolution. Ag-impact Mailing List, 16 April 2002.

Sharma, R. 2005. Overview of reported cases of import surges from the standpoint of analytical content. Import Surge Proj. Working Pap. 1. FAO, Rome.

Shaw, S., and R. Schwartz. 2005. Trading precaution: The Precautionary Principle and the WTO. UN Univ. Inst. Advanced Studies, Japan: UNU-IAS.

Shennan, C., T.P. Gareau, and J.R. Sirrine. 2005. Agroecological approaches to pest management in the US. p. 193-211. *In* J. Pretty (ed) The pesticide detox: Towards a more sustainable agriculture. Earthscan, London.

Shiva, V. 1991. The violence of the green revolution: Third world agriculture, ecology and politics. Third World Network, Penang.

Silberston, A. 2003. Anti-dumping rules—time for change? J. World Trade 37(6):1063-1081.

Singer, H.W. 1950. The distribution of gains between investing and borrowing countries. Am. Econ. Rev. 40:473-485.

Singh, S. 2006a. Organic produce supply chains in India. CMA, IIM, Ahmedabad.

Singh, S. 2006b. Leveraging contract farming for agricultural development in India: Status, issues, and strategies. Working group on agricultural marketing reforms constituted by the Nat. Dev. Council, Gov. India, New Delhi.

Sorby, K., G. Fleischer, and E. Pehu. 2003. Integrated pest management in development: Review of trends and implementation strategies. ARD Working Pap. 5. World Bank, Washington DC.

Srinivasan T.N. 2000. The TRIPS agreement: A comment. Available at http://www.econ.yale.edu/~srinivas/TRIPS.pdf.

Steinfeld, H., P. Gerber, T. Wassenaar, V. Castel, M. Rosales, and C. de Haan. 2006. Livestock's long shadow: Environmental issues and options. FAO, Rome.

Stern, N. 2007. The economics of climate change: The Stern review. Cambridge Univ. Press, UK.

Stiglitz, J. 2006. Making globalization work. Houghton Mifflin, NY.

Suparmoko M, 2000. The Impact of the WTO Agreement on Agriculture in the Rice Sector. Paper presented at UNEP Workshop on Integrated Assessment of the WTO Agreement on Agriculture in the Rice Sector, Geneva, 2002.

Swinbank, A. 1999. The role of the WTO and the international agencies in SPS standard setting. Agribusiness 15(3):323-333.

Tabor, R., M.H. Sawit, and H.S. Dillon. 2002. Indonesia rice policy and the choice of a trade regime for rice in Indonesia. Roundtable Workshop, Inst. Dev. Econ. Finance, Univ. Indonesia, 11 Mar 2002.

Tallontire, A. 2001. Fair trade and development. Natural resources and ethical trade program. NRI, Univ. Greenwich, UK.

Tauli-Corpuz, V. 1999. Indigenous peoples caucus statement. Roundtable on Intellectual Property and Traditional Knowledge, WIPO, Geneva, 2 Nov 1999.

Tebtebba Foundation. 1999. No to patenting of life! Indigenous peoples' statement on the TRIPS Agreement of the WTO. Tebtebba Foundation, Philippines.

Ter Weel, P., and H. van der Wulp. 1999. Participatory integrated pest management. Policy and Best Practice Doc. 3. Min. Foreign Affairs, Dev. Cooperation, The Hague, The Netherlands.

Tharakan, P.K.M. 1999. Is Anti-dumping here to stay? The World Econ. 22(2):179-206.

The Washington Post. 2007. NAFTA anxiety: Trade agreement has not produced expected gains in Mexico's economy. Wednesday, 10 Jan 2007.

Tozun, N. 2001. New policy, old patterns: A survey of IPM in World Bank projects. Global Pesticide Campaigner 11(1):1,14-15

TWN. 2005. Asian Regional workshop on bilateral free trade agreements: Summary and recommendations. Third World Network, Penang.

UNCTAD. 2003. Report of the UN Panel of Eminent Persons on Commodities. UNCTAD, Geneva.

UNCTAD. 2004. Development and globalization: Facts and figures. UNCTAD, Geneva.

UNCTAD. 2005. Trade and development report, 2005. UNCTAD, Geneva.

UNCTAD. 2006a. The emerging biofuels market: Regulation, trade and development implications. UNCTAD/DITC/TED/2006/4. UNCTAD, Geneva.

UNCTAD. 2006b. Tracking the trend towards market concentration: The case of the agricultural input industry. Available at www.UNCTAD/DITC/COM/2005/16. UNCTAD, Geneva.

UNCTAD. 2007. World economic and social prospects, 2007. Available at http://www.unctad.org/en/docs/ wesp2007_en.pdf. UNCTAD, Geneva.

UNDP. 1999. Human Development Report 1999, New York, Oxford University Press.

UNEP. 2005. Integrated assessment of the impact of trade liberalization: A country study Vietnam rice sector. UNEP, Nairobi.

USAID. 2004. USAID agricultural strategy: Linking producers to markets. Available at http://www.usaid.gov/our_work/agriculture/ag_strategy_9_04_508.pdf. USAID, Washington DC.

US Embassy Beijing. 1996. Agricultural pesticide use in China. Available at http://www.usembassychina.org.cn/sandt/mu5pest.htm.

Van den Berg, H., and J. Jiggins. 2007. Investing in farmers: The impacts of Farmer Field Schools in Integrated Pest Management. World Dev 34:663-686.

Vietnam News/Vietnam Economic Times. 2006. VND1 trillion earmarked for bio-technology. 24 Feb 2006.

Wailes, E.J., and A. Durand-Morat. 2005. Impacts of WTO policy on U.S. rice policy. Selected paper presented at the World Trade Organization Impacts on U.S. Farm Policy Conference. New Orleans. June, 2005.

Wen, D. 2005. China copes with globalization. Int. Forum Globalization, San Francisco.

Weisbrot, M., and D. Baker. 2002. The relative impact of trade liberalization on developing countries. Available at http://www.earlham.edu/~jond/2003%20Stuff/relative_impact_of_trade_liberal.htm. Center for Econ. Policy Res., Washington DC.

World Bank. 1999a. Mozambique agricultural sector public expenditure project, #039437. Rep. #22286. Proj. Inform. Doc. World Bank, Washington DC.

World Bank. 1999b. Peru agricultural research and extension project (AREP). P047690 Rep. #19841, Proj. Appraisal Doc. 1999/10/25. World Bank, Washington DC.

World Bank. 2002. Biotechnology for poverty alleviation and economic growth: Challenges and options for the World Bank. World Bank, Washington D.

World Bank. 2005. Going, going, gone: The illegal trade in wildlife in East and Southeast Asia. Available at http://www.edis.org/dbtw-wpd/exec/dbtwpcgi.exe?AC=GET.RECORD&XC=dbtw-wpd. World Bank, Washington DC.

World Bank. 2006a. India-Bangladesh bilateral trade and potential free trade agreement. World Bank, Dhaka.

World Bank. 2006b. State of the carbon market. Carbon Finance Unit, World Bank, Washington DC.

Xiaoyun, L., A. Changsheng, J.B. Tschirley, S.E. Webb, and A. Morton. 1997. Sustainable agriculture and rural development in China, Part 1: The agro-ecosystem and China's rural economy. Available at http://www.fao.org/sd/EPdirect/EPan0009.htm. SD Dimensions, FAO, Rome.

Xie, J. 1993. The principles and ways in the design of man-made ecological systems and their application to the multiple development of mountain areas. Inst. Ecol., Kunming.

Ye, X., S. Al-Babili, A. Kloeti, J. Zhang, P. Lucca, P. Beyer et al. 2000. Engineering provitamin A (carotene) biosynthetic pathway into (carotenoid-free) rice endosperm. Science 287:303-305.

Yeats, A. 1998. Just how big is global production sharing? Policy Res. Working Pap. 1871. World Bank, Washington DC.

Yussefi, M., and H. Willer. (ed) 2003. The world of organic agriculture 2003—statistics and future prospects. IFOAM, Bonn.

Zanardi, M. 2004. Anti-dumping: What are the numbers to discuss at Doha? The World Econ. 27(3):403-433.

4

Agricultural Change and Its Drivers: A Regional Outlook

Coordinating Lead Authors:
M. Monirul Qader Mirza (Canada), Rajeswari S. Raina (India)

Lead Authors:
Anna Matysek (Australia), Ma Shimming (China)

Review Editor:
Thelma Paris (The Philippines)

Key Messages

1. Demographic changes will have a significant impact on labor supply and the nature and capacity of the agricultural labor force. The population of the ESAP region constitutes about 56% of the global population. The demography of the region is changing rapidly. In India and some other developing ESAP countries, the younger generation will continue to dominate the labor force. In China, Japan and the Republic of Korea, however, the labor force will be characterized by an increasingly ageing population. Economic development and rural-urban wage differentials will encourage rural to urban migration. Increasing migration in response to local stress, opportunities for education and employment will contribute to acceleration in rural depopulation. The implication is that in the developing ESAP countries, the rural labor force will increasingly be relatively less educated, largely female and consequently have less access to AKST.

2. Poverty, malnutrition, social and political problems will continue to be significant in ESAP despite the prospects for economic growth, with South Asia performing the worst unless corrective measures are implemented. Although there will be economic growth and improvements in nutrition status, within a decade South Asia will be home to nearly half of the malnourished children in the world. ESAP is an epicenter of domestic and regional political tensions and will need concerted regional cooperation to ensure peace, prosperity and opportunities for economic growth. Increasingly, ESAP countries will realize the importance of the capacity of the public sector to respond to the needs of the poor, provide for social security where it is lacking, generate public and private investments in infrastructure and social development programs, enable pro-active civil society participation in economic decision-making and engage actively with emerging threats to impoverished rural livelihoods in order to promote political stability, economic growth and social justice.

3. Globalization, economic and trade liberalization pose both challenges and opportunities for AKST in addressing the development goals. Although the contribution of agricultural GDP to national incomes is declining, the agriculture sector will remain significant to economic growth, employment and rural livelihood. The opportunities from globalization and trade liberalization include enhanced access to markets and increased flow of commodities, technology, labor and/or knowledge and skills. One of the major challenges in developing ESAP countries will be the capacity to address in a timely manner the issue of maintaining access to appropriate quantity and quality of food for under-privileged sections of the population. In the developing ESAP countries, the existence of small-scale and marginal farmers will demand capacities to address scale issues and in developed ESAP countries the removal of trade barriers/subsidies will present potential opportunities. The capacity of governments to influence trade and AKST outcomes will be affected by the magnitude of global private sector control over agribusiness and trade. Increasing integration of the ESAP region with the global economy, especially through technology/knowledge sharing and greater infrastructural investments will enhance employment and income in the ESAP region.

4. Degradation of natural resources and environmental systems will have adverse implications for achievement of each of the development goals. Natural resources, especially freshwater and arable land, will increasingly be subject to serious pressure from competing sectors. Along with a continuing increase in agricultural production, intensive agriculture and overuse of agrochemicals will worsen current trends of degradation of soil and water quality as well as loss of biodiversity in many parts of ESAP. Rapid urbanization and industrial expansion will place increasing pressure on demands for land and water. Water transfers to intensive irrigation based agriculture and urban areas are placing substantial ecological and political pressure on water resources and in the absence of new technological and policy options, this trend will become more severe in the future. Industrial and agricultural effluents will affect water quality across the region. In addition, intensive agriculture will further reduce the areas available for fisheries and livestock production. AKST has the technologies but may require appropriate institutional arrangements to enable sustainable natural resource management.

5. Climate change and climate variability will emerge as threats to the agricultural sector in most of the ESAP region. However, some parts of the region may benefit from climate change. The IPCC projections show that occurrences of natural hazards are likely to increase globally. The frequency and magnitude of these events in developing countries that are already vulnerable to these hazards and dependent on agriculture is of particular concern. The outcomes of these hazards may be food insecurity and worsening poverty, increases in average temperature, changes in precipitation patterns, sea level rise and resulting inundation in the coastal areas, increases in soil and water salinity, and new and more favorable environments for pests and diseases; these will have ramifications for agricultural productivity and livelihoods. There are many technological and institutional options to help mitigate and adapt to climate change. Climatic change may have some beneficial effects for agriculture in some parts of the ESAP region.

6. The agricultural sector will continue to compete with other sectors for energy and a range of other goods and services. Biofuels and renewable energy sources will provide additional energy supply and utilization opportunities within a wider portfolio of energy sources. Rapid economic growth will substantially increase energy demand in the ESAP region over the next few decades. The gap between energy demand and supply is growing and is likely to be partly met by biofuels and other renewable energy sources. Developing countries of the ESAP region will invest in and utilize alternative energy sources to meet increasing local demands. Biofuel production may result in competition for land, water and energy within economies, with implications for local food security. There are options for collaboration between AKST organizations (public and private) and wider energy systems R&D that can help address the development goals.

7. Existing national and international institutions, and educational and R&D organizations will be inadequate to address the multiple roles and functions of agriculture. The balance between public and private sector investments and capacities for innovation will continue to shift towards commercial interests. Current linear R&D models and technology transfer approaches will be increasingly inadequate to address emerging concerns/challenges. Increasing capacities within public sector AKST organizations will depend on the development of flexible institutional arrangements. Multiple and varied demands of the agriculture sector will increase the pressure on current educational and R&D organizations to evolve to work with a diverse range of partners. There will also be demands for the skill base of AKST to include social, political and legal knowledge. Local and traditional knowledge systems (e.g., medicinal and aromatic plants) will become mainstream in parts of some ESAP countries, but will continue to decline in other areas (for example, tribal and mountainous communities). Availability and affordability of advanced information and communication technologies will enhance effectiveness of AKST. Increased investment in science and technology and enhanced innovation capacity will play an increasingly important role in providing adaptive responses for agriculture to stressors such as climate change, increasing natural hazards, HIV/AIDS, avian flu, SARS and malaria.

8. The capacity of political, administrative and scientific systems to learn from past evidence of the impacts of AKST and to adapt to emerging challenges will be crucial for the achievement of the development goals. There are both uncertainties and predictable aspects to the evolution of drivers that influence the future of agriculture and AKST. Some ESAP countries will assess the capacity of existing systems and their flexibility in responding to past challenges, while many will continue to address crises on an ad hoc basis. Proactive investment in learning and capacity building within local, national and international organizations will facilitate flexibility and adaptation. Poverty and underemployment, demographic shifts, trade and economic liberalization, natural resource degradation, climate and other stresses are all crucial challenges in many ESAP countries. The interrelated nature of these challenges suggests that achievement of the development goals will also depend on policy choices and decisions beyond the immediate agricultural sector.

4.1 Context for Looking Forward

The economic and social structures of South and South East Asia and the Pacific islands are unique. The economies of most of the countries are highly dependent on agriculture. Only a few countries are industrialized and a few others are in transition. In recent years, the contribution of the agriculture sector to overall GDP in all ESAP countries has declined as a result of increasing contributions from the manufacturing and services sectors (ADB, 2006b). However, the agriculture sector still provides employment to the largest section of the population in ESAP. Agricultural sector employment in South Asian countries is in the range of 43-52%. Overall employment in the agriculture sector in South East Asia (except for Malaysia) is in the range of 37-62%, which is slightly lower than in South Asia (ADB, 2006b). The ESAP region's agricultural sector is characterized by small peasantry, with small and marginal farms accounting for almost 86% of the farming sector in some countries like India (Hobsbawm, 2006).

Over the recent decades the agriculture sector provided impressive yields and great services to reduce hunger and poverty in many parts of the ESAP region, especially in South and South East Asia. Despite successes in food grain production, the sector has been largely unable to reduce hunger and poverty in Asia. Despite the steady growth of ESAP agricultural trade, intra-regionally and globally, the region is marked for the highest incidence of malnutrition, especially among children. Notwithstanding the contribution of agriculture to employment, income and food, it is now a major polluter of land, water and atmospheric systems in Asia. Some of the worst cases of gender disparities and marginalization of indigenous and tribal people are evident in ESAP. Some specific features of the ESAP region are:

- Land degradation has become a serious threat to agriculture. A global assessment of the extent and form of land degradation showed that 57% of the total dryland area in China and India are degraded (UNEP, 2006).
- The gap between water demand and supply is increasing due mainly to increased demand from agriculture, rapid urbanization and industrialization.
- Natural hazards—floods, droughts and coastal inundations have become regular threats to agriculture. Floods and droughts damage enormous quantities food crops every year. Storm surge and tidal inundations cause crop losses in the coastal areas. Climate change is emerging as a new threat to cropping in terms of excessive flood related inundation, shrinking cropping seasons, and temperature related yield losses.
- Globalization has multi-dimensional effects on agriculture: withdrawal of subsidies makes agricultural products less competitive with highly subsidized developed country agriculture, however deregulation and liberalization also bring more opportunities for investment.
- In spite of the green revolution and per capita increases in income driven by globalization efforts, food insecurity remains a major problem in many ESAP countries. There are many causes: the poorest of the poor have not yet attracted targeted policies for education and healthcare; the population growth rate exceeds the growth rate of food production; there are growing disparities among the poor and the rich; increasing prices of food commodities are putting them out of reach of the poor because of limited income; and gender disparity.

Given this context the key question for decision makers who have to invest in or promote agriculture and AKST is: what will be the future outlook for agriculture in the ESAP region and in particular, what will be the role of AKST in meeting the development and sustainability goals of reducing hunger and poverty; improving nutrition, health and rural livelihoods; and facilitating social and environmental sustainability?

To analyze this question based on currently available knowledge, the key issue was deconstructed into a series of more specific questions:

- What will be the future of food systems, agricultural products and services?
- What are the major uncertainties of the drivers and projections?
- What are the implications for AKST in the future?
- What will be the implications for development goals?

4.1.1 Approaches of scenarios development and impact assessment

The specific questions identified above can be addressed by either developing plausible futures based on socioeconomic, technological and political assumptions, or by extrapolating agricultural or related variables, based on assumptions made about the baseline period. Each driver might take different shapes in the future based on these assumptions. Recently many studies developed scenarios to help decision makers see different plausible futures with reference to climate change, ecosystems, environment and agriculture. These scenarios considered various timelines for the projections (Table 4-1).

Conventional impact assessment of agricultural science and technology (S&T) has been conducted mainly by social scientists, especially economists, in national and international agricultural research organizations or policy making agencies. The explicit purpose of such assessments is to understand how S&T as a major driver of agricultural growth (productivity and production) brings returns and thereby legitimizes investments made already as well as the scope for further investments in S&T. Global assessments such as the Millennium Ecosystem Assessment (MA) and the Intergovernmental Panel on Climate Change (IPCC) (Table 4-1) have been conducted to enable a better understanding of drivers and processes of change in the ecosystem and global climate regimes in order to identify options for action to address the drivers of change and/or the processes. These assessments have been conducted by large groups of people with a variety of expertise and experience, drawing upon a variety of natural science and social science disciplines as well as regional/local experiences.

4.1.2 Assessment approach for ESAP

This chapter presents an assessment of agricultural change and its drivers from the existing literature (national and international) based on historical trends and future projections of key drivers of change and expected changes in future policies and politics (See Global Chapter 5). The assessment

Table 4-1. *Approaches for scenario development and impact assessment.*

Scenario	Focus	Time-line	Approach
IPCC SRES 2000	Climate Change	2100	Designed four different plausible worlds: *A1*: rapid economic growth; *A2*: fragmented world; *B1*: convergence with global environmental emphasis; and *B2*: local sustainability. The scenarios consistently describe the relationship between emission driving forces and their evolution over different timeline.
MA 2005	Ecosystem	2050	Developed four plausible scenarios by combining qualitative storyline development and quantitative modeling of driving forces. The four scenarios are: *Adapting Mosaic*-recognizes extensive value of ecosystem services for human well-being; *Techno-garden*-proactive policies towards economic value of ecosystem services; *Global Orchestration*-technology development to fix damaged ecosystem; *Order from strength*-security and protection within national boundary and ecosystems are less important.
GEO 2002	Environment	2032	Developed four scenarios through consultation and experience of other scenarios groups. *The Markets First*: market driven development; *The Policy First*: strong actions at national level for specific social and environmental goals; *The Security First*: high inequality and conflict caused by socioeconomic and environmental stresses; and *The Sustainability First*: A world with the emergence of new development paradigm in responses to the challenges of sustainability.
FAO 2003	Agriculture	2030/2050	Looked at different driving forces that lead the growth of global agriculture and food consumption. Discussed future prospects for food, nutrition, agriculture and major commodity group in the future.
IFPRI 2002	Agriculture	2020	Examined alternative regional and global scenarios (optimistic and pessimistic) based on a number of driving variables. These variables are also affected by policy decisions on investment in agricultural research, irrigation, clean water, and health, population programs and economic policies.

Source: Authors' elaboration.

is based on a list of key drivers of change of agriculture and AKST relevant to the ESAP region. A driver is defined as "any natural or human induced factor that directly or indirectly causes a change in AKST" (MA, 2005). The IAASTD conceptual framework (ESAP Chapter 1) illustrates this as the mutual interaction of direct and indirect drivers, as well as the effect of each of these drivers on innovation, knowledge and learning, mediated through actors/networks and processes/rules and norms. However, given the nature of the drivers and the complex relationships between drivers of change in the ESAP region, a classification/listing of direct and indirect drivers of change would be an academic exercise of little relevance to decision-making. What is more important is to explore how the individual drivers will evolve in the future; how the drivers of change relate to each other; and how these inter-relationships and changing contexts will shape AKST in future.

A lacuna in current global decision-making and negotiations on globalization or climate change or poverty or any such international processes is that they "take place in compartmentalized sectors such as trade/finance/development aid/health," and do not question or assess the inter-relationships and impacts of each of these on the other global processes (see WCSDG, 2004). In this chapter we present some trends in the major drivers of change that are important for plausible decision-making in the future. We refer here to a large body of work on trends or projections of each driver of change. Much of this work has been attempted for different purposes, by different authors/agencies with different ideological orientations and values. Wherever possible we refer to certain time points such as 2015, 2020, 2030, or 2050 to allow for some comparability across the different drivers discussed here.

4.2 Drivers of Agricultural Change

4.2.1 Demographic change

4.2.1.1 Population growth

The Asian population reached 3.7 billion in 2000. The population of the ESAP region constitutes about 56% of the global population. It is projected that the ESAP population will increase continuously to reach 4.8 billion in 2025 and 5.0 billion in 2050, which will be 56% and 54% of the world's population respectively (UN, 2001). South Asia has 40% of the region's population and one of the highest rates of population growth (ADB, 2001a). The slowing down of population growth is due to rising levels of education, increased female participation in the work force and greater use of contraceptives. Countries such as the Philippines and Bangladesh continue to maintain high birth rates—with challenging implications for job creation, food security and environmental stress.

Together the combined populations of China and India currently constitute about 38.5% of global population, and 73% of the ESAP population. The demography of the ESAP region is changing rapidly. The population of India will exceed that in China by 2035 (UN Economic and Social Council, 2004). India's population is projected to reach 1.25 billion by the year 2015 and 1.53 billion by 2050 (UNDP, 2003) (Figure 4-1).

FAO projections show a continuing slowdown in the growth of the world's population. The average annual population growth rate has fallen to 1.35% from a peak of 2.04% in the late 1960s; it is expected to fall to 1.1% in the period 2010 to 2015 and to 0.8% during 2025 to 2030. Absolute global population has fallen from a growth peak of 86 million per year in the late 1980s to current annual additions of around 77 million, but is projected to decline to 67 million on average between 2025 and 2030 and 43 million between 2045 and 2050 (FAO, 2003).

The SRES storylines constructed for the Asia region reflect a range of socioeconomic scenarios. The population projections for Asia are 1.54 billion people in 2050 and 4.5 billion people in 2100 (IPCC, 2007). Overall, despite variations in the different population projections, the future is one of increasing population pressure in South Asia. In terms of sheer numbers, the population of South Asia is likely to be a major concern driving the decisions made for agricultural and rural development both globally and within nations.

4.2.1.2 Demographic factors

Demographic factors have shaped agricultural expansion and growth over the centuries. The Green Revolution in the ESAP region is the prime example of how population growth or the rhetoric of population growth and associated projected food shortages led to a planned and rapid increase in food production and productivity. Over the next 20-25 years South Asia will be the most populated part of the world and by 2025 will be home to about 46% of the population in the Asian region (Hussain et al., 2006).

The two key elements of demographic transition, fertility rate and population growth rate, have implications for economic growth (and prospects) in rural areas, agricultural growth rates, education levels (especially female literacy and educational attainments) and per person incomes (Hussain et al., 2006). Crucial among the determinants in Asia are fertility rate and the prospects for a well-nourished population.

Fertility rates (births per woman) in South Asia (SA) and the East Asia and Pacific (EAP) countries are expected to change from the current rates of 3.1 and 2 (in 2000-2005) to 2.1 and 2.0 (in 2045-50) (World Bank, 2005). The slower than expected decline in fertility rate in SA is because of a slow expansion and relatively poor access to medical/health care and prevailing weaknesses in child health care systems.

Maternal mortality will continue to be highest in South Asia as a regional cluster—the only exceptions being Bangladesh and Sri Lanka where the figures are expected to be lower. The marginal increase in fertility rates in East Asia and Pacific is explained by the decreasing infant and maternal mortality rates in a large number of Pacific island countries and improvements in health services in general. Fertility decline in rural areas in general will be slower than in the urban regions.

Child malnutrition will be difficult to eradicate even by 2050—and in the SA region it is likely that child malnutrition may increase from current levels given the degradation of ecosystems and increasingly limited access to ecosystem services for the poor (MA, 2005). Even with estimates of increasing health and sanitation and better access to and

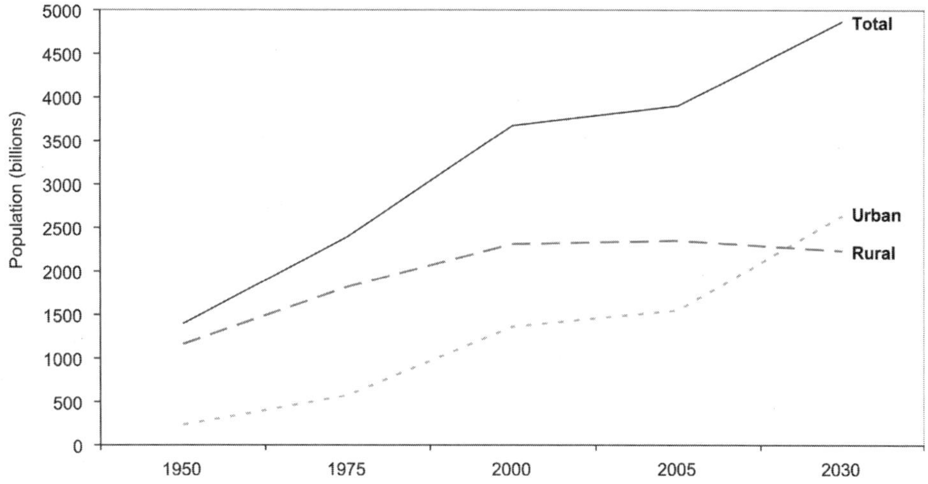

Figure 4-1. *Asian population trends. Source: DESA, 2006*

availability of food, 46% of children in South Asia will still be malnourished in 2020 (Rosegrant and Malik, 1995). In 2020, South Asia will account for 48% of the world's malnourished children. Progress in improving nutrition is expected to continue, though more slowly than in the past.

Average per capita food consumption in developing countries is projected to rise by 6.3%, from 2680 kcal in 1997-99 to 2850 kcal in 2015. The proportion of undernourished people is anticipated to fall from 20% in 1990-92 to 11% by 2015 and 6% by 2030. However, the 2015 target of halving the total number of undernourished people will probably only be reached in 2030 when the numbers are expected to fall to 440 million. This delay is a result of rapid population growth associated with lagging economic growth as well as the fact that many countries are starting from extremely low national average food consumption levels. The proportion of the global population living in countries with per capita food consumption under 2200 kcal per day will fall to only 2.4% in 2030. However, in South Asia, the fraction could fall by 40% from 1997-99 to 2030 and in East Asia the number could halve by 2030 (FAO, 2003).

The number of malnourished children under the age of five in the developing world is projected to decline by 21% between 1997 and 2020. Although child malnutrition is expected to decline by 31% in South Asia, India will still be home to 44 million malnourished children in 2020, representing 34% of the total in the developing world. China will have the largest decline, with a 54% reduction in the number of malnourished children by 2020 (IFPRI, 2002).

4.2.1.3 Age structure

The demographic data show ageing in developed countries and an increasing proportion of younger people in most of the developing countries. In India and some other ESAP countries, the younger generation will continue to dominate the labor force for decades. But in China, Japan and the Republic of Korea, family planning policies and improvements in health care will contribute to an ageing of the populations. An ageing population is now one of the crucial social problems being addressed in China (Yuankai, 2007). In the rural areas in particular, the Government is being forced to acknowledge the need to support millions of senior citizens. Over 60% of China's aged live in rural areas. The number of China's rural senior citizens (i.e., over 60 years old) will reach 120 million in the next 20 years.

Asia is one of the world's fastest ageing regions, with the percentage of elderly projected to double between 2000 and 2030 (Kaneda, 2006). Japan, Australia and New Zealand are expected to see rapid rates of ageing and by 2050 the proportion of people over 60 is projected to be 25% of the population. Ageing presents challenges to agricultural sector productivity and innovation adoption and raises concerns of poverty among the rural elderly and women. Asian developing countries have relatively large youthful populations that require strategies to balance their aspirations with opportunities. A population age structure dominated by youth in developing countries suggests the potential for labor migration to regional developed countries. It also has implications for strong national urban development.

The outlook for the future in the Asian region is for the younger population to increase slowly to 685 million in 2040, when they would account for 14% of the total population. While the growth in the young adult population will continue over the next two decades in the region, in most OECD nations their numbers will decrease (Hugo, 2005). For agriculture, the implication of an increasingly younger population who have development aspirations and who are malnourished (4.2.1.2) is worsened by the emerging gender ratios in the region.

4.2.1.4 Gender composition

A declining labor force (and in particular a declining male labor force) will be available for agriculture, particularly on small to medium farming enterprises and the available labor force will be dominated by women. Due to out-migration of male heads of households, the pressure of maintaining the farm and household will fall on the women left behind despite the constraints they face such as lack of access to land, new seeds and technical knowledge (Paris et al., 2005). The current trends in declining fertility rates are associated with a reduction in family size. Along with improvements in the population gender ratio and the potential for increased matriarchal households and stress on family labor, these trends present a mixed picture for agriculture (United Nations, 2001). The region demonstrates a decreasing trend in agricultural employment rates with increasing importance of non-farm work for income security.

Over the past half-century, women's participation in the labor force has increased steadily in many Asian countries, particularly in the rapidly growing economies of East and Southeast Asia. In 1999, half or more of women in the age groups 15-64 were employed in all sectors, including agriculture, in nine Asian countries (ADB, 2001a) (Figure 4-2).

The female contribution to the overall economy is high throughout the ESAP region, particularly in terms of labor input into agriculture. Bangladesh, Bhutan, Cambodia, China, India, Myanmar, Nepal, Pakistan and Viet Nam have particularly high percentages of women employed in the agricultural sector, with estimates ranging between 60 and 98%. Indeed, in most Asian countries the number of women employed in agriculture as a percentage of the EAP is higher than that of men (Garcia et al., 2006). Female members from farming households will play increasingly important roles as key agents of change by serving as extension agents to other women farmers and laborers. Experiences in Bangladesh demonstrated that including women in training in specific technologies in which they were actively engaged and in the overall production system contributed to more effective decision-making. Several successful women-led projects included promotion of post-harvest and storage technologies and video-centered learning. Rural women will become the direct recipients of village level training courses which deal with knowledge-intensive agricultural technologies (crops, livestock, fisheries, agroforestry) crop and livestock (Paris et al., 2005).

4.2.1.5 Rural-urban and inter-regional migration

Rural to urban migration still dominates migration flows in most Asian countries because of the high proportion of the population living in rural areas (DESA, 2006). According to the 2005 Revision of World Urbanization Prospect, the global proportion of urban population increased from 13% in 1900 to 49% in 2005 and is expected to reach 60%, or 4.9 billion people, by 2030. While urbanization continues to be on the rise, rural populations are still significant and are currently growing. However, a gradual decline in rural populations is expected to commence in 2019, reaching slightly less than the current 3.3 billion by 2030. In 2005, 71% of all rural dwellers lived in Asia, primarily in India, China, Indonesia and Bangladesh. At the same time, the Asian urban population has reached over 1.5 billion. This is projected to rise to over 2.6 billion by 2030 (Figure 4-1).

Two other estimates show slightly different trends. Between 2005 and 2030 the Asian urban population is expected to rise by 2.12% annually (DESA, 2006). At the same time, rural populations are anticipated to decline by 0.2% annually. There are likely to be 2.2 billion rural Asians by the year 2020 and this rural population will have much lower access to health and education, and a lower level of general well-being (ADB, 2001b).

Since 1950 countries such as Australia, China, Fiji, Indonesia, Japan, Korea DPR, Republic of Korea and the Philippines have recorded negative population growth in rural areas. Yet many developing countries in the region still have a larger percentage of rural population compared to urban population. In particular, countries that depend heavily on agriculture as economic drivers register less than 30% of their total population as urban, e.g., Bangladesh, Bhutan, Cambodia, India, Lao PDR, Nepal, Papua New Guinea, Sri Lanka, Samoa, Solomon Islands, Vanuatu and Viet Nam. For China, the urban population is projected to reach around 60.5% of the total population and hence exceed rural population by 2030 (UNDP, 2003).

Urbanization is likely to pose a major labor market concern as well as environmental concerns (Dyson et al., 2004). Countries such as China, Thailand, Sri Lanka and Malaysia

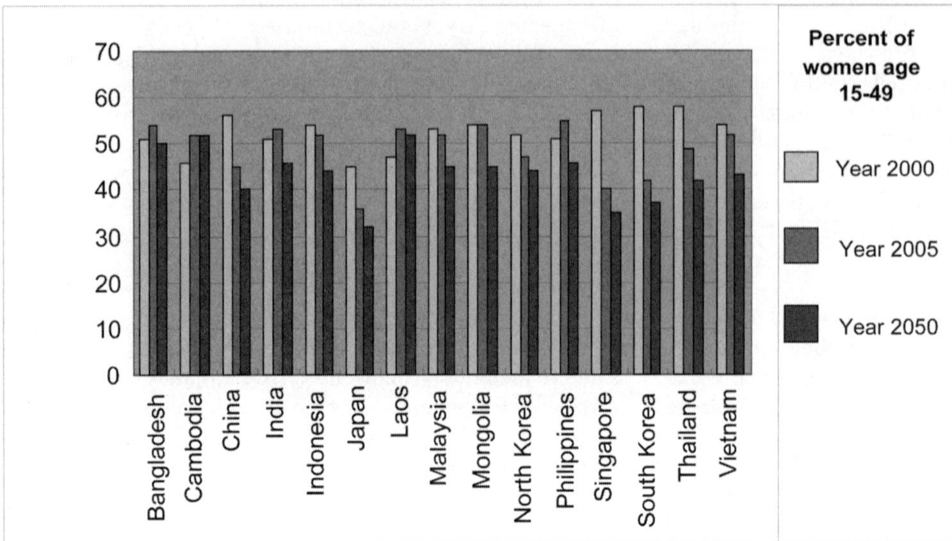

Figure 4-2. *Women age 15-49 in some Asian countries and regions. Source: UN, 2001.*

will face caring for an ageing population, while India will have to grapple with the problems of educating and finding jobs for a significantly younger population (Hussain et al., 2006).

Besides domestic rural-urban migration, international and intra-regional migration is on the rise in ESAP though there is a marked difference among educated and unskilled migrants in destination countries. Increasing movements in search of economic mobility brings mixed migration impacts; skilled and semi-skilled labor is lost, but capital gains are acquired through remittances.

Lower costs of transport and information have increased both domestic and international labor mobility, such that workers are more sensitive to international wage differences and no longer consider only wage differentials between rural and urban areas within their own country. The migration pattern between Cambodia and Thailand for example, has been shown to be affected by cross border wage differentials, job availability and work opportunities in each country (Acharya, 2003). Implications for agriculture are that labor shortages can be offset by migration of human capital from other countries. Continued regionalization and reductions in the cost of transport imply a move toward a common labor market in the ESAP region.

Likely regional implications for ESAP will be a smaller, more educated and wealthier rural class characterized by an agricultural sector adopting more efficient farming processes. Skilled individuals who pursue off-farm employment may provide remittances back to rural communities and provide skills to assist in the incorporation of new farming technologies and farming practices. Rural communities that incorporate mechanization as a substitute for labor will provide individuals with more time to devote to higher paid work, thus assisting in the alleviation of poverty.

The projected future of demographic change is one of increasing population pressure in the region with malnourished children concentrated mainly in South Asia; an increasingly larger younger population in developing countries; an ageing population in developed countries; a declining and mostly female rural labor force; increasing urbanization and inter-regional migration.

4.2.2 Economic drivers

Many of the economies of the ESAP region are economies in transition and are highly agriculture dependent. A typical pattern that emerges as economies undergo demographic transition is one where total agricultural output increases but dependence on agriculture tends to decline. The recent historical development pattern in most of the ESAP region has been a declining dependence on agriculture and an increasing dependence on associated structural shifts toward manufacturing and services (ADB, 2005).

Economic growth is itself a function of many variables, including demographics, factor endowments, international trade, savings and investment, institutional capacity and technical progress.

4.2.2.1 Gross domestic product

Projections of GDP are often difficult to compare between studies, since regional aggregations typically vary with respect to country inclusions. For this reason, care should be exercised in drawing broad conclusions from the studies reported here for regional growth rates.

The scenario approach adopted in the IPCC SRES (IPCC, 2000) resulted in four main "marker" projections for economic growth within each of the scenario families: A1, A2B1, A2B1 and B2 (Box 4-1). Growth rates for Asia were projected to span a range of 3.9% (A2) to 6.2 (A1) per annum over the period 1990-2050. Longer term projections for 1990-2100 indicated a GDP growth rate range for Asia of 3.3 (A2) to 4.5% (A1) per annum.

The Asian Development Bank (Roland-Holst et al., 2005) projected real GDP growth rates by country over the period 2005-2025. These projections are underpinned by assumptions about continued rapid productivity growth and capital accumulation (Figure 4-3). The greatest growth potential relative to the baseline scenario occurs under assumptions that all tariffs and non-tariff barriers within Asia, as well as all export subsidies in Asia are removed (Roland-Holst et al., 2005). This scenario results in aggregate income differentials relative to the baseline in 2025 of between 8.1% for Japan and 116.6% for Malaysia.

OECD-FAO (2006) growth projections are more aggregated with respect to regional reporting and indicate an average annual growth rate of 3.13% for Asia for the period 2006-15. Oceania is projected to grow by 3.09% per year on average over the same timeframe. This compares to projections undertaken in 2003 by FAO indicating a 2.2% per annum growth rate for Oceania over the period 1999-2010, 5% for South East Asia and 3.3% for South Asia (FAO, 2003).

The World Bank (2007) projections of income growth per person (2010-30) were developed for central and high growth scenarios. In the central scenario, mid-term GDP growth projections for East Asia and the Pacific and South Asia are projected to be higher than for any other region. Income growth per person in East Asia and the Pacific is projected at around 4.4% per annum on an annual average basis for 2010-2030 in the central scenario and 6.8% in the high growth scenario. The corresponding figures for South Asia are around 3.4% and 5.1% respectively.

4.2.2.2 Agricultural productivity

Output can be increased either by using more factors of production, using those factors more intensively, or by increasing the productivity of those input factors. In all cases, improvements in education are important.

As discussed above, and in line with regional population projections, labor supply in developing ESAP countries is expected to continue to grow over the next two decades. However labor supply growth in Australia and Japan is expected to slow as the population ages markedly. The case is particularly severe for Japan, whose labor supply declines over the entire projection period. China, South Korea and Chinese Taipei are all expected to have negative growth in labor supply by 2030-50 (UNDP, 2003; Matysek et al., 2006). However, significant improvements in education and literacy rates are expected to raise labor productivity, particularly in South and Southeast Asia (Table 4-2). This could have either beneficial or negative implications for agriculture, depending on the level of development of the country and its economic structure, inter alia.

Box 4-1. SRES storyline summaries.

A1. The A1 storyline and scenario family describes a future world of very rapid economic growth, global population peaking mid-century and declining thereafter, and the rapid introduction of new and more efficient technologies. There is convergence in regional incomes, capacity building and increased cultural and social interactions. There are three groups within the A1 family that describe alternative directions of technological change in the energy system. The three A1 groups are distinguished by technological emphasis: fossil intensive (A1FI), non-fossil energy sources (A1T), and balanced across all sources (A1B) (i.e., not relying too heavily on any one particular energy source). The food system dynamics of this scenario family are characterized by rapid increases in the volume of trade in food and feed; fast increases in agricultural productivity and rapidly increasing per person consumption of livestock products in line with rising per person incomes.

A2. The A2 storyline and scenario family describes a much more heterogeneous world. Themes include preservation of regional cultural identities. Economic development is regionally oriented, trade barriers remain and per capita economic growth is slow. Fertility patterns across regions thus converge only slowly, such that global population increases substantially. Technological change is more fragmented with slower diffusion than in other storylines. Food system dynamics are characterized by only moderate increases in agricultural trade, slow improvements in crop and livestock productivity and slow increases in per person consumption of livestock products.

B1. The B1 scenario family is characterized by a globalizing, convergent world with the same population dynamics as A1, but with more rapid changes in economic structures toward a service and information based economy. There is a strong focus on energy efficiency due to high fossil fuel prices and rapid introduction of clean technologies. Agricultural trade volumes increase rapidly and productivity growth in the sector is high. However per person consumption of livestock products is lower than in A1.

B2. The B2 storyline describes a world with increasing global population, but at a rate lower than A2, intermediate levels of economic development, and less rapid and more diverse technological change than in the B1 and A1 storylines. The scenario emphasizes local solutions to economic, social and environmental sustainability and a focus on non-material quality of life factors. The food system is characterized by moderate increases in agricultural trade volumes, productivity, and increases in per person consumption of livestock products.

Source: IPCC, 2000.

Progress in agricultural mechanization has varied significantly across the ESAP economies. In recent times, mechanization has increased most strongly in South and Southeast Asia and China. By contrast, the mechanization trend has been stagnant or in slight decline in Australia and New Zealand (FAO, 2003). The use of capital intensification as a proxy for technological change and productivity improvements can be misleading. For example, although mechanization intensity in Australian broad acre cropping is expected to continue to decline over time relative to levels observed two decades ago, this reflects the adoption of new techniques such as no-tillage agriculture, which not only has beneficial implications for reducing land degradation but is typically productivity enhancing and more profitable (in the absence of herbicide resistance) (McTainsh et al., 2001; D'Emden et al., 2006).

Other factors that will influence the level of agricultural productivity in the future include the level of public investment in agricultural research and rural infrastructure, improvements in inputs such as fertilizers, irrigation and genetically modified crops, environmental degradation and climate change. This is not an exhaustive list of the influences on agricultural productivity, nor are the listed factors of equal importance with respect to their potential effects on productivity.

In all Asian economies, the predominant source of growth over the coming decades will be attributable chiefly to improvements in capital and total factor productivity rather than labor inputs. However, it is difficult to disentangle the relative contributions of capital and labor productivity.

Higher levels of education tend to promote mechanization. Substitution toward capital inputs from labor in farming is expected to further increase the importance of R&D in agriculture. Several countries within the ESAP region including Australia, Philippines, India and China will be able to exploit their comparative advantage in research and jointly benefit from each other. Their commitment toward the continued development of their rural sectors will most likely benefit from joint R&D projects that focus on the development and adoption of AKST.

Improvements in total factor or partial factor productivity in the ESAP region have resulted from increased investment in agricultural research and irrigation infrastructure. Returns to these investments have been highest in areas with significant land shortages but good institutional structures (Pingali and Heisey, 1999).

The growth rate of areas harvested for agricultural production will decline from now to 2020, along with decreased availability of arable land, increasing population pressures and land degradation (IFPRI, 2001). Given these limitations, yield growth becomes an important determinant of productivity growth. Of particular interest for this region is yield growth rates for cereals (including rice and wheat), which are projected to decline significantly in South Asia, Southeast Asia and East Asia over the period to 2020 (Figure 4-4).

Agricultural productivity growth will be crucial to ensuring adequate food supplies, however this will be challenged over the coming decades by resource depletion, environmental degradation and increased market pressure to use food crops for ethanol feedstock. Agricultural research and

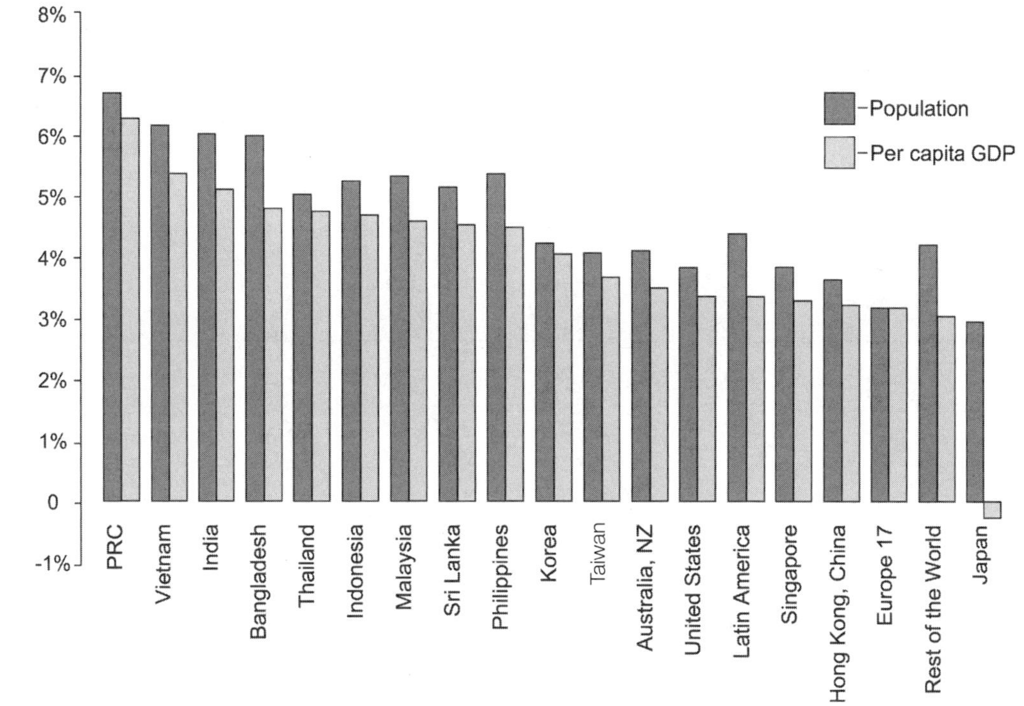

Figure 4-3. *Baseline growth projections by Asian Development Bank. Source: Roland-Holst et al., 2005.*

Table 4-2. *Sources of productivity.*

		2005-2010	2010-2015	2015-2020	2020-2025
East Asia, excluding Japan					
GDP		6.7	6.0	5.5	5.3
Contribution of:	**Labor**	0.4	0.3	-0.1	-0.1
	Capital	3.4	3.6	3.6	3.5
	TFP	2.8	2.2	2.1	2.0
Southeast Asia					
GDP		6.9	6.8	6.1	5.5
Contribution of:	**Labor**	1.0	0.8	0.6	0.4
	Capital	1.8	2.1	2.4	2.6
	TFP	4.4	4.2	3.4	3.0
South Asia					
GDP		7.0	6.1	5.7	5.3
Contribution of:	**Labor**	1.0	0.9	0.8	0.7
	Capital	2.2	2.2	2.2	2.2
	TFP	3.8	3.0	2.6	2.4
Developing Asia					
GDP		6.8	6.2	5.6	5.3
Contribution of:	**Labor**	0.6	0.5	0.2	0.1
	Capital	2.9	3.2	3.2	3.2
	TFP	3.2	2.7	2.4	2.2

Source: Roland-Holst et al., 2005.

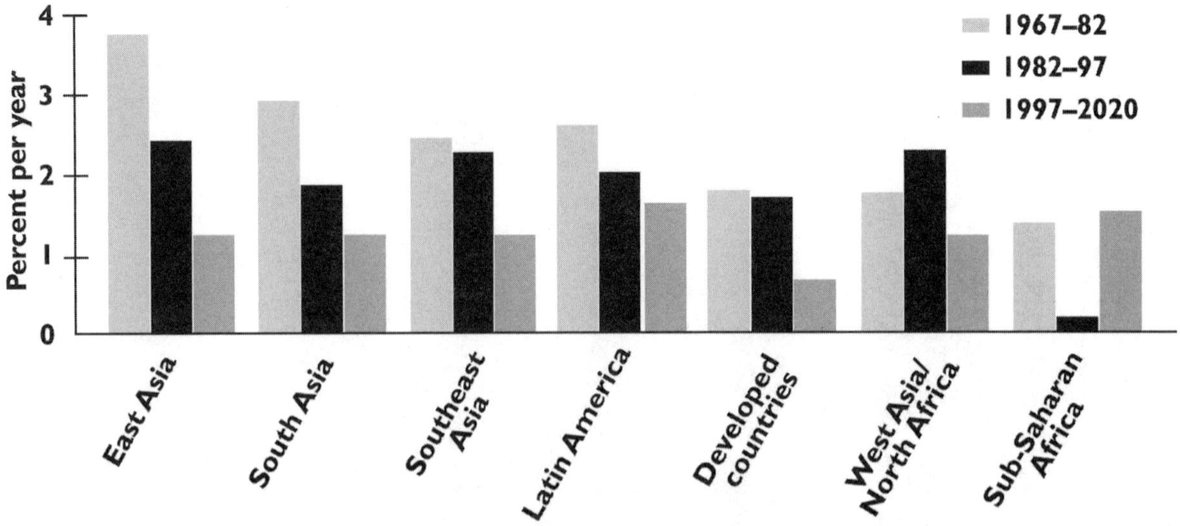

Figure 4-4. *Yield growth rates by region, all cereals. Source: Rosegrant et al., 2001.*

development will be important in ensuring continued productivity gains of the magnitude required to limit local food security concerns.

It is estimated that in Asia and Oceania, there are around 12 million ha under certified organic production. Most of this land area is in Australia, with only a fraction of organic agriculture production attributable to Asia (Willer and Yusefi, 2005). Organic agriculture is of increasing interest to both NGOs and agribusinesses because it offers a new sustainable development business opportunity. In addition, organic agriculture is considered to offer potentially higher incomes due to lower energy requirements, decreased water and pesticide inputs, and because it buffers yields against drought (Pimental et al., 2005).

4.2.2.3 Commodity prices

Theoretically it is expected that over time the prices of primary food commodities will decline relative to prices of manufactures as a result of (1) wage and productivity differentials between low income countries that typically produce primary commodities and higher income countries that typically produce manufactures, and (2) lower income elasticity of demand for primary commodities (Prebisch, 1950; Singer, 1950). This globally worsening terms of trade portends progressive squeezing of the rural economies in ESAP where a majority of the population still subsists on agriculture.

History reveals a downward trend in the prices of agricultural commodities as productivity in the agricultural sector and technological advances have been made, with some fluctuations caused by weather conditions. Particularly prior to the current resurgence of globalization since the 1990s, most developing countries kept their domestic food prices below global food prices and artificially raised the price of manufactured goods above international prices. Gradual changes and reductions in tariffs, as well as new import and export regimes have changed these price regimes in the ESAP region.

Commodity and energy prices are interlinked. High energy prices translate into higher commodity prices vis à vis manufactures. For example, the highly energy intensive production of fertilizers can affect the prices of agricultural products by raising input costs (ADB, 2006b). High oil prices may also result in increased demand for biofuels as consumers substitute least cost fuel sources. This may have implications for local food security in some regions (see 4.9.1) if demand for biofuels is strong enough to displace traditional agricultural food crop production.

Projected real agricultural commodity prices are expected to fall by 2015, relative to 2001 levels. ADB (Park and Zhai, 2006) presents a range of commodity price projections to 2015 under three scenarios. The scenarios centre around a baseline, with assumptions about higher and lower agricultural TFP growth, energy efficiency and energy reserves. The disaggregated impacts of the softening and tightening variables are given alongside the aggregate impacts on commodity prices. Changes in agricultural productivity are associated with roughly equivalent percentage changes in world agricultural commodity prices. Energy security and environmental constraints represent two significant uncertainties that could affect projections of agricultural commodities prices into the future (see 4.2.7, 4.2.8 and 4.2.9).

4.2.2.4 Globalization, economic growth and agricultural markets

Industrialization in agriculture has resulted in the coordination, production and distribution of produce to a larger market. Improvements in transport and information networks have a tendency to increase mobility between suppliers around the world. The higher degree of access to a variety of goods has assisted in integration along the supply chain including between abattoirs, supermarkets and other retailers. This means that demand for farm produce is no longer restricted to nearby markets. Rather, industrialization has increased the accessibility of goods and services to include distant consumers and suppliers as well. These trends will accelerate in developing countries into the future. Globalization is also likely to result in greater product diversification to supply new and varied markets. Technological change in agricultural production, improvements in rural infrastructure and diversification in food demand patterns will trigger product diversification toward high value food products (Pingali, 2004). Demand for organic agricultural products has been growing faster than overall food markets, particularly in developed countries for some time, and the expected continuation of this growth trend into the future could have substantive positive implications for local land management, biodiversity, food security and rural livelihoods (El-Hage Scialabba and Williamson, 2004; UNCTAD, 2006).

The effects of trade liberalization could have significant growth benefits translating into poverty alleviation in developing Asia, reflecting the importance of agricultural incomes. The removal of tariffs and agricultural subsidies on agriculture and food is expected to significantly improve market access for developing countries (Anderson and Martin, 2006). Estimates of a 48% increase in exports of agriculture and food from China, a 17% increase from Indonesia, 13% increase from the Philippines, 88% increase from India, 24% increase from Bangladesh, 35% increase from other East Asia and a 10% reduction from Viet Nam are reported (Hertel and Keeney, 2006). These improvements in market access are not without potential downsides; globalization involves the transportation of products over much greater distances and thus also increases the energy inputs required to move products from source to markets (see 4.2.9). There are visible trends of environmental degradation following in the wake of increasing globalization and trade liberalization, especially when production of high-value crops or commodities draw upon limited water resources, use chemicals or pesticides in excessive doses, add to energy consumption in agriculture, and worsen employment opportunities for women (unskilled) and marginal populations, forcing them to resort to further exploitation of fragile resources like ground water, forest land, protected species of wildlife, etc. (MA, 2005; Nijam, 2005; UNEP, 2006) (see 4.2.5).

Moreover, if improvements in market access are limited, e.g., barriers to trade in agricultural products remain in place in the large EU and United States markets, this could have serious negative implications for developing countries including those in the ESAP region. A significant question moving forward in the liberalization negotiations is whether developing countries should aim for a reduction in barriers in developed countries or pursue maintained or expanded preferential treatment (Binswanger and Lutz, 2000).

4.2.3 Implications of growth for agriculture

The implications of economic growth prospects on poverty reduction will be significant over the period to 2030 (Table 4-3). Different assumptions about capital and labor mobility will influence future economic growth prospects in the ESAP region. Estimates of poverty reduction in Asia in the future will depend on key assumptions including (1) the nature of growth and redistributive policies, (2) the benchmark of abject poverty (less than $1 per day and less than $2 per day), and (3) the inclusion of China and the impact of its polices on poverty reduction (in contrast to India—where little effort or impact is evident). The projected decline in poverty in Asia as a whole is positively affected by the inclusion of China (and its policies) and negatively influenced by the relative inability of some developing countries to make the necessary investments for economic growth (Roland-Holst et al., 2005). If the ESAP member countries can sustain the rate of growth they experienced during the period 2000-2004, then it is likely that all the countries will be able to meet their income and poverty eradication targets by 2015 (ADB, 2005). Further, if these rates of growth are sustained over the next decade, then these countries can completely eradicate abject poverty of less than $1 per capita per day by 2025 (ADB, 2005).

4.2.3.1 Food consumption and demand

Projections indicate that as population growth rates decline and countries become richer, demand for food will continue to grow, albeit at a declining rate. The annual average rate

Table 4-3. *Projections of poverty by region.*

Region	Population (%)					
	less than $1 per day			less than $2 per day		
	2003	2015	2030	2003	2015	2030
East Asia and Pacific	11.5	2.8	0.8	40.2	15.5	6.7
China	13.9	3.6	1.1	41.2	16.5	7.3
Rest of East Asia and Pacific	6	1.1	0.2	37.7	13.5	5.4
South Asia	33.2	16.2	8.1	79.5	60.2	46

Source: World Bank, 2007.

is projected to fall from 2% for the period 1989 to 1997-99 to around 1.4% on an annual average basis over the period 2015 to 2030 (FAO, 2003). This declining growth rate is expected to be more pronounced in developing countries, where the food demand growth rate is projected to fall from an annual average of 3.7% over the past 30 years to a mean of 2% a year over the next 30 years. This differential is attributable to the fact that daily food consumption in many large developing countries such as China is approaching that of industrialized countries. Once this level converges, the growth rate in total food demand will slow. Notably however, this trend is not expected in India because cultural traditions will hold the country's demand for meat and animal feed to well below that in China (FAO, 2003).

Rapid population and income growth in East Asia have been key drivers behind the increasing demand for world food commodities (ADB, 2006b). Population and income growth in South Asia over the coming decades will contribute to this growing food demand, as will movement toward adequate food consumption levels and improvements in nutrition (FAO, 2006b).

Growing projected demand for agricultural commodities in the ESAP region suggests there will be significant challenges in achieving the Millennium Development Goals. For example, while future economic growth is projected to raise per person incomes by 1.9% a year between 2000 and 2015, it is unlikely that the absolute impoverished population can be halved, with projections suggesting a more limited reduction from 1.27 billion in 1990 to 0.75 billion in 2015 (FAO, 2003).

To satisfy the projected increases in food requirements, an additional billion tonnes of cereals will be needed each year by 2030. These projections imply that developing countries could be required to import 265 million tonnes of cereals annually (FAO, 2003). Trade liberalization will therefore be an important factor in ensuring adequate food supplies in the region as regional production falls short of regional food demands (Park and Zhai, 2006).

Developing Asia is increasingly influencing international commodity markets and prices. The rapid industrialization and structural change witnessed in China will continue to have enormous implications for global commodities demand and prices. Agricultural and light manufactures demand will increasingly give way to heavy industrial raw materials (ADB, 2006b). This process is expected to be repeated in India over the next few decades, with additional significant implications for energy and other raw materials demand, and further into the future, food demands that can only be met through imports from lower income countries (ADB, 2006).

Increasing wealth over the past few decades within the ESAP countries has facilitated a change in consumption patterns toward higher value food products and imports. Wage income will increase at the fastest rate in East and South Asia of anywhere in the world (World Bank, 2007). As disposable incomes increase, demand for starchy staples is expected to decline, while demand for livestock products, fruits, vegetables and processed food products is projected to increase (OECD-FAO, 2006).

The implications of economic growth for food demand will depend on the relative wealth of countries. For low income Asian countries, changing diets will result in an increase in per person consumption of wheat and rice and a decline in consumption of maize and other coarse grains by 2020 (IFPRI, 2002). With additional growth and improvements in per person wealth, food demand will shift again.

A driver of changing import demand for cereals has been significant improvements in agricultural productivity as a consequence of the Green Revolution (Francks et al., 1999). Rapid urbanization has also contributed to changing dietary profiles. Urbanization often generates additional demand for higher value processed food and tropical beverages such as coffee (ADB, 2006b). In China and India, this has also led to significant shifts in food import demands. This trend has been particularly apparent in China, where wage rates have risen faster than in many other developing countries and this has also created opportunities for agricultural exports from other low income countries (World Bank, 2007). In particular, changing dietary patterns have exerted an influence on the level and variety of imports for food products such as meats, vegetables, edible oil and oil seeds (ADB, 2006b).

The fraction of the global population living in countries with per capita food consumption under 2200 kcal per day is projected to fall to around 2.4% in 2030. In South Asia, this fraction is expected to fall by 40% between 1997-99 and 2030 and in East Asia by 50% (FAO, 2003).

Daily calorific intake increases through time in both regions, however the composition of that intake varies between South and East Asia (Table 4-4). The key differences are that while demand increases for roots and tubers in South Asia, the trend in demand is downward for those commodities in East Asia. Meat consumption grows faster over the projection period in East Asia than in South Asia, and South Asian diets are far more heavily weighted toward milk and dairy products than in East Asia.

The composition of aggregate cereals demand is expected to substantially change over the period to 2020. In Asia, the aggregate demand for rice is projected to decline by 4% and wheat demand to decline by 1% between 1997 and 2020, while aggregate demand for maize as a percentage of cereal consumption will rise 6% (IFPRI, 2001). However, per capita demand projections for cereals reveal a slightly different picture. Per person demand for rice in Asia is projected to remain constant over the period 1997 to 2020, while per person wheat demand is expected to rise 9% and per person maize consumption is projected to decline by around 16%. These shifts are due to rapidly growing incomes and urbanization (IFPRI, 2001).

At the sub-regional ESAP level, it is expected that in North India, Northern and Southern China, the current demand for rice will continue well into the future (FAO, 2006b). A gradual shift in demand from rice to wheat is taking place in Southeast Asia—especially in Thailand, Philippines, Malaysia and Indonesia. Although this trend is just starting in Viet Nam, and not yet evident in Myanmar, Laos PDR and Cambodia, it is expected that there will be an increasing shift in the ESAP region as a whole from rice staple diets to rice and wheat staple diets (FAO, 2006b). Pacific countries such as Samoa, Tuvalu and Solomon Islands are projected to reveal an increasing preference for rice and wheat based diets in place of traditional starchy

Table 4-4. *Changes in the commodity composition of food demand (expressed in kcal/person/day)*

	1969/71	1999/01	2030	2050
South Asia				
Cereals, food	150.4	157.1	167	169
Roots and tubers	16.9	23.5	31	36
Sugar (raw sugar eq.)	20.3	25.6	30	32
Pulses, dry	14.5	10.1	8	7
Vegetable oils, oilseeds and products (oil eq.)	4.6	9.7	15	18
Meat (carcass weight)	3.9	5.5	12	18
Milk and dairy, excl. butter (fresh milk eq.)	37.0	67.6	106	129
Other food (kcal/person/day)	84	141	180	200
Total food (kcal/person/day)	**2,066**	**2,392**	**2,790**	**2,980**
East Asia				
Cereals, food	152.2	186.7	176	162
Roots and tubers	96.6	65.8	61	53
Sugar (raw sugar eq.)	5.7	11.6	17	20
Pulses, dry	4.8	2.0	2	2
Vegetable oils, oilseeds and products (oil eq.)	3.5	10.6	15	17
Meat (carcass weight)	9.2	39.8	62	73
Milk and dairy, excl. butter (fresh milk eq.)	3.7	11.3	21	24
Other food (kcal/person/day)	98	322	405	440
Total food (kcal/person/day)	**2,012**	**2,872**	**3,190**	**3,230**

Source: FAO, 2006b.

roots and tubers. A notable exception is Bangladesh, which is expected to continue to maintain a heavy dependence on rice based diets well into the future.

While East Asian cereal demand will double and significantly exceed its production levels, a surplus of 73 million tonnes will be produced in the developed world between 1997 and 2020 (IFPRI, 2002). Unlike many other regions projected to expand cereal cultivation areas, Asia will face limited development opportunities since approximately 80% of the potentially arable land is already under cultivation. In addition, rapid urbanization will consume vast areas of potentially arable land.

Yield growth rates will therefore be more important for ESAP countries than some other regions. However despite the importance of improvements in cereal yield, rates of yield increase are expected to continue to decline to around 1.25% annually in South Asia, 1.2% in Southeast Asia and 1% in East Asia over the period 1997-2020 (IFPRI, 2002).

Other crop markets will face similar futures. Aggregate roots and tubers demand in the developing world will increase by 55% between 1997 and 2020 but supply is expected to increase by only 51%. This is expected to result in a decline of exports in roots and tubers out of Southeast Asia. Southeast Asia will increase the regional surplus of edible oils, with production growth exceeding demand growth by 7 million tonnes between 1997 and 2020. East Asia will increase its edible oil imports from 4 million tonnes in 1997 to 10 million tonnes in 2020 (IFPRI, 2002).

Under a pessimistic scenario, Indian and Chinese cereal production declined 15% relative to the baseline scenario, resulting in significant trade deficits for both countries (IFPRI, 2002). However, the analysis indicates that world markets are capable of absorbing these large increases in imports without major price consequences. Under this scenario, Indian kilocalorie consumption in 2020 declined by 171 kcal per capita per day relative to the baseline scenario and Chinese kilocalorie consumption declined by 264 kcal per capita per day. The number of malnourished children is projected to increase by 2 million relative to the baseline scenario in both countries.

Access to water and sanitation remains a major concern in the ESAP region and governments may need to provide potable water as a basic input to ensuring food safety and health. Expansion of domestic markets for processed foods and beverages along with growth in agricultural trade has led to increasing awareness of food safety and quality in the region. Despite acceptance of international food safety regulations (e.g., Hazard Analysis and Critical Control Point) many ESAP countries need to intensify efforts to implement and enable quality assurance systems. With the exception of a few countries like Japan, the consumers in ESAP are largely indifferent to questions concerning the safety of biofortified and genetically engineered food (Hoban, 2004), and reveal widespread acceptance of relatively low cost pharmaceutical products (drugs and vaccines) from genetically modified or cloned animals.

4.2.3.2 Implications of GDP growth for agriculture

In the past few decades, agriculture has experienced some stagnation in growth compared with historical rates. The sector has also experienced lower investment in recent years and this trend is likely to continue in future. Other issues that may affect the agriculture sector include withdrawal of subsidies, less priority for R&D and an aging labor force. The manufacturing sector, which has grown strongly in the past few decades, is by contrast receiving many incentives including tax exemptions, low import duties and prioritized training. This trend will likely intensify, especially in the large emerging economies of ESAP. In ESAP as in much of other developing regions of the world, the impact of an increasingly globalizing and industrializing food system will be evident as diverging rural worlds and increasing concentration of power in the hands of a few transnational actors (Pimbert et al., 2001).

The expected rapid growth of many of the ESAP countries will have significant implications for agriculture. Structural transformation of these economies is expected—while the absolute output of agricultural production from this region is expected to grow over time, the relative importance of agriculture will decline as manufactures and services become relatively more important sources of GDP. The largest contributors to the rise in developing country service exports over the past two decades have been East Asia and the Pacific and Europe and Central Asia (World Bank, 2007).

Although the agriculture sector is expected to continue to be a major employer in the region, labor will continue to move to other sectors and increasingly so as wage differentials grow in relation to other sectors. Although there is a shift toward off-farm employment in South East Asia, agricultural output in the region has not fallen; in fact, agricultural production capacities in Indonesia, Malaysia and other countries in the ESAP region have increased in the past two decades (FAO, 2006a). This is partially explained by the adoption of increased capital investments in technology to offset labor shortages (Mahmoud and Shively, 2004).

It can be expected that trade and investment liberalization and greater globalization will enhance allocative efficiency and enhance specialization along the lines of countries' comparative advantage. Globalization, or domestic policies that lead to more efficient allocation of resources, reduces production costs and thereby leads to significant increases in output and income growth. These effects are documented in the literature on the economic effects of trade liberalization (see Schneider et al., 2000; Anderson and Martin, 2006). However, there are predictions that agriculture in particular may be negatively affected by the pressures of globalization given that 60% of the farming community is small-scale/marginal peasantry who may suffer heavy losses during adjustments to global changes (Ghosh, 2005; WCSDG, 2004).

4.2.4 Sociopolitical drivers

4.2.4.1 Economic liberalization and regulation

Economic liberalization and increasing globalization have benefited those countries and populations who have capital, entrepreneurial ability, education and skills (WCSDG, 2004). The poor populations of the world, illiterate and unskilled, with limited capital assets, have lost out. Since a majority of this population is in Asia, especially South Asia, this region will face a crisis of employment and income generation, especially in rural and remote areas, amidst rapidly expanding urban growth and flourishing international trade in services and manufacturing. (WCSDG, 2004). Most countries in the ESAP region will meet their MDG targets of reducing poverty by half between 1990 and 2015 (ADB, 2005). Some countries like China, Thailand and Sri Lanka have already halved or more than halved their population living in abject poverty.

What are the chances that economic liberalization in an era of globalization will bring more effective redistribution in unequal societies in ESAP countries such as India, the Philippines and Bangladesh? Overall it does appear that globalization and economic liberalization may create a new political cleavage between "cosmopolitans" who have skills and assets to adjust easily to changes in global markets (and consequently increasing political clout domestically) and "provincials" who are less mobile with lower labor market skills (Bowles and Pagano, 2006).

Economic liberalization and new labor and knowledge market regulations may enhance private investment—but to what extent these drivers will change investment in R&D is not known. In a region where fundamental labor regulations that protect basic rights of workers are "conspicuous by their absence" or are never enforced (ADB, 2005), there is a significant opposition to any reform in the regulations governing input markets, trade or tariffs. This opposition is largely voiced by NGOs who state that poor farmers cannot afford to pay for or purchase seeds and other technologies (see Chapter 3).

The impacts of globalization are likely to vary widely, from posing severe production and consumption constraints on small-scale farmers and the agricultural sector in general in areas that are not well-endowed with resources or that are without social security, to rapid growth in markets, commodity production and trade in other areas. Latecomers into processes of globalization are likely to be penalized because of a lag time in bringing domestic institutional structures and productive sectors into line with new norms for tariffs, markets, and standards, as well as to compete with other countries that have been pre-disposed to such norms or have fewer people affected by these new norms (Nayyar, 2006). Given the uneven impacts of globalization, it is likely that developing countries may invest less of their scarce public sector resources on relatively long term investments like AKST and may address more immediate issues like subsidies for small-scale farmers to access inputs or price support mechanisms. These in turn may have negative consequences for agriculture in developing and developed countries.

4.2.4.2 Political stability

In most developing countries (including the ESAP region) the failures or lack of development commitment of the State are passed on as the ill effects of globalization (see Bardhan et al., 2006; Przeworski and Meseguer Yebra, 2006). Given that the world over, developed countries (except the USA and UK) have enhanced and expanded the size of their government sector (ratio of government expenditures to GDP), it may be expected that the ESAP region will also respond in a similar way in the future—with governments playing a major role in poverty alleviation, macro-economic management, social insurance and environmental protection (Bardhan et al., 2006). Politically, there may also be increasing diversity of institutions (policies) among nation states. The key message for AKST institutions and organizations is that the capacity of economic liberalization and globalization to dismantle the barriers to economic opportunity faced by the poor depends critically on the capacity of public bodies to respond to the voices of the poor.

One of the major responsibilities of public bodies in the developing countries in Asia will be to find the resources and appropriate allocation of these resources to enable rural regeneration, as well as provide the much needed social security in rural areas. The Asian crisis of the late 1990s showed the vulnerability of the rural economy to urban-centered financial and economic crisis (Gerard and Ruf, 2001). When remittances from migrants decline and many who lose their jobs return to the rural economy, agriculture becomes the mainstay of the rural economy. The burden, without government support or any risk coverage, falls on the rural economy, particularly on rural women (Nathan and Kelkar, 1999; Cook et al., 2003). Despite the existence of useful models of social security nets or investments by the State, rural Asia is likely to be lacking in these kinds of investments, because the pressure for other productive investments will outrun the demand for these in the near future. The Indian National Rural Employment Guarantee Act (NREGA, which provides a minimum assurance of up to 100 person-days of employment per household), the Chinese continued ownership of land under the household responsibility system and the Bangladeshi microcredit systems (where NGOs, with financing support from the financing authorities, reschedule loan repayments, provide new loans and often undertake relief measures) play the role of safety net. With privatization of health and education services continuing at the current pace, there will be a large gap between these minimum incomes and what is needed to meet the social needs of Asia's impoverished rural populations (Ahmed et al., 1991). In parts of Asia, chronic poverty and the lack of safety nets drives people to organize around paths of violence.

The ESAP region is a hotbed of political crises of various sorts—largely domestic (ranging from secessionist parties, naxalite movements, domestic and cross-border terrorism, communal tensions, totalitarian regimes and anti-democratic legislations) in countries ranging from Sri Lanka, India, Nepal, Thailand, Bangladesh, Myanmar, East Timor, China, Viet Nam, Indonesia, Papua New Guinea and some other Pacific countries. Political instability caused by such forces will reduce and even disable trade in agricultural goods and discourage learning and technology dissemination in these regions. There is significant cross-border trade and regular economic activities occurring along a soft and self-negotiated border among people living in different (political) nations but similar agroecological terrains. The exchange of agricultural produce or other rural goods across soft borders, e.g., in India's Northeastern states, contributes to increasing economic prosperity and sharing of ecological and cultural resources (see Hazarika, 2000; Baruah, 2005). An important political message from these fungible border zones across countries in the ESAP region is for nation states to recognize that people and their fundamental rights and access to regional or local natural resources and cultural activities can be nurtured as a major instrument of peace in the ESAP region.

In the coming decades, a strengthening India-ASEAN relationship may add to political stability in the region in a wider sense. This depends on India's capacity to generate economic growth within its borders and on its capability to enhance national resilience of ASEAN's regional member states, and thereby to promote regional resilience (Prasad, 2006). In the ESAP region, India will play a much greater role in ensuring regional integration, promoting relationships in the ESAP region with global powers (e.g., the USA) and allaying member country fears about Chinese expansion and potential control over their own markets (Tammen, 2006). Indian political parties, irrespective of their ideological differences, are committed to (1) peace and friendly neighborhood relationships (2) industrialization as the mainstay of economic growth in future, (3) rural regeneration, and (4) infrastructure and energy investments. However, political strategies toward regional cooperation (especially in trade and labor mobility), domestic policies for poverty alleviation, health and education services and rural industrialization will depend significantly on whether the moderate and secular political parties or the right wing parties are in power in future.

Politically the Chinese Communist Party is keen to improve the image of China in an increasingly globalizing world. The political agenda is to make sure that domestic development is endogenously driven, will enhance transparency in decision-making and will combat corruption at all levels (Economic Daily, 2007). An important commitment for the future will be that the country will adhere to peace and peaceful processes for development in the region. Political stability in the ESAP region will be significantly affected by the power play between the USA and China as well as the interests of the USA its allies in the region (Christensen, 2007).

The impact of political stability on agriculture and AKST will mainly be in terms of enabling investments in cross-border infrastructure or natural resource management (such as waterways, mountain ecosystems, roads, etc.). These are likely to bring long term improvements in the livelihoods, market opportunities and economic growth for people in these political borderlands in ESAP. Increasingly, the ESAP countries will realize that it is the capacity of public organizations to respond to the voices of the poor, provide social security where it is lacking and engage constantly with emerging threats to impoverished rural livelihoods that will lead to political stability and economic growth.

4.2.4.3 Deregulation

Deregulation has been attempted in many ESAP countries such as India, China, Thailand, and Indonesia; developed ESAP countries (Australia, New Zealand and Japan) have taken the lead. While the impact of fiscal reform on AKST may not be direct, it is important to see that countries that have made structural changes have attracted a significant amount of private investment in AKST—especially in food processing and retailing, biotechnology, and specialized product development like organic agriculture.

Overall the macroeconomic policy reforms in South Asia began by liberalizing trade. The current scenario promises that this trend will continue well into the future with implementation focusing on deregulation and privatization (Kemal, 2007). The degree of openness in the economy will continue to be high in Sri Lanka; India will be the most closed (relatively) for some time to come (World Bank, 2006a). The latter is to be expected until employment growth rates match the growth rates of the economy; unemployment in an economy that is increasingly deregulated will remain a major concern for the Government.

In India, many argue that deregulation with trade liberalization spells doom for the agricultural sector (Ghosh, 2005; Patnaik, 2005). Specifically for AKST, deregulation will imply that poor farmers will increasingly lack the resources to buy essential inputs, access relevant S&T inputs and information and participate in crucial export-market driven agricultural developments. This may lead to an overall increase in hunger and poverty among the already poor in rural areas. Given the growth and diversification patterns in Indian and Chinese manufacturing and trade (including domestic trade) it appears that the apprehension that import liberalization might lead to a large-scale demise of domestic industries is unwarranted (Mani, 2005; Veeramani, 2007). Domestic industry in the Asian region has been able to and will continue to compete and survive by specialization in narrow product lines (Veeramani, 2007).

There is a likelihood of increasing concentration of agricultural input and output actors with a few multinationals converging to control a major share of the global agricultural markets. Given the rate of growth of supermarkets and the increasing openness in Asian economies to Foreign Direct Investment in food retailing, it is estimated that by 2010 there will be only 10 major global retailers of food (Vorley, 2001; Reardon et al., 2003). All food grain trade in the region will be controlled by a few major transnationals like Cargill, ADM (Archer Daniel Midlands), Conagra, Monsanto, Nestle and Atria (who now control over 90% of global foodgrain trade (Shrivastava, 2006). Cargill will steadily increase its investments in the oilseeds and edible oils market and it promises to increase its share in the Asian vegetable oil market in the future.

Implications for the agriculture sector arising from increasing deregulation of the economy include massive growth of private investment in agribusiness, especially in commodity markets and retail trade, and increasing standardization of agricultural produce from the region. What this implies for the diversity of food systems in Asia is not known yet, though it is likely that some provisions such as "geographical indicators" may enable development of a market niche for select agricultural products like basmati rice, jasmine tea or tussar silk. A direct positive impact of deregulation will be in increasing the linkages of the agricultural sector with other manufacturing and service sectors, thus expanding resources and facilities for growth.

4.2.4.4 Infrastructure

Infrastructure constraints affect economic growth in the ESAP region. If economic growth is considered important and is held as a key to poverty reduction, then all the ESAP countries will invest heavily in infrastructure provision and improvements. Currently there is a major gap between levels of infrastructure investment and access to basic infrastructure between the East Asian economies and South Asian economies. Significant improvements in infrastructure investments can add up to 0.85% per annum to economic growth in China (2005-2014), 0.80% in Indonesia, 1.32% in India, and 0.45% in Bangladesh (Ianchovichina and Kacker, 2005). It is estimated that the Asian economies will have to invest at least 6.5 to 7% of their GDP on infrastructure provision during 2005-2010, without which there will be increasing infrastructure constraint to economic growth (Fay and Yeppes, 2003; Jones, 2006). Currently only China and Viet Nam seem to be investing at these rates. Countries like India, Indonesia and Philippines have fallen behind their own target investment levels, with the marginal increase in private capital investment in infrastructure not compensating for the decline in public investment over the 1990s and early 2000s (Jones, 2006). In all these economies, the overall macroeconomic orientation seems to follow the trend from the 1990s, with increasing foreign direct investment (FDI) in infrastructure development, more relaxed norms and less formal approval regimes, special incentives for technological improvements or export oriented units (in industrial investments) encouraging private infrastructure investments.

Liberalization has had a direct impact on infrastructure development in the ESAP region. Investment levels have been high since the early 1990s in the entire region, with countries investing an average of 30% of GDP in various investments, with much of this (ranging from about 1-14%) share going into infrastructure development (World Bank, 2006b). The growth of rural infrastructure, especially rural roads, has been shown to have a positive impact on the growth of private extension in South India, electronic commerce and crop advisory services in the Deccan Plateau states of Andhra Pradesh, Maharashtra and Karnataka (Dhan Foundation, 2005; Prahalad, 2005). Another key infrastructural investment is in the water and sanitation front, creating immense opportunities for services and achievement of the broader MDGs (Farrington, 2006).

A major development that will transform infrastructure and development opportunities across Asia has been the recent Intergovernmental Agreement on the Asian Highway Network (adopted in 2005) and the Intergovernmental Agreement on the Trans-Asian Railways (see www.unescap.org). These UNESCAP-led pan Asian infrastructure investments will lead to direct road access across South-East-Central and West-Asian countries and will also provide a land link to Europe, as well as dry ports which can consolidate and distribute produce, create employment locally and provide port access to some of the hinterland production centers, especially in transporting perishable or delicate products. These projects will enable overall infrastructural improvements with shared technical standards across countries and sharing or collaboration with financial organizations.

Much of the infrastructural investments, however, continue to take place in urban or peri-urban areas and coastal China, metropolitan areas in Thailand, Indonesia, Viet Nam and India, leaving rural infrastructure relatively unattended (World Bank, 2006a). Urbanization is a major driver of infrastructure—with the likelihood of 50% of the East Asian population being urban in 2025 and 40% of this urban population likely to be poor (in 2025), there is an urgent need for public sector investment in urban infrastructure and delivery of essential services such as piped water, electricity, communication and roads (Jones, 2006). The ESAP region reveals a wide disparity in basic needs infrastructure such as water supply and sanitation—ranging from 93% access to rural sanitation in Thailand to 13% in rural Solomon Islands (World Bank, 2005). On average the investment in and access to basic infrastructure for water and sanitation is marginally better in East Asia and Pacific countries compared to South Asia. The South Asian countries are likely to invest more in infrastructure provision, especially in energy and energy trade across the border (Jones, 2006).

The growth of infrastructure, especially massive rates of growth of investments in urban or peri-urban areas promises little for rural areas and agriculture where the lack of infrastructure will continue to be a major hurdle for further growth in yield, incomes and overall development. Private investment may always find it attractive to invest in areas of quick and assured returns—something that agriculture does not promise in the ESAP region. In South Asia, which is already starved of essential water and sanitation services, the demand on rural environmental services due to rising urban infrastructure investments may become untenable.

4.2.4.5 Regional cooperation

Regional cooperation in infrastructure and service delivery is bound to increase in the near future throughout the ESAP region. Investment in water and sanitation programs in many countries, including some small countries like Nepal and Sri Lanka, seems to play an important role in strengthening local democracy by bringing people's participation in the delivery and monitoring of water services. While increasing tensions over water along international borders seems to be a feature in all the ESAP border regions, there are several regional networks and cooperation agreements being confirmed or implemented. Despite increasing conflicts and contrariness in Government behavior around key river basins—the Ganges, Brahmaputra, Mekong, Indus, etc., platforms for negotiation such as the Indus River Treaty, Ganges Water Treaty and the Mekong River Commission have worked well and have the potential to evolve and expand into further infrastructure arenas as well as pro-active regional cooperation platforms. While the South Asian Association for Regional Cooperation (SAARC), Asia Pacific Economic Cooperation (APEC), Bay of Bengal Initiative for MultiSectoral Technical and Economic Cooperation (BIMSTEC), and Association of Southeast Asian Nations (ASEAN) are examples of regional cooperation, the Greater Mekong Sub-region, initiated and facilitated by Asian Development Bank (ADB) is an innovation in international cooperation especially in infrastructure development and benefit sharing. The unique features of the GMS are its geography (with each country sharing at least three border areas), its economics (bordered by China and Thailand—both dynamic economies), sponsorship (ADB as neutral facilitator and sponsor) and budget (ADB—from national allocations). The countries can opt into a "6 – x" agreement or choose not to enter the agreement/investment (Jones, 2006). Other regional cooperation initiatives that have emerged in this pattern and are expected to enhance agricultural trade are the South-Asia sub-regional Economic Cooperation, and the Central and South Asia Trade and Transport Forum. In sectors such as fisheries that are marked by heated conflicts within countries and between countries sharing seas/coastlines, regional cooperation and active academia-Government interactions with a wide range of stakeholders along with experiments with several institutional and policy options are emerging (Gupta, 2006; Salayo et al., 2006).

Regional cooperation in Asia has thus far focused on trade and economic cooperation, peace and security, and "less on deeper aspects of integration" (WCSDG, 2004). The SAARC regional cooperation appear to be wanting in several key areas of cooperation because of a lack of political commitment. More cooperation will be required to bring in a new "social charter" for regional cooperation addressing poverty and injustice, growing inequality, social disparity and environmental security (Najam, 2005; Sobhan, 2005). However, recent developments in ASEAN point toward deepening regional integration over the coming decade (Sobhan, 2005). It is important to recall that monetary cooperation among Asian countries increased substantially after the Asian financial crisis of 1997/98.

The tension between the two developed economies (Japan and Australia) in the ESAP region and their differing views on and expectations from Chinese economic growth is likely to increase in the future. Cooperation in monitoring financial health of the Asia Pacific economies includes arrangements such as the Manila Framework Group and the ASEAN surveillance group. Most importantly the discussion on an Asian Monetary Fund has evolved (with Japanese initiation) in Chiang Mai, Thailand, into two liquidity funds—the ASEAN Swap Arrangement and the Bilateral Swap Arrangement. The Japanese and Australian economies have also contributed to setting up the Pacific Economic Cooperation Council and the Asia Pacific Economic Cooperation (APEC) in the past thirty years (since the signing of the Nippon Australia Relations Agreement—NARA—in 1976); these have become more meaningful since the financial crisis of the 1990s.

The ESAP region embraces Japan with its mistrust of Chinese growth and Australia and its expectations of access to Chinese markets and mobilizing investments from China, both of which will continue to increase (Terada, 2006). As emerging industrial powerhouses and major investment markets, growth patterns in both India and China are likely to influence regional cooperation and rivalries.

Globalization and increasing intra-regional trade have played a significant role in enhancing regional cooperation in two sub-regions of the world—Latin American and the

Caribbean (LAC) and the Asia-Pacific region (Monata, 2006; Nayyar, 2006). The trend toward increasing trade relationships and regional cooperation between Asia and Latin America is likely to increase in the near future. It is likely that trade and economic cooperation between LAC and ESAP will increase at the cost of ESAP-EU and ESAP-USA cooperation (Tammen, 2006).

Chinese investments and Chinese and Indian cooperation with African countries is another trend that will increase significantly in the near future, bringing the ESAP region the status of preferred development partner for the African countries. Chinese overseas investments are expected to grow to US$60 billion in 2010-2015 (GIBS Review, 2006).

The call from the EU seeking ASEAN investors to invest in the new EU member States is an example of negotiating stakes for ASEAN in the EU (bilaterals.org, 2007). In today's hierarchy, the U.S. dominance is unchallenged, but U.S. preeminence is declining in relative terms and may in two to four decades eventually dissipate (Tammen, 2006).

It is predicted that smaller trading groups will bring in much greater intra-group gains, though globally their share may not gain much. The projections for SAARC trade are:

- Complete elimination of tariffs will increase intraregional trade by 1.6 times the existing levels. The volume of intra-SAARC trade will increase from the present figure of US$5 billion to around 14 billion in 2015. An increase in overall trade does not necessarily represent an increase in the ratio of inter-SAARC trade to total trade by SAARC members.
- Dynamic gains are 25% more than static gains.
- Smaller member countries tend to gain relatively more than larger ones. The latter is stated as an empirical argument, but it is a theoretical one as well and is borne out by the experiences of other FTAs elsewhere in the world (SFG, 2005).

Overall, regional cooperation presents a positive scenario for agriculture for all of ESAP. Major benefits that arise from regional cooperation include the potential for redressing the negative impacts of globalization on small farmers, rural women and other marginal production systems like coastal fisheries.

4.2.4.6 *Agricultural trade*

Trade in agriculture and allied products grew significantly in the region during 1991-2004 (FAO, 2006a). The ESAP region as a whole will continue to be a net importer of agricultural products (including forestry and fisheries). Current trends in net surplus production and trade surplus in sub-regions such as Southeast Asia and the Pacific islands are likely to continue (FAO, 2006a). South Asia and the developed Asian economies, particularly Japan and Singapore, will continue to be net agricultural trade deficit sub-regions. Many of the developing economies are likely to expend less of their foreign exchange reserves on import of cereals/other agricultural commodities, thus revealing relatively stronger agricultural trade positions. The impact of AKST on agricultural trade is evident in the fact that in ESAP countries in 2002-2004, food/agricultural imports, especially cereal imports, accounted for less than 10% of foreign exchange reserves compared to 1969-71, when cereal imports were 40% to 120% of foreign exchange reserves in some countries (FAO, 2006a). If WTO regulations on domestic subsidies in agriculture are accepted or enforced, some of the current agricultural exporters may become importers—of food and of labor to cultivate the food. The strength and resilience of multilateral treaties and multilateral organizations, as well as domestic policies to maintain economic growth and social justice will be tested in the context of agricultural trade in ESAP during the period 2010-2020.

The current trend is towards Preferential Trade Agreements (PTAs). Both India and China have increased their share in global trade and in various bilateral agreements and PTAs within the ESAP region. SAARC, SAFTA, PICTA and several other bilateral agreements have added to the flow of goods—especially agricultural and manufactured goods within the ESAP region. This trend is expected to continue well into the next two decades. China's growth in trade (9% of the global increase in exports and 8% of the global increase in imports during 1995-2004) compares favorably to India's (2% of increase in global exports and imports during the period) (FAO, 2006b; World Bank, 2006a).

The ASEAN + 4 (China, Korea, Japan and India) promises to be a powerful alignment for agricultural trade and economic development (Batra, 2006). In 2003, the intra-block trade for ASEAN+4 was 44% of the total value of trade in these countries, which is significantly higher than the intra-block trade in any other PTA (Batra, 2006). The sharp decline in tariffs and non-tariff barriers has been a major factor fueling this intra-regional trade. Some protection for selected commodities are often negotiated among members, especially among bilateral/preferential trade agreements, e.g., palm oil (Malaysia), rubber (Thailand), fruits and vegetables (Thailand, China), sugarcane, wheat, oil seeds (India).

One of the fears in most ESAP economies is that farmer livelihoods will diminish with increasing imports of critical crops and with removal of tariff barriers. With globalization the domestic and international policies that govern the barriers to economic opportunity for the poor will change (Bardhan et al., 2006). The key question is whether a responsible public sector in the Asia-Pacific countries and regional cooperation will help the rural poor and small-scale farmers weather the crisis and provide investments and incentives for better economic opportunities.

Regional and sub-regional collaboration around environmental concerns should follow increases in agricultural trade in ESAP. Major trade and economic cooperation agreements, environmental, biodiversity and ethnic factors call for sub-regional cooperation in ensuring sustainable development, especially for poor people in small island Pacific countries (UNESCAP, 2006). The Pacific countries are likely to face the greatest threat from climate change, with potentially disastrous consequences (See 4.2.8). The success of agricultural trade and its translation into sustainable economic development depends on how individual Governments handle social security for the rural and urban poor who will face a crisis in production and consumption and how the tradeoffs between agricultural production for trade and environmental costs are addressed in the wake of increasing input prices and decreasing real prices of agricultural commodities.

4.2.4.7 NGOs and civil society

The ESAP region is perhaps second only to the LAC region in terms of the intensity and capacity of NGOs and their capacity on a wide range of issues that are important to social issues. What originated as voluntary work in the immediate post-independence decades (1960s and 1970s) in the Asian countries became organized nongovernment organizations (NGOs) (Tandon, 2000). Starting as small trusts and locally based civil society groups, they have taken on different organizational formats: for instance the Asia Pacific Research Network (APRN) is a collection of 37 member organizations, and has a mandate to exchange information on local and international issues that shape society in the region. These organizations are expected to grow into international and multilateral organizations in the near future similar to the way in which the World Social Forum evolved. Increasingly there is a trend among NGOs in the region to focus on building an Asia-Pacific community (Yamamoto, 1995).

From being the effective or people-friendly implementation arm and partner of the State in development programs, the NGO sector in Asia will increasingly partner with actors in arenas such as agriculture, health, population, gender and empowerment, urban planning, water management, microcredit and insurance. A relatively new path that the NGOs are treading now, as partners of the corporate sector and in leading the environmental movement, promises to grow into a powerful driver of change in the ESAP region (Korten, 1997; Yamamoto and Ashizawa, 1999; Barkenbus, 2001). There are also civil society networks that work towards building effective working relationships among countries. The South Asian Perspectives Network Association (SAPNA) is an example, pointing out future directions in development policies in the South Asian countries (Wignaraja, 2005). The argument for this engagement is that South Asian countries will need a new praxis and management of knowledge systems in order to address the demand for development with equity. An emerging trend in the Asia-Pacific region is that of NGO innovation in non-formal education (NFE) (UNESCO, 2003). Advances in NFE and NGO leadership in empowering local people (e.g., NFE and access to credit in Korean villages, functional literacy in rural China/Bangladesh) have major implications for pluralistic agricultural extension practice and AKST in general (Sulaiman and Hall, 2005; UNESCO, 2005).

The Asia-Pacific NGOs have not partnered with science—agricultural science in particular—in learning for, planning and implementing knowledge-based agricultural development. Their role in agricultural science and technology has been limited to technology dissemination. Though there are cases where NGOs have helped scientists to learn about local contexts, generate new/modified technologies and find new ways of working, these cases are rarely acknowledged by formal public sector agricultural science (Rhoades, 2000). The trend of NGOs to partner with research and non-research actors in the agricultural innovation system may be strengthened in the coming decades (Hall et al., 2004). While the NGO arm of corporate social responsibility is appreciated widely and is even considered an essential partnership as a check on unhindered exploitation and profiteering, there is increasing concern that NGOs funded by/co-opted by corporate sector may lose their capacity to articulate social and ethical issues in development when corporate strategies neglect such implications.

Overall, the social and political drivers of change present a mixed bag of positive and negative impacts on the future of agriculture in ESAP. For AKST, while increasing investments and learning opportunities emerge from more social interactions and political changes, it is likely that many of these investments may be private sector investments in AKST given that public resources may be increasingly diverted to social security nets or other essential infrastructure investments. Social and political drivers portend a change in the nature of regional cooperation and of the actors who will ensure that AKST continues to generate useful technologies that are accessible to and utilized by farming communities and other rural producers. These new actors with new social and scientific skills will be from the private sector and NGOs, and therefore public sector AKST (constituting a large component of KST in ESAP) must equip itself to partner with these new actors.

4.2.5 Education, culture, ethics and health

Important social drivers such as education, health and cultural norms (their resilience, capacity for modernization and global human rights and value systems) can shape future AKST in the region. This shaping occurs primarily by bringing more educated and healthy people to generate and utilize knowledge in the agricultural sector and by absorbing global advances in S&T into local cultures or adapting local habits and practices, perceptions of risk, etc., to accept modern technologies or ways of working.

Standards of living in 2050 are expected to decline in response to demographic transitions in countries such as New Zealand, Singapore, Japan and Australia that are already undergoing a transition toward increasingly older populations (Ross, 2006). In countries such as Malaysia, Indonesia, and Philippines where the population is relatively young now, the impending transition (combined with appropriate saving responses) is likely to have large positive impacts on standards of living for at least 50 years or more (Ross, 2006).

4.2.5.1 Employment

Employment opportunities are closely related to overall demographic composition and location; the ESAP population will be predominantly urban and engaged in service or manufacturing activities by the 2030s. A decreasing share of the economic pie will come from the agriculture sector (World Bank, 2007) reflecting the changing and increasing employment opportunities in other sectors, which may absorb rural unemployment and surplus agricultural labor force (see 4.2.1).

Asia's labor force will increase by 14% (245 million people) between 2005 and 2015. Though China will contribute to this increase, China's share will be limited because of its internal fertility rate and population growth rate. By 2015 China will add 7% additional labor to its current labor force. The corresponding figures for Bangladesh (25%) and Philippines (24%) are far higher (ADB, 2005). Labor force participation rates will tend to be lower in South Asia compared to East Asia and Pacific countries. Yet, the developing countries of the region will reap a "demographic dividend" as the share of the young working population increases (except in the developed ESAP countries and others such as Sri Lanka and Thailand), adding to national incomes (ADB, 2005). Their success will depend on the availability of education, employment opportunities, infrastructure and capital investments that provided to employ this young labor force.

Policies that will have the highest impact on full and productive employment in the ESAP region are growth-promoting policies (ADB, 2005), primarily:

1. Policies to improve incomes in the rural economy and urban non-formal sector;
2. Policies to shift productivity gains into higher real wages and aggregate demand; and
3. Industrial policies that provide government with a major role in co-coordinating and monitoring industry.

Increasing economic liberalization and reduction in tariff rates projected to begin in 2010 causes concern about the demise of the domestic industry and widespread unemployment in the manufacturing sector in the Asian countries. Given the likely evolution of specialization in industry (as the driver of growth), there will not be any worker displacement/redundancy; with specialization, workers will move within industry rather than between or out (Veeramani, 2007). Employment opportunities and incomes are likely to be highly differentiated in rural areas, with globally competitive farm entrepreneurs (Rural World 1) standing to gain at the cost of the falling fortunes of family farmers (Rural World 2) and the struggle for survival of the poor peasants and laborers (Rural World 3) (Pimbert et al., 2001). There will be increasing demand for labor/employment policies to ensure that different segments of the rural population can survive the pressures of globalized agricultural and food systems.

4.2.5.2 Education

Education, especially access to primary and secondary education, will continue to enable the increasing migration of rural educated youth to urban or rural non-farm sector employment (IFAD, 2001; ADB, 2005). A decline in fertility rates allows increased participation of women in the workforce. China for instance, has gender disparity in educational levels and in opportunities for women in the labor market (Hussain et al., 2006; World Bank, 2007). This is a significant issue for the country since China has the highest female labor force participation in the world.

Gender and urban biases in education (see IFAD, 2001; ADB, 2005; UNESCO, 2006) will continue to be major problems in achieving the targets set for poverty reduction and better rural livelihoods in the Asia-Pacific region. Though gender gaps in primary education have been reduced in several countries, there is significant gender bias in secondary and higher education, as well as employment opportunities for women. These gaps are likely to grow in future unless addressed in a focused and perhaps regional manner (IFAD, 2001; Fennel, 2006). Agricultural education investments are likely to decline in formal universities and agricultural universities (Byerlee and Echeverria, 2002). But investments in private and public sector higher education and research as well as investments in Farmer Field Schools, training programs at various levels of participatory research and extension, and most importantly in functional education and non-formal education for sustainable development are likely to increase in all the ASEAN, APEC and SAARC countries (UNESCO, 2006). Investment in informal education in the Asia Pacific region is increasingly seen by donor agencies and governments as a mechanism for (1) enhancing skills and capacities for better livelihoods and incomes, (2) enabling employment opportunities, especially non-farm rural employment, (3) reducing the gender bias and thereby poverty in rural areas and in agriculture, and (4) increasing capacities for technology uptake, especially through functional education (IFPRI, 1995; Ooi, 2001; UNESCO, 2006) (Box 4-2).

It is commonly agreed that formal education to improve literacy and numeracy improve farm productivity since such skills increase the sources from which farmers may obtain information. Higher education may lead to increased rural to urban migration meaning that policies targeting improved farm productivity through education may also instigate human capital outflow to off-farm employment. This shift of skilled labor to other sectors will be increasingly seen as a source of growth opportunities for manufacturing sector growth.

Several gaps in our knowledge about food systems have to be addressed to ensure democratic and environmentally sustainable food systems in the future (Pimbert et al., 2001). AKST organizations are increasingly acknowledging their need to educate themselves about diverse contexts and their implications for S&T.

The trend to understand local knowledge systems and their role in shaping or utilizing the outputs of AKST will be strengthened in the ESAP region. Unfortunately democratic participation of the relevant actors/rural poor in shaping formal S&T systems will not be realized in the near future in ESAP countries. Farmer participatory research will continue to be conducted after the technologies have been proven in laboratory and field station trials and farmers will continue to be seen as tail-end adopters of technologies. These methods will continue for at least several more decades despite social science research showing the success of different approaches to learning, technology generation and utilization (Fujisaka, 1994; Biggs and Matsaert, 2004; Hall et al., 2004; Biggs, 2006). The basic problem within AKST remains an awareness of little social science outside of agricultural economics; there is no inclusion of disciplines such as anthropology, which lead to emphasis on local contexts, and poverty-relevant sciences. This situation may continue well into the 2020s in the ESAP region unless challenged by dynamic developments (Cernea, 1991; Raina, 2005).

4.2.5.3 Indigenous knowledge

Basic education helps when farmers want to make the transition from traditional to modern agricultural practices. Yet AKST actors—public sector R&D organizations, private firms and private R&D, NGOs/CSOs, policy makers and donors—have made little attempt to explain these education-led changes in AKST uptake other than the usual technology adoption studies. Though few attempts have been

Box 4-2. Future investment in human resources in Asia-Pacific region.

The IFPRI vision 2020 highlights the need for investment in human resources in the Asia-Pacific region. Accordingly the countries will need to:

- Introduce new programs and strengthen existing ones to target the poor and disadvantaged at household and intra-household levels based on effective policy research.
- Emphasize maternal and child health and nutrition programs.
- Improve access to clean water and sanitation.
- Provide safety nets for the poor and landless rural households affected by the new economic policy reforms.
- Invest more in schooling, especially for girls.

Source: IFPRI, 1995.

made to validate or incorporate indigenous knowledge into knowledge systems useful for locally adapted agricultural and food systems, there is increasing interest among formal scientists to understand indigenous knowledge and use it for agricultural/rural development (Rajashekharan, 1993; Joshi, 1997; Talawar and Rhoades, 1998; Rhoades, 2000). Given rapid urbanization and the corporatization of agriculture, this is a trend that will prevail in the region for at least 10-15 years.

Indigenous knowledge and agroecological approaches to understanding and scientifically validating indigenous knowledge have been much analyzed and discussed for over twenty years now (See Chapter 2). In the ESAP region many donor agencies and national governments will continue to support research to better understand and integrate indigenous knowledge with modern scientific knowledge or ways of cultivation (e.g., KIT, the Netherlands; DFID, UK; GTZ, Germany; Aga Khan Foundation, India; USAID, USA). Given the projections that the majority (48%) of the world's poor will live in South Asia by 2020 (ADB, 2005) and a majority of these will be people living in mountainous or remote rural areas relying entirely on indigenous knowledge systems, markets and other local institutions (IFAD, 2001; ADB, 2005), the demand for devoting greater investment and R&D attention to these people, their crops and livelihoods is increasing.

Several traditional knowledge systems have been revived in recent years in agriculture and allied knowledge, for example herbal medicine and ayurveda. These traditional knowledge systems have important economic implications (bioprospecting, IPM, health care, fisheries, etc.) and thereby pose challenges to IPR and knowledge piracy often indulged in by Western pharmaceuticals (Pushpangadan, 2000). The demand for investments in nonformal education, traditional health care, organic agriculture, IPM, etc., will continue to grow, bringing opportunities to acknowledge, revive and provide entrepreneurial growth opportunities to repositories/practitioners of traditional knowledge.

The ESAP region is acknowledging the role of indigenous knowledge, women and their traditional knowledge of agricultural and natural resource management practices, and rural community histories in enabling environment friendly, location specific agricultural development. The role of CSOs/NGOs in ensuring green development and a sustainable growth pattern in Asia is increasingly being pushed into the policy arena (Barkenbus, 2001). CSOs will also play an increasingly strategic role in the campaign for the right to food and human rights for marginalized and tribal peoples for whom the pressures for survival will increase with increasing globalization. Consequently, the demands to "de-globalize" and invest in building local capacities for sustainable agricultural and food systems that can feed the resource poor people of Asia will increase (see Foodjustice.net, 2003—statement by NGOs in 14 Asian countries, five years after the World Food Summit; Tyler, 2006).

4.2.5.4 Human and ecosystem health

By 2020, the ESAP region will be home to large numbers of the poorest and under- and malnourished people in the world. There will be more malnourished children in South Asia than in sub-Saharan Africa (Rosegrant et al., 2001). The region will also see unprecedented growth in industrialization and urbanization, with the urban population expected to increase by 352 million people between 2005-2015 (UNESCAP, 2005). Municipalities in 2025 will face a tenfold increase in their solid waste burden (Bass and Steele, 2006). Consumer demand in China alone is expected to rise in the next decade (ending 2015) to the equivalent of four USAs—this includes demand for cement, timber, coal and steel (Bass and Steele, 2006). The health of both human beings and ecosystems will depend on how this urban population is fed and provided the goods and services it needs.

The ESAP region will need to invest heavily in environment friendly and socially and ethically just development. The environmental technology business in Asia will reach over $212 billion by 2015 (ADB, 2006a). Japan with the launch of the 3R initiative in March 2005 and China with its commitment of resources to renewable energy (the highest in the world) have already given due policy attention to developing a "Resource Saving Society" (UNESCAP, 2005). China has pledged to generate 15% of its energy from renewable energy sources by 2020 (up from the current level of 7%) (Bass and Steele, 2006). In addition to national strategies, one of the key elements of regional cooperation will be focusing on engaging the public and private sectors in Asian economies to build capacity for generating and utilizing environmental technologies. Generation and sale of power to public and private players has already begun in Bhutan to the benefit of neighboring states, especially India. The ESAP region, especially Australia, New Zealand, Japan, China, India and some of the bigger and faster growing economies now propose to collaborate and help develop action plans for sustainable development in the Pacific countries. These Pacific countries (Melanesian, Polynesian and Micronesian regions) are some of the smallest countries in the world, contain unique biogeographical features and are inhabited by some of the oldest ethnic populations (UNESCAP, 2006).

Despite a lack of quantitative data, it is clear that land-improving investments are creating a number of "bright spots" in the developing world (Scherr and Yadav, 1997; Lele, 2006). Investments to prevent land degradation will lead to further rehabilitation of ecosystems in the ESAP region throughout the next few decades. In addition to investments in organic agriculture, diversification into higher value perennial crops (in all ESAP countries), conservation farming (in Thailand, Philippines and other East Asian countries, also promoted actively by the CG centers like IRRI and CIMMYT through their Rice-Wheat Consortium across the five countries in the Indo-Gangetic Plains), water management (in all ESAP countries), agroforestry (in India, China, Thailand, and some East Asian and Pacific countries), resource conserving and pro-poor mechanization (by NAEF in Nepal), favorable property rights (Cambodia, Laos, Viet Nam), and several community-based NRM projects (in almost all ESAP countries) are ongoing and will increase in scope and scale in the near future (Scherr and Yadav, 1997; Seth et al., 2003).

A sector that has immense human health and ecosystem health/sustainability implications is the fisheries sector in the ESAP region. The sector has been experiencing a biological decline due to over-fishing, the spread of virus in the shrimp industry and other diseases. Conflict resolution among countries sharing the seas in the region and between poor fishing villages along the coast and the large-scale fishing industry (large gear operators) financed by global or domestic capital is now a major concern in the South Asian and Southeast Asian countries (Salayo et al., 2006). International instruments such as the Code of Conduct for Responsible Fisheries of the FAO, the World Summit on Sustainable Development and the International Plan of Action for the Management of Fishing Capacity have addressed these issues, specifically prescribing codes for safeguarding the right to livelihoods for millions of fisherfolk and the critical issue of building and maintaining ecosystem health.

Human ecosystems are increasingly polluted; air and water pollution are very high in some cities in Asia. In China, it is estimated that by 2025 pollution may reach intolerable levels. Eight of the ten most polluted cities in the world are in China, accounting for a loss of about 3-6 million life years (Bass and Steele, 2006). Air pollution is likely to pose severe constraints to agricultural production and productivity (Marshall et al., 1997). By 2020, Asian emissions of SO_2, NO_x and NH_3 will be equal to or greater than the combined emissions from Europe and North America (Galloway, 1995). Crop production adds to the N-load in the environment primarily due to increased use of fertilizers. A nitrogen balance model for east Asian countries projects that the N-load in 2020 due to food production-consumption and energy production will be 1.3 to 1.6 times times greater than it was in 2002 (Shindo et al., 2006). The trend of high N concentrations in groundwater in eastern and northern China, the Republic of Korea and Japan will continue. It is predicted that N-pollution (NH_3) due to fertilizer use and domestic animal wastes from China will surpass N-emissions and environmental accumulation levels in the United States by 2020 (Galloway et al., 1996). Anthropogenic reactive nitrogen in Asia, most of which is accumulated in the environment, will dramatically increase from current levels of about 67.7 Tg N yr^{-1} in 2000 to 105.3 Tg N yr^{-1} by 2030 (Zheng et al., 2002).

Along with increasing demand for food and energy supplies, Asia lacks effective measures to improve the efficiency of fertilizer nitrogen use and to prevent NOx emissions from fossil-fuel combustion. Environmental N pollution may be lessened through the substitution of synthetic nitrogen fertilizers with high-efficiency nitrogen sources yet to be developed. There is a need for open communication on current knowledge and uncertainties among government, industry and consumers (Hansen, 2002).

The tradeoffs between trade liberalization and the environment and poverty caused due to depleting natural resources are increasingly being recognized in Asia (UNEP, 2001). Generally, the economic benefits from trade liberalization are high, but the environment costs are also high (see Chapter 3 and 4.2.4). It will be increasingly important to have an assessment of the environmental and human livelihoods costs incurred for trade and find ways of financing and technically supporting conservation measures to reverse the damage, along with compensation for affected livelihoods, e.g., small-scale fisheries with depleted fish stocks due to deep sea trawling that destroys fingerlings necessary for fish breeding and growth.

Many ESAP countries, desirous of conserving their agricultural biodiversity (plants and animals) attempt to comply with the Cartagena Protocol and CBD. Conservation responses range from mandatory assessment of the impact of all programs, technologies and development interventions on the biological resources and access of the major stakeholders (such as farmers, herders, etc.) to these resources and natural environments in some countries to a ban on genetically modified crops and organisms in others.

Increasing awareness of the erosion of genetic variability (due to monocultures of major cereals and selective breeding of crops and livestock) and of the inherent value of indigenous and traditional knowledge systems, and the increasing importance of bioprospecting and patenting for industrial and pharmaceutical applications will lead to increasing investment and legal assurances in the conservation of biodiversity. To enhance local involvement and incentives to conserve agricultural biodiversity, governments, corporate sector and CSOs may encourage learning platforms as active repositories of indigenous practices of seed storage, cultivation and conservation. Alternative cultivation systems, such as ecological agriculture, and ecotourism (around the theme of genetic wealth) that recognize the relevance of gene pools in local ecosystems and application in local diets/cultures may also be explored.

A key dimension of environmental decision-making is the costs involved—often directly translated into closure, relocation or rehabilitation of polluting industries and workers. In agriculture, the tradeoff between environmental costs and the costs of setting aside land or entire production systems, are very high, complex, and involve millions of households that live entirely on agriculture. Many countries with shared resources will increasingly feel threatened by environmental security issues, adding to political tensions in ESAP (Nijam, 2005).

Many resource conserving technologies and production practices that are environment friendly are unused for want of enabling institutional and policy arrangements (Rhoades, 1999; SaciWATERs, 2002; Raina and Sangar, 2004). Decisions made about the environment are also burdened by static assumptions and compartmentalized analyses. In a static sense, environmental regulations add to project cost. In a dynamic sense, if conventional waste products are reused or captured within production systems, there are no additional costs for disposal (Porter and van der Linde, 1995) and total value of output is larger. In a production system that involves synergies, decision makers will have to look beyond the single output/end product to the range of joint outputs and their mutual dependencies. In such systems, the increase in productivity can cover for the increased costs in utilizing waste products. Local value added in tea plantations can be much higher if the shade trees (e.g., silver oak) are replaced by arhar (red gram/pigeon pea), which has additional value. Local employment opportunities in rural areas are fairly high in environmental biotechnology, e.g., production of bioplastics from agricultural wastes. There is significant opportunity for industrial developments to incorporate local value addition or processing activities, involving a dynamic mutually beneficial environment-production relationship (Hatti-Kaul et al., 2007). These developments have major implications for multifunctional agricultural systems and AKST.

AKST, being almost static, single commodity focused and weak in natural resource and social systems capacities, is not equipped in the near future to handle such multiple products/functions, synergies and dynamic systems. For instance in industrial agriculture (producing biofuel or bioplastics) and the pharmaceutical livestock farms (producing specific drugs—as isolates or as part of animal products), the entire chain of activities and relationships of supplies, finance, infrastructure, services, etc., required will need to be assessed not by an individual scientist or a research institute, but by a coalition of actors.

A major message is that little is known about balancing the benefits and the pains of adjustment or change in farming practices and policies that will be essential to reduce sensitivity of the sector to pollution. Even less is known to enable better linkages between agriculture and other sectors of the economy, where environment-friendly production practices and waste reduction or utilization become built-in synergies within the system.

Overall, the commitment in ESAP to invest in local capacities for development demands a convergence of policies and programs in education and employment. ESAP will need skills and competencies as well as a diversified basket of employment strategies designed to meet the needs of globally competitive producers and small-scale farmers and landless laborers. The region will also need knowledge and investment inputs to ensure that its tribal and indigenous knowledge systems are developed and utilized for environment-friendly development. Most of all, education in general and AKST in particular, has to consciously acquire the disciplinary and social competencies that are necessary to understand the tradeoffs between economic growth and the environment.

4.2.6 Science and technology

4.2.6.1 Research investment

There is little information or projections available for future investments in agricultural research in the ESAP region. Despite estimates of attractive rates of return to investments in agricultural research and several impact assessment studies, there is little available on possible trends/estimates of potential agricultural research investment in future. The determinants of agricultural research investment, especially in the public sector, seem not to respond to any conventional economic arguments/findings and are influenced by a myriad of complex factors (Tabor, et al., 1998; World Bank, 2005). Rather, investments in agricultural R&D in the ESAP region seem to be arbitrary and entirely politically driven or to follow a precedent approach at best, like any other public bureaucracy. "The sheer complexity of agricultural development, especially the lack of obvious and standard approaches for investing in agriculture, has led to rather fickle support from international and national policy makers" (World Bank, 2005). Despite a recent resurgence in national/international policy realms and donor commitment to invest in agriculture, there is little attention being given to increasing investment in agricultural research.

It was argued that as part of a wider global phenomenon, the ESAP region faces and will continue to face a relative decline in the amount of public funding available for agricultural research and extension (Anderson and Purcell, 1996). However, within the ESAP region the recent resurgence of public investment in agricultural research (accounting for 32.7% of global public agricultural research expenditures in 2000, compared to just 20% in 1981) is likely to continue. In 2000 China and India together accounted for 31% of the public agricultural R&D investment in developing countries. They will contribute hugely to future public agricultural R&D investment (Pardey et al., 2006). This revitalization of public R&D investment in India and China is largely a result of government policies to revitalize public sector research and enable commercialization of agricultural technologies, especially increasing linkages with the private sector. The decline in Japan's public sector commitment to agricultural R&D is more than compensated by its massive increase in the 1990s in private agricultural R&D investments, a trend that is likely to continue. Japan and Thailand will be the next large investors in domestic agricultural science.

The developing countries will continue to expand their share in global public sector agricultural R&D investments. The ESAP region, mainly because of China and India, will have an increasing share of this developing country investment in formal public sector AKST (Pardey et al., 2006). But the increasing share of the developing countries (from 45.4% in 1981 to 55.7% in 2000), may mean that in future the developing countries cannot rely on technological spillover from the rich countries or international research communities. The S&T linkages between rich and poor countries will get more attenuated as the funding in developed countries gets reoriented from agricultural productivity to environmental impacts of agriculture, food quality, medical, energy and industrial applications of agriculture (Pardey et al., 2006). The productivity obsession to the neglect of other aspects of agricultural-environmental-social systems has been the bane of formal agricultural R&D in the developing countries (Roling and Wagemaker, 1998). Given the past record of Western science-determined research in much of public sector R&D in developing countries, especially Asia, changing priorities in the West might bring a positive turn in the funding patterns, bringing spillovers of knowledge that can enable better livelihoods for the rural poor and their environments in developing countries.

Almost all formal AKST in ESAP is organized based on some Western, especially US, model of agricultural research. Given the overview of social and political drivers and increasing opportunities for education in the ESAP region (though rural areas will still lag behind) there will be new private and NGO sector actors involved in AKST. Then, with a committed program of institutional reform, this shift in research interests and allocations between rich and poor countries can push Asian economies to invest more heavily on environmental impact, food quality and gender sensitive technologies, and integrate more with research for non-farm rural employment, and other needs that are currently under-researched.

Internationally and in the developed countries the slowdown in public spending has been more than compensated by increasing private funding of agricultural research. But the developing countries of the ESAP region have made little gains in private funding of agricultural research, though their share in total private sector R&D investment continues to be highest among the developing countries. Not much has changed since the finding that private sources fund less than 7% of total agricultural research spending in India, Bangladesh, Pakistan and Indonesia (Umali, 1992). Private investment in agricultural R&D is less than 10% of total agricultural R&D in the ESAP countries (Pray, 2001). More accurately, private investment in the region is only 8% of total investment in agricultural R&D (CGIAR Science Council, 2005; Pardey et al., 2006).

Private R&D in biotechnology has grown significantly in Asia (Chaturvedi and Rao, 2004; La Vina and Caleda, 2006; see Biospectrum various issues and APBioNet). In Asia, the magnitude of private sector presence in R&D will continue to be significantly different in different sectors, with health-biotech R&D (pharmaceuticals, vaccines and diagnostics research) being located largely in the private sector (both domestic and FDI led investments in R&D in multinational corporations [MNC] in Asia). Private R&D in agricultural biotechnology in the near future will be located largely within MNCs with strong product lines as they focus on buying all market-end firms/facilities in ESAP. Much of the non-MNC private biotech R&D is funded by domestic public or international donor funds, fuelled by the promise of profits (La Vina and Caleda, 2006).

Contrary to conventional wisdom that private investment in agricultural research and extension will increase in Asia-Pacific countries and that the role of the Government is to focus investment on basic research, human capital and infrastructure, and to provide an environment conducive to private research (Tabor et al., 1998; Chang and Zepeda, 2001), there is an increasing emphasis on the role of the government in research and other enabling investments in the agriculture sector (Byerlee and Echeverria, 2002; Hall et al., 2002; World Bank, 2006c). Much of private investment in agriculture is currently in the industrialization and capitalization of the sector. Given the trends of increasing urbanization and expected growth of value-addition and food retailing in ESAP, private investments will continue to be made in the off-farm parts of the agricultural production and marketing chain, in seed, fertilizers, herbicides, machinery, processing, retail marketing, etc., where the private sector can commercialize technologies and knowledge outputs (Reardon et al., 2003). Increasingly, the argument for pro-poor agricultural/rural innovation is to look for complementary investments in other organizations besides public sector and national agricultural research systems (NARS) that can play a major role in enabling generation and utilization of knowledge (including technology) (Hall et al., 2004; Biggs, 2006; World Bank, 2006c).

This also reaffirms an interest in the NGOs or non-profit research entities investing in the agricultural sciences, with the current figure of 1% of the ESAP agricultural research investment being made by them likely to increase in future, as partnerships with public and private R&D increases. In India, for instance, there are organizations like MSSRF, MANAGE, MYRADA, CISED, Mitraniketan, BASIX, and several of the NGOs who host ICAR-KVKs (the extension units under the ICAR—called Krishi Vigyan Kendra). These NGOs are at the forefront of not-for-profit research in agricultural, horticultural and livestock systems. The lack of reform in the public sector agricultural research organizations is one of the factors that promote the role of NGOs or not-for-profit firms in agricultural research (Raina, 2003a).

Changes in dietary profiles in response to rapid urbanization, growth of incomes and expansion of value-chains and food retail in ESAP show that in 2020 the demand for livestock and dairy products will be more than double the current level. While integrated crop-livestock systems and access to common property resources for herders and pastoralists may become crucial for poverty alleviation, there may be increasing investments in high-tech animal production systems. Much of this investment is expected to come from the private sector, especially from MNCs in food processing and retail.

Overall, investments in agricultural R&D have proven that predictions of a decline in investment (Anderson and Purcell, 1996) have in fact been reversed in the ESAP countries. Agricultural R&D investments have grown since the mid-1990s, a trend that may continue into the future, focused on non-conventional but increasingly crucial issues for ESAP, such as green production systems, reduced environmental pollution, farmer friendly markets, open source software/genetic material/protocols and gender relationships. Increasing market opportunities will also bring increasing investments from private R&D.

4.2.6.2 *Research organizations and institutions*

The expansion of research organizations in agriculture and allied sectors was a phenomenon of the 1970s and 1980s in most of the ESAP countries. The current interest in these countries is on changing or reforming institutions or the rules/norms that govern these research organizations (Byerlee and Echeverria, 2002; Hall et al., 2002; Huang et al., 2002; Raina, 2003a). It is expected that this resurgence of interest in funding agricultural R&D and in institutional reform in agricultural R&D will lead to greater effectiveness of knowledge to achieve the MDGs. The lessons from experiments on new models or approaches to funding research organizations and corresponding changes in their institutions/rules will be applied increasingly in the ESAP region (World Bank, 2006b).

It is only in the late 1990s that the distinction between the agricultural research organizations themselves and the institutions or rules/norms that govern these organizations, was made in the social science literature on agricultural research in the ESAP region (Biggs and Smith, 1998; Raina, 1999, 2003ab; Biggs and Matsaert, 2004; Hall et al., 2004). In the ESAP countries, the linear compartmentalized (into research, extension and adoption) organization of knowledge in AKST will continue for some time because it is deeply entrenched in existing formal agricultural research organizations and policy making organizations (Hall et al., 2004; World Bank, 2006c; Biggs, 2007; Raina and Sulaiman, 2007b). But recent trends show increasing donor interest in non-linear systems of knowledge generation and utilization, as well as the institutions or rules/norms that will promote new non-linear ways of working in R&D organizations and extension organizations (IDRC, 2006; World Bank, 2006c). These will increasingly be applied to the agricultural sciences and existing formal R&D organizations.

The CGIAR organizations in the ESAP region are experimenting with institutional reforms, partly in response to the pressure to prove their efficacy in reducing poverty and the rapidly declining rate of donor support to the CGIAR system (Lele, 2004; CGIAR Science Council, 2005). Recent experiments that will give important insights and incentives to initiate institutional reform include increasing pressure for agricultural diversification, the private-plant breeders consortium in ICRISAT, regionally differentiated research strategies in AVRDC, the natural-resource based no-tillage technology systems in the CIMMYT-IRRI sponsored Rice-Wheat Consortium in South Asia, the internal processes to set research priorities that suit the spatial dimensions of poverty in CIMMYT and innovation systems research to enable pro-poor livestock and fodder innovation systems in ILRI (Joshi et al., 2003; Hall et al., 2004; Erenstein et al., 2006; Raina and Sulaiman, 2007a). Different theoretical frameworks and approaches like the Sustainable Livelihoods Framework, the Innovation Systems approach, the Network models, etc., will increasingly be tried by different CGIAR organizations in the ESAP region to enable changes in ways of working in science and among its partners to achieve goals of sustainable development. The Asian Vegetable Research and Development Centre (AVRDC), for instance, has a regionally differentiated strategy for vegetable research and development that emphasizes research on more nutritious vegetables in the East Asia region through a network approach (the AVRDC-ASEAN Regional Network on Vegetable Research and Development), and enhancement of vegetable cultivation area (by over 9 million hectares) in the Indo-Gangetic Plains of South Asia by 2010 (AVRDC, 2002). The organizations involved in these emerging networks and new research strategies may continue to be governed by linear knowledge flows and will not be in a position to influence or change policy decisions and practices to enable innovation unless proactive investments are made to enable learning and change within these organizations.

Pressure for institutional reform and reorganization is now increasing within the university system in Asia—especially India, Thailand, China (much experimentation and change) and Sri Lanka. Universities are being forced to reconsider their traditional roles (as sources of ideas, basic scientific knowledge and teaching resources) and embrace new ones like making contributions to regional development through innovation. They are being asked to transform themselves from "knowledge containers" to "entrepreneurial universities" (EDB, 1995; Government of Japan, 2002; National Knowledge Commission, 2006). These new functions of the University evolve as they interact and learn from/with several key development partners. Universities evolve with the demands that society places on them (David, 2004). If the agricultural growth drivers and social and economic drivers of change place sufficient pressure on the university system, the decline of university research in Asia (Byerlee and Echeverria, 2002) may be avoided. It is unlikely that the universities will invest in research that is less visible and pro-poor—they are more likely to address areas like IT, textiles, architecture, biopharmaceuticals and other more visible areas with an articulate demand.

In ESAP the response to new institutions and policies like the Trade Related Intellectual Property Rights (TRIPS) under the WTO regime and the relative reluctance of countries to buy into the intellectual property rights (IPR) regimes will be conditioned by the major fear of negative impacts (of these institutions as curbs on technology/knowledge) on millions of lives and livelihoods. Many address the issue of IPRs as a knowledge generation incentive. Little is asked about how future institutional reform must address issues of knowledge utilization in the field, given that public agricultural research has generated vast amounts of technologies that find no application at all. Notwithstanding the overstated generalization that patent regimes and intellectual property rights will enhance commercialization of technology and knowledge in agriculture, the ESAP region will face continuing opposition to IPRs in agriculture or any form of private appropriation of biological material, technology or knowledge for commercialization (Shiva, 2000). There is also some questioning of how the TRIPS agreement may or may not detract R&D from addressing relevant knowledge generation and use questions in the Third World, especially the ESAP region (Connett-Porceddu, 2006; Musunga, 2006).

ESAP is also home to the "open source biological information sources" that are available (over a million life science patents and appropriate software make it transparent and accessible to users) and growing rapidly to meet the needs of ESAP's scientific community and industry (Herrera, 2005). The problem is that even in the medium term, there will only be a few groups (limited often to closely networked actors) that have the capacity to share or utilize this open source data base (Herrera, 2005; Connett-Porceddu, 2006; BioForge website). The developed countries have led the evolution of a policy framework for exclusionary patent rights; the evolution of norms/institutional arrangements for sharing an open and inclusionary source will now be taken up by several networks and technical cooperation programs in the ESAP region. In terms of ideal institutional arrangements, ESAP will soon have to decide whether IPRs or Open Source Biological Software or a flexible combination of the two will help more effective knowledge utilization among peasants. Given that open source software can be effective in incremental innovations and may not be useful for radical innovations (Bonardi and Warin, 2007), it may be ideal for developing countries with large peasant populations and rural non-farm employment demands to invest in open source software. The question that will be asked increasingly will be about their potential for generating local ownership and appropriate institutional reform in public and private R&D in individual countries (Connett-Porceddu, 2006).

The institutions or rules/norms that govern the public sector agricultural research organizations and their ways of generating and evaluating technologies is increasingly receiving attention in the ESAP region (Raina, 2003b; Hall, 2005; World Bank, 2006c; Raina and Sulaiman, 2007b). Formal AKST organizations in the region will be increasingly called upon by Governments and the public to prove their efficacy in improved livelihoods and incomes for the rural poor, poverty reduction, overall quantity and quality of food systems and a sustainable environment with significant reductions in agricultural pollution. Increasingly emerging challenges will also demand institutional reform within the agricultural R&D organizations; these challenges may be pest or disease outbreaks (avian flu, anthrax, etc.), climate change impacts, water scarcity, trade or other market standards, energy or fuel crises in rural areas, food processing and retail demands. New or modified ways of working and institutional changes will be necessary to achieve these goals.

4.2.6.3 *Biotechnology, transgenic crops and pesticides*

Much as the chemical and mechanical inputs into agriculture have moved from household level and village artisan based sources to public and private funded sources of knowledge and artifacts, the biological components will also move to industrial/manufactured goods markets and private sources of knowledge and product generation (Rieuvenkamp, 2003). Asian biotechnology investments are predominantly in public sector organizations, despite the fact that private research in the biosciences and biotechnology has grown rapidly over the past two decades. Most Asian bioscience capacities are therefore in a "public-sector bind" with

- A precedence and affiliation towards conducting research almost as an end in itself without worrying about application, commercialization or utilization;
- Significant compartmentalization of R&D into different sectors (Ministries); and
- Little capacity for biotechnology commercialization and the partnerships necessary for commercialization (Chaturvedi, 2002; Peczon and Manalo, 2004; Wong et al., 2004; Hidayat, 2006; La Vina and Caleda, 2006).

A common feature that marks the public sector R&D in the biosciences is inertia and resistance to change. This inertia has more to do with internal institutional rigidities than with lack of technological capabilities or an increase in concern about environmental safety as evidence by the fact that public research organizations in Asia have access to several biotechnology results that have potential applications in society. Though there are only a few notable public sector enterprises in biotechnology, it is likely that increasing institutional reform will enable venture capital led biotechnology enterprises to emerge from public R&D and there will be increasing collaboration between public-private R&D and private industry.

Some countries like Japan, Singapore, China, India and Malaysia are home to fast growing and thriving biotechnology enterprises—national biotechnology strategies in each of these countries reveal ways to continue this expansion. Some countries like South Korea have taken a pro-active approach to development of biotechnology and transgenics for application in various sectors. The Korea BioVision 2010 is a plan put forth by the Ministry of Commerce Industry and Energy (MOCIE) highlighting the role of transgenics in several fields (Feddema, 2003). It is projected that following this strategy the country will be seventh in the world of biotechnology.

While China has an edge in commercializing agricultural biotechnology and some specialized areas like stem cell research, India has established its advantages in industrial biotechnology, animal/veterinary health products and diagnostics. The growth of these specific sectors has also been conditioned by the domestic policies enabling their growth, e.g., the Indian pharmaceutical industry. Increasingly, there is pressure on these enterprises to shift their attention from catering to the global MNCs/global market in order to focus on domestic development needs (Frew et al., 2007).

Asia will continue to be the "new promised land" for the large diversified MNCs in the pharmaceutical sector (Ekchart et al., 2005). Besides cost savings due to availability of cheap labor, industry now seeks quality manpower to work in the biotechnology sector. In the enterprise domain, the most globally integrated and tantalizing bio-innovation success in Asia is that of Singapore. The Government of Singapore, through its Economic Development Board (a statutory board under the Ministry of Finance), spearheaded the development of biotech enterprises in the country (Tsui-Auch, 1999; Pownall, 2000; Ekchart et al., 2005; La Vina and Caleda, 2006). The strategic research and health biotechnology spearheaded by this Board provide a model and potential applications of this model of biotechnology development may emerge in other Asian countries.

Increasing applications of biotechnology in agriculture and health may give pro-poor benefits but these are contingent upon new and modified ways of working, involving new directions in science, new partners, finding and sustaining non-research partners and other complementary skills, the participation of poor people in identification and selection of problems and pro-active policies and institutional arrangements (Raina, 2003; Sahai, 2003; Hall, 2005; Chathaway et al., 2006; IDRC, 2006).

Transgenic crops continue to receive substantial funding from public and private sector research as well as several quasi-public-private organizations in ESAP. Several multinational firms as well as international research institutes and universities have been investing heavily in biotechnology and transgenic research in Asian countries (Pray and Naseem, 2005). The Asia Pacific Association of Plant Tissue Culture and AgroBiotechnology have reasons for approving the release and utilization of GM crops in the region (Sahai, 2003). The ESAP region, with the exception of China, is wary of release of transgenic crops and livestock into the production environment. Japan still has and perhaps will continue to hold its ban on all transgenic agricultural crops and commodities though it is a leader in biotechnology applications in the health and environment sectors. Overall the ESAP region is weak in evaluation and risk analysis required for the release of genetically modified organisms and research policy in ESAP countries will continue to ignore the alternatives proposed to assess risk and estimate the costs and benefits of/from GMOs using non-monetary and pluralistic approaches (Jasanoff, 2000). Pro-active and democratic decision making processes in S&T, especially concerning choice of technology, may take time to emerge in ESAP, despite the demand for discursive processes, ecological and democratic values in S&T in the region (see Dryzek, 1998).

There is a need to address gaps in communication and articulation capacities that arise from a discontent with and mistrust about technological changes that have evolved over time. These are often called the "violence of science" (Shiva, 2000; PAN, 2002). This distrust is of two kinds: (1) a blue, nostalgic mode that craves for the old world uncorrupted by modern medicine, chemical fertilizers and polluted water, and (2) a green, forward-looking mode that demands a cleanup of the current mess of lifestyles, disrupted ecosystems, social tensions and pollution (Bauer and Gaskell, 2002). Asia will now increasingly invest in media research to assess these nuances of discontent with biotechnology and to see how it is articulated (as blue or green objections to modern technology, as wider political economy arguments, as governance or institutional reform processes).

4.2.6.4 *Pesticide and herbicide use*

Pesticide and herbicide use in the ESAP region is not as intensive as it is in much of North America or Europe. Yet, the damage to water bodies and ecosystems at large is rated as severe (UNESCAP, 2005). There are some tracts that use a heavy dose (of over 80% of recommended dose) of pesticides and herbicides (Huang et al., 2002). The pesticide industry expects that world demand for pesticides will soar until 2009 when many of the current pesticides become cheaper because patents have expired (http://freedonia.ecnext.com/coms2/summary_0285-284519_ITM). This trend may be sustained in the developed regions. In Asia fungicides may increase in use because of their effectiveness against soybean rust (see 4.2.7)

The alternatives to heavy pesticide and herbicide use come from at least three different scientific and social perspectives, all receiving acclaim, policy attention and NGO-led mobilization or campaigns in the ESAP region.

- The emergence of genetically modified crops with inbred resistance to specific pests and diseases; examples include cotton engineered for resistance to the American boll worm, brinjal/egg-plant engineered to resist the fruit borer and rice engineered to resist the tungro virus and blight. Developments in this area are likely to continue.
- An increasingly popular alternative is integrated pest management (IPM or a more stringent version called non-pesticide management, NPM), which uses a combination of physical, chemical, biological agents and modern weather based or traditional knowledge inputs.
- Organic agriculture use no chemical inputs (including synthetic fertilizers) and soil and water systems are detoxified; cultivation processes and processing of produce are certified by some EU-based or American certification process.

While agricultural research systems have and will continue to invest heavily in pesticide research and in biotechnology and in IPM (ranging from no pesticides to biopesticides), there is very little resource allocation, capacity building, and conviction in the public sector NARS to work on organic agriculture. In India organic agriculture is largely funded by international donor agencies and conducted largely by university departments and NGOs, with little information or scientific support from public sector agricultural R&D organizations (See www.csa-india.org). While organic agriculture is big business with private investment and research in Australia, in countries like China, Indonesia, Thailand and Philippines, organic agriculture receives significant public sector R&D attention. They have established organic standards and certification systems and have policies to promote organic agriculture for export and for domestic markets.

Japan is and will continue to be the largest consumer of organic products in Asia with a growth rate of 21% projected till 2020. Several countries, including China, Thailand and Bhutan have declared their own national strategies for organic agriculture (FAO, 2004). The growth of organic agriculture and markets and other facilitation services like certification agencies must be viewed in conjunction with the overall growth expected in the Asian retail market. Food retail in Asia is expected to greatly increase, accounting for 41% of the global food retail trade in 2020 (Partos, 2007).

4.2.6.5 *Innovation or generation and utilization of knowledge*

A recent development in the Philippines, the emergence of a 5-year development plan by the Benguet State University (BSU) to develop an "organic agricultural college," is an innovation intended to produce human capacities for learning about, researching, and extending technical support and market services for organic agriculture in the country. In Thailand, the Prime Minister has requested experts to ensure that Thai farmers produce the finest and best quality produce (www.bilaterals.org, 1 Nov 2006). These examples reflect two major concerns that Asian agricultural research organizations are being asked to address: (1) the capacity to utilize in the field the knowledge that is generated, and (2) the capacity to learn about and respond to emerging challenges in the agriculture sector/other sectors that may affect agriculture. The ESAP region plans to build the capacity of a system of inter-related organizations and individuals to enable the generation and utilization of knowledge and prepare for the changes that may take place.

The ESAP region has witnessed a renewed interest and scholarship in agricultural innovation. Questions being asked include (1) why is there a widening gap between research and utilization of knowledge, and (2) why is there so little that policy makers learn about field conditions (SDPI, 2005ab;

Tyler, 2006). Social science researchers in Asia are increasingly realizing that these questions are not about research but about innovation. Conceptually innovation signifies the process of change when knowledge (including technology) is generated and used in economically productive ways in agricultural or rural societies (Douthwaite, 2002; Biggs and Matsaert, 2004; Hall et al., 2004; Raina, 2004). In order to enable innovation systems to emerge many Asian countries are emphasizing the need for institutional change within the formal agricultural research organizations, especially their capacity to work effectively in partnership with other organizations.

Despite well-intentioned recommendations (based on agricultural innovation systems analysis in four developing countries) to facilitate or strengthen the relationship of agricultural R&D with a range of relevant partner organizations (IDRC, 2006; World Bank, 2006c), developing countries in ESAP are not likely to follow the recommendations made. Future trends in innovation are likely to be bolstered by some recent decisions, such as increasing investments in rural infrastructure, attention to increasing access to rural credit and other services for the rural poor, investments in new networks and relationships to address specific development concerns such as HIV/AIDS, pollution, biotechnology, and markets for rural value added products. Active institutional reform within public sector AKST organizations for innovation and development in ESAP looks remote, given the institutional inertia and the overwhelming euphoria of the green revolution that refuses to wear off despite evidence of rural poverty. There is also the deeply entrenched technological determinism that drives public sector funding of R&D worldwide and the relative reluctance of R&D policy makers to learn lessons from past experiences (Raina and Sulaiman, 2007b).

The diversity of the ESAP region demands that innovation strategies have different goals and occupy different geopolitical niches and ecological systems within developed countries, and different domestic concerns in developing countries (ranging from poverty reduction to reclamation of degraded land, controlling arsenic pollution in water and increasing phytosanitary innovation capacity to meet international trade regulations). Even in the largely globalized food system of today, agricultural and food systems adapted to local cultures and ecosystem carrying capacities can be developed and function fruitfully (Hendrickson and Heffernan, 2002). It is likely that with increasing donor interest (World Bank, IDRC, GTZ, AKF and Japanese Government among the major ones) the ESAP agricultural R&D, extension, industry, farmers, rural banks, rural service providers, input suppliers, environmentalists and other local markets actors will pay more attention to building locally relevant innovation capacity.

4.2.7 Natural resources—land use and water use, land cover change

Sixty percent of ecosystems are degraded or used unsustainably and could further degrade significantly until 2050 (MA, 2005). There are a number of natural resource management concerns that are likely to affect the ESAP region in future. They include conversion of forest and coastal lands for agriculture and aquaculture, flood control and loss of natural fish habitat, increased use of fertilizers and pesticides and their impacts on the natural environment, overexploitation of inland and marine fisheries, land degradation, competition between urban and agricultural water supplies, and degradation of water bodies/wetlands, water pollution and loss of biodiversity. These degradation processes will affect agricultural productivity and livelihoods, especially in marginal and vulnerable areas. Expansion of agriculture in developing countries could lead to 10-20% of current grassland and forestland lost by 2050 (MA, 2005).

Fragmentation as a result of inheritance, population pressures or land scarcity inhibits efficiency in agricultural management efforts. Land and ownership fragmentation poses a problem since irregular farm shapes and diverse ownership structures make it difficult to approve and implement large scale technologies such as irrigation. Irrigation requires large up front costs to purchase fixed capital such as control units and water pumps. Extending irrigation networks is considerably cheaper since extensions use existing fixed capital. Evidence from China and South Asia indicates that land fragmentation is detrimental to land conservation and economic gain, thereby discouraging farmers from adopting agricultural innovations (Niroula et al., 2005) and have negative impacts on land productivity (Nguyen et al., 1996; Jha et al., 2005; Niroula et al., 2005).

Over the recent decades the agriculture sector provided services to reduce hunger and poverty in many parts of the ESAP region, especially in South and Southeast Asia. Despite successes in food grain production, the agriculture sector has recently been facing a large number of problems as natural resources have come under serious pressure from competing sectors. Rapid urbanization and industrial expansion are creating huge and increasing demands for land and water.

Globally, there is enough land, soil and water and enough potential for further growth in yields for increased production to be feasible. At present some 1.5 billion ha of land is used for arable and permanent crops and a further 2.8 billion ha are suitable for rainfed production. A significant fraction of potential land is either locked up in other valuable uses or unsuitable for cultivation due to low soil fertility, high soil toxicity, high incidence of human and animal diseases, poor infrastructure, and difficult terrain. Land expansion is expected to account for 20% of production growth with the remaining 80% resulting from intensification practices such as higher yields, increased multiple cropping and shorter fallow periods (Bruinsma, 2003).

In South Asia, almost 98% of suitable land is already in use. Thus, there is little capacity for expansion in area and it is projected that more than 80% of the increase in production will have to come from yield increases. Furthermore, about a third of the harvested area in developing countries in 2030 is expected to be irrigated land. However, by 2030, East Asia is expected to use 75% of their irrigable area and South Asia (excluding India) is expected to exploit almost 90%. This will result in a 14% increase in water withdrawals for irrigation in developing countries by 2030 and hence 20% of developing countries are expected to face water shortages (FAO, 2002).

A future policy environment is needed that promotes sustainable farming methods and reduces environmental damage while maintaining or even increasing production and associated costs. These methods include no-till and conservation agriculture, integrated pest management and plant nutrient systems and organic agriculture. To reduce the population of undernourished, FAO recommends giving increased priority to agriculture, increasing national food production and reducing inequality of access to food (FAO, 2002). China has initiated new guidelines in response to increased pressure on resources (see Box 4-3). Intensive agriculture and overuse of fertilizers have led to degradation of soil quality and fertility in many parts of ESAP.

Physical responses to land degradation problems have originated from the competing influences of fiscal and market incentive programs (GEO, 2002). In the past, policies were only concerned with increasing supply, however recently focus has increasingly shifted to integrated water resources management. No direct reference to the future of land and water resources is made in the four scenarios of the GEO which looked at a 2032 timeline; however, the following conclusions could be drawn:

Market First: Advances in technology and structural changes in economies might slow the trend in land degradation and water scarcity management.

Policy First: Rate of land degradation could fall due to implementation of more stringent land conservation measures in response to changing tax and subsidy structures. Water conservation, uses and management would improve in future.

Security First: Pressure on land and water resources would increase due to rapid environmental changes, wider inequality in economic and social developments and erosion of traditional livelihoods and communities.

Sustainability First: Land and water resources would be managed in a better way due to emergence of new environmental and development paradigms and be supported by new, more equitable values and institutions.

Box 4-3. Chinese National Development Program Compendium of Science and Technology for Mid-long Term (2006-2020).

Chinese agriculture is facing increased pressure on natural resources such as land and water with the increasing population and rapid economic development. The key issues urgently to be resolved for a sustainable agriculture in the near future include the food security, eco-safety, increase of farmer's income and sustentation of science and technology due to decline of arable land, land degradation and desertification, water pollution and scarcity, use of agriculture technology. The Chinese government recognized a range of such resource management problems and constituted guidelines to facilitate the development of the agricultural science and technology.

The newly released "National Development Program Compendium of Science and Technology for Mid-long Term (2006-2020)" in February 2006 developed the strategies of sustainable agriculture for the next fifteen years through enhancing development of agricultural science and technology. The four strategies that will be implemented include:

- By developing water saving and improving land productivity to substitute the resource scarcity;
- By changing agricultural patterns in a sustainable manner to achieve a win-win both in improving ecosystem function and facilitating agricultural production;
- By extending agro-product processing chains, promoting industrialized agriculture and agribusiness to increase farmer's income and
- By using the innovation of biotechnology and information technology to promote the efficiency and upgrading of the traditional technology, enhance the technological transfer and research on basic science in order to improve the ability of the agricultural science.

Source: State Council, 2006.

While water availability is decreasing, water demand for agriculture, industry and households is increasing in the region as a result of population growth and economic development. In 2025, water availability per capita in the region will be between 15 and 35% less than that in 1950 (ADB, 2001c) (Figure 4-5).

The amount of water used to meet domestic and industrial demands in Asia is increasing rapidly, with rates of increase between the years 1995 and 2025 projected to range from 70 to 345% (ADB, 2001c). As economies grow, people typically begin to overuse water, resulting in increased domestic demand. China, India, Indonesia, Malaysia, the Philippines and Viet Nam are typical countries in that water consumption needs have increased with increased industrialization (ADB, 2001c). Economic growth in the region will require more water for industry and people. In urban areas, which by 2030 will be home to about half of Asia's population, water stress will become more severe. The increase in water demand in all sectors is expected to create conflicts among sectors and within each sector over water allocation. In Asia, many countries share international river basins as sources of water. Increases in domestic water stress would also bring more potential for conflicts among countries. On the other hand, such competition for water also provides opportunities for cooperation on allocation and sharing of water resources (Kataoka, 2002).

The gap between water demand and supply is increasing due mainly to increased demand from agriculture, rapid urbanization and industrialization. Water transfers to intensive irrigation based agriculture and urban areas are placing substantial ecological and political pressure on water resources and this trend will become more severe in the future. Industrial and agricultural effluents are affecting water quality across the region and threaten public and aquatic health.

Water will also be major constraint to the achievement of food security in many developing countries. As of 1997, the cereal harvested irrigated area was approximately 176 million ha in developing countries but is only expected to rise by 29 million ha by 2020 (IFPRI, 2001).

Increasingly in the ESAP region, there is concern that

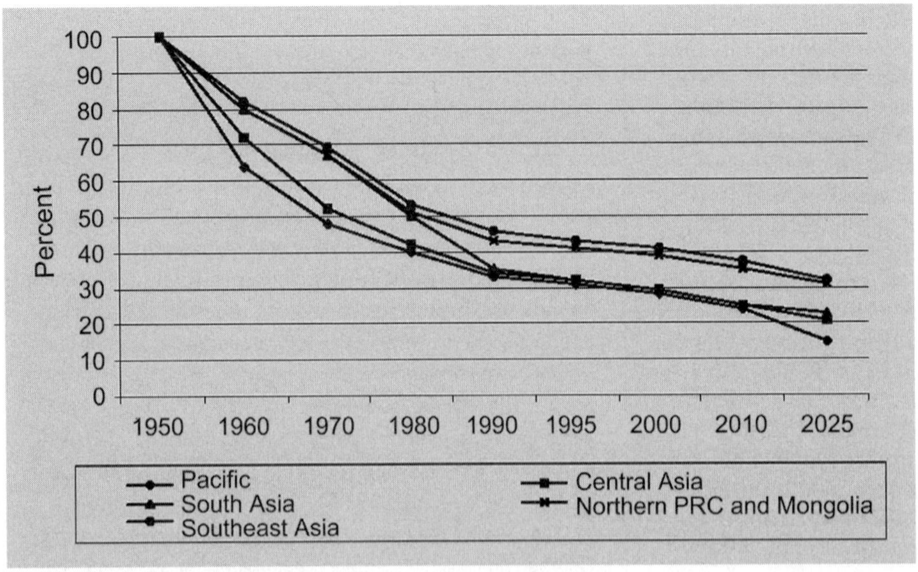

Figure 4-5. *Decline in water resource per capita (1950-2025). Source: ADB, 2001c.*

conservation practices and environmental movements in general pose a development cost, mainly due to the tradeoff between a development option that can lead to economic growth and an environmental option which can reduce or mitigate the degradation.

4.2.8 Climate change, natural hazards and adaptation

The ESAP region is highly vulnerable to climate variability and change. Climate change and natural hazards, such as floods, droughts, pest attacks and diseases are important drivers of change for agriculture and AKST. The Southwest monsoon weather system which lasts from June to October dominates agricultural production, employment and human well-being. Increasing trends in temperature have already been observed in various regions of ESAP with some spatial variability. Interseasonal, interannual and spatial variability in rainfall trends have been observed in the last few decades. A decreasing trend in annual rainfall was observed in Northeast and North China, parts of northeast India, Indonesia, the Philippines and some areas in Japan. Annual rainfall shows increasing trends in western China, Changjiang Valley and along the southeastern coast of China, Bangladesh and along the western coasts of the Philippines (IPCC, 2007).

A significant acceleration in warming over that observed in the 20th century has occurred in the ESAP region (IPCC, 2007). Warming similar to the global mean is expected over Southeast Asia. However, more significant warming is projected for South Asia and East Asia.

Significant warming in future is likely over the Himalayan Highland including the Tibetan Plateau and arid regions of Asia (Gao et al., 2003). Extreme heat waves and intense precipitation are projected for South Asia, East Asia and Southeast Asia (Emori et al., 2000; Lal, 2003; Hasumi et al., 2004). Tropical cyclones/typhoons already cause significant damage to infrastructure and coastal agriculture in the ESAP region. Since generation of cyclones and typhoons are highly related to sea surface temperature, a 2 to 4°C rise in sea surface temperature could increase cyclone/typhoon intensities in East Asia, Southeast Asia and South Asia by 10-20% (Knutson et al., 2004).

4.2.8.1 Climate change and agriculture

Agricultural production in ESAP will be threatened by climatic variability and increased frequency of extreme weather events in coming decades (IPCC, 2007). Impacts to agricultural systems as a result of climate change could include degradation of land, destruction of existing crops, loss of biodiversity, changes in crop and livestock production and increased health problems due to nutritional impacts and vector distribution. However, advancements in AKST could drive new initiatives that promote adaptation to climate change.

A large area in ESAP is under rainfed agriculture. For example, in India, 60% of the land area under cultivation is rainfed. Any change in temperature, humidity and climate will have impacts on crop production. Modeling analyses suggest substantial decreases in cereal production in Asia by the end of this century. Rainfed wheat production may substantially decrease (Fischer et al., 2002) and there will be significant regional differences in wheat, maize and rice yields. In East and Southeast Asia, crop yield could increase by 20% while South Asia may experience a 30% decline in crop yield (Murdiyarso, 2000). Temperature alone could be detrimental to crop yield; in the Philippines rice grain yield declined by 10% for each 1°C increase in growing-season minimum temperature in the dry season between 1979 and 2003 (Peng et al., 2004). Sea level rise and resulting intrusion of saline water could cause significant damage to rice crops in many coastal areas especially in the mega-delta region of ESAP. Currently available rice and other crop varieties are highly vulnerable to salinity and a small change in salinity can cause significant damage to crops. Climate change may introduce positive benefits for rice crops in some areas in ESAP, especially in Northeast China (Wang et al., 2005).

4.2.8.2 Floods

Flooding occurs annually in many parts of ESAP with positive and negative impacts. Bangladesh is one of the most highly flood vulnerable countries in the world: 21% of the land area is flooded annually. India is also highly vulnerable to flooding: 40 million ha (12% of land) is flood prone. Recent episodes of flooding in Indonesia, Malaysia and Viet Nam demonstrate the vulnerability of these countries.

Future climate change will likely alter flood patterns in the ESAP region. Projected increases in precipitation in eastern India and Bangladesh may trigger more flood episodes. With a 2°C rise in global mean temperature, the mean flooded area in Bangladesh could increase by 23 to 29% (Mirza, 2005). The maximum monthly flow of the Mekong is estimated to increase by 35-41% in the basin and by 16-19% in the delta with lower values estimated for the years 2010-2038 and higher values for years 2070-2099 compared to 1961-90 levels. The thawing volume and speed of snow cover in spring is projected to accelerate in Northwest China and the western part of Mongolia while the thawing time could advance, increasing water sources and potentially leading to floods in spring (IPCC, 2007).

Damage to crops by flooding is highly dependent on three related factors: timing, magnitude and frequency. For example in Bangladesh, the crop calendar has evolved around the onset and retreat of the monsoon. If floods (flash floods) occur earlier than expected they can cause substantial damage to standing crops. However, if floods occur very late, farmers cannot replant as the growing period would be severely limited by the onset of winter.

In Asia, climate change may negatively affect the fisheries sector (IPCC, 2007). Rising air temperature would lower availability of oxygen for fish species at higher elevations. In the plains, the timing and amount of precipitation could also affect the migration of fish species from river to floodplain for spawning, dispersal, and growth (FAO, 2003). Future changes in ocean currents, sea level, sea water temperature, salinity, wind speed and direction, strength of upwelling, the mixing layer thickness and predator response to climate change have the potential to substantially alter fish breeding habitats and food supply for fish and ultimately the abundance of fish populations in Asian waters (IPCC, 2007).

4.2.8.3 Droughts

Droughts are very common in many countries in the ESAP region. India is one of the most vulnerable countries to drought, particularly the states of Gujarat, Haryana, Rajasthan and Punjab. Bangladesh is also highly vulnerable to drought, especially those occurring in the monsoon period, which severely affect rice crop production. In China, the areas affected by drought have exceeded 6.7 million ha since 2000 in Beijing, Hebei Province, Shanxi Province, inner Mongolia and North China (Zhou, 2003; Yoshino, 2000). North Korea and Australia are also vulnerable to drought.

Climate change could influence variations in temperature, precipitation and evapotranspiration and exacerbate water shortages and drought frequency in the ESAP region (IPCC, 2007). Since over 80% of water supplies in the region are used for agriculture, crop productivity will be severely affected if access to water is diminished (García et al., 2006). Decreased precipitation and increases in evapotranspiration can lead to deficiencies in soil moisture as well as stream flows that provide for irrigation (Wang, 2005). The retreat of glaciers and decline in snow melt in much of South Asia could place significant pressure on water availability in the dry summer months and instigate the gradual transformation of already water stricken areas into arid deserts (IPCC, 2007; Barnett et al., 2005).

4.2.8.4 Pest attack and diseases

The predicted warming trends linked to climate change are expected to influence pest and disease frequency and damage extent. For example, a decline in frost events in New Zealand has led to an increase in the tropical grass webworm and caused severe damage in northern agricultural regions of the country (IPCC, 2007; UNEP, 2006). Changing pest and disease patterns due to climate change will likely affect how food production systems perform in the future. This will have a direct influence on food security and poverty levels, particularly in countries with a high dependency on agriculture (IPCC, 2007).

4.2.8.5 Adaptation

The agricultural sector could minimize damage and take advantage of changing environments by implementing adaptation measures and coping mechanisms such as the development and introduction of high temperature tolerant crop varieties (Agarwal et al., 2003). Currently available crop varieties are not capable of tolerating high temperatures, especially increased nighttime temperatures. Yield responses from current rice varieties are high in response to increases in nighttime temperature (Peng et al., 2004). The development of salt tolerant crop varieties could reduce shortfalls of crops under sea level rise scenarios. Current rice varieties and especially the HYVs have two major limitations: their stems are short in length and they cannot sustain long inundation periods during flooding resulting in severe losses of crops. In the Philippines a gene has been identified that enables rice to survive complete submergence. The development could benefit millions of farmers whose rice crops are constantly vulnerable to flooding especially in the low lying mega-deltas of ESAP.

As water resources availability will be highly variable over time and space, technological development and diffusion are necessary for water conservation and increasing irrigation water application efficiency (roughly 40% in India and in Bangladesh). The introduction of low cost drip irrigation can increase efficiency; the water saved could be used to bring additional lands under irrigation. Weather information, such as advance warning on floods, droughts, pest attacks, etc., can help farmers to better prepare for extremes.

Climate change can introduce new pathogens and expand the territory of animal diseases. Improved understanding is needed of the mechanisms that spread pathogens, their hosts and the potential damage to animal health. Some adaptation measures (IPCC, 2007) for the agriculture sector in Asia will require substantial inputs of technical, financial and human resources (Table 4-5).

Table 4-5. *Climate change and adaptation measures in agriculture.*

Sectors	Adaptation Measures	AKST Challenges
Agriculture cropping	*Choice of crop and cultivar:* Use of more heat/drought tolerant crop varieties in areas under water stress Use of more disease and pest tolerant crop varieties Use of salt-tolerant crop varieties Introduce higher yielding, earlier maturing crop varieties in cold regions *Farm management* Altered application of nutrients/fertilizer Altered application of insecticides/pesticides Change planting date to effectively use the prolonged growing season and irrigation Develop adaptive management strategy at farm level	• Identification of appropriate gene • Lack of resources for the development of varieties • Time-lag between development; field trial, acceptability of farmers and onset of climate change • Onset of new pests and diseases • Needs extensive research on nutrients and fertilizer requirements of new crop varieties • Changing planting date could have effect on yield • Resources and technology require at grass roots level
Livestock production	Breeding livestock for greater tolerance and productivity Increase stocks of forages for unfavorable time periods Improve pasture and grazing management including improved grasslands and pastures Improve management of stocking rates and rotation of pastures Increase the quantity of forages used to graze animals Plant native grassland species Increase plant coverage per hectare Provide local specific support in supplementary feed and veterinary service	• Breeding less climate sensitive livestock will be a formidable challenge • Less climate sensitive grass and pasture varieties need to be developed • Many native grassland species are not nutritious for animals • Need resources, advanced technology for feed and veterinary service
Fishery	Breeding fish tolerant to high water temperature Improved fisheries management capabilities to tackle climate change	• Cross breeding with fishes from arid region is a possibility but its effects on local varieties will be unknown for long period • Technology and resources will be major obstacle
Development of agricultural bio-technologies	Development and distribution of more drought, disease, pest and salt-tolerant crop varieties Develop improved processing and conservation technologies in livestock production Improve crossbreeds of high productivity animals	• Will emerge as technological challenge for poor countries • Faster technological transfer is required • A new nexus between technology owners may emerge to take advantage of climate change
Improvement of agricultural infrastructure	Improve pasture water supply Improve irrigation systems and their efficiency Improve use/store of rain and snow water Improve information exchange system on new technologies at national as well as regional and international level Improve sea defense and flood management Improve access of herders, fishers and farmers to timely weather forecasts	• Improved water store, supply and irrigation need new technologies and replacement of the old • Dissemination of information on technology requires to build institutional capacity and educating farmers • Improved sea defense and flood management have potentials but they have certain limits

Source: Authors' elaboration modified from Cruz et al., 2007.

4.2.8.6 Food security and human migration

Food security, hunger and famine are not exclusively related to climate affected crop losses; but natural hazards do exacerbate these issues. Climate change and possible increases in natural hazards could increase hunger and malnutrition in many nations in ESAP including Bangladesh, China, India, North Korea, Viet Nam and the Philippines. A sizeable proportion of these populations (urban and rural) suffer from food insecurity, especially rural marginal farmers. Although as a result of globalization per capita income has increased, inequality is on the rise. Prices of essentials, particularly food, are increasing. A threefold increase in global cereal prices by 2080 as a consequence of decline in net productivity due to climate change is projected (Parry et al., 2004). Subsistence producers who grow crops like sorghum and millets in dry conditions could be at greatest risk of a potential drop in productivity and from the danger of losing crop genetic diversity (IPCC, 2007). In the near term, an additional 49 million people would be at risk of hunger if the world follows the SRES A2 emissions scenario without any carbon fertilization (Parry et al., 2004). By 2050 (mid-term) and 2080 (long-term), the additional people at risk of hunger could increase to 132 and 266 million, respectively.

Rural to urban migration may increase if sufficient income sources are not available in rural areas. Countries of the ESAP region could face substantial food shortages unless they succeed in adapting to environmental changes. The situation does not look optimistic given the recent stagnancy in agricultural productivity. Higher temperatures, increased rainfall, drier summer months and saline water intrusion will decrease agriculture productivity in the short to mid-term. In the long-term, technological breakthroughs may alter the situation. However, this will be highly dependent on the development, deployment, and diffusion of new technologies. The commitment of individual countries to reduce emissions and enable mitigation and adaptation to climate change will be crucial.

4.2.9 Energy

4.2.9.1 Energy crisis in agriculture

Access to adequate, reliable and affordable supplies of modern energy sources, such as hydrocarbons or electricity, is minimal and traditional energy sources for food production, such as fuelwood, biomass and human and animal power, are also diminishing. Since global energy use is unsustainable in the long term, the energy sector is undergoing a shift toward energy efficiency and conservation in addition to the development of renewable and recyclable energy sources. Rural areas have the advantage of transitioning to more sustainable energy systems by employing techniques such as organic farming, improved water and soil management, integrated pest management, mechanization and biotechnology. Technological and institutional challenges remain in building the capacity of rural areas to adopt more sustainable measures, which often involve high initial investments in capital, labor and training. If rural populations are excluded from the shift to sustainability there is a risk of massive emigration to urban centers (Dutkiewicz, 1999).

Regional groundwater exploitation has escalated at the expense of the energy economy. South Asia as a whole spends 5 to 6 billion USD per year to pump approximately 210 km^3 of water, mostly for irrigation (with 27-35% of the power being subsidized). Economic losses in the electricity sector due to agricultural power subsidies are estimated at 5.4 billion USD in India (Shah et al., 2003). While farmers will soon be faced with water availability issues, such as declining water levels, high rate of well-failure, salinity and reduced well-yields, irrigation will face high energy costs and unreliable electricity supply. Energy demands for agriculture activities can be influenced by climate change, in the form of increased electricity requirements for irrigation pumping during warmer weather to maintain soil moisture (IPCC, 2007).

4.2.9.2 Bioenergy

The general emerging pattern is to move from traditional bioenergy (wood fuels, charcoals, etc.) to modern fuels as household income rises (Barnes and Floor, 1996). However, with rising oil prices and falling oil supplies, concerns over greenhouse gas emissions and political instability in many oil rich countries, there is a renewed interest in bioenergy, mostly liquid biofuels but also for electricity generation. Improving the efficiency and reducing the harm of traditional bioenergy remains a challenge and needs to be addressed.

Bioenergy can take the form of solid biomass or liquid biofuels (Bird Life International, 2005; Tustin, 2006). In ESAP, the main energy crops include sugar, coconut, cassava, castor kernel and oil palm. Since supporting local farmers is good politics for national leaders, policymakers are directing resources towards the biofuel cause (Yuit and Wall, 2006). It can benefit commercial plantations and similar agriculture in the rural areas. The rural dwellers can also benefit from the use of by-products from bioenergy production. However, increasing food prices, potential deforestation and depletion of water resources could emerge as byproduct environmental problems. In addition, a variety of energy inputs used during the cultivation of feed stocks and production of biofuels can produce greenhouse gases that contribute to global warming.

World primary energy demand projections suggest that the supply of non-hydro renewables as a percentage of global electricity supply/electricity generation will triple from 2% in 2002 to 6% in 2030. While wind power will see the biggest increase from 0.3% in 2002 to 3% in 2030 and is expected to succeed biomass as the largest source of non-hydro renewable electricity generation, it is anticipated that electricity generation from biomass will triple between now and 2030. Furthermore, the demand for biomass and waste fuels will rise by 1.3%, of which 0.7% is attributed to traditional biomass (IEA, 2004) (Table 4-6).

Both Thailand and India have launched national poli-

Table 4-6. *Biofuel policy initiatives in ESAP.*

Country	Source of Biofuel	Suggested Blending (%)	Year of National Biofuel Policy/Act
China	Ethanol	10	2005
India	Ethanol, Jatropha oil, Pongemia oil	5 (with up to 10 for public transit)	2003
Japan	Ethanol	3	2003
Malaysia	Palm oil	5	2005
Thailand	Tapioca, Ethanol	10	2005
The Philippines	Coconut oil	10	2005

Source: Raju, 2006.

cies to promote gasohol, which is a blend of 10% ethyl alcohol and 90% gasoline. The Thai gasohol program started in 1985. As of December 2005 the country had more than 4,000 stations serving alternative fuels and an import ban on methyl tertiary-butyl ether (MTBE), which is the petrol-based fuel additive that ethanol replaces, will be mandated in 2007. The government has pledged a renewable energy target of 8% of total energy consumption with 24% of the target as liquid biofuel. Initiatives are also under way in the Philippines and Indonesia to implement similar gasohol policies (Bhandhubanyong, 2005; Yuit and Wall, 2006). In addition, the promotion of biodiesel produced from coconut oil is under way in the Philippines, with Thailand, Malaysia and Singapore expected to follow suit. The Indonesian government is focusing on biodiesel production from palm oil. An anticipated 5.6 billion of the 22 billion USD pledged for biofuel production and distribution initiatives will be spent on palm oil production. Similarly, the Malaysian biodiesel policy is expected to produce up to 500,000 tonnes of a biodiesel blend of 5% palm-oil-derived and 95% petroleum-derived diesel (Yuit and Wall, 2006). These are technical potential estimates and the utilization of full technical potential is dependent on economic viability.

4.2.9.3 Bioelectricity or electricity from biomass

Given growing global energy demands, a question of interest relates to the potential for biomass to produce electricity. A sustainable, economically competitive global bioenergy supply is around 270 EJ per year, which is approximately 70% of the total world energy consumption in 1990 (Coombs et al., 1992).

Rice husks in Southeast Asia have significant potential for electricity generation. It was estimated that 28.5 million tonnes of rice husk are produced in Southeast Asia per year. Assuming energy content of 3,000 kcal per kg of rice husk and accounting for typical boiler and steam turbine efficiencies, a steam turbine consuming 5.4 kg kW^{-1} will potentially produce 2778 MWe or 24.35 x 106 MWh (at 100% load factor). Since the total average energy consumption per hour for the Southeast Asian countries is 16,628 MWe, rice husk could theoretically supply 13.6% of the total electricity consumption. However, this figure depends on capacity of husking mills and associated costs (Himpe, 1997).

A recent study analyzed the global bioenergy potential for the period 2050-2100 based on forecasted future development paths and land-use patterns using four storylines of the IPCC SRES Emission Scenarios. The resulting potential for abandoned land ranges from about 130 to 410 EJ yr^{-1} in 2050 up to 240 to 850 EJ yr^{-1} in 2100. While the potential at low-productive land is negligible, "at rest" land could potentially provide approximately 35 to 245 EJ yr^{-1} in 2050 and from about 35 to 265 EJ yr^{-1} in 2100. At a regional level, South Asia has an average of approximately 3% of world potential for abandoned agricultural land and 5% of world potential for "at rest" land for the year 2050 (Hoogwijk, 2004). A contrasting study of bioenergy potential in the USA concluded that 709 million ha would be needed to meet the country's gasoline demands for a business-as-usual scenario in 2050. This figure decreases to 46 million ha through improvement of biofuel conversion efficiency and increases in feedstock yield (Greene, 2004).

4.2.9.4 Competing land uses and implications for food security

While the intent of biofuels projects would be to make use of existing agricultural land or abandoned and/or low quality farmlands, the clearing of virgin forest as well as agriculture and forest lands may be necessary to meet projected energy demands if biofuels are to be relied upon to the extent projected in some of the estimates reported above. As a result, any emission reductions provided by the use of biofuels will be lessened due to the significant loss of carbon sequestration capacity when virgin forest is cut down (Yuit and Wall, 2006).

Furthermore, while forests themselves provide a source of biomass (in the form of timber harvest waste, unmarketable lumber, trees removed during land clearing operations, wood residues produced by sawmills, forest thinning material, and leaves and other forest litter), overexploitation of this resource will result in damage to forest ecosystems and a subsequent loss in biodiversity. This is a concern especially for biodiversity rich continents such as Asia (Bird Life International, 2005; Kampman et al., 2005). Unless alternative sources of energy are developed, forest policy must incorporate energy needs into afforestation and forest preservation strategies in order to meet projected demands for biofuel.

Energy security is also linked with food production since a predominant use of traditional biofuels is cooking. Therefore, the adequate supply of traditional biofuels has an important bearing on nutritional security, especially in rural areas and low income households (Mahapatra and Mitchell, 1999; Kampman et al., 2005).

A study in eastern India investigated the increased pressure on regional forests to provide fuelwood, which is the major traditional biofuel in rural eastern India. Dwindling supplies are influencing the use of crop residues, leaf litter, dung, and kerosene to meet energy needs. The mean per person consumption of fuelwood, dung, leaf fuel and crop residues by farm households is 0.46, 0.08, 0.12 and 0.04 tonnes respectively. Other reasons for using dung include higher livestock numbers, insufficient labor to gather fuelwood, and accessibility of biogas plants. Leaf fuel is gaining recognition since it is essentially free from the legal, social and political constraints associated with forest biomass. However, intensive use of dung, agricultural wastes and leaf litter may deprive the soils of much needed organic nutrients. The study also evaluated the hypothesis that dwindling forest biomass supplies will motivate tree planting. The analyses concluded that on-farm production of fuelwood was not influenced by scarcity of forests. However, agroforestry has the potential to limit deforestation and improve agricultural productivity by freeing up labor hours normally dedicated to fuelwood collection (Mahapatra and Mitchell, 1999). Thus, the promotion of tree planting on-farm and provision of community land to meet fuelwood demand should be deliberated (FAO, 1998, 1999; Slingerland and van Geuns, 2005). However, the influence of community agroforestry on conservation depends on secure land tenure and associated land ownership rights (Contreras-Hermosilla and Fay, 2005).

Competing uses of land for biofuel feedstock production could have impacts on food security. A major constraint of the biofuel industry is land availability and the competition between biofuel feedstock and food crops for this land (Gratzl and Fawer-Wasser, 2006). Studies predict an exponential decrease in global per person agricultural land in future (2025) due to urbanization, degradation, and biofuel plantations among other things (Lal, 2006). Per person arable land area will be <0.1 ha by 2025 in some densely populated countries of Asia under a scenario of medium population growth (Engelman and LeRoy, 1995). Competition is heightened when soils are degraded due to poor agricultural practices or natural processes (Swift and Sanchez, 1984; FAO, 1999). The prices of food and fodder crops would increase even if large-scale cultivation of biofuel feedstocks is promoted irrespective of economic value of food crops and biofuel feedstocks (Kampman et al., 2005). This could have significant socioeconomic effects for ESAP where food security is already an issue (Yuit and Wall, 2006). Furthermore, monocultures and block cropping, which could arise from increasing biofuel demands, are associated with declines in biodiversity (Bird Life International, 2005). Ecosystems may also be harmed by more intensive forms of biomass farming as a result of changes in the water table, increased pesticide use and encroachment on wildlife habitats (Kampman et al., 2005).

Countries should concentrate on region specific biofuel feedstocks (e.g., *Jatropha curcas*, sugarcane, palm oil, sun flower, etc.) depending on soil and climate conditions, as well as energy crops that can be grown with minimal chemical inputs and with high water efficiency (Bird Life International, 2005; Raju, 2006; Yuit and Wall, 2006). However, in order to avoid pressure on lands suitable for food crop agriculture, initiatives are required to bring degraded and low quality lands under crops rich in nonedible oils (Lal, 2006; Yuit and Wall, 2006). For example, Jatropha is such a crop that is cultivated in almost all tropical and subtropical countries, and is an important feedstock for biodiesel as the plant grows in poor soil and withstands long periods of drought. The Jatropha system promotes the planting of Jatropha hedges to protect gardens and fields against roaming animals and reduces erosion. The oil from the seeds can then be used as a bioenergy source. The Jatropha system and/or plantations are currently employed or are under consideration in Cambodia, India, Indonesia, Laos, Nepal, Papua New Guinea, Sri Lanka and Thailand (Henning, 2004; Yuit and Wall, 2006).

4.2.9.5 Outsourcing of biofuel feedstock production

Industrialized nations are looking to ESAP to increase production of biofuels to meet their environmental targets in terms of emissions of greenhouse gases and reduce dependency on fossil fuels (Yuit and Wall, 2006). There is a potential imbalance between countries with a high and rising demand for biofuel use and countries where biofuels can be produced at lower cost (Kampman et al., 2005; Slingerland and van Geuns, 2005; Henniges and Zeddies, 2006). Many developing countries are currently considering or already expanding biofuel production and processing capacities. This may create challenges as it becomes increasingly efficient to produce bioenergy, since competition for land and other resources will arise between food and fuel production (von Braun, 2005). Bioenergy should be seen as a positive step forward only if it is conducted in a sustainable fashion with equal emphasis placed on conservation efforts, energy efficiency and climate change policy (Yuit and Wall, 2006).

4.2.9.6 Pollution and health impacts

Traditional bioenergy derived from the combustion of wood and agricultural residues for heating and cooking may impart negative health impacts from indoor air pollution (De La Torre Ugarte, 2006). On a more global scale, the burning of biofuels is linked to large pollution plumes. The Mediterranean Intensive Oxidant Study (MINOS) investigated long-range transport of pollutants and concluded that air pollution over the eastern Mediterranean between 1 and 12 August originated in ESAP. During the Asian Summer Monsoon, convection carries polluted air into the upper tropospheric anti-cyclonic circulation and is then transported in the upper troposphere over the Mediterranean. This plume is characterized by enhanced concentrations of biofuel combustion tracers, such as methonal (Scheeren et al., 2003). However, the introduction of cleaner cooking technologies can reduce the contribution of traditional biofuels to air pollution. Traditional biofuel black carbon emissions in India have essentially remained unchanged from 1985 to 1995 due to the gradual introduction of clean technologies (Venkataraman et al., 2005). This is despite an increase in traditional biofuel consumption. The use of modern biofuels, such as biodiesel and ethanol, is expected to reduce harmful pollutants from vehicle exhaust compared to petroleum-based fuels. An extensive survey of emissions from a range of biodiesels suggests that burning this fuel reduces health risks associated with petroleum diesel. Biodiesel emissions show decreased levels of polycyclic aromatic hydrocarbons (PAH) and nitrated polycyclic aromatic hydrocarbons (nPAH), which have been identified as potential cancer causing compounds (US EPA, 2002). If wastes from modern biofuel plants are not processed as per environmental requirements, they could be sources of pollution. Biofuels will play a major role in the energy management issues that ESAP countries will confront in the near future, involving wider policy regimes, institutional and technological choices.

4.3 Major Uncertainties of the Drivers and Projections

Major uncertainties affect individual drivers as well as the interaction between different drivers. It is important to note the difference between risk and uncertainty. The definition of risk refers to a quantifiable change in or likely future of some variable (life of an individual, catch of fish, agricultural production, etc.). Therefore, risks are associated with certain assumptions that decision makers can make about these variables. Unlike risk which is amenable to a certain (limited) extent of quantification, uncertainties are beyond quantification because there are gaps in assumptions (including those about number and nature of relevant determinants), missing information, and poor systems of analysis. These are variables whose future trends or changes may bring with them questions that decision-makers may not be asking at present. In some cases, crucial data may be missing; in others the capacity to understand or project nonlinear changes may be lacking. Yet it is important to keep decision makers informed about the existence of uncertainties (Table 4-7).

Table 4-7. *Major uncertainties and likely impacts in the ESAP region.*

Key Uncertainties	Drivers of change directly affected	Implications for agriculture, food systems, products and services	Implications for development and sustainability goals in the ESAP region
Climate change[1] Rise in: sea level temperature precipitation (2015-2075)	Demographic changes Economic growth Agricultural growth Trade Investment	• Resurgence of tropical diseases • High morbidity rates • Reduced labor availability • Unpredictable employment opportunities • Factor productivity declines • Food prices increase • Unstable markets • Declining crop productivity • Cost of production increase • Animal and crop diseases increase • Preservation and storage crucial • Deciduous forests incapable of regeneration • Desertification increases • Unpredictable production estimates, quality standards, etc. • Futures markets collapse • Higher market regulation with increasing loopholes • Capital diverted to survival (food, health) • Returns to investment decline	• Increase in poverty, hunger and malnutrition • Inequality, civil strife increase • Economic growth unsustainable • Natural resource degradation • S&T becomes emergency driven and legitimacy falls • Governance and decision-making become more centralized • End of capitalism[2]
Regional conflicts (water/energy) (2015-2030)	Water Energy Trade Economic growth	• Irrigation water pricing • Production costs increase • Urbanization slows • Off-shore fishing collapses • Migration increases • China and India—efforts to contain domestic inequities • India—global collaboration for basic needs • Nuclear energy increases • Hydel power collapse • Non-conventional energy increases • Gains in off shore oil gas • Energy prices soar • Trade declines—tariffs increase • Regional cooperation collapses • GDP and agricultural growth declines • Public investment in maintenance/ compensation • Private capital shifts to EU/LAC/ Africa	• Water conflicts—worsening law and order • Hunger and food crisis severe—global aid resolves some of it • Limited employment • Gender relationships, female labor options worsen • Economic development grinds to a halt • Investments in water/ energy saving, construction, agricultural and industrial production • South Asian trade blocks collapse—China or China + Australia dominate East Asian trade • China regional leader • Civil liberties curtailed
Global conflicts (2020-2030)	Fiscal / Political stability Globalization	• China withdraws investments in the USA • USA[3] attempts to contain Chinese growth • ESAP divided—Chinese vs. US allies • Collapse of WTO • EU vs. USA—sub-regional trade blocks • Labour/capital mobility constrained • Global trade declines • Markets/ investments shift to ESAP and LAC • Regional and sub-regional instability • Regulations increase	• USA and EU unemployment increase • Worst global economic depression • Economic growth limited to some pockets • China dominates ESAP • Intra-regional alliances increases • Global defense expenses escalate

Table 4-7. *continued*

Key Uncertainties	Drivers of change directly affected	Implications for agriculture, food systems, products and services	Implications for development and sustainability goals in the ESAP region
Confederation of Asia-Pacific States (2020-2050)	Political stability Civic space Energy/Climate change Economic growth	• Increasing domestic strife • Lack of faith in the UN to resolve local conflicts, detract authoritarian nation states in Asia • Transnational constitutionalism • Civil society and private sector led negotiations • Pan-Asian identity • Political maturity—South Asian Parliament and Boao Forum ideas[4] • Federalism and democratic governance • Fungible borders, shared problems and solutions • Regional protocols, policies and S&T for systems understanding and solutions • Intra-regional cooperation and trade increase	• Asia-Pacific Union/ Parliament emerges • Yeng$ (China-Japan-Australia-Singapore-Brunei lead Asia-Pacific currency), Asian Monetary Fund • Unrestricted labor and capital mobility • Trans-Asian transport improved • Trans-national integration and governance given prime importance • Human rights and ecological values politically accepted.

[1]IPCC, 2007

[2]Industrial capitalism "will walk out of the economy silently, on its toes" after it has wreaked havoc on humanity and the ecosystem Beck, 1992.

[3]Tammen, 2006; Christensen, 2007

[4]Muni, 2004

The implications of these uncertainties for AKST are many and varied. Climate change and consequent variations in crop/animal production and productivity may increase investments in AKST but have a diminishing success rate over time if climate change continues to worsen. Increasing resource constraints and deteriorating law and order situations could impose restrictions on access to, participation in, and utilization of, technology and knowledge for millions of people.

Several other changes in the relationships among the drivers of change will also shape the nature and intensity of AKST in the region. For instance, increasing production problems in dryland agriculture and fisheries could lead to migration, consequent displacement and strife and may carry an entire generation of location specific knowledge to urban or other centers. Displacement will in turn affect the access to education and S&T training, thus reducing further the availability of technically qualified and trained human resources for AKST.

4.4 Relevance and Implications for Agriculture and AKST

4.4.1 Drivers of change and implications for agriculture

Given the pattern of evolution of the key drivers of change, the future of agriculture and food systems in ESAP reflects the continuing social, economic and environmental importance of agriculture. Agriculture in ESAP will be influenced by the economic and political choices made by individual Governments, developing and developed, and will in turn play a crucial role in shaping some of these macroeconomic decisions in the region. The latter may include options for domestic social security nets, regional and global trade choices, impacts on and investments to mitigate global climate change processes and may also lead to internal negotiations and alignments within the region among countries willing to adopt different paths to address social and environmental sustainability. The overall sociopolitical power play in the region is an important dimension that underpins this discussion.

Overall, the trends or projections of the key drivers of change draw upon on a diverse set of policies and institutional arrangements prevalent in the ESAP region and they reflect some of the inconsistencies and the diversity that exists in these policies and regional contexts. The major implications of the key drivers of change for agriculture and food systems in ESAP are the following.

1. Given the decreasing share of agriculture in the economy of the region, and its overriding importance in providing employment, food, nutrition and health, social and environmental well-being, it is clear that *agriculture can provide solutions only if appropriate macroeconomic policies are in place.* Despite impressive overall economic growth and a declining share in the overall economy (GDP), agriculture will continue to play a major role in the economy of ESAP countries, mainly because it will continue to be the mainstay for a large proportion of the rural population in almost all the developing ESAP countries and will be a major contributor to environmental degradation as well as a focus of environmental remediation in both developing and developed ESAP countries.

South Asia will be home to half the Asian population

in the next 20-25 years, with an increasingly younger work force and larger proportion of women engaged in cultivation and agricultural labor. Rural women will play greater roles not only as unpaid family workers but more importantly as de facto farm managers. Thus in the future AKST will have to empower rural women by enhancing their skills and knowledge on all aspects of production and processing which will raise agricultural productivity and lead to economic development.

Despite projected increases in food production and a marginal improvement in per capita food consumption, South Asia will account for almost half of the world's malnourished children (with India hosting one in every three undernourished children in the world, and China recording the highest reduction in child malnutrition by 2020), and will continue to have low access to education, health and basic environmental services like drinking water and sanitation.

While the diversity of agricultural production and market access will increase, the role of small and marginal farmers in production systems will continue well into the future, until overall rural out-migration due to education and more urban jobs will allow these populations dependent on small unviable holdings to move to higher paid jobs and more predictable incomes. This has implications for investments and subsidies in the region, to protect small producers, to provide some social security, or to enable through investments, more opportunities for employment in a liberalized Asian/global context.

Globalization and increasing market liberalization will lead to increasing regional preferential trade agreements, especially for agriculture, with all ESAP countries gradually withdrawing production (input and price) subsidies and other trade distorting practices; some studies point to a trend of increasing inequality and of a need for redistributive policies that may reduce widening income gaps in ESAP.

2. Agriculture will gain from and contribute to economic growth in the region, *but the economic gains will be far less than the contributions of the sector to factors of growth and growth processes.* The ESAP region is expected to be the fastest growing region in the world over the next two decades (national and per capita income growth rates in East Asia and Pacific countries will be higher than those of South Asian countries). There will be increasing trade in agriculture which will be accompanied by a steady fall in prices of agricultural commodities and an overall increase in prices of inputs (especially rising energy prices), with increasing total factor productivity and substitution of labor for capital in agriculture.

Demand for agricultural commodities will not only increase dramatically over time, but the composition of that demand will change significantly as per person wealth increases. The ESAP region is gradually diversifying its farm production in favor of higher valued commodities including fruits, vegetables and meats. The change in diets and declining terms of trade for cereals in Asia will lead to diversification of farm production into higher value products. The decline in the terms of trade and falling prices will also mean that for countries in the ESAP region to maintain a comparative advantage in agricultural commodities, they must offset through higher productivity by increasing farm sizes and increasing the mechanization of farming processes. Further agricultural development in ESAP must exploit comparative advantages in a more globalized economy. This will mean further industrialization and product diversification, leading to the creation of larger, more technologically advanced farming industries.

The demand for rural employment, improved livelihoods and political stability, along with increasing agricultural diversity, post-harvest and value-addition processes, specialization (organic agriculture for instance) and urbanization of food markets, will pose macroeconomic challenges as well as opportunities to move people out of direct dependence on agriculture to other non-farm or urban employment. Agriculture will be integrated into industrial, environmental and health sector growth in a wider, more diverse form.

Agriculture will be at the receiving end of most of the negative consequences of globalization and trade liberalization with elimination of tariffs in the ESAP region, marking a wider rural-urban disparity, increasing concentration of food markets/retail and grain trade in the hands of a few global players, varying levels of investment (public and private), improved transport and communication facilities, along with increasing restrictions on economic activity due to IPRs or other trade policies, increasing disillusionment, political instability, intra and inter-regional tensions (over water, trade, subsidies, environmental compliance, oceans and fishing rights, etc.), and increasing marginalization of indigenous and tribal people within these countries. Concerted action by Governments will be necessary to ensure that social safety nets and adequate investments and benefits flow to the agriculture sector and into poorer regions and communities within ESAP.

3. *Environmental and social costs of agricultural growth and overall economic growth will pose additional challenges.* With increasing evidence of impending climate change and marginalization of significant proportions of rural Asians from mainstream development processes, there is a requirement in ESAP for an increasingly diverse portfolio of policies, institutions and organizations to address these diverse complex problems.

Agriculture in ESAP will be increasingly constrained by worsening environmental degradation (land, water and air pollution) caused by population pressures, agricultural production practices, urban and industrial wastes, and perceptible climate change impacts. The lack of legal and institutional mechanisms as well as lack of compliance to arrest environmental degradation and improve mitigation investments and practices will continue in most developing ESAP countries.

Increasing land degradation, decreasing access to sufficient quality irrigation, increasing incidence of pests and diseases (with added problems epidemics like the avian flu and other livestock diseases), and varying impacts of global warming, will most likely result in increasing rural distress and migration in many ESAP countries. This will add to environmental degradation and urbanization pressures.

The role of nonconventional actors will become increasingly important in ESAP agriculture (the socially responsible among the corporate sector and the CSOs); especially in enabling environmental sustainability through efforts to invest in organic agriculture, and indigenous knowledge systems, and to improve livelihoods in marginalized and remote parts of the region, and through their capacity to mobilize community-wide compliance to environmental legislation (at the local, national and global levels).

Many ESAP countries—both developed and developing—will become increasingly conscious of the tradeoffs between resource depleting subsidies given to farmers, polluting industries, and trade. There is evidence of increasing efforts at the national and regional/global levels from Governments and international bodies to address the options for reducing waste and inefficiencies, and for environmental remediation and conservation policies.

The region is also experimenting with and rapidly adopting several conventional resource conserving technologies (watershed management, water harvesting, zero-tillage, IPM,) as well as new technologies and institutional arrangements that promote these technologies, like biotechnology, bioenergy/biofuels, biopharmaceuticals, bioplastics (environmental/green chemistry).This is despite a lack of consensus among the major actors (the State, private industry and CSOs) about the national and local processes and choices available for technology adoption and the associated effects on environmental, economic and social well-being.

Thus, the key drivers and their evolution convey the following messages for the future of the agricultural sector and food systems in ESAP:

- Agriculture and food systems need well-informed and more integrated support in the form of several direct and enabling policies;
- While existing and future terms of trade will ensure that the ESAP economy will draw benefits from agriculture, specific policies or increased investments to reduce the losses to rural and agricultural populations may be necessary;
- The environmental costs of agricultural production and the tradeoffs between environment and development will have to be addressed.

4.4.2 Relevance and implications for AKST to achieve development goals

Given the evolution of the drivers of change and their relevance to agriculture, we now explore the nature of AKST needed to address the development and sustainability goals of reducing hunger and poverty, improving nutrition, health and rural livelihoods, to facilitate social and environmental sustainability.

The ESAP region shifted fairly rapidly from a combination of subsistence agriculture and commercial cultivation (catering to colonial industries and trade), to an overwhelmingly food security driven commercial production of staple cereals. There was also a significant increase in production of milk, fruits, vegetables, poultry and fish. These were accompanied by the establishment and development of formal organizations and research programs for agricultural technology generation and diffusion, aided initially by global NGOs (e.g., Rockefeller, Ford), and later by international development organizations and national governments. Overall, AKST in the ESAP region has been crop/commodity focused (i.e., cereal crops, livestock, apiculture, aquaculture) and has paid some attention to natural resources as factors of production that contribute to agricultural growth (See Chapter 2). Formal organized AKST has focused mostly on irrigated cereal producing tracts and less on arid ecosystems or drylands, coastal or mountain ecosystems, and other marginal production systems. AKST in the ESAP region has generated and enabled the utilization of several environment-friendly production technologies, such as land and water conserving technologies and IPM. Public sector R&D continues to grow in ESAP countries, along with emergence of new technologies such as biotechnology and concerns about institutional reform to enable increasing access to and utilization of technology and knowledge.

Given this context it is likely that in future, AKST in the ESAP region will be characterized as follows:

- Public sector R&D dominated AKST systems will continue in many developing ESAP countries and in some CGIAR centers;
- Private investment in AKST and commercially appropriable technologies will continue to increase in the developed ESAP countries and in East Asia;
- Given the young age of much of the population, and opportunities for education and information technology in Asia, there will be an increasingly important role for a university education system that caters to formal AKST, with associated private and public goods;
- ESAP is also likely to witness increasing collaboration between the corporate sector and NGOs, both in technology dissemination and in new farming arrangements such as contract farming or cooperative farming that gives scale advantages to small farmers;
- Institutional reform in public sector NARS to enable effective AKST systems that can enable innovation and development will continue to be a hurdle in many ESAP countries.
- AKST will engage proactively with many marginal agricultural production systems and indigenous knowledge systems.
- Allocations in AKST on environmental problems will increase and so will corresponding investments in appropriate production systems, and resource conserving technologies.

Based on the projections for the drivers of change and messages about increasing hunger, poverty and environmental degradation in parts of ESAP, the AKST requirements for achievement of the development and sustainability goals include:

- Reducing hunger and poverty:
  - (a) Increasing public sector investment in formal AKST organizations to address declining productivity growth rates in staple cereals in Asia.
  - (b) Greater exchange of and participation in international knowledge and technology networks and markets, in response to increasing globalization.
  - (c) Increasing attempts to understand, analyze and improve agricultural production and livelihoods in dryland, pastoral, coastal and mountain ecosystems, and other marginal agricultural production systems.

(d) Open debate about subsidies to protect small holders versus investments along with AKST to enhance their production and market capacities.
(e) Facilitation of other employment options for rural landless labor in areas where capital substitution for labor occurs.
(f) Increasing recognition of the role of complementary investments in rural and meso- or macro-variables like rural infrastructure, rural banks, post-harvest systems, and transport, that enable more effective generation and utilization of agricultural knowledge and technologies, and more non-farm rural employment.
(g) Application of agricultural knowledge and technologies to complement social security systems or targeted food distribution systems in areas of abject poverty.
(h) Increasing partnerships between public AKST organizations and CSOs—for innovation and for specific monitoring of local agroecological systems.
(i) More attention from national and international policy circles to institutional arrangements that restrict poor people's (especially women's) access to knowledge and technologies, and appropriate corrective measures from local AKST to enable utilization of knowledge that can enhance the income and well-being of the poor.

- Improving nutrition, health and rural livelihoods:
  (a) Increasing investments in AKST for a diverse range of crops/commodities/livestock, forestry and aquaculture systems.
  (b) Increasing food safety and standards, and adoption of sanitary and phyto-sanitary systems that enable rather than distort trade in agricultural/food products.
  (c) Formal AKST investments in alternative methods of production (e.g., organic) that are more environmentally friendly.
  (d) Increasing interactions and partnerships in formal AKST, with a wide range of organizations involved in research, development, finance, transport, storage, packaging and other products or services, in public, private or civil society sectors.
  (e) Substitution of all resource degrading and resource depleting inputs/chemicals/pesticides, with new technologies or production practices such as biotechnology or IPM, wherever all required safety regulations are in place and adequate monitoring mechanisms are available.
  (f) Integration of AKST with specific programs that address child malnutrition or other targeted nutrition programs—biofortified rice, vegetables, organic grain, high-protein local foods, preserved or processed foods, for example.
- Environmental sustainability:
  (a) Improved AKST assessment methods and systems of monitoring and evaluation that can forewarn of environmental consequences.
  (b) Increased domestic and global investments in AKST organizations for research on the implications of climate change for agriculture, and for technologies for adaptation and mitigation.
  (c) Increased investments in cost sharing schemes for environmental services provision—enabled by resource-conserving technologies in developing countries—especially in parts of the Pacific islands and South Asia.
  (d) Promotion of industrial and environmental biotechnology, including biofuels and bioplastics that can enable both remediation of the environment and increase rural non-farm employment.
  (e) Greater development of partnerships and international agreements in AKST that can promote regional and global networks for specific ecosystems and recognize ways to minimize tradeoffs between the environmental and production imperatives of each agroecological system.

Current trends in many developing ESAP countries reveal that the capacities to make such investments or partnerships in AKST and the political willingness to make the choices (subsidies vs. investments, IPR vs. open source knowledge systems) to enable better utilization of AKST by the poor are limited. The capacity of the agriculture sector and AKST to reduce hunger and poverty and enhance environmental sustainability depends significantly on reforms within the sector, as well as on several other macroeconomic variables and political processes. AKST in ESAP needs significant capacity development to learn from and work with a wide range of organizations and processes, in order to enable innovation for development.

References

Acharya, S. 2003. Labour migration in the transitional economies of South-East Asia. Econ. Social Comm. Asia and the Pacific. ADB, Manila.

ADB. 2001a. Growth and change in Asia and the Pacific: Key indicators 2001. Oxford Univ. Press, NY.

ADB. 2001b. Asian environment outlook 2001. ADB, Manila.

ADB. 2001c. Water for all: The water policy of the Asian Development Bank. Available at http://www.adb.org/Documents/Policies/Water/water-policy.pdf. Dep. Econ. Social Affairs. ADB, Manila.

ADB. 2005. Key indicators of developing Asian and Pacific countries. ADB, Manila.

ADB. 2006a. Regional cooperation for sustainable development in Asia Pacific: Asian Development Bank's perspective and experience. Speech by Rajat M. Nag, Spec. Adv. Pres. Reg. Cooperation. Available at www.adb.org/Documents/Speeches/2006/sp2006003.asp. ADB, Manila.

ADB. 2006b. Asian development outlook 2006: Routes for Asia's trade. ADB, Manila.

Agarwal, M., C. Sahi, S. Katigar-Agarwal, S. Argawal, T. Young, Gallie et al. 2003. Molecular characterization of rice Hsp101: Complementation of yeast Hsp104 mutation by disaggregation of protein granules and differential expression in *indica* and *japonica* rice types. Plant Mol. Biol. 51:543-553.

Ahmed, E., J. Dreze, and A. Sen (ed) 1991. Social security in developing countries. Clarendon Press, Oxford.

Anderson, K. and W. Martin. 2006. Agriculture, trade reform and the Doha agenda. *In* K. Anderson and W. Martin (ed) Agriculture, trade reform and the Doha development agenda. World Bank, Washington DC.

Anderson, J.R., and D.L. Purcell. 1996. International policies for agricultural research investments. Conf. Proc. Global Science Policy for the Twenty-first Century, Melbourne, 16-28 Aug 1996.

AVRDC. 2002. Empowering small scale farmers for knowledge-based agriculture. ARDC Strategy 2010. AVRDC, Shanhua, Taiwan.

Bardhan, P., S. Bowles, and M. Wallerstein. 2006. Globalization and egalitarian redistribution. Oxford Univ. Press, New Delhi.

Barkenbus, J. 2001. APEC and the environment: Civil society in an age of globalization. Asia Pacific Issues, No. 51, East-West Centre, Hawaii.

Barnes, D., and W. Floor. 1996. Rural energy in developing countries: A challenge for economic development. Biomass Bioenergy 12(5):347-361.

Barnett, T.P., J.C. Adam, and D.P. Lettenmaier. 2005. Potential impacts of a warming climate on water availability in snow-dominated regions. Nature 438:303-309.

Baruah, S. 2005. Durable disorder: Understanding the politics of northeast India. Oxford Univ. Press, New Delhi.

Bass, S. and P. Steele. 2006. Managing the environment for development and to sustain pro-poor growth, challenges and risks to development in Asia. Asia 2015 Promoting growth, ending poverty. Available at www.asia2015conference.org.

Batra, A. 2006. Asian economic integration: ASEAN +3+1 or ASEAN +1s? Working Pap. 186. ICRIER, New Delhi.

Bauer, M.W., and G. Gaskell (ed) 2002. Biotechnology—The making of a global controversy. Cambridge Univ. Press and Science Museum, London.

Beck, U. 1992. Risk society: Towards a new modernity. Sage, New Delhi.

Bhandhubanyong, P., 2005. Development of ethanol as a transportation fuel in Thailand. Paper presented at the Int. Symposia on Alcohol Fuels (ISAF) XV, San Diego, 26-28 Sept 2005.

Biggs, S.D., and G. Smith. 1998. Beyond methodologies: Coalition-building for participatory technology development. World Dev. 26:239-248.

Biggs, S., 2006. Learning from the positive to reduce rural poverty: Institutional innovations in agricultural and natural resources research and development. Impact assessment workshop. CGIAR Systemwide PRGA and CIMMYT. Mexico, 19-21 Oct 2005.

Biggs, S., and H. Matsaert. 2004. Strengthening poverty reduction programs using an actor oriented approach: Examples from natural resources innovation systems. AgREN Network Paper No. 134. ODI, London.

Biggs, S. 2007. Building on the positive: An actor innovation systems approach to finding and promoting pro-poor natural resources institutional and technical innovations. Int. J. Agric. Resourc. overn. Ecol. 6(2):144-164.

Binswanger, H., and E. Lutz. 2000. Agricultural trade barriers, trade negotiations, and the interests of developing countries. Pap. for UNCTAD X, Dec. 1999. Available at http://www.unctad-10.org/pdfs/ux_tdxrt1d8.en.pdf.

Bird Life International. 2005. Bioenergy fuel for the future? A BirdLife International position paper on bioenergy use in the EU. Available at http://www.birdlife.org/eu/pdfs/2006/2005_bioenergy_position.pdf.

Bonardi, J.-P., and T. Warin. 2007. Open source software development, innovation and co-ordination costs. Available at http://www.middlebury.edu/~econ. Econ. Disc. Pap. 07-01. Middlebury College, VT.

Bowles., S., and U. Pagano. 2006. Economic integration, cultural standardization and the politics of social insurance. p. 289-305. *In* P. Bardhan et al. (ed) Globalization and egalitarian redistribution. Oxford Univ. Press, New Delhi.

Bruinsma, J. (ed) 2003. World agriculture: Towards 2015/2030—An FAO perspective. FAO, Rome.

Byerlee, D., and R.G. Echeverria (ed) 2002. Agricultural research policy in an era of privatization. CABI Publ., U.K.

Cernea, M.M. 1991. Using knowledge from social science in development projects. World Bank Disc. Pap. 114. World Bank, Washington DC.

CGIAR. 2005. Summary report on system priorities for CGIAR research 2005-2015. Science Council, CGIAR, Rome.

Chang, H.S., and L. Zepeda. 2001. Agricultural productivity for sustainable food security in Asia and the Pacific: The role of investment. *In* L. Zepeda (ed) Agricultural investment and productivity in developing countries. FAO, Rome.

Chathaway, J., J. Tait, and D. Wield. 2006. The governance of agro and pharmaceutical biotechnology innovation: Public policy and industrial strategy. Tech. Analy. Strat. Manage. 18(2):169-185.

Chaturvedi, S. 2002. Measuring developments in biotechnology: International initiatives, status in India and agenda before developing countries. Disc. Pap. 30/2002, RIS, New Delhi.

Chaturvedi, S., and S.R. Rao. 2004. Biotechnology and development: Challenges and opportunities for Asia. Academic Foundation, New Delhi.

Christensen, T.J., 2007. The state of US-China diplomacy. Remarks before the U.S.–China Economic and Security Review Commission, Washington DC, 2 Feb 2007.

Connett-Proceddu, M. 2006. Intellectual property and promoting R&D for public interest in the Asian region: Inclusionary and distributive innovation system options. CAMBIA-BIOS Initiative, UNCTAD, ICTSD, HKU, IDRC regional dialogue on intellectual property rights, innovation and sustainable development, Hong Kong, 8-10 Nov.

Contreras-Hermosilla, A., and C. Fay. 2005. Strengthening forest management in Indonesia through land tenure reform: Issues and framework for action. Forest Trends and the World Agroforestry Center, Nairobi.

Cook, S., N. Kabeer, and G. Suwannarat. 2003. Social protection in Asia. Ford Foundation and Har-Anand Publ., Delhi.

Coombs, J., D.O. Hall, R.P. Overend, and W.H. Smith (ed) 1992. Biomass and Bioenergy. Vol. II, No. 1-6.

Cruz, R.V., H. Harasawa, M. Lal, S. Wu, Y. Anokhin, B. Punsalmaa et al. 2007. Asia. p. 469-506. *In* M.L. Parry et al. (ed) Climate Change 2007: Impacts, adaptation and vulnerability. Contribution of Working Group II to the Fourth Assessment Report of the Intergovernmental Panel on Climate Change. Cambridge Univ. Press, UK.

D'Emden, F.H., R.S. Llewellyn, and M.P. Burton. 2006. Adoption of conservation tillage in Australian cropping regions: an application of duration analysis. Tech. Forecast. Social Change 73(6):630-647.

David, Paul A. 2004. Understanding the emergence of "open science" institutions: Functionalist economics in historical context. Indust. Corpor. Change 13(4):571-589.

De La Torre Ugarte, D.G. 2006. Developing bioenergy: Economic and social issues. *In* P. Hazell and R.K. Pachauri (ed) Bioenergy and agriculture promises and challenges. Available at http://www.ifpri.org/2020/focus/focus14/focus14.pdf. IFPRI, Washington DC.

DESA. 2006. World urbanization prospects: The 2005 revision. United Nations, NY.

Dhan Foundation 2005. Annual report, 2005—Culture building for nurturing institutions. Dhan Foundation, Madurai.

Douthwaite, B. 2002. Enabling innovation. Zed Books, London.

Dryzek, John 1998. The politics of the earth—environmental discourses. Oxford Univ. Press, UK.

Dutkiewicz, R. (ed) 1999. The challenge of rural energy poverty in developing countries. World Energy Council, London.

Dyson, T., R. Cassen, and L. Visaria. 2004. Twenty-first century India: Population, economy, human development and the environment. Oxford Univ. Press, UK.

Eckhart, K.V., K. Naruse, and M.Y. Tang. 2005. The Asian frontier: Big pharma's new promised land. Life Sci. Monitor 1:6-10.

Economic Daily. 2007. Xinhua News Agency, 27 February.

EDB. 1995. Singapore: Biotechnology a growing industry. Econ. Dev. Board, Singapore.

El-Hage Scialabba, N. and D. Williamson. 2004. The Scope of organic agriculture, sustainable forest management and ecoforestry in protected area management. Environ. Nat. Resourc. Pap. 18, FAO, Rome.

Emori, S., T. Nozawa, A. Numaguchi and I. Uno. 2000. A regional climate change projection over East Asia. p. 15-18. *In* Preprint Vol. 11th Symposium on Global Change Studies, Long Beach, 9-14 Jan 2000.

Engelman, R. and P. LeRoy. 1995. Conserving land: Population and sustainable food production. Popul. Action Int., Washington DC.

Erenstein, O., W. Thorpe, J. Singh, and A. Varma. 2006. Crop-livestock interactions in the Trans-Gangetic Plains, India. CIMMYT, ILRI and RWC, New Delhi.

FAO. 1998. Proc. of the regional expert consultation on modern applications of biomass energy. Kuala Lumpur, 6-10 Jan 1997.

FAO. 1999. Poverty alleviation and food security in Asia—land resources. RAP Publ. 1999/2. FAO, Bangkok.

FAO. 2002. Aquatstat Information system on water in agriculture. www.fao.org/waicent/faoinfo/agricult/agl/aglw/aquastat/dbase/index.htm. FAO, Rome.

FAO. 2003. FAO agricultural commodity projections to 2010: Medium term prospects, Rome.

FAO. 2004. Production and export of organic fruit and vegetables in Asia. Commodities Trade Tech. Pap. 6. FAO, Rome.

FAO. 2006a. The state of food and agriculture in Asia and the Pacific. FAO, Reg. Off. Asia Pacific, Bangkok.

FAO. 2006b. World agriculture: Towards 2030/50: Prospects for food, nutrition, agriculture and major commodity groups. Interim report. Global Perspect. Study Unit. FAO, Rome.

Farrington, J. 2006. Promoting growth and ending poverty in Asia. Asia 2015 promoting growth, ending poverty. Conf. overview paper. Available at www.asia2015conference.org.

Fay, M. and T. Yeppes. 2003. Investing in infrastructure: What is needed from 2000-2010? Policy Res. Working Pap. 3102. World Bank, Washington DC.

Feddema, R. 2003. South Korea plans its biotechnology future. Biotech. Dev. Monitor 50:388.

Fennel, S. 2006. Future policy choices for the education sector in Asia. Session 3, Asia 2015—promoting growth, ending poverty. London, Mar 2006.

Fischer, G., M. Shah, and H. van Velthuizen. 2002. Climate change and agricultural vulnerability. IIASA, Vienna.

Foodjustice.net 2003. The right to food: NGOs/CSOs statement at the Asian regional consultation on the World Food Summit Five Years Later. Bangkok, 28-29 Aug 2001. Available at www.foodjustice.net/modules.php?name=Content&pa=showpage&pid=17&cid=7.

Francks, P., J. Boestel, J. and H. Kim. 1999. Agriculture and economic development in East Asia: from growth to protectionism in Japan, Korea and Taiwan. Routledge, London.

Frew, S.E., R. Riziae, S. Rammut, M. Ray, A.S. Daar, and P. Singer. 2007. India's health biotech sector at a crossroads. Nature Biotech. 25(4):403-417.

Fujisaka, S. 1994. Will farmer participatory research survive in the international agricultural research centres? Gatekeeper Ser. 44. IIED, London.

Galloway, J.N. 1995. Acid deposition—perspectives in time and space. Water Air Soil Pollut. 85(1):15-24.

Galloway, J., Z. Dianwu, V.E. Thomsun, L.H. Chang, and S. Penkett. 1996. Nitrogen mobilization in the United States of America and the People's Republic of China. Atmos. Environ. 30(10-11):1551-1561.

Gao, X.J., D.L. Li, Z.C. Zhao and F. Giorgi. 2003. Climate change due to greenhouse effects in Qinghai-Xizang Plateau and along Qianghai-Tibet Railway. Plateau Meteorol. 22(5):458-463 (In Chinese with English abstract).

García, Z., J. Nyberg, and S.O. Saadat. 2006. Agriculture, trade negotiations and gender. Available at ftp://ftp.fao.org/docrep/fao/009/a0493e/a0493e.pdf. FAO, Rome.

Gerard, F., and F. Ruf (ed) 2001. Agriculture in crisis: People, commodities and natural resources in Indonesia, 1996-2000. CIRAD, Montpellier; Curzon Press, Richmond.

Ghosh, J. 2005. Trade liberalization in agriculture: An examination of impact and policy strategies with special reference to India. HDR Off. Occas. Pap. 2005/12. UNDP, NY.

GIBS Review. 2006. Doing business in 2007. Oct Vol. 10. Available at http://www.gibsreview.co.za/home.asp?PID=11&ToolID=2&reviewid=199&itemid=202.

GEO. 2002. Global environment outlook: GEO 3. UNEP, Nairobi.

Government of Japan, 2002. Biotechnology strategy guidelines. Biotech. Strategy Council, Tokyo.

Gratzl, B. and M. Fawer-Wasser, 2006. New report from Bank Sarasin on the sustainability of biofuels. Bank Sarasin Media Release, Basel, 25 Jul 2006.

Greene, N., 2004. Growing energy: How biofuels can help end America's oil dependence. Nat. Resourc. Defense Council (NRDC), NY.

Gupta, M.V. 2006. Challenges in sustaining and increasing fish production to combat hunger and poverty in Asia. WorldFish Centre Q. 29(1&2).

Hall, A.J. 2005. Capacity development for agricultural biotechnology in developing countries: an innovation systems view of what it is and how to develop it. J. Int. Dev. 19(5):611-630.

Hall, A.J., R. Sulaiman, N. Clark, M.V.K. Sivamohan, and B. Yoganand. 2002. Public-private sector interaction in the Indian agricultural research system: An innovation systems perspective on institutional reform. p.155-176. *In* D. Byerlee and R. Echeverria (ed) Agricultural research policy in an era of privatization. CABI Publ., UK.

Hall, A.J., B. Yoganand, R.V. Sulaiman, R.S. Raina, C.S. Prasad, G.C. Naik et al. (ed) 2004. Innovations in innovation: Reflections on partnership, institutions and learning. CPHP South Asia and ICRISAT, Patancheru NCAP, New Delhi.

Hansen, J. (ed.) 2002. Air pollution as a climate forcing: A workshop. East-West Centre, Hawaii.

Hasumi, H., and S. Emori (ed) 2004. K-1 coupled model (MIROC) description. K-1 Tech. Rep. 1. Center Climate Syst. Res., Univ. Tokyo.

Hatti-Kaul, R., U. Törnvall, L. Gustafsson, and P. Börjesson 2007. Industrial biotechnology for the production of bio-based chemicals: A cradle to grave perspective. Trends Biotech. 25:119-124.

Hazarika, S. 2000. Rites of passage: Border crossings, imagined homelands and India's East and Bangladesh. Penguin, New Delhi.

Hendrickson, M. and Heffernan. 2002. Opening spaces through relocalization: Locating potential resistance in the weaknesses of the global food system. Sociol. Ruralis 42(4):347-369.

Henniges, O., and J. Zeddies. 2006. Bioenergy in Europe: Experiences and prospects. Available at http://www.ifpri.org/2020/focus/focus14/focus14_09.pdf. 2020 Vision Focus 14 Brief 9. IFPRI, Washington DC.

Henning, R.K. 2004. Integrated rural development by utilization of Jatropha curcas L. (JCL) as raw material and as renewable energy—economy & dissemination strategy. Available at http://www.jatropha.de/. The Jatropha Website.

Herrera, S. 2005. Profile—Richard Jefferson. Nature Biotech. 23(6):642-643.

Hertel, T. and R. Keeney. 2006. What is at stake: The relative importance of import barriers, export subsidies and domestic support. *In* Agriculture, trade reform and the Doha development agenda. World Bank, Washington DC.

Hidayat, D. 2006. Indonesian S&T policy and development and S&T Indicators. Int. Conf. S&T Policy Res. Statistical Indicators. NAM S&T Centre and NSF, Colombo, 8-10 Nov 2006.

Himpe, F. 1997. Technical and economic feasibility of cogeneration projects from rice husk. p. 16-18. *In* Proc. regional expert consultation on modern applications of biomass energy. Kuala Lumpur, 6-10 Jan 1997. Reg. Wood Energy Dev. Prog. in Asia (RWEDP) Rep. 36.

Hoban, T.J. 2004. Public attitudes towards agricultural biotechnology. ESA Working Pap. 04-09. FAO, Rome.

Hobsbawm, E. 2006. War, peace and hegemony at the beginning of the 21st Century.

Hoogwijk, M. 2004. On the global and regional potential of renewable energy sources. Ph.D. thesis. Univ. Utrecht, Netherlands.

Huang, J., R. Hu, C. Pray, and S. Rozelle. 2002. Reforming China's agricultural research system. p. 245-264. *In* D. Byerlee and R. Echeverria (ed) Agricultural research policy in an era of privatization. CABI Publ., UK.

Hugo, G. 2005. Migration in the Asia-Pacific region. Global Comm. Int. Migration (GCIM), Geneva.

Hussain, A., R. Cassen, and T. Dyson. 2006. Demographic transition in Asia and its consequences. IDS Bull. 37:79-87.

Ianchovichina, E., and P. Kacker. 2005. Growth trends in the developing world: Country forecasts and determinants. Policy Res. Working Pap. 3775. World Bank, Washington DC.

IDRC. 2006. Innovation, technology and society under the IDRC, IPS program. IDRC, Ottawa.

IEA. 2004. World energy outlook 2004. IEA, Paris.

IFAD. 2001. Rural poverty report. IFAD, Rome.

IFPRI. 1995. Vision 2020—A vision for food, agriculture and the environment. IFPRI, Washington DC.

IFPRI. 2001. 2020 Global food outlook: Trends, alternatives and choices. 2020 Vision for food, agriculture and the environment food policy report. IFPRI, Washington DC.

IFPRI. 2002. Sustainable food security for all by 2020. Conf. Proc., Bonn, 4-6 Sept 2001. IFPRI, Washington DC.

IPCC. 2000. Special report on emission scenarios. Cambridge Univ. Press, UK.

IPCC. 2007. Climate change 2007: Impacts, adaptation and vulnerability. Working Group II Contribution to Intergovernmental Panel on Climate Change Fourth Assessment Report. Cambridge Univ. Press, UK.

Jasanoff, S., 2000. Technological risk and cultures of rationality. *In* Incorporating science, economics and sociology in developing sanitary and phytosanitary standards in international trade. NRC, Nat. Acad. Press, Washington DC.

Jha, R., H.K. Nagarajan and S. Prasanna. 2005. Land fragmentation and its implications for productivity: Evidence from Southern India. Australia South Asia Res. Centre, Australian National Univ., Canberra.

Jones, S. 2006. Infrastructure challenges in East and South East Asia. Paper presented at Asia 2015: Promoting growth ending poverty. Available at www .asia2915conference.org

Joshi, L. 1997. Incorporating farmers' knowledge in the planning of interdisciplinary research and extension. PhD thesis. Sch. Agric. Forest Sci., Univ. Wales, Bangor.

Joshi, P.K., A.. Gulati, P.S. Birthal, and L. Tewari. 2003. Agriculture diversification in South Asia: Patterns, determinants, and policy implications. IFPRI, Washington DC.

Kampman, B. E., L.C. den Boer and H.J. Croezen. 2005. Biofuels under development: An analysis of currently available and future biofuels, and a comparison with biomass application in other sectors. CE, Delft.

Kaneda, T. 2006. A critical window for policymaking on population ageing in developing countries. January. Population Reference Bureau, Washington DC.

Kataoka, Y. 2002. Overview paper on water for sustainable development in Asia and the Pacific. Asia-Pacific forum for environment and development first substantive meeting, Bangkok, 12-13 Jan. 2002. Available at http://www.apfed.net/apfed1/pdf/5_1.pdf.

Kemal, A.R. 2007. Macroeconomic policies in South Asia. South Asian J. 15:23-44.

Knutson, T.R., and R.E. Tuleya. 2004. Impacts of CO_2 induced warming on simulated hurricane intensities and precipitation: sensitivity to the choice of climate model and convective parameterization. J. Climate. 17:3477-3495.

Korten, D. 1997. The post-corporate world. Kumarian Press and B.K. Publ., San Francisco

La Vina, G., M. Antonio, and M.J. Caleda. 2006. Biotechnological innovations in Asia: Policy and governance challenges. IDRC, Ottawa.

Lal, M. 2003. Global climate change: India's monsoon and its variability. J. Environ. Studies Policy 6 (1):1-34.

Lal, R. 2006. Perspective: Managing soils for feeding a global population of 10 billion. J. Sci. Food Agric. 86:2273-2284.

Lele, S. 2006. Land use, nature conservation and the stability of rainforest margins in Southeast Asia. Book review. Conserv. Soc. 4(1):169-172.

Lele, U. 2004. The CGIAR at 31: An independent meta evaluation of the consultative group on international agricultural research. World Bank, Washington DC.

MA. 2005. Ecosystems and human well-being: Synthesis. Millennium Ecosystem Assessment. Island Press, Washington DC.

Mahapatra, A.K., and C.P. Mitchell. 1999. Biofuel consumption, deforestation, and farm level tree growing in rural India. Biomass Bioenergy 17:291-303.

Mahmoud, C. and G. Shively. 2004. Agricultural diversification and integrated pest management in Bangladesh. Agric. Econ. 30(3):187-94.

Mani, S. 2005. The Dragon vs. the elephant: Comparative analysis of innovation capability in the telecom industry of China and India. Econ. Polit. Weekly 40(39):4271-4283.

Marshall, F., M. Ashmore, and F. Hinchcliffe. 1997. A hidden threat to food production: Air pollution and agriculture in the developing world. Gatekeeper Ser. 73. IIED, London.

Matysek, A., M. Ford, G. Jakeman, A. Gurney, and B.S. Fisher. 2006. Technology and climate change. ABARE Res. Rep. 06.6, Canberra.

McTainsh, G., J. Leys, and K. Tews. 2001. Wind erosion trends from meteorological records. Australia: State of the environment technical paper series (land). Ser. 2. Available at www.deh.gov.au/soe/2001/land/index.html. Dep. Environ. Heritage, Canberra.

Mirza, M.M.Q. 2005. The implications of climate change on river discharge in bangladesh. p. 103-136. *In* M.M.Q. Mirza and Q.K. Ahmad (ed) Climate change and water resources in South Asia. Taylor and Francis, UK.

Monata, C. 2006. Latin America and Asia-Pacific regions toward 2015-2020—Closer neighbours or partners? Seminar: Beyond P4: Outlook for further regional integration in Asia-Pacific: A Latin American perspective. New Zealand Inst. Int. Affairs, Wellington.

Muni, S.D. 2004. A South Asian parliament. South Asian J. 6:49-59.

Murdiyarso, D. 2000. Adaptation to climatic variability and change: Asian perspectives on agriculture and food security. Environ. Monitor. Assess. 61:123-131.

Musungu, S.F. 2006. Intellectual property rights and biotechnology in developing countries: Key research and knowledge gaps. Conf. Bioinnovation policy in Asia. Bangkok, Nov 2006. IDRC, AIT and IDRC, Bangkok.

Nathan, D., and G. Kelkar. 1999. Agrarian involution, domestic economy and women: Rural dimensions of the Asian crisis. Econ. Polit. Weekly 34(18).

National Knowledge Commission (NKC). 2006. Report of the National Knowledge Commission. Gov. India, New Delhi.

Nayyar, D. 2006. Development through globalization? UNU-WIDER Res. Pap. 2006/29. UNU-WIDER, Helsinki.

Nguyen, T., E. Cheng, and C. Findlay. 1996. Land fragmentation and farm productivity in China in the 1990s. China Econ. Rev. 7(2):169-180.

Nijam, A. 2005. Broadening the security debate: Human and environmental dimensions. *In* Sustainable development—Bridging the research/policy gaps in southern contexts. Vol.1 Environment. SDPI and Oxford Univ. Press, Islamabad.

Niroula, G.S. and G.B.Thapa. 2005. Impacts and causes of land fragmentation, and lessons learned from land consolidation in South Asia. Land Use Policy 22:358-372.

OECD-FAO. 2006. Agricultural outlook 2006-2015. FAO, Rome.

Ooi, P.A.C. 2001. From passive observer to pest management expert: Science education and farmers.p. 105-118. *In* L.G. Soon et al. (ed) Curriculum development for plant-pest management in Asia-Pacific. FAO, Bangkok.

PAN (Pesticide Action Network). 2002. Malaysian Government bans the herbicide paraquat. Environ. News Service, 23 Sept 2002.

Pardey, P., N. Beintema, S. Dehmer, and S. Wood. 2006. Agricultural research: A growing global divide? ASTI Initiative. IFPRI, Washington DC.

Paris, T., A. Singh, L. Joyce, M.M. Hossain. 2005. Labours outmigration, livelihoods of rice farming households and women left behind: A case study of Eastern UP. EPW XL(25):2522-2529.

Park, C.Y., and Zhai, F. 2006. Asia's imprint on global commodity markets. ERD Working Pap. 90. Asian Dev. Bank, Manila.

Parry, M.L., C. Rosenzweig, A. Iglesias, M. Livermore, and G. Fischer. 2004. Effects of climate change on global food production under SRES emissions and socioeconomic scenarios. Global Environ. Change 14:53-67.

Partos, L. 2007. Chinese food market to reach number two global position by 2020. Available at http://www.foodnavigator.com/news/printNewsBis.asp?id=58687.

Patnaik, U. 2005. Theorising food security and poverty in the era of economic reforms. Social Sci. 33(7-8):50-80.

Peczon, B.D. and A.J. Manalo. 2004. Public-private sector partnerships in the Philippines: Issues, modalities and lessons. Paper presented at BIO 2004, San Francisco, June 2004.

Peng, S., J. Huang, J.E. Sheehy, R.C. Laza, R.M. Visperas, X. Zhong et al. 2004. Rice yields decline with higher night temperature from global warming. PNAS 101(27):9971-9975.

Pimbert, M.P., J. Thompson, W.T. Vorley, T. Fox, N. Kanji and C. Tacoli. 2001. Global restructuring, agri-food systems and livelihoods. Gatekeeper Ser. 100. IIED, London.

Pimental, D., P. Hepperly, J. Hanson, D. Douds, and R. Seidel. 2005. Environmental, energy and economic comparisons of organic and conventional farming systems. BioScience 55(7):573-582.

Pingali, P. 2004. Agricultural diversification: Opportunities and constraints. Agric. Dev. Econ Div., FAO, Rome.

Pingali, P.L., and P.W. Heisey. 1999. Cereal crop productivity in developing countries. Econ. Pap. 99-03. CIMMYT, Mexico.

Porter, M. and C. Van der Linde. 1995. Towards a new conception of the environment-competitiveness relationship. J. Econ. Perspect. 9(4):97-118.

Pownall, I.E. 2000. An international political economic view of the biotechnology industry. Electronic J. Biotech (EJB) 3(2). Available at http://www.scielo.cl/pdf/ejb/v3n2/art02.pdf.

Prahalad, C.K. 2005. The fortune at the bottom of the pyramid: Eradicating poverty through profits. Pearson Education, Wharton School Publ., NJ.

Prasad, Malla V.S.V. 2006. Political and security cooperation between India and ASEAN: Implications for economic cooperation. *In* Kumar et al. (ed) India-ASEAN economic relations: Meeting the challenges of globalization. ISEAS, Singapore.

Pray, C.E. 2001. Public-private sector linkages in research and development: Biotechnology and the seed industry in Brazil, China and India. Am. J. Agric. Econ. 83(3):742-747.

Pray, C.E., and A. Naseem. 2005. Intellectual property rights on research tools: Incentives or barriers to innovation? Case studies of rice genomics and plant transformation technologies. AgBioForum 8(2&3):108-117.

Prebisch, R. 1950. The economic development of Latin America and its principal problems.

Przeworski, A., and C. Meseguer Yebra. 2006. Globalization and democracy. p. 169-191. *In* Bardhan et al. (ed) Globalization and egalitarian redistribution. Oxford Univ. Press, New Delhi.

Pushpangadan, P. 2000. The Pushpangadan model of benefit sharing. Nat. Botanical Res. Inst., Lucknow.

Raina. R.S. 1999. Patronage and evaluation in the Indian Council of Agricultural Research. Evaluation 5(3):278-302.

Raina, R.S. 2003. Innovation capability for agricultural biotechnology: Policy issues. Biotech. Dev. Monitor 50(3):29-31.

Raina, R.S. 2003a. Institutions and organizations: enabling reforms in Indian agricultural research and policy. Int. J. Tech. Manage. Sustain. Dev. 2(2):97-116.

Raina, R.S. 2003b. Disciplines, institutions and organizations: Impact assessments in context, Agric. Syst. 78(2):185-212.

Raina, R.S., and S. Sangar. 2004. Institutional and policy requirements for wider application of conservation agriculture. Pap. presented Nat. Conf. Conserv. Agriculture—Conserving resources, enhancing productivity. CASA and RWC-IGP, Nat. Agric. Sci. Complex, New Delhi, 22-23 Sept 2004.

Raina, R.S. 2005. Social sciences for agricultural and rural development: Lessons for rural innovation. Conf. Globalization and Chinese Agriculture, China Agricultural Univ., Beijing, 17-18 Sept 2005.

Raina, R.S., and R.V. Sulaiman. 2007a. Report on livelihood patterns, impact pathways and innovation systems in the Indian Indo Gangetic Plains: Synthesis of existing information. Rice-Wheat Consortium, CIMMYT, New Delhi.

Raina, R.S., and R.V. Sulaiman. 2007b. From technology development and dissemination to learning approaches: Institutional change for rural development. p. 173-208. *In* V. Ballabh (ed) Institutional alternatives and governance of agriculture. Academic Foundation, New Delhi.

Rajasekaran, B. 1993. A framework for incorporating indigenous knowledge systems into agricultural research, extension, and NGOs for sustainable agricultural development. Studies Tech. Social Change No. 21. Iowa State Univ., Ames.

Raju, K.V. 2006. Biofuels in South Asia. Overview—editorial introduction. Asian Biotech. Rev. 8(2):1-9.

Reardon, T., C.P. Timmer, C.B. Barrett, and J.A. Berdegue. 2003. The rise of supermarkets in Africa, Asia, and Latin America. Am. J. Agric. Econ. 85(5):1140-1146.

Rhoades, R.E. 1999. Evolution of agricultural research and development since 1950: Toward an integrated framework. Gatekeeper Ser. 12. IIED, London.

Rhoades, R.E. 2000. Agricultural anthropology: New disciplinary blood in an international research organization. p. 159-162. *In* P.J. Higgins and J.A. Paredas (ed) Classics of practicing anthropology 1978-1998. Soc. Appl. Anthropol., Oklahoma City.

Rieuvenkamp, G. 2003. Looking back and looking forward. Biotech. Dev. Monitor 50:2-5.

Roland-Holst, D., J.P. Verbiest, and F. Zhai. 2005. Growth and trade horizons for Asia: Long term forecasts for regional integration. Available at http://www.adb.org/Documents/ERD/Working_Papers/WP074.pdf. ERD Working Pap. Ser. 74. ADB, Manila.

Roling, N., and Wagemaker (ed) 1998. Facilitating sustainable agriculture. Cambridge Univ. Press, UK.

Rosegrant, M., and S. Malik (ed) 1995. A 2020 Vision for food, agriculture, and the environment in South Asia: A synthesis. IFPRI, Washington DC.

Rosegrant, M., M.S. Paisner, S. Meijer, and J. Witcover. 2001. Global food projections to 2020: Emerging trends and alternative futures. IFPRI, Washington DC.

Ross, S. 2006. Demographic change in the Asia-Pacific: The effect on average living standards. J. Asia-Pacific Econ. 11(3):229-248.

SaciWATER. 2002. Higher education on water resources in South Asia: Towards capacity building for IWRM. South Asia Consortium Interdisciplinary Water Resources Studies. Hyderabad.

Sahai, S. 2003. Genetically modified crops: A resource guide for the Asia-Pacific. Consumers Int. Asia-Pacific Off., Kuala Lumpur.

Salayo, N.D., M. Ahmed, L. Garces, and K. Viswanathan.2006. An overview of fisheries conflicts in South and Southeast Asia: Recommendations, challenges and directions. NAGA WorldFish Center Q. 29(1-2).

Scheeren, H.A., J. Lelieveld, G.J. Roelofs, J. Williams, H. Fischer, M. de Reus et al. 2003. Impact of monsoon outflow from India and Southeast Asia. Atmos. Chem. Phys. 3:1589-1608.

Scherr, S.J., and S. Yadav. 1997. Land degradation in the developing world—issues and policy options for 2020. Vision 2020 Brief 44. IFPRI, Washington DC.

Schneider, K., B. Graham, C. Millsteed, M. Saunders, and R. Stuart. 2000. Trade and investment liberalisation in APEC. ABARE Res. Rep. 2000.2, Canberra.

SDPI. 2005a. Sustainable development—Bridging the research/policy gaps in southern contexts. Vol. I: Environment. SDPI, Islamabad and Oxford Univ. Press, Karachi.

SDPI. 2005b. Sustainable development—Bridging the research/policy gaps in southern contexts. Vol. II: Social policy. SDPI, Islamabad and Oxford Univ. Press, Karachi.

Seth, A., K. Fischer, J. Anderson, and D. Jha. 2003. The review panel report: The rice-wheat consortium—an institutional innovation in international agricultural research on the rice-wheat cropping systems of the Indo-Gangetic Plains. CIMMYT, New Delhi.

SFG. 2005. Second freedom—South Asian challenge 2005-2025. Strategic Foresight Group, Mumbai.

Shah, T., C. Scott, A. Kishore, and A. Sharma. 2003. Energy-irrigation nexus in South Asia: Improving groundwater conservation and power sector viability. IWMI, Colombo.

Shindo, J., K. Okamoto, and H. Kawashima. 2006. Prediction of the environmental effects of excess nitrogen caused by increasing food demand with rapid economic growth in eastern Asian countries, 1961-2020. Ecol. Modelling 193(3-4):703-720.

Shiva, V., 2000. Stolen harvest: The hijacking of the global food supply. Southend Press, Cambridge.

Shrivastava, A. 2006. Globalized world: Who gains who loses? Available at http://www .infochangeindia.org/analysis157.jsp. InfoChangeIndia News Features.

Singer, H. 1950. The distribution of gains between investing and borrowing countries. Am. Econ. Rev. 40(2):473-85.

Slingerland, S., and L. van Geuns. 2005. Drivers for an international biofuels market. Disc. Paper, CIEP Future Fuel Seminar, Clingendael Inst., 9 Dec 2005.

Sobhan, R. 2005. Charting a roadmap for South Asian cooperation. South Asian Survey 12(3):3-19.

State Council. 2006. National Development Program Compendium of Science and Technology for Mid-long Term (2006-2020). (in Chinese) Available at http://www.gov.cn/jrzg/2006-02/09/content_183787.htm. State Council of the People's Republic of China, Beijing.

Sulaiman, R.V., and A. Hall. 2005. Extension policy at the national level in Asia. Plant Prod. Sci. 8(3):308-319.

Swift, M.J., and P.A. Sanchez. 1984. Biological management of tropical soil fertility for sustainable productivity. Nature Resourc. 20:1-10.

Tabor, S.R., W. Janssen, and H. Bruneau. 1998. Financing agricultural research: A sourcebook. ISNAR, The Hague.

Talawar, S., and R. Rhoades 1998. Scientific and local classification and management of soils. Agric. Human Values 15:3-14.

Tammen, R.L. 2006. The impact of Asia on world politics: The China and the India options for the United States. Int. Studies Rev. 8(4):563-580.

Tandon, R. 2000. Voluntary action, civil society and State. Mosaic Books, New Delhi.

Terada, T. 2006. Thirty years of the Australia-Japan partnership in Asian regionalism: evolution and future directions. Aust. J. Int. Affairs 60(4):536-551.

Tsui-Auch, Lai Si. 1999. Functional versus sectoral industrial policy: A comparative study of the biotechnology development in Hong Kong and Singapore. Business Sch., Nanyang Tech. Univ., Singapore.

Tustin, J. (ed) 2006. Annual Report 2005. IEA Bioenergy, Rotorua, New Zealand.

Tyler, S. (ed) 2006. Communities, livelihoods, and natural resources—action research and policy change in Asia. Intermediate Tech. Publ., London and IDRC, Ottawa.

Umali, D.L. 1992. Public and private sector roles in agricultural research—theory and experience. Disc. Pap. 176. World Bank, Washington DC.

UN. 2001. World population prospects: The 2000 revision. UN Population Div., NY.

UNCTAD. 2006. UNCTAD trade and environment review. UNCTAD, Geneva.

UNDP. 2003 Human development report: 2003. UNDP, Oxford Univ. Press, New York.

UNEP. 1997. World atlas of desertification. 2nd ed. Arnold, London.

UNEP. 2001. Reference manual for the integrated assessment of trade-related policies. UNEP, Geneva.

UNEP 2006. GEO yearbook 2006. EarthPrint, Hertfordshire, UK.

UN Economic and Social Council. 2004. World demographic trends. Commission on population and development. Thirty-Eighth Session. E/CN.9/2005/8. UN, New York.

UNESCAP. 2005. State of the environment in Asia and the Pacific 2005—Synthesis. UNESCAP, Bangkok.

UNESCAP. 2006. Policy issues for the ESCAP region. Strengthening pacific island developing countries and territories through regional co-operation. Available at http://www.unescap.org/62/English/E1361e.pdf. UNESCAP, Bangkok.

UNESCO. 2003. Fifteen non-formal case studies collected by ASPBAE. Available at http://portal.unesco.org/education/en/ev.php-URL_ID=31362&URL_DO=DO_TOPIC&URL_SECTION=201.html. UNESCO, Paris.

UNESCO. 2005. UNESCO science report 2005. Available at www2.unescobkk.org/elib/publications/ICT_NFE/. UNESCO, Paris.

UNESCO. 2006. Innovations in non-formal education: A review of selected initiatives from the Asia-Pacific Region. UNESCO, Bangkok.

US EPA. 2002. A comprehensive analysis of biodiesel impacts on exhaust emissions Available at http://www.epa.gov/otaq/models/analysis/biodsl/p02001.pdf. US EPA, Washington DC.

Veeramani, C. 2007. Specialization patterns under trade liberalization: Evidence from India and China. Available at http://www .gdnet.org/middle.php?oid=237&zone=docs&action=doc&doc=11301. Global Development Network.

Venkataraman, C., G. Habib, A. Eiguren-Fernandez, A.H. Miguel, and S.K. Friedlander, 2005. Residential biofuels in South Asia: Carbonaceous aerosol emissions and climate impacts. Science 307:1454-1456.

Von Braun, J. 2005. The world food situation—an overview. CGIAR Ann. Gen. Meeting, Marrakech, 6 Dec 2005. IFPRI, Washington DC.

Vorley, W.T. 2001. The chains of agriculture: Sustainability and the restructuring of agri-food markets. WSSD Paper. IIED, London.

Wang, G. 2005. Agricultural drought in a future climate: results from 15 global climate models participating in the IPCC 4th Assessment. Climate Dynam. 25(7-8):739-753.

WCSDG. 2004. A fair globalization—creating opportunities for all. World Comm. Social Dimens. Globalization, ILO, Geneva.

Wignaraja, P. 2005. Pro-poor growth and governance in South Asia—Decentralization and participatory development. Pakistan Dev. Rev. 44(4):1159-1171.

Willer, H., and M. Yuseffi (ed) 2005. The world of organic agriculture. Statistics and emerging trends, 2005. IFOAM, Bonn.

Wong, J., U. Quach, H. Thorstendottir, P.A. Singer, and A.S. Daar. 2004. South

Korean biotechnology—a rising industrial and scientific powerhouse. Nature Biotech. 22:DC42-DC47.

World Bank. 2005. Agricultural growth for the poor—an agenda for development. World Bank, Washington DC.

World Bank. 2006a. Dancing with the giants: India, China and the global economy. World Bank, Washington DC.

World Bank. 2006b. Agricultural investment sourcebook. World Bank, Washington DC.

World Bank. 2006c. Enhancing agricultural innovation: How to go beyond strengthening research system. ARD, World Bank, Washington DC.

World Bank. 2007. Global economic prospects: Managing the next wave of globalization. World Bank, Washington DC.

Yamamoto, T. (ed) 1995. Emerging civil society in the Asia Pacific community: Integrative survey report and conference report of the Osaka Symposium, JCIE Publ., Tokyo.

Yamamoto, T., and K.G. Ashizawa (ed) 1999. Corporate NGO partnership in Asia Pacific. JCIE Publ., Tokyo.

Yoshino, M. 2000. Problems in climatology of dust storm and its relation to human activities in Northwest China. J. Arid Land Studies 10:171-181.

Yuankai. T. 2007. Growing old. Beijing Rev. No. 2, Jan 11.

Yuit, G.C.H., and J. Wall. 2006. Alternative energy: Beware the hidden costs. The Straits Times (Review) 22 July 2006.

Zheng, X., C. Fu, X. Xu, X. Yan, Y. Huang, and S. Han et al. 2002. The Asian nitrogen cycle case study. AMBIO 31(2):79-87.

Zhou, Y.H. 2003: Characteristics of weather and climate during drought periods in South China. J. Appl. Meteorol. Sci. 14:S118-S125.

5

Development and Sustainability Goals: AKST Options

Coordinating Lead Authors:
Medha Devare (India/USA), Carl B. Greenidge (Netherlands), Govind Kelkar (India)

Lead Authors:
Elenita Dano (Philippines), Wais Kabir (Bangladesh), Rasheed Sulaiman V. (India)

Contributing Authors:
J.E. Thies (USA) and Meriel Watts (New Zealand)

Review Editors:
Shelley Feldman (USA), Digna Mazanilla (Philippines)

Key Messages

1. If the current trend in soil, water and forest degradation is to be arrested, productivity and environmental sustainability must be addressed in an integrated fashion. While part of the solution may involve the application of technologies, a comprehensive solution depends on appropriate institutional arrangements and policies to ensure an integrated approach to natural resources management. One option is to provide systems of incentives that promote sustainable resource use.

2. If the poverty associated with rainfed agriculture is to be reduced, policies and resources will need to be extended to agriculture in arid, semi-arid and upland areas. Policies might include encouraging the use of locally adapted varieties and conservation agriculture, such as moisture conservation and zero or minimum tillage.

3. A variety of agricultural technologies will need to be applied in an integrated, site-specific manner if agricultural, environmental and development goals are to be realized. Organic, traditional and local systems and integrated nutrient and pest management can capitalize on the expanding domestic and global demand for food produced in ways that minimize costs to human and environmental health and reduce fossil fuel dependency. Such technologies are especially advantageous to rural producers constrained by access to credit, external inputs and extension services. Transgenics, nanotechnology and precision agriculture may positively affect natural resource and human welfare by reducing pesticide use and providing inexpensive vaccine delivery, but should be deployed under site-specific scientific and social monitoring within a stringent biosafety framework.

4. To improve agricultural performance and rural livelihoods, availability and access to information and communication technologies that facilitate the rapid dissemination and exchange of information will need to be enhanced. Frameworks for information exchange among farmers, extension workers, researchers and policymakers need to be flexible, adaptable, inter-related and science-based with a capacity to incorporate as well as develop new knowledge streams. This may be accomplished through decision tools, e-learning modules, and market information systems accessible through mobile technologies and computer kiosks as evident among producers in Bangladesh, China, India and the Philippines.

5. Integrating the complementary expertise of actors involved in AKST will be necessary if the rural sector is to respond and benefit from the growing complexity of regional agricultural development. To benefit from the opportunities offered by public-private partnerships, potential patterns of interaction need to be identified and current constraints overcome. Stakeholder meetings, institutional reforms, funding and other fiscal policy measures offer opportunities for promoting such collaboration.

6. If community livelihoods and the rural economy are to be enhanced, discriminatory barriers that limit the participation of women, indigenous peoples, certain caste groups, religious and ethnic minorities need to be identified and eliminated. This may be accomplished through the development and implementation of policies and governance structures, anti-discrimination legislation and by ensuring access to public services and markets.

7. The capacity of women agricultural producers, as holders of AKST, needs to be strengthened if household and community livelihoods are to be improved. Given the feminization of agriculture and women's limited access to resources and markets, this may be accomplished by empowering women to secure access to and manage land, knowledge and technologies that recognize their economic contribution to agricultural production, ensure wage parity and providing women with vocational and technical training.

8. To help ensure equitable access to environmentally sustainable and appropriate technologies, differences in technological capacity and the specific needs of countries in the development of Intellectual Property Rights (IPRs) policies will need to be recognized. IPRs may create barriers to trade and limit the capacity of developing countries to move up the science and technology ladder. The appropriateness and fairness of the patent system when applied to biodiversity and traditional and local knowledge are urgent concerns in the conservation of natural resources and access of the poor to them.

9. The capacities of various actors require strengthening if the opportunities arising from global trade integration, the information and communication revolution are to work to the benefit of the rural economy. The capacities of farmers, researchers, local governments, extension workers, financial institutions, local entrepreneurs and market agents, agroindustry and NGOs may be enhanced through training, professional exchange and vocational education. Efforts in the areas of research, policy and governance, extension and training could include: (1) traditional and emerging technologies, (2) international regulations, IPRs, trade negotiations, institutional reforms, (3) support systems not limited to production such as organizational, marketing, entrepreneurship to farmers, producer groups and NGOs and (4) the non-farm rural sector.

10. If economic development is to become sustainable given increasing global competitiveness, member states will need to develop their external trade relations taking into account that the latter can affect the achievement of development goals. This can be accomplished by (1) configuring and phasing sub-regional integration in ways that enable the development of a global, non-discriminatory trading system, (2) ensuring that Agreement on Agriculture (AoA) reforms take into account their impact on the divergent agricultural sector/s in the region, (3) improving the quality and coherence of regional trade policies and decision-making and monitoring the impact of competitive trade regimes that address environment, health and labor and (4) assessing and adopting alternative mechanisms, including "marketization" (e.g., carbon trading), to respond to conflicting interests. Such broad-based interventions will reduce extensive rural unemployment and sociopolitical dislocations that may result from increased competitiveness, technological obsolescence and trade rivalries.

11. If hunger and poverty are to be reduced through accelerated agricultural growth, rural investment needs to be increased and priorities changed. Such changes should address (1) the mix of agricultural activities (e.g., in favor of rainfed crops and those grown by the poor), (2) agricultural research, extension and science and technology infrastructure and support infrastructure (e.g., farm to market roads), (3) enhancement of the value chain including postharvest technologies, agroindustries and markets. In addition to public funding and donor support, investment in the foregoing areas can be enhanced by innovative means such as competitive (contract) research grant schemes, commodity cesses or levies.

12. Global consensus is necessary to achieve food security and natural resource conservation given the challenges posed by climate change and increasing biofuel use. Climate variability and change are threats to the agricultural sector in most of the ESAP region, while agriculture in high to mid-latitude parts of the region may benefit from climate change. The increasing use of biofuel crops such as oil palm, *Jatropha*, sugarcane and traditional food crops such as batata and cassava will increase land and water pressures, pose threats to natural ecosystems such as forests and potentially have negative impacts on food security and prices. A major challenge is to ensure that the development of biofuels meets sustainability goals.

13. To achieve environmental sustainability and economic development, the region needs to capitalize on the emerging global knowledge economy through enhanced capacity of national innovation systems. This involves establishing and strengthening links between networks in the knowledge economy. The state can play a critical role as sponsor or champion of this process by identifying actors and organizations, encouraging collaboration and developing enabling institutions and policies to build an effective system.

14. Policies that address the linkages between agricultural and non-farm rural employment need to be developed to reduce the poverty associated with limited rural employment opportunities. These might include a focus on local value addition opportunities such as agroprocessing and non-timber forest products as well as wage employment programs to enhance rural infrastructure.

15. If rights over competing use of water are to be equitably resolved, coherence is needed among administrative functions and policies. Resolution mechanisms might include the establishment and strengthening of interministerial coordination, multistakeholder consultations/management and multi-sectoral dialogue.

5.1 Context

The pursuit of development and growth in the ESAP region has generally been undertaken without sufficient consideration to sustainability and in some cases with poorly supported assumptions about the sharing of benefits that result from economic growth. Appropriate decision-making processes need to consider equity and sustainability issues while also assessing the gains to be garnered from productivity and growth. Factors that can influence the achievement of broad social and political goals ought not to be restricted to science and technology or AKST. Rather, achieving such goals will depend on resolving social issues that are shaped by factors that can alter relations of power and control and affect entitlements and access to resources. Thus, facilitating innovation is not only a question of developing and transferring concrete, science-based innovations, but also about facilitating innovative processes.

Technological advances in realizing development goals for much of ESAP unfold through social dialogue and interaction that have implications for the dynamics of policy making. Social goals include reducing poverty which we understand to mean a human condition characterized by low income, lack of voice and sustained deprivation of capabilities, choices and power that are necessary for the enjoyment of fundamental human rights. Poverty corresponds to the inability to access the full range of rights, standards of social equality and non-discrimination, as well as to be protected by the state and other development actors, including civil society organizations, community management bodies and corporations (Narayan et al., 2000; Hulme and Mckay, 2005).

This chapter begins with a discussion of the institutional and organizational context in which humans strive to produce and survive in ESAP. Parts of the region are characterized by social exclusion and inequality, particularly of women who constitute the majority of agricultural workers. Exclusion also characterizes access to the fruits of such economic growth including public services, markets and governance structures. Coupled with the challenges of climate change, water scarcity and petroleum dependence, countries in the region need to undertake effective measures for inclusive and equitable growth and put in place context-specific regimes for intellectual property rights and ethical and fair trade. They also need to capitalize on the emerging global knowledge economy and enhance the capacity of AKST actors and institutions to meet the broad goals of improved agricultural growth and capacity, sustainability and livelihood options.

In the section on technologies, we argue for an integrated approach to agriculture, using best management practices that blend traditional knowledge and organic practices with conventional and emerging technologies to help improve rural livelihoods and human health. This integrated approach will help to ensure consistency with the goals of greater productivity on the one hand and sustainability and equity, on the other. We recognize the potential of biotechnology, nanotechnology and precision agriculture to improve human welfare and preserve natural resources when these technologies are deployed appropriately, with site-specific scientific and social monitoring and within a stringent biosafety framework. We suggest that the development of organic agriculture can also supply a growing regional demand while helping to improve rural livelihoods and human health. Despite the inroads of information technology in many parts of Asia, information asymmetry remains a serious challenge, even recognizing that the information and communication technologies gap between expert and layperson has shrunk. Improvement in the availability and outreach of information is still required to strengthen adaptability and science based capacity to build and support new knowledge streams.

The chapter concludes with a discussion of institutions and policies and trade and markets, covering a range of issues, including public-private-community partnerships and networks, organizational reforms and enabling environments including decentralized and community-based entities with experience in science and technology generation and management. Further, technical and funding assistance from development partners can offer opportunities to galvanize efforts that reduce the mismatch between economic growth and poverty, vulnerability and inequality throughout the ESAP region. The section closes with an examination of trade policies and the steps required to ensure their consistency with development and sustainability goals.

5.2 Emerging Challenges

To feed the anticipated world population of approximately 9 billion people by 2050, 99% of whose growth will be in developing countries, food production will have to be trebled on dwindling arable land and freshwater resources (Population Reference Bureau, 2002). Options for increasing food production while reducing dependence on water resources and contributing to sustainable environmental practices are limited (Serageldin, 1999; Chrispeels, 2000; Vasil, 2003). Per capita available arable land area will be reduced by 50% over the next 50 years, with prime land being lost to industrialization and urbanization in a number of rapidly growing ESAP countries (Krattiger, 1998). Further, yield gains from the input and seed-oriented technologies utilized since the 1970s are stagnant or declining and unlikely to be able to keep pace with increasing population pressure in the region (Bouis, 1993; Cassman et al., 1995). The conversion of marginal areas into productive land, expansion of low-cost irrigation and trait modification and improved management through transgenic and nanotechnologies offer only limited potential (Crosson and Anderson, 1992; Peng et al., 1994; Carruthers et al., 1997; Rosegrant, 1997; Huang et al., 2002) while increasing the productivity of agricultural systems in ESAP continues to be complicated by climate change, animal and human diseases, social inequalities and the need for changes in institutional and governance practices.

5.2.1 Natural resources

The ESAP region's natural resources (soil, forests and water) are being degraded in intensive as well as non-intensive agricultural systems, from both the over- and under-utilization of inputs. In the developed countries in the region, high rates of inorganic and organic fertilizer have led to soil and water contamination (NRC, 1989; Conway and Pretty, 1991), while in the developing countries, population pressure, land constraints and poor soil management have reduced soil fertility with attendant biodiversity and functionality losses (Stoorvogel and Smaling, 1990; Tandon, 1998; Henao and Baanante, 1999). In the recently released National Development Programme Compendium of Science and Technology for Medium-Long Term (2006-2020), the Government of China recognized a range of resource management problems including eco-safety, land degradation and desertification, water shortage and water pollution. China's agricultural challenges also include serious surface water pollution arising from the improper use and disposal of some 3×10^9 tonnes of manure annually and the excessive application of pesticides and fertilizers on land and water.

Forests cover almost 30% of the land area in Asia, are critical for providing products and services and contribute to the cultural and spiritual heritage of much of the region's population (CFAN, 2004). Throughout the developing countries of ESAP, intensive timber harvesting, expanding commercial agriculture and the conversion of forests to rubber or oil palm plantations have contributed to forest degradation, primarily in Indonesia, Malaysia, Myanmar and Thailand (FAO, 2003a). Land and forest degradation, climate change, air and water pollution, water shortage and the loss of biodiversity are now the focus of new social and environmental movements (Krishna, 2004) that seek improved agricultural efficiencies and new policy measures.

Water shortages are common throughout rural and urban areas of the region. Although China has 28% of the world average per capita availability of water, it is estimated to have a shortage of 30×10^9 km^3 per year. While there may be sufficient water in much of the region for domestic as well as agricultural and industrial use, the poor are generally systematically excluded from access (UNDP, 2006). Among the poor, women and girls face a distinct disadvantage since they often sacrifice education and skill development to collect water. Moreover, lacking informal rights to land, women are excluded from irrigation systems management. These deep-seated inequalities, such as gender division of household labor, norms on women speaking in public and constraints on women's mobility, adversely affect women's decision-making in community management, farmers' associations and water user groups.

Addressing inequities in water access will require clear targets for reducing asset-based land and water inequalities and establishing regulatory systems for public-private-community partnerships to provide clean water and sanitation to households at affordable prices. The National Slum Dwellers Federation in India, Orangi Pilot Project in Pakistan and Total Sanitation Campaign in Bangladesh (adopted in Cambodia, Chain and India) are examples of community-government partnerships that have led to rapid increases in access to sanitation and water. However, efforts to address competing uses of water are frustrated by the separation of administrative responsibilities for water and agriculture. In many cases, building on local artisans and local joint management would help to expand water development technologies. There is also a need for process monitoring to understand the logic of adoption and build coherence among administrative functions and policies to equitably resolve rights over competing water uses. Resolution mechanisms might include establishing and strengthening interministerial coordination, multi-stakeholder consultations/management and multi-sectoral dialogue.

The marine environment, too, is being degraded because of economic development (e.g., oil development and logging, dams), population growth and human activity, particularly in coastal areas. Marine ecosystems in the Vietnam Sea play an important role in the economic development of East and Southeast Asia, but are subject to pollution from several nations, resulting in habitat loss and a declining resource base (Thanh, 2005). Resolving such problems, which are increasing despite conservation efforts, requires not only national action, but also regional cooperation and modified institutional arrangements that include the international aid regime. For instance, the sustainable harvesting of fish stocks might be most effective if governments of the region cooperated to:

- Invest in integrated ecological socioeconomic assessment of fisheries to set sustainable goals and identify priority actions
- Improve tenure regimes to clarify roles and responsibilities for access and management of fisheries resources, as well as capacities for enforcement
- Pursue adaptive management based on sustainability goals

As current methods to prevent the collapse of fish stocks have not proved effective, action is needed to conserve stocks, such as big eye and yellowfin tuna in the Pacific and to increase marine food security. In pursuit of those goals, three major initiatives are needed in the medium and long term:

- Careful review and comparison of fisheries agreements, particularly the means employed to attract fleets and to raise revenues for small island states.
- Establish networks of "fish parks" so that protected areas of the ocean can be increased from the current level of only 0.5% to 20-30%, as proposed by the international academy of sciences (Balmford et al., 2004) and
- Complement fish reserves or parks with arrangements to disseminate monitoring data for internet use to engage citizens in policing activities (Clover, 2004).

5.2.2 Climate change

A daunting challenge in the ESAP agricultural sector is climate change and its predicted effects on productivity and livelihoods. Water is already one of the greatest constraints to agricultural productivity in much of the ESAP region and is likely to become more so with global warming-induced reductions in precipitation and accelerating glacier recession in the Himalayas (Duan et al., 2006; Singh et al., 2006; Kulkarni et al., 2007). The consequences of such changes are water scarcity as well as catastrophic flooding in mountain countries like Nepal, which are at risk of glacial lake outburst floods (Rai and Gurung, 2005). Unseasonal rains, debilitating droughts, excessive floods, devastating cyclones and storms and other extreme weather conditions seriously threaten agricultural, livestock and aquaculture production systems. Drought has consequences for 500 million farmers largely living on rainfed agriculture in India, Myanmar, Pakistan, Nepal, Thailand, Philippines, Australia and the Pacific Islands (UNESCAP, 1995). The Intergovernmental Panel on Climate Change (IPCC, 2007) projects a decline in grain harvest for tropical regions of 11-46% by 2050, largely in response to global warming, with effects likely to be felt by large and small farmers throughout the ESAP region.

Reducing CO_2 emissions requires multiple approaches, including reductions in energy consumption, more efficient use of available energy, application of renewable energy sources and carbon sequestration (Han et al., 2007). At the field and farm level, new agricultural approaches and models are essential for higher productivity, energy efficiency and sustainability in view of climate change threats (Seguin et al., 2007). For instance, a field or farm-level decision to eliminate tillage operations can reduce production costs by 50%, save labor and decrease erosion, agrochemical contamination and fuel consumption (Huang et al., 2002). Moisture conservation technologies along with crop and cropping system diversification are important to lower risk in a variable climate scenario, improve production and sustain livelihoods, particularly in marginal areas (Lantican et al., 2003).

At the regional level, governments in ESAP also need to adopt strategies to prepare farmers, particularly those in the most vulnerable (drought and flood-prone) sectors, for the long-term impacts of climate change and variability. Increased investment in rainfed agriculture throughout the region is one necessary intervention, as is raising awareness and support for farmers and affected populations to formulate mitigation and adaptation measures. Modern technologies in conjunction with traditional knowledge of water management and crop selection can support community-based adaptation efforts and account for the specific needs and conditions of particular populations. This is especially important in the low-lying deltaic areas of Bangladesh and the small-island developing states of the Pacific. With the exception of a few cases in Southeast Asia, irrigation intervention excluded women from access to water resources (Lambrou and Piana 2006a), thus generating especially adverse impacts of drought and climate variability for women. Yet, gender specific impacts are seldom noted in discussions related to the effects of climate change suggesting that effective policy implementation to respond to climate change requires accounting for its differential effects on women and men (Lambrou and Piana, 2006ab).

As forest degradation is endemic throughout the region, it is useful to examine recent positive responses to curb deforestation and support reforestation. In China, for instance, the introduction of large scale tree-plantation schemes between 2000 and 2005 may suggest strategies for other areas of the ESAP region (UN, 2007). Forests currently store a substantial stock of carbon, amounting to 826 billion tonnes in trees and soil (Brown, 1998), an amount that exceeds the stock of carbon currently in the atmosphere. Sequestering carbon in forests through activities that expand forest cover through plantations and agroforestry plantings and minimize deforestation, forest fires and soil disturbance can help mitigate the accumulation of greenhouse gases in the atmosphere (Adams et al., 1993; IPCC, 1996; Adams et al., 1999; IPCC, 2000; Roper, 2001). Mitigation strategies to address fossil fuel combustion and land use changes, particularly deforestation for agronomic purposes are also required since they have the greatest impact on the global carbon cycle (Janzen, 2004).

5.2.3 Biofuels/bioenergy

Biofuel as an energy source is traditionally used in many communities across the region, albeit on a small scale and often only by individual households. During the Japanese occupation in Southeast Asia, for example, rural communities with no access to kerosene used oil extracted from *Jatropha curcas*, coconut and castor beans for light and cooking. More recently, communities across the Pacific (the Marshall Islands and the Bougainville), use coconut oil to fuel vehicles (Cloin, 2007). Within these contexts, biofuels show promising potential to increase access to energy for the poor. Governments in the ESAP region are now assessing biofuels, or bioenergy as a solution to reducing dependence on expensive petroleum and curbing emissions of greenhouse gases that cause climate change. While many of the current sources of biofuels are derived from corn, root crops, coconut, groundnut, sugarcane, sweet potato and palm oil (Ohga and Koizumi, 2007), tree crops such as *Jatropha* and *pongamia* are also receiving significant attention. A new generation of biofuels produced from agricultural and timber waste is being explored but is as yet not commercially viable.

Biofuel production promoted by regional governments is based on large scale industrial production that supplies energy for urban areas and industrialized countries, giving rise to competition between biofuel and food production on already strained natural resources. Even for biofuels produced from secondary food crops, such as China's announced shift to ethanol production from cassava, sweet potato and sorghum (Sun, 2007) and non-grain oil crops such as *jatropha* that can grow in marginal lands, massive production of biofuels requires the conversion of agricultural, forest and public lands to grow these crops commercially. Commercial production of biofuels that require area expansion is therefore likely to substantively increase the agricultural demand for water, which is already at 78% of the available fresh water supply in Southeast Asia (ASEAN, 2006).

Further, the increasing demand for oil palm is expected to create additional pressure on the region's steadily declining forests. In Indonesia, for example, the government plans to convert some 1.5 million ha of land to oil palm plantations and another 1.5 million for *jatropha* plantations, beginning with approximately one million hectares of land not productively used, including forestry concessions that have been abandoned by concessionaires (Haswidi, 2006). This conversion can potentially increase forest and land fires associated with oil palm plantations and lead to a serious loss of biodiversity, ecological degradation and recurring transboundary haze that endangers human health and economic security. Apart from climate change concerns, deforestation threatens the survival of indigenous peoples, forest dwellers and the rural poor who depend on forests for their food, livelihood and cultural identity. Current debate on the environmental impacts of biofuels in ESAP also extends to biofuel production in input-intensive monoculture systems and genetically modified crops developed to accelerate the growth of biofuel crops and trees with little risk assessment analyses.

With increasing market demand, crops traditionally grown for food, feed and oil are expected to shift to biofuels. This will likely increase food prices, especially if supplies remain the same, as is already evident in the current corn price trends worldwide. Rising grain prices and the diversion of local carbohydrate and protein sources to the energy market could represent a market opportunity, but also a threat for the region's poorest people (Brown, 2006). High feedstock prices could drive small livestock and poultry producers out of business, depriving millions of families of their livelihood. The joint announcement by Malaysia and Indonesia to allocate 40% of their combined annual palm oil output for biodiesel production is expected to increase the price of edible oil, making it expensive for both food and energy users. In response, countries like the Philippines have adopted measures to ensure that ethanol production does not compete with food production by requiring prohibitive distancing between sugar mills for food/feedstock production and biofuel processing. Root crops in particular often serve as secondary or even survival crops for many rural and indigenous communities across ESAP, so using them for biofuel production could directly affect food affordability.

Biofuels have the potential to provide food security and livelihoods for the poor in ESAP if their production is based on the sustainable use of local resources and the provision of off-farm income, especially to women, and the improved management of energy production and consumption. Supplying energy to urban areas and industrialized countries may offer short-term economic gains for developing countries in the region, but with high costs for the environment and for the capacity of countries to produce food that is available, accessible and affordable to poor people.

5.2.4 Urban-periurban agriculture

Urban and periurban agriculture (UPA) can contribute to reducing poverty and enhancing food security. It involves the production of crops, livestock, fish and related goods and processing and marketing activities in and around cities and towns. UPA is distinguished from rural agriculture by its integration into the urban economic and ecological system. Its objectives include income generation, contribution to urban dwellers' food needs and environmental sustainability (van Veenhuizen, 2006). UPA can increase food availability and enhance the freshness of perishable foods reaching urban consumers. Case studies show differences in nutrition, especially among children, when poor urban families farm. UPA also has the potential to efficiently recycle nutrients from municipal solid wastes and waste water (Hussain et al., 2002).

Despite potential efficiencies, there is a bias against urban and periurban agriculture that has resulted in overlooking, underestimating and under-reporting UPA practices. Concerns arise over competition for land, water, labor energy resources and its incompatibility with urban life (smells, noises, pollution). Though many of these concerns arise in rural agricultural production, two important differences in the urban environment are: proximity to greater numbers of people and high existing stresses on the natural resource base. In the absence of appropriate management and monitoring of resources, the negative environmental and health effects of UPA are critical concerns.

In order to enhance the positive consequences of UPA for urban livelihoods, resource utilization and the environment, research must bridge the gap between UPA research and practice and between urban planning and policy. Other constraints to UPA in urban areas of the ESAP region include limited and declining land, polluted irrigated systems and wastewater contamination. As municipalities play an important role in urban planning, their involvement is critical for promoting UPA for its potential contributions to improving the urban environment.

5.2.5 Human, animal and zoonotic diseases

HIV/AIDS: The current situation of HIV/AIDs-infected and affected people in the region remains grim; 4.3 million people were newly infected in 2006, with East Asia host to one of the fastest rates of infection (United Nations, 2007). Almost half of those living with HIV in the region are women, with an increasing number being married women and youth between the ages of 15 to 24 years (Mehta and Gupta, 2005). Gender inequality and economic imbalances continue to drive the "feminization" of the HIV epidemic (United Nations, 2007). In the absence of accessible, affordable health care and social acceptance, HIV has the potential to affect not only individuals, but entire communities and economies—as witnessed in large parts of Africa.

Key interventions to control infection have expanded in recent years; however, they are not sufficient to arrest infection rates. Country-specific and region wide consultations (i.e., public hearings, women's court) are held with some success and include HIV-positive people's networks, local and international NGOs, experts and practitioners of HIV control initiatives and international development partners. Additional interventions could include such enabling strategies as ensuring employment opportunities for HIV infected and affected people, the implementation of policies to protect women's inheritance and property rights (Bhatla et al., 2007); dignity-based access to health care, education, skill development programs, training and an inclusive community-based approach to overcome the local world of shame and social stigma often experienced by HIV infected/affected people.

Avian flu: The 2004 avian flu outbreak of the deadly H5N1 strain of avian influenza in East Asia resulted in the extermination death of 140 million domesticated poultry and birds (WHO Global Influenza Program Surveillance Network, 2005; Webster, 2006). The outbreak also proved that avian flu may be transmitted from birds to humans, with 130 human cases, including 70 fatalities, of avian flu in Vietnam, Thailand, Cambodia and Indonesia (Webster, 2006).

The damage caused by the epidemic made apparent that, despite intermittent outbreaks since the 1990s, most ESAP countries are not adequately prepared to cope with this disease. Few governments have rapid response mechanisms to warn of possible culling campaigns or offer appropriate compensation (Martin, 2006). FAO has provided guidance on establishing mechanisms and programs for surveillance, monitoring and diagnostics and early warning systems (FAO, 2004a).

These outbreaks have severely tested the capacity of national agricultural extension systems across the region. The Thai experience of 2004 showed weakness in the capacity of extension systems to timely and efficiently report outbreaks. Communication systems need to be improved to ensure that reports of local outbreaks are transmitted to appropriate authorities at the national level to ensure immediate surveillance and diagnostics to arrest further spread to other areas (FAO, 2004a). Public funds must be mobilized to improve the capacity of local extension workers to conduct surveillance and detection, as well as to handle outbreaks. Such training programs need to target poultry-raisers, especially those small backyard producers who comprise the majority of the poultry sector in the developing ESAP countries. Regular capacity building programs on preparedness for veterinary diseases need to include avian flu outbreaks, even in countries that have yet to be affected by the virus.

Public awareness and education on avian flu, especially among segments of the population involved in handling poultry, is also critical. Indonesia was severely hit by the 2004 outbreak and has now invested heavily in public education, while Thailand implemented a national public education program focusing on preventing virus transmission (Olsen, 2005). Greater awareness is needed regarding the risks and costs of avian flu outbreaks need improvement, as well as coordination between appropriate authorities and law enforcers. Prevention measures such as fencing poultry populations to keep them away from wild birds suspected of carrying highly pathogenic avian flu virus need to be supported, possibly through incentive mechanisms to encourage cooperation from those for whom fencing costs may be prohibitive, especially in countries where free-range raising of chickens and ducks is traditionally practiced (Martin, 2006).

Governments also need to allocate emergency funds to provide immediate and fair compensation for culled flocks, especially to backyard poultry raisers whose livelihoods depend primarily on small-scale poultry raising (Martin, 2006). For example, during the 2004 outbreak, Thailand allocated a budget of US$132.5 million for direct compensation to affected farmers (Tiensin, 2005). Countries may also consider investing in vaccination of poultry against avian flu, but while vaccination may be a possibility for large-scale poultry raisers, it is largely inaccessible to backyard raisers dependent on government support. Investment must also take into account the resources needed to develop extension worker capacity to properly and safely handle vaccines from the source to the farms. In considering various options it is crucial to account for the impacts on animal and human health of veterinary vaccines and antibiotics (Ho, 2007).

5.2.6 Institutional and governance practices

As agriculture in the ESAP region is changing rapidly, organizations must continuously evolve appropriate institutions and governance systems to respond to new challenges. Designing, evaluating and implementing institutional reforms will be a continuing challenge for managers in Research,

Development, Training and Extension (RDTE) and related systems if agriculture is to meet sustainability and development goals. It may be important to implement institutional changes in rules, norms, conventions and habits within RDTE and related organizations to enhance exchange among various stakeholders. A learning-based approach appropriate to each specific context is one creative option to meet this need.

Developing appropriate policies is important, but having sound policy does not ensure compliance with guidelines for improved performance. First, many of the small countries in the region have limited capacity to implement policies as exemplified by the gaps in the capacity of several nations to implement policies related to biosafety, IPR and food quality standards. Secondly, quite often the policy only prescribes what to do, without taking into account what needs to be done to get the policy implemented. Ideally, the policy should also facilitate change, through a process of experimentation, reflection and learning so that it develops the capacity of the various stakeholders to identify bottlenecks, experiment with alternative ways of working and evaluate performance. The actors thereby learn what needs to be changed or modified and how to develop better policies.

The implementation of appropriate agricultural policies and programs requires collaboration among a large number of organizations that involves the creation of appropriate forums for partnership (e.g., working groups, steering committees) and the development of mechanisms to ensure collective decision-making. Organizations with a tradition of working in isolation may find it difficult to adjust to inclusive governance mechanisms and may invite opposition. Yet, developing skills to manage transitions can contribute to a collective ownership of change.

5.2.7 Actors and organizations

Agricultural development in the region depends upon the performance of a large number of actors and organizations; those involved in RDTE, as well as those involved in the generation and distribution of inputs, supply of credit, value addition and marketing and the development and implementation of policies with attention to the context that shapes the interaction among these different constituencies. Though many people are present in the different ESAP countries, there is a wide diversity in number, capability and performance, which has implications for planning agricultural development interventions. First, importing models of technological change that may have been successful elsewhere, may not adequately address agricultural development in this region indicating the need for country and region-specific interventions and approaches. Second, the development or application of new technologies need not be the starting point for agricultural development. Instead, complementing new technologies with institutional innovations may more adequately address the complex interaction of agricultural development and environmental sustainability.

Donor funds and expertise from CGIAR centers and the FAO, for example, have facilitated the development of infrastructure, extension services and human resources in most ESAP countries. However, many of the small countries in this region do not have adequate capacity to adopt emerging technologies, exploit the potential of ICTs, respond to new challenges from integration of trade and enforce protocols related to biodiversity and biosafety. Research and development efforts need to be strengthened in biotechnology, postharvest technology, IPRs and food legumes and minor fruits and vegetables. Currently, international donors including the World Bank, Asian Development Bank, FAO, IFAD and international networks such as CGIAR, ACIAR, GFAR and APAARI support select countries in agricultural development along with regional networks. Increasing the funding levels of these international and regional networks could greatly increase their reach.

There has also been a decline in public support for research and extension leading to vacancies and reduced operational funding. Concurrently, private sector investments in agricultural research and extension have increased. Apart from private input companies and agribusiness firms, several other actors are increasingly intervening in research and extension, including producer cooperatives, farmer associations and NGOs. Media (print, radio and television) have increased their role in the dissemination of information on agricultural practices and technologies, development programs and market arrivals and commodity pricing. The internet is an emerging source of agricultural information, although its use varies widely within and between ESAP countries.

Though a large number of organizations with complimentary skills and expertise exist, they have yet to collaboratively address new and existing agricultural challenges. Making new knowledge, information, or technology available and put into productive use occurs best when different actors in innovation systems share knowledge and work in partnership. Though efforts have been made to promote such collaboration, they have yet to alter how organizations function. New forms of interaction, organization and agreement between multiple actors ought to be encouraged if agriculture is to be environmentally and economically sustainable in the ESAP region.

5.2.8 Social exclusion and gender inequality

Social exclusion derives from exclusionary relationships based on power and hierarchy and intersects with other aspects of social disadvantage, such as gender, caste, ethnicity and religious minority status. Exclusion is a multidimensional process that prevents individuals or groups from access to institutions of governance, public services like health care and education and economic resources as well as factors of production (Beall et al., 2005). Further, exclusionary relationships are nurtured and maintained through systemic violence and the denial of rights that contributes to sustained poverty.

Most of the region's poor live and work in rural areas, making agriculture and land rights central to the struggle against poverty. To date, debate over land rights has brought rural and urban populations and industrial interests and farmers into violent confrontation as inequalities in decision-making have denied villagers access to traditional lands in bids to facilitate industrial expansion. In China, where the average land available per capita is only 40% of the global average, some 227,000 ha are lost to urban and industrial construction annually.

Reviews carried out between 2000 and 2006, note that throughout the ESAP region, countries have improved their development policies, consequently scoring higher on indicators relating to the MDGs (UN, 2007). However, they also reveal that highly unequal countries—India, Pakistan and Philippines—have had less success than more equal ones such as China and Viet Nam (UNDP, 2006). The most noticeable policy improvements are related to gender and efficiency of resource mobilization for poverty reduction even though large groups of rural women and indigenous peoples continue to be excluded from economic and social institutions (Narayan et al., 2000).

Of the estimated 300 million indigenous/tribal peoples in the world, about 70% live in Asia and are generally among the most marginalized communities in almost every country with high levels of landlessness, illiteracy, malnutrition and no access to health services. Many of these communities experience social discrimination, economic exploitation and political marginalization (www.ifad.ord/pub/factsheet/ip/e.pdf). Among indigenous/tribal peoples, women are further marginalized as they have little representation or voice in village councils. This is true even in matrilineal communities as in Meghalaya, India and among the Mosuo in Lijiang, China. Even when new state-sponsored organizations like Joint Forest Management councils are created, indigenous/tribal women continue to be excluded (Kelkar and Nathan, 2003).

Widespread gender-based inequalities in access, control and ownership of productive resources characterize relations in the region in ways that hinder development and are inextricably linked to poverty outcomes (Mason and King, 2001). Exclusion from skills, capabilities, assets and recognition, for example, is one consequence of labor market restructuring. Thus, it is hardly surprising that 70 to 80% of informal workers are women who are employed in low-paid jobs with long working hours. Further, their exclusion from formal employment and forms of informal work has denied women the dignity of livelihood.

Cultural norms that govern women's work and mobility play a significant role in women's work relations. Increasing the participation of women and indigenous peoples in community governance is a matter of drawing upon their labor, skills and knowledge to enhance resource management and AKST and their development choices and rights. Sustainability depends on allocating resources to excluded groups in ways that enable them to negotiate with government, the private sector and civil society. Gender specific approaches can overcome the marginalization of women in the governance of community resources. The capacity of women agricultural producers, as holders of AKST, also needs strengthening if household and community livelihoods are to be improved. Given the feminization of agriculture and women's limited access to resources and markets, this may be accomplished by empowering women to secure access to and the management of land, knowledge and technologies in ways that recognize women's economic contribution to agricultural production, ensure wage parity and provide women with vocational and technical training.

Thus, if community livelihoods and the rural economy are to be enhanced, discriminatory barriers that limit the participation of women, indigenous peoples, certain caste groups and religious and ethnic minorities need to be identified and eliminated. This may be accomplished through the development and implementation of policies and governance structures, anti-discrimination legislation and ensuring resource equity.

5.3 Existing and Emerging Technologies in the ESAP region

5.3.1 Local and traditional knowledge and practices

Through millennia, local and traditional knowledge played an important role in maintaining and improving the livelihood of farming and indigenous communities—from producing food and providing shelter, to achieving control of their lives. Traditional knowledge systems are often the basis for local-level decision-making not only in agriculture, but also in health care, food preparation, education, natural resource management and a host of other activities in rural communities. The valuable contributions made by these systems to global knowledge will continue only with appropriate and adequate support, based on the recognition of their dynamic nature and interdependence with the environment where they evolved.

Traditional knowledge is both cumulative and dynamic, building upon the experience of earlier generations and adapting to the new technological and socioeconomic changes of the present (Johnson, 1992). Traditional knowledge includes indigenous indicators to determine favorable times for each phase in crop production, land preparation and plant propagation practices, seed storage and processing, sowing and intercropping, seedling preparation and care, crop harvesting and storage, food processing and marketing and pest management and plant-protection methods (Grenier, 1998).

5.3.1.1 Impact

Traditional knowledge of crops and medicinal plants contribute much to our understanding of cropping systems in the ESAP region and may be adapted for non-traditional systems. In Nepal, hill farmers have a ranking system for the nutrient value of manure—from bat to buffalo—from different animals in terms of its use as fertilizer which corresponds well with scientific findings based on macronutrient content (Tamang, 1993). Modern plant breeding also owes much to the landraces bred, conserved and developed by traditional communities over the millennia. These local varieties have been a continuous source of genes used in the development and improvement of high-yielding varieties, with conservation and innovations involving plant genetic resources practiced by farming communities for about 100 years (Rerkasem and Rerkasem, 2002; Nagarajan et al., 2006). For instance, a centuries-old seed management system allows farmers in Nepal to grow and protect their seeds (Timsina, 2000; Upreti and Upreti, 2002).

Traditional knowledge of crops has also contributed greatly to modern medicine and biopesticides in the region and beyond. For example ground neem (*Azadirachta indica*) seeds have been widely used in India as a natural insecticide and fertilizer, and the leaves are used to protect grains stored in local containers (Arnason et al., 1985; Prijono and Hassan, 1993; Gajalakshmi and Abbasi, 2004). Neem, currently marketed as a modern biopesticide, is effective against malaria and internal worms and is reportedly used by 500 million Indians in brushing their teeth, as well as for making soap. Another example is *Conospermum*, commonly called smokebush in Australia, used traditionally by Aboriginal peoples for a variety of therapeutic purposes. In the late 1980s the plant was found to contain a substance called conocurvone, reported to have potential in treating AIDS (Nader and Mateo, 1998; Vermani and Garg, 2002). It is estimated that in 1985, US$43 million worth of plant-based medicines (many of which were used first by indigenous peoples) were sold in developed countries (Principe, 1989). As advances in biotechnology broaden the range of life forms containing attributes with commercial applications, the full market value of traditional knowledge will certainly increase.

Similarly, local and traditional knowledge and practices in animal agriculture systems include methods of animal breeding and production, specific uses of traditional fodder and forage species, animal-disease classification and traditional ethno-veterinary medicine (Grenier, 1998). A recent study in India revealed that traditional health control and treatment systems were effective in curing a number of ailments in animals including dysentery, arthritis, dog bites, coughs and colds, anoestrus, wounds, bloat and diarrhea (De et al., 2004). While modern veterinary medicines were found to provide relief more rapidly than traditional treatment and preparations, the latter are cheaper and locally available to farmers, especially in remote areas.

Aquaculture systems integrated into cropping systems and utilizing indigenous knowledge and traditional practices—such as the rice-fish systems in Bangladesh—can contribute to food security and food diversity (Pretty and Hine, 2000). Rice-fish systems are also traditionally practiced across Southeast and East Asia where they have provided numerous benefits to farming households beyond additional and diversified sources of food and nutrients. For example, the introduction of larva-eating fish into rice fields in Quanzhou County, Jiangsu Province, China has increased rice yields, reduced pesticide use and significantly reduced the incidence of malaria (Pretty and Hine, 2000). The diversified and integrated nature of rice-fish farming has also provided a natural system of protection against crop pests and diseases.

Indigenous peoples in the region have practiced sustainable forest utilization and management techniques for centuries. Indigenous knowledge in forestry includes management of forest plots and their productivity, knowledge and sustainable use of forest plants and animals and understanding the interrelationships between tree species, improved crop yields and soil fertility (Grenier, 1998). The indigenous Hani communities in Mengsong, Xishuangbanna, in China's Yunnan province have developed a system of classifying forests according to their function and products. In Sri Lanka, forestry has traditionally been combined with cropping and animal rearing, such as in the classical Kandyan Forest Gardens where the agricultural system simulates a tropical rain forest involving small land units averaging about one hectare and consists of a mix of as many as 30 perennial and semi-perennial trees and shrubs (UNESCAP, 2002). The farmers who practice this system are reported to enjoy a relatively better living because of returns from both cash crops and subsistence products and improved health and longevity.

Indigenous practices of importance include the traditional use of fire in forest management and conservation, facilitating crop production and pest control and preventing unplanned burns (Jackson and Moore, 1998). For instance, annual fires as practiced by local people in Sumbawa, Indonesia are intended to maintain grasslands in a condition that favors grazing animals and enables wild herbivore populations to be maintained at higher levels than they would be without fires. Similarly, local communities in Nepal use fire extensively to promote the regeneration of grass for animal agriculture.

Traditional knowledge systems continue to contribute in the development of agriculture throughout the ESAP region, largely because they are often inexpensive to implement and can be paid for in goods or services and they are readily available and accessible even to those who do not have cash incomes (Fernández, 1994). In most cases indigenous knowledge systems are socially desirable, economically affordable, environmentally sustainable and involve minimal risk to rural farmers and producers. Since they evolved gradually within the community and are under its control, local and traditional knowledge systems are considered more directed towards self-reliance and self-sufficiency than are modern technologies imposed from outside (Fernández, 1994).

5.3.1.2 Challenges

Traditional agriculture is considered labor intensive which may be a disadvantage or advantage depending on the social circumstances. The labor requirement may keep people from engaging in other economic activities, but it can also provide employment for rural people who might otherwise migrate to urban areas. Indigenous knowledge and traditional agricultural systems have much to offer to the future development of sustainable agricultural systems in ESAP countries. The role they play should be in response to specific local needs and conditions to maintain food security and improve micronutrient content of food with minimal human and environmental health consequences, taking into account gender equity and equitable access to resources. Government policies must not ignore people's knowledge and must recognize that traditional knowledge plays an essential role in development (Warren, 1991). An enabling environment that supports the conservation, use and promotion of traditional knowledge, as well as its interaction with modern agricultural systems is needed to maximize the potentials of these knowledge systems.

Increasing population and the accompanying pressure on productive resources such as water and land brought about by the expansion of modern agriculture and industrialization, pressures from modernization and cultural homogenization are posing serious challenges to the viability of traditional agricultural systems and knowledge in many ESAP countries (Doane, 1999). An appropriate response might lie in providing monetary and infrastructural support to indigenous peoples and practitioners of traditional agriculture. Protection of traditional knowledge should also involve equity considerations anchored on the principles of conservation, preservation of traditional lifestyles, and adequate recognition and compensation.

5.3.2 Organic agriculture

While there is no single definition of organic agriculture, the term is generally understood to represent a suite of farm management practices rooted in using natural rather than synthetic materials to enhance soil fertility and combat plant disease (Raynolds, 2000). Organic agriculture fosters processes and interactions occurring in natural ecosystems, encouraging internal stability rather than relying on external control measures.

Through its emphasis on building soil structure and fertility by the incorporation of organic matter, organic agriculture can play a role in the rehabilitation of degraded agricultural lands where erosion, compaction, salinization and desertification have significantly reduced yields and threaten future agricultural use. The incorporation of organic matter into the soil improves water-holding capacity, which is essential in areas of poor/erratic rainfall. While some evidence suggests higher above and below-ground biodiversity and improved biochemical characteristics in organic systems (Bengtsson et al., 2003; Melero et al., 2006), neutral outcomes are also reported (Franke-Snyder et al., 2001). Species richness has been shown to increase among some organism groups while others are unaffected (Bengtsson et al., 2003). The impacts of organic agriculture on biodiversity depend on factors such as crop rotation and tillage practices, quantity and quality of organic amendments and the characteristics of the surrounding landscape. Although some studies suggest reduced environmental losses of nutrients like nitrate nitrogen in organic systems (Kramer et al., 2006), most evidence suggests that nitrate losses are not reduced in high-yielding organic systems when contrasted to conventional production systems (Kirchmann and Bergstroem, 2001; Torstensson et al., 2006).

Organic animal agriculture practices include soil management techniques based on appropriate stocking rates and grazing regimes that minimize compaction, provision of good quality drinking water and organically grown feed, free access to graze and range on a wide variety of pasture and browsing species, adequate housing to ensure animal welfare and the use of natural health remedies (BIO-GRO, 2001).

5.3.2.1 Impact

Overall, organic agriculture's emphasis on the sustainable use of local, often free, resources makes this system particularly important for the rural poor in the ESAP region, especially those in marginal areas with access to organic inputs. Organic agriculture is particularly beneficial for women-headed households because of women's greater difficulty in accessing financial resources to purchase seeds, fertilizers and pesticides. It also eliminates the exposure to pesticides, which have had a significant impact on the health status of the rural poor (UNESCAP, 2002).

In addition to its potential for marginalized areas in the region, organic agriculture is developing into a lucrative export market-driven sector—as well as a thriving domestic market sector, with produce increasingly being sold at local farmers markets in Australia and New Zealand as well as many other ESAP countries. There is also growing demand for organic food in the ESAP urban centers and the premiums paid for this food could offer an opportunity for increased incomes for the rural poor, especially in areas largely untouched by Green Revolution technologies (IFAD, 2002). Overall, the organic market in Asia is currently growing at the rate of 10-20% annually, making it the fastest growing sector in the region's agricultural market. A number of countries in the region have adopted policies to promote organic agriculture to take advantage of the growth in market demand. The area under organic management in Asia is around 2.9 million ha, while in the Pacific (largely Australia and New Zealand) it is around 11.8 million hectares (Yussefi and Willer, 2003).

Similar to traditional agriculture in the region, animals in organic systems are incorporated into a mixed animal agriculture/cropping enterprise, often with the addition of forestry. Yields from organic dairy systems in New Zealand average around 5% lower than their conventional counterparts but with higher returns and lower input costs resulting in little change in overall returns to the farmer (Christensen and Saunders, 2003). Organic aquaculture in the ESAP region has lagged behind the development of other kinds of organic agriculture. New Zealand is one of the largest producers of organic fish and mussels outside Europe; other major producers include Indonesia, Viet Nam and Thailand (shrimp) and Australia (salmon). One constraining factor is the requirement that food for the farmed species be obtained from sustainably managed fisheries, derived from locally-available fishery products not suitable for direct human consumption, free from synthetic additives and contaminants and only fed to farmed aquatic species with naturally piscivorous feeding habits (FAO, 2002). Removal of this constraint would considerably increase the production of farmed organic aquatic plants and molluscs.

Organic systems may require more labor than other systems, which can be negative or positive, depending on the availability of labor and income-generating activities. The question of yields from organic versus conventional systems remains controversial, but there is clearly a niche for organic crops in the ESAP region, particularly when combined with nitrogen-fixing legumes, livestock systems that supply reliable organic inputs, or where sufficient crop residues are available. There are a number of organic conversion projects throughout the region that have reported yield increases and social benefits after conversion (FAO, 2002). In Asia a positive relationship was noted between organic farming and improvements in rural livelihoods, including positive effects on employment, income and household food security (UNESCAP 2002a). For example, an improvement in the health status of members of the Nayakrishi Andalon organic movement in Bangladesh, which includes more than 65,000 families, was observed.

Organic agriculture can reduce fossil fuel dependence through reductions in petrol-based inputs. This may not hold true if organic inputs and products are transported over great distance, but this is an unlikely scenario in most of this region, particularly for small-scale farmers. Calculations on comparative energy use in OECD countries indicate that energy consumption on organic farms is 64% that of conventional farms (Haas and Kolke, 1994; Lampkin, 1997), while other research in Iran and Switzerland puts this figure as low as 30-50% (Zarea et al., 2000). In a three-year comparative study on organic and conventional strawberry production in China, 98% of the energy inputs in the organic systems were from renewable sources such as animal manure and biogas, whereas 70% of the energy inputs in the conventional system were non-renewable sources such as electricity, chemical fertilizers and pesticides (FAO, 2002). In New Zealand, the mean annual energy input was considerably lower under organic management systems than under conventional management (Nguyen and Haynes, 1995). While fossil fuel consumption can be substantially reduced in organic systems, these energy savings must often be balanced against productivity reductions (Dalgaard et al., 2001). For organic systems with substantially lower yields than conventional alternatives, enterprise energy efficiency (energy output per unit energy input) can be lower than the efficiency of conventional systems (Loges et al., 2006).

5.3.2.2 Challenges

Among the biggest obstacles to organic agriculture in the ESAP region are those associated with insufficient quality and quantity of organic inputs and the cost of the organic certification. Shortages of organic soil amendments are common throughout the region (Husain and Raina, 2004), especially where high population pressure, cropping intensity, or small land holdings preclude rotations with N-fixing legumes and there are competing uses for animal manures (e.g., for cooking fuel). Some of the most common organic inputs such as cereal stover are of poor quality, having low nutrient concentrations and macronutrient ratios that are not optimal for crops. As organic agriculture does not require the purchase of expensive inputs it becomes more accessible to poor rural women who are unable to obtain credit. However, as organic systems need to be developed over a number of years to reach maximum productive capacity, women's often insecure access to long-term control over land may be a hindrance (FAO, 2002).

The present scale of organic agriculture in the region is still considered small, partly because development and uptake by farmers have been hampered by a lack of supportive government policy in many countries (UNESCAP, 2002). While organic certification paves the way for producers to take advantage of the growing market demand for organic products, the costs involved work to the disadvantage of resource-poor farmers. The weak bargaining power in setting the price for agricultural produce in general, poor farm-to-market support infrastructure and lack of clear policies in the marketing of organic products all contribute to the limited access of poor farmers even to the domestic organic markets. Government policies in promoting organic agriculture therefore need to address these problems to ensure that poor rural producers will directly benefit from the socioeconomic potentials of this system. Measures should also be adopted to ensure that the expansion of organic agriculture cultivation in response to market demands does not sacrifice local food security and the environment.

Despite its potential and proven benefits, the impetus for private sector research in organic agricultural research has been largely absent for several reasons, including the presumption that organic agriculture is lower-yielding, the relatively low market share of organic products in ESAP countries and the reliance on inexpensive, rather than expensive inputs. It is posited that if funding levels were to increase, organic production could be increased substantially, improving the social conditions of the rural poor of Asia, thus going a long way toward meeting development and sustainability goals.

5.3.3 Conventional technologies and practices

Agricultural technologies and practices generated by formal institutions and research centers, which might involve combining indigenous knowledge, organic practices and relatively new innovations or technologies, are termed "conventional" or "modern" and have contributed to substantial gains in global agricultural productivity. The best known of these technologies, developed and disseminated in the 1960s and 70s after a decade of food shortages and famines, is known throughout Asia as the Green Revolution and depends almost entirely upon plant breeding to produce high-yielding varieties, mineral fertilizers, irrigation, synthetic pesticides for weed, disease and pest control, animal breed improvement and intensification of feeding and mechanization.

The Green Revolution led to the introduction of stronger-stemmed and higher yielding "modern" varieties of the major cereal crops rice, wheat and maize, fueling an explosion in their yields on lowland, intensively irrigated land. Cereal production in Asia more than doubled from about 313 million tonnes in 1970 to about 650 million tonnes in 1995 (IFAD, 2002) and enabled double and triple cropping in areas that previously produced only one or two crops per year (Umetsu et al., 2003; Gupta and Seth, 2007). A great majority of all recent gains in crop yields are attributable to these conventional breeding-induced improvements targeting the physiological yield potential of crop plants and their tolerance to biotic and abiotic stresses (Khush, 2005; Reynolds and Borlaug, 2006).

Other conventional technologies include those of Integrated Pest Management (IPM) and Integrated Nutrient Management (INM), which draw substantially upon indigenous knowledge and organic practices. IPM is an environmentally sensitive approach to pest management that relies on a suite of pest management options. INM is a suite of practices designed to integrate the use of organic and inorganic sources of crop nutrients, so that agronomic productivity increases in an environmentally sustainable manner, without compromising soil resources. INM relies on a number of factors, including appropriate, balanced nutrient application, and soil and nutrient conservation practices such as low- and no-till farming, terracing, mulching and green manuring.

5.3.3.1 Impact

There is a body of evidence that suggests millions of lives in the ESAP region were saved, and that the view that the rich became richer and the poor even poorer as a consequence of the Green Revolution is over-generalized (Acharya, 1982; Lipton and Longhurst, 1989; Paul, 1989; Kar, 1991; David and Otsuka, 1993; Edmundson, 1994; Osmani, 1998). While pesticide dependence undoubtedly had negative implications for environmental quality and human health, the increased productivity engendered by Green Revolution technologies reduced the need to bring natural ecosystems under cultivation by about a billion hectares of land for 1999 levels of productivity globally (Borlaug and Dowswell, 2005). Adoption of the Green Revolution package was largely focused on irrigated and high yield-potential areas to the neglect of marginal areas (IFAD, 2002) where a large percentage of the rural poor live. However, adoption since 1977 has increasingly been in rainfed and abiotic stress-prone areas—particularly in the case of wheat (Byerlee and Moya, 1993; Heisey et al., 2002; Lantican et al., 2003).

Advances in animal breeding and health have increased both the quantity and quality of animal protein available to consumers in the region. Livestock plays a vital role in economic development and in the life of Asian farmers. Animal products such as meat, milk and eggs provide daily cash income to farmers and also provide much required nutrition to the rural population. It also contributes to draft power and the dung adds not only to soil fertility but also is used as a material for house construction and cooking fuel. For the poor, livestock has multiple uses. This perspective needs to be kept in focus when designing and promoting livestock sector interventions.

Although it is commonly viewed narrowly as the "intensive culture of salmon and shrimp to provide high value products for luxury markets," conventional aquaculture actually comprises diverse systems of farming plants and animals in inland and coastal areas, many of which have relevance for the poor (Edwards, 2000). The ESAP region was responsible for about 91% of the world's aquaculture production in 2003, with China alone accounting for almost 71% (FAO, 2005). Nine of the top ten aquaculture production countries are in this region, with China, India and the Philippines topping the list (FAO, 2006). Thus, aquaculture contributes to both the overall growth of many states and to the livelihoods of the poor in many parts of Asia where it is traditional practice. It is a major foreign exchange earner in addition to being a food source and employing the poor. In Thailand, for example, shrimp farming is dominated by small-scale farmers (Apu and Middendorp, 1998).

Conventional aquaculture and fisheries offer options for employment for landless laborers. These options include cage and pen culture, enhanced fisheries in large communal water bodies (e.g., grass carp in Vietnam, Chinese carps in Nepal) and farming molluscs and seaweeds in coastal bodies (Indonesia and the Philippines). Inland aquaculture systems may have the greatest potential for expansion because aquaculture can be integrated with existing agricultural practices of small farmers, resulting in increased overall production through the synergies of integrated practices. Aquaculture can utilize irrigation systems as well as saline areas and swamps which would otherwise be of marginal use to agriculture (Edwards, 2000).

In general, technologies generated by formal institutions are designed to suit a wide range of environments and may prove beneficial in changing agroecological environments. High input, mechanized, increasingly consumer- and market-driven systems have substantially increased the productivity of agricultural and animal systems in the ESAP region while resulting in substantial labour savings. This often translates to less time expended in physically demanding work for the women who are the primary agricultural workers in the ESAP region and a decreased need for the entire family, including children to be involved in field work. This may be a positive or negative effect depending on the level of income generated by the farm and the availability of other means of income generation.

5.3.3.2 Challenges

Conventional practices—particularly those placing high dependence on pesticides and improved seed—may have unintended economic, social and environmental effects (Bautista, 1997; Whitten and Settle, 1998). Although risk of crop failure is generally lower with such practices, the risk of disease outbreaks is higher. The economic outlay required for seed, pesticide and fertilizer inputs to maximize success is high and translates to a greater risk undertaken by farmers with much to lose if crops should fail. These technologies too often did not focus on ways to achieve increased food production in a resource-efficient manner that is environmentally benign.

Regardless of how the Green Revolution is viewed, there is no doubt that yields throughout the ESAP region have begun to stagnate or drop despite increased inputs, reflecting declining soil biological activity, fertility and structure (Harrington et al., 1989; Abrol and Gill, 1994; Kijne, 1994; Singh and Paroda, 1994; Huang et al., 2002). The rice yield in Asia declined sharply during the 1980s from an annual growth rate of 2.6% in the 1970s to 1.5% in the 1990s (Pingali and Rosegrant, 1994; Gruhn et al., 2000; Gupta and Seth, 2007). Clearly productivity cannot be sustained in the long term in such intensive cereal monocultures by using increasing levels of chemical inputs. Additionally, the adverse livelihood-related and environmental consequences resulting from the mismanagement of Green Revolution technologies are now well recognized. The negative effects of insecticide use on human health, the environment and on beneficial insects have been documented (IRRI, 1994; Way and Heong, 1994; Pingali and Roger, 1995). Many farmers entered a spiral of debt precipitated by rising costs of inputs and falling prices for outputs: rice farmers in the Philippines were found to be economically better off before they shifted away from the mixed cropping enterprises to the high-yielding monocultures (UNESCAP, 2002).

Much of the previous effort in livestock farming focused on breed improvement and development of vaccines. Competition from both developed and developing countries and more sophisticated and changing sets of domestic and international trade norms and standards are increasing pressure on developing country producers. Throughout the region, the cost of feed is an important constraint to increased livestock production, along with poor-quality and fluctuating feed supplies (FFTC, 2007). Deterioration of common property resources and an increasing shift towards cash crops have adversely affected fodder availability and the use of indigenous feed resources, especially for the poor, landless and pastoralists. Poor people need technical, infrastructural and institutional support to benefit from livestock market opportunities.

Similarly, a number of social, economic and institutional issues must be confronted if aquaculture is to fulfill its potential in the ESAP region. Since the 1990s NGOs, researchers and environmental groups have focused on the environmental and social impacts of aquaculture. The criticisms, centered on concerns about mangrove destruction, pollution and social conflict generated by aquaculture, were convincing enough to cause funding to India and Thailand to be halted in the 1990s. Recent solutions to address these issues tend to favor capital-intensive technical interventions, sidelining small farmers. Furthermore, they suggest farm-level solutions, which may reduce environmental impact at that level, but overlook interactions with activities in related sectors. Policies for aquaculture need to be cross-sectoral, integrated and wide-reaching (Dene, 2005), with policy guidelines that specifically target the poor to encourage development and explore participatory community management in aquaculture (Edwards, 2000).

As the resource systems in question are both complex and dynamic in their biophysical and human aspects, it is not always possible to understand how a system works or to predict the outcome of management actions (Arthur, 2005). In such circumstances, the standard approach of government guidelines based on "best practice" in management is unhelpful since not only are best practices uncertain or unknown but the resources to implement them are also lacking. An "adaptive learning" approach takes these constraints as a starting point and seeks to build on whatever knowledge is available with the aid of planned management experiments and the development of knowledge sharing networks which seek to reduce uncertainties. This approach has yielded fruitful results in the rice-fish systems of West Bengal, India and in the fisheries, including reservoirs, in Lao PDR, Cambodia, Vietnam and Thailand (by MRAD Ltd., WorldFish Centre, Mekong River Commission, Indian Central Inland Fisheries research Institute and the State Government of West Bengal).

The challenge for post Green Revolution crop, livestock and aquaculture systems in the ESAP region is to improve productivity without the negative ecological and social side effects experienced during and after the Green Revolution. It needs to address the problem of diminishing supplies of oil and escalating prices of fuels and petrochemical products such as fertilizers and pesticides, finding and incentivizing ways to minimize these inputs—for health, environmental, as well as economic reasons. Similarly, incentives for improved water use efficiency, rain harvesting measures, training, credit and infrastructure (e.g., for cheaper fuel and energy) for increasing organic inputs are critical as the natural resource base in the region becomes more oversubscribed and degraded. More attention and investment in the public and private sectors needs to be devoted to integrated pest and nutrient management (IPM and INM) technologies. These approaches hold promise for optimizing agricultural productivity and environmental sustainability while minimizing adverse effects on human health by combining low input approaches with the judicious and timely use of reduced chemical applications.

5.3.4 Transgenic technology or "the gene revolution"

Humans have been knowingly or unknowingly modifying the genetic makeup of plants for thousands of years, but transgenic technology to produce crops, pharmaceuticals (pharming), food vaccines and genetic use restriction technology (GURT) is one of the newest and most controversial developments. Transgenic technology uses genetic engineering to produce crops with a variety of properties, including herbicide tolerance and insect resistance, micronutrient enhancement and vaccine production. Despite arguments for a cautious approach as with any extensive change in agricultural practices (NAS, 2003), land planted to transgenic crops is expanding rapidly (James, 2005). Often fourteen countries in which more than 50,000 ha are planted to transgenic crops, nine are "resource-poor", of which three are in the ESAP region: China, India and the Philippines (James, 2005).

Transgenic crops can increase agricultural production (Peng et al., 1999; Taylor et al., 2001; Regierer et al., 2002) e.g., through decreased loss to pests. By replacing chemical sprays to control pests, insecticide-resistant crops such as those with insecticidal genes from the bacterium *Bacillus thuringiensis* (Bt) can reduce or eliminate adverse effects of such insecticides on human and environmental health (Jeyaratnam, 1990; Gray et al., 1993; Gray, 2000; Huang et al., 2002; Qaim and Zilberman, 2003). On small-scale farms in China and India Bt cotton yields were significantly higher with pesticide use reduced by up to 70% (Huang, 2002; Qaim and Zilberman, 2003; Hossain et al., 2004; Wu and Guo, 2005). Herbicide-resistant crops presuppose the availability of affordable herbicide and represent an economic risk for farmers who are dependent on seed and chemicals. From the perspective of a mechanized agroecosystem, use of herbicide-tolerant crops allows reduced and zero-till practices to work more effectively. The resulting decrease in soil disturbance is beneficial for retaining soil organic matter, improving soil structure, reducing soil compaction and improving soil water relations.

Transgenic technology is being used to develop crops resistant to abiotic stresses such as drought, soil acidity and salinity, although the value of these modifications in the field has yet to be established (de la Fuente-Martinez, 1997; de la Fuente-Martinez and Herrera-Estrella, 1999; Liu et al., 1999; Zhang and Blumwald, 2001; Zhang et al., 2001; Garg et al., 2002; Singla-Pareek et al., 2003). Work on improving storage stability and manipulating ripening and processing-related factors aims to provide improved storage stability, delayed ripening and other changes to increase flexibility in distribution and/or facilitate juicing or other processing for greater economic benefit. Much of this work has been limited to transgenic tomatoes (Fromm et al., 1993; Grierson, 1994; Picton et al., 1995; Kalamaki et al., 2003ab; Powell et al., 2003) and potatoes (Greiner et al., 1999). Advances in resistance to abiotic pressures may allow crops to be grown in marginal, low productivity areas, while increased storage stability and delayed ripening will benefit those with few resources to invest in refrigeration and other equipment to increase the shelf life of agricultural produce. Micronutrient-enriched staple crops have also been developed to target the most vulnerable—resource-poor women and children (Combs Jr. et al., 1996; Bouis, 2000).

There is increasing interest in the potential for transgenic plants to produce pharmaceuticals and vaccines through molecular farming (pharming). Vaccines can be used to prevent or combat many of the diseases that cause illness and death in low-income countries, but are expensive, must be refrigerated and administered by trained personnel and require sterile delivery through needles that are often difficult and expensive to obtain. Although research in this area is in the early stages, vaccines against some infectious gastrointestinal diseases have been produced in potatoes, bananas and corn (Thanavala et al., 1995; Lamphear et al., 2004). Transgenic plants are also being evaluated for a variety of non-food applications, including bioremediation, modification of fiber content and biodegradable plastics (Haigler and Holaday, 2001; He et al., 2001; Poirier, 2001; Scheller et al., 2001).

Genetic use restriction technologies (GURT) are based on regulating gene expression to restrict plant propagation from a second generation of seed. Unlike variety-specific (V-GURT; "terminator technology") which results in sterile seed, specific trait, or T-GURT would enable farmers to save their own seed, but lack access to the added traits in the absence of payment for chemical activators. In addition to their use restriction properties discussed above, GURTs represent an environmental containment strategy through their ability to eliminate the spread of transgenic seed (V-GURT) or transgenes to wild plant relatives (T-GURT).

Potential productivity advantages from GURTs include T-GURTs enabling a farmer to restrict the expression of a trait when there is an advantage in doing so; for example, at a specific phase of development or during periods of biotic or abiotic stress (FAO, 2001). Productivity gains from V-GURTs include the ability to safeguard the integrity of adapted maternal breeds or to prevent preharvest sprouting. As with any growth regulator applied to crops, environmental or human health issues may be associated with the use of chemical activators (i.e., tetracycline, copper, steroids) and hence the effects of these need to be addressed.

Transgenic crops and GURT offer a promising means to increase agricultural productivity in cropping systems. However, these technologies have the potential to affect human and animal health, have substantial social and economic impacts on grower communities and can significantly alter agricultural ecosystems through effects on the environment. Despite human health concerns, several studies with animal models and a range of transgenic crops have failed to show any toxic, allergenic, or nutritional effects of the transgenic crop tested (Noteborn et al., 1995; Hammond et al., 1996; Harrison et al., 1996; Hashimoto et al., 1999ab; Folmer et al., 2000; Momma et al., 2000; Sidhu et al., 2000; Teshima et al., 2000; Ash, 2003; Donkin et al., 2003; Stanford et al., 2003; Hammond et al., 2004; Kan and Hartnell, 2004). These findings in no way negate the need to apply rigorous standards to health risk assessments of individual technologies; in addition, comparative risk assessments with alternative control regimes will be needed.

Transgenic and genetic use technologies have the potential to increase economic returns via improved crop yields under stress conditions and reduced chemical inputs, while preventing the spread of transgenes in the case of GURT applications. Transgenic technology can significantly affect the cultural and economic situations of producers—as can conventional plant breeding—but at a faster rate than the latter. The threats of biodiversity reduction through "genetic pollution" and "superweed" creation are scenarios with far-reaching negative consequences for livelihoods and cropping systems. Further, the technologies are expensive and commit farmers to regular financial outlays for improved seed or chemicals each season that may not be achievable.

Potential environmental effects of transgenic technology include: (1) adverse effects on non-target organisms, (2) gene flow into wild plant communities or soil organisms and (3) development of resistance by target pests. Non-target entomophagous insects and parasitoids are invaluable in integrated pest management approaches employed to control outbreaks of insect pests. Most of the insecticides used for the control of rice stem borers and leaffolders have been found to harm beneficial insects, while multiple Bt rice lines show no significant non-target effects (Chen et al., 2006). An evaluation of direct toxicity or indirect food chain-related effects on a large variety of insects and crops indicates no adverse impacts (Lopez and Ferro, 1995; Orr and Landis, 1997; Pilcher et al., 1997; Riddick and Barbosa, 1998; Volkmar et al., 2000; Zwahlen et al., 2000; Bourguet et al., 2002; Cowgill et al., 2002; Al-Deeb et al., 2003; Cowgill and Atkinson, 2003; Dutton et al., 2003; Jorg et al., 2003; Volkmar et al., 2003). Although most of the evidence suggests that transgenic crops do not have direct effects on beneficial insects, adverse effects of Bt proteins on beneficial insects via compromised food quality of their prey have been reported (Hilbeck et al., 1998; Schuler et al., 1999; Meier and Hilbeck, 2001; Ponsard-Sergine et al., 2002) and transgenic corn resulting from event 176 had adverse effects on Lepidoptera (butterflies), arguing for case-by-case evaluations (Losey et al., 1999; Jesse and Obrycki, 2000; Hellmich et al., 2001; and Stanley-Horn et al., 2001; Zangerl et al., 2001).

Although a variety of transgenic crops have demonstrated changes in microbial, protozoan and nematode populations in soil (Donegan et al., 1995; Di Giovanni et al., 1999; Donegan et al., 1999; Griffiths et al., 2000; Lukow et al., 2000; Marroquin et al., 2000; Cowgill et al., 2002), the data are difficult to interpret and tie to ecosystem function and a large number of studies have shown no effect on these soil populations or their processes (Al-Deeb et al., 2003; Blackwood and Buyer, 2004; Devare et al., 2004; Wu et al., 2004; Fang et al., 2005; Devare et al., 2007).

Gene flow between crops and their wild relatives is common and, between plants capable of hybridizing, inevitable if grown within the crop's pollen dispersal range (Ellstrand et al., 1999). Pollen-mediated crop-to-crop transgene flow in rice can be maintained at negligible levels with short spatial isolation (Rong et al., 2007). Insect resistance to Bt has been demonstrated in laboratory trials, but despite Bt crops being grown on more than 15 million ha worldwide, an increase in the frequency of resistance caused by exposure to Bt crops in the field has not yet been reported (Fox et al., 2003; Tabashnik et al., 2003). Resistance has been slowed in pest populations through high doses of Bt toxin expressed in plant tissues. This decreases the likelihood of survivorship, ensuring that insects are not exposed to sublethal doses that might promote development of resistance. This high dose strategy has been combined with the use of refugia which serve to maintain susceptible insect populations that delay resistance (Roush, 1994; Tabashnik, 1994).

5.3.4.1 Impact

In 347 Chinese rice fields farmers growing non-transgenic rice, small and poor farm households benefited from 6-9% higher transgenic rice yields and reduced pesticide use by 80% (Huang et al., 2005). The negative impacts of insecticide use on the environment and on the control of pests by beneficial insects have been extensively documented (Way and Heong, 1994; Pingali and Roger, 1995). Recent reports on several farming systems incorporating these insect resistant (IR) crops confirm increases in beneficial insects and a return of songbirds (FAO, 2004; Morse et al., 2004). IR crop growers are likely to also see savings in fuel consumption and decreased greenhouse gas emissions (Phipps and Park, 2002).

Micronutrient-enriched crops can do much to improve human health in the region. While the consumption of green-leafy vegetables and unpolished or brown rice can mitigate the effects of vitamin A deficiency which is common throughout the ESAP region and causes over 1 million deaths and partial or complete blindness in 0.25-0.5 million of the poorest children each year (Conway and Toennissen, 1999; Hunter, 2000), these are not always consumed in the region for a variety of reasons including lack of irrigation, poverty and lack of knowledge (Mishra, 1996; Pee et al., 1998; Khadka, 2001; Faruk et al., 2003). Rice, chosen because it is a staple in the ESAP region and engineered to overproduce pro-vitamin A or beta-carotene can greatly reduce Vitamin A deficiency (Ye et al., 2000). Rice is also being targeted to address iron deficiency, identified as a contributing factor in over 20% of post-partum maternal deaths in Asia and Africa (Conway, 1999; Goto et al., 1999; Lucca, 1999). Transgenic crops can potentially have direct positive effects on health and nutrition through the elimination of toxins, e.g., cassava (Siritunga et al., 2003) and increases in nutrients or antioxidants (Regierer et al., 2002; Bovy et al., 2002).

5.3.4.2 Challenges

Transgenic technologies can increase crop yields and lower pesticide use, but can also threaten livelihoods and cropping systems if biodiversity is reduced and farmers do not profit after committing scarce resources to these expensive agricultural packages. In many ESAP countries where seed saving is common practice, farmers can save traditional seeds and other public varieties, but would have to expend resources in the purchase of transgenic seed. While economic loss from the inability to save seed may be recouped by gains from increased yield and reduced input costs, transgenic crops have the potential to confer an economic advantage only in the presence of the stress they are engineered against. If the risk and debts incurred by growers of transgenic crops are to be minimized, access to low-interest loans needs to be greatly improved and market and infrastructure instabilities effectively minimized.

Most regulatory frameworks dictate that transgenic crops not be grown in areas where wild relatives are endemic and often advocate a high dose strategy and the use of refugia. These precautions are considered to limit gene flow and delay resistance development in target insect populations and will need to be enforced. This is likely to be a challenge in many ESAP countries that do not have strong regulatory and monitoring infrastructures. Thus, the risk of out-crossing is likely to be increased in these countries. A case in point is the "stealth seeds" phenomenon in India, which involved farmers multiplying, crossing and selling Bt cotton seed which was viewed as desirable but expensive to legally obtain. Although conventional breeding has already led to significant reductions in biodiversity via the replacement of land races with hybrids, such a result is of at least equal concern for transgenic crops. ESAP countries urgently need to invest in and enforce stringent biosafety protocols to assess the risk of gene flow and mitigate the cultural and biodiversity implications of "genetic pollution" and the potential ecological consequences accompanying the creation of "superweeds". To be effective, these protocols will need to include local agronomic, socioeconomic and ethical considerations.

5.3.5 Nanotechnology

Nanotechnology is defined as the "the design, characterization, production and application of structures, devices and systems by controlling shape and size at nanometer scale." (UK Royal Society and Royal Academy of Engineering, 2004). There are growing concerns regarding the safety and the long-term impacts of nanotechnology on food and agriculture and its long-term prospects in large-scale replication and production of biological material such as proteins. The specific genetic modification of life-forms through nanotechnology to give enhanced properties is also a subject of heated debate (ETC Group, 2004; UK Royal Society and Royal Academy of Engineering, 2004). International civil society groups monitoring developments in nanotech project that the impacts of nanobiotechnology in food and agriculture will surpass that of the Green Revolution and farm mechanization in a couple of decades.

The agricultural application of nanotechnologies is primarily focused on the seed and pesticide industries. In the area of seed development, researchers are experimenting on techniques that use nanoparticles to manipulate native DNA or insert foreign DNA into plant cells without being passed on the next generations, thus avoiding cross-pollination concerns (Dalke, 2003). Thai researchers at Chiang Mai University drilled a hole through the membrane of a rice cell to insert a nitrogen atom that would stimulate a rearrangement of DNA, resulting in grain color change and potentially other characteristics such as maturity period (Ranjana, 2004). The world's major agrochemical companies have already started commercially marketing pesticides formulated at the nanoscale, which are claimed to be more easily dissolved in water, more stable, optimally target pests and be completely absorbed in the plant's system (ETC Group, 2004). Some of these pesticides are emulsions containing nanoscale droplets and microencapsulated formulations.

The convergence of nanotechnologies with information and communication technologies also has wide applications in agriculture. One of these applications is precision farming, the site-specific farm management involving a bundle of new information technologies applied to the management of large-scale, commercial agriculture (ETC Group, 2004). The US Department of Agriculture (USDA) is developing a "Smart Field System" that automatically detects, locates, reports and applies water, fertilizers and pesticides through nanotechnologies. Companies and the public sector in the US are experimenting on the potential of "smart dust," which involves the development of autonomous matchhead-size sensors with the ability to detect light, temperature and vibration, communicate with other sensors or "motes" in the vicinity and self-organize into ad hoc computer networks capable of relaying data using wireless technology (ETC Group, 2004). Smart dust or nanosensor technology is already being applied in engineering and microclimate sensing. Other emerging nanotechnology applications with long-term implications for AKST include the development of nanowater which involves the use of nanotubes to filter pollutants and saline particles from water for human consumption and agricultural uses.

Despite the fact that some products of nanotechnology have already reached the commercial stage, there are few studies on the potential health and environmental impacts of nanotechnologies. Nanoparticles can be inhaled, ingested or pass through the skin. Once in the bloodstream, there are concerns that nanoparticles can elude the body's immune system and penetrate the blood-brain barrier (ETC Group, 2002). The increased reactivity of nanoparticles could harm living tissue, perhaps by giving rise to "free radicals" that may cause inflammation, tissue damage or tumors (ETC Group, 2005).

Buckyballs (precursor of nanotubes) can cause rapid onset of brain damage in fish (Oberdörster, 2004). Researchers at the US National Aeronautic and Space Administration (NASA) reported that when injecting commercially available carbon nanotubes into the lungs of rats caused significant damage (Raloff, 2005). In other studies, researchers reported substantial DNA damage to hearts and aortic arteries of mice exposed to carbon nanotubes and increased susceptibility to blood clotting in rabbits inhaling nanotech buckyballs (ETC Group, 2004). Buckyballs clump together in water to form soluble nanoparticles and even in very low concentrations can harm soil bacteria, raising concerns about how these carbon molecules will interact with natural ecosystems (ETC Group, 2004). In recognition of the knowledge gaps and the health concerns arising from available toxicological studies, factories and research laboratories should treat manufactured nanoparticles and nanotubes as if they were hazardous and reduce them in waste streams. The use of free nanoparticles may need to be prohibited in environmental applications such as groundwater remediation (UK Royal Society and Royal Academy of Engineering, 2004).

5.3.5.1 Impact

Research and development in nanotechnologies have been receiving substantial investments from both the public and private sectors. Most of the world's major seed and agrochemical companies have a substantive stake in nanotech research and development. The European Commission estimates the current global investments in nanotechnologies at around €5 billion, with 40% from the private sector (UK Royal Society and Royal Academy of Engineering, 2004). Civil society groups monitoring development in nanotechnologies have placed the combined investments of the public and private sectors in 2004 at $\$10 \times 10^9$ (ETC Group, 2004). The projected value of nanotech products by 2011-2015 is estimated at around US$1 trillion. Patents involving nanotechnologies have jumped from 521 in 1995 to 1,976 in 2001 (UK Royal Society and Royal Academy of Engineering, 2004). In the ESAP region, Japan, China and India are the leading investors in nanotechnology. Japan invested US$800 million on the technology in 2003, while India has allocated US$22.8 million under its current 5 year plan (Barker et al., 2005). The substantial financial and technological investments required in nanotechnology applications limit the capacity of many developing countries in the region to tap its potentials in agriculture.

Some nanotechnology applications such as the development of nanowater have great potential to improve water for human and agricultural uses. The Indian Institute of Technology (IIT) is about to commercialize its nanofilter technology in water purification at the household level, while US universities and the International Water Management Institute (IWMI) are advancing research to remove arsenic from groundwater in Bangladesh to render it potable (Barker et al., 2005). Future developments in the search for cheaper and renewable energy sources through the use of nanotechnologies may also have strategic implications on AKST in the region.

5.3.5.2 Challenges

An imminent concern over nanotechnology for the ESAP region is its impact on trade in agricultural commodities. Nanotech products in the global market such as synthetic textiles and nanotech rubbers are projected to provide stiff competition and affect world prices, posing a threat to cotton and rubber industries and the livelihoods of millions of farming families in the region.

Most of the social and ethical concerns surrounding nanotechnologies revolve around control, transparency and governance. While governments in industrial countries have substantially invested financial resources in nanotechnology research and development, the private sector has a significant advantage through products already in the market and/or the pipeline. More policy attention is required on regulation and standards at the national and international levels and on controversial social and ethical issues such as their role in the modification or production of living material.

ESAP governments that decide to adopt nanotechnology for agricultural development will need to take into account its potential risks to human, animal and environmental health, as well as its socioeconomic and ethical implications. Adequate precautionary measures will need to be put in place, from the production process to commercialization and consumption. Biosafety measures must be formulated and adopted for nanotechnology research and development activities with the active participation of all stakeholders to ensure that socioeconomic and ethical concerns are addressed. Countries may consider adopting a legally-binding international standard or protocol on the assessment of nanotechnology products and processes to ensure safety for human and animal health and the environment.

5.3.6 Precision agriculture

Precision agriculture is a comprehensive system designed to optimize agricultural productivity while minimizing production costs, fertilizer, pesticide and water inputs and adverse environmental effects through the application of crop information, advanced technology and management practices (NRC, 1997). The main ideas behind precision agriculture are understanding spatial variability of soil properties, crop status and yield within a field; identifying the reasons for yield variability; making farming prescription and crop production management decisions based on variability and knowledge; implementing site-specific field management operations; evaluating the efficiency of treatment; and accumulating spatial resource information for further management decision making (Wang, 2001). Simply put, this translates to using the appropriate inputs at the optimal times in the appropriate ways.

Precision agriculture may include the integration of geographic information systems (GIS) or remote sensing technology with farm management and technologies to improve crop and livestock production in terms of product quality, environmental issues and the welfare of people and livestock (Cox, 2002). The suite of technologies currently used include GIS hardware and software, variable-rate application equipment for seed, fertilizer and pesticide, grid soil sampling, low-volume irrigation, soil fertility and weed population sensors, yield monitoring capability and remote sensing imagery. Attention to soil quality, efficient water management, IPM, INM and efficient postharvest management are all important in precision farming (Persley and Doyle, 1999).

5.3.6.1 Impact

Most formal and informal agricultural research and extension systems in the ESAP region currently provide blanket fertilizer and pesticide recommendations for large productions areas. Yet, on-farm studies in double and triple-rice cropping systems in India, Indonesia, Thailand and Vietnam present evidence of huge field-to-field variation in native soil nitrogen (N) supply where the variation was not associated with soil organic matter content, total N, or other measures of soil N availability (Olk et al., 1999). Given this variation, site-specific nutrient and pesticide management is more efficient and could have positive economic and human and environmental health benefits (Cassman, 1999). Although there is a potential for benefits from precision agriculture, these have not yet been well-documented (Auernhammer, 2001). As most of the technologies involved are expensive as well as data- and knowledge-intensive, their implementation and adoption is likely to be slow and variable in developing countries of ESAP.

5.3.6.2 Challenges

Adoption of precision agriculture has been relatively slow even in developed ESAP countries such as Australia, Japan and New Zealand. Causes of slow adoption in Australia include: (1) cost of adoption, (2) lack of perceived benefit from adoption, (3) unwillingness to be early adopters and (4) lack of a technology delivery mechanism (Cook et al., 2000). These obstacles are likely to impede adoption in developing countries, but lack of reliable information and data (GIS coverage, satellite imagery, soil maps) as well as expertise, equipment and small land holdings can also represent important barriers.

As with the dissemination of almost any new technology-oriented intervention, precision agriculture is likely to be adopted first by resource-rich farmers in areas with high yield potential, with poorer farmers and particularly women, benefiting later, if at all. However, it may be possible to improve rural livelihoods in rainfed and marginal areas in the ESAP region by disseminating elements of precision agriculture that do not call for sophisticated technologies like GIS, but rely on quick, easy to use, cheap tests and measures of soil, crop and pest infestation parameters. Such low-investment, low-technology interventions could improve production efficiencies through a combination of conservation agriculture practices such as IPM and INM and efficient postharvest management. Public and private sector investment will be needed to develop scientific capacity and technology transfer and support mechanisms (Cassman, 1999). It will also require educating extension agents and farmers to use locally-adapted seed, diagnose limiting factors, predict yields and input requirements and modify management regimes accordingly.

5.3.7 Information and communication technologies (ICT)

Limited access to information has been a major hurdle facing low-income farmers, extension agents, civil society organization workers and others in the agricultural sector throughout the ESAP region. Possibilities for ICT application in agriculture include facilitating the access of rural communities to information on efficient farm management and the market through radio and TV shows as well as computer kiosks. GIS and related tools provide information on land use/land cover, water quality, productivity, tidal influence and coastal infrastructure, and can increase the efficiency and sustainability of coastal fisheries and shrimp farming (Rajitha et al., 2007). New and emerging agricultural technologies depend heavily on advances in ICT and would not be possible without applications that support high-throughput genetic and genomic work and the manipulation, analysis and interpretation of large sets of data. Further, ICT has resulted in "knowledge management" (KM), the creation, dissemination and utilization of knowledge by combining organizational dynamics and knowledge engineering with ICT (Flor, 2001). Much of the KM experience has been limited to the private sector, but organizations such as the World Bank, FAO and CGIAR have also launched initiatives.

The proliferation of ICT in the form of radios, telephones, televisions and computers and more recently, GIS, remote sensing and the use of information technology in climate forecasting and modeling is benefitting ESAP agriculture. Improved climate forecasting is essential if effective coping strategies are to be devised (Verdin et al., 2005). ICT can facilitate the rapid transfer of technology between regions within ESAP, or be utilised to develop a Decision Support System (DSS) for sustainable land management at the farm, village, watershed and regional scale (Rais, 1997; Craswell et al., 1998).

5.3.7.1 *Impact*

The effective dissemination and adoption of ICT in the region can be critical to economic and social development in this information age—akin to electrification in the industrial age (Castells, 1999). The correlation between access to information and poverty is widely acknowledged, for reasons outlined by Flor (2001): information leads to resources; information leads to opportunities that generate resources; access to information leads to access to resources; and access to information leads to access to opportunities that generate resources. Data from the Human Development Report (UNDP, 1999) database on four major ICT indicators (internet hosts per 1000 persons, telephone lines per 1000 persons, personal computer ownership and television ownership) were compared with the human development index (HDI) and the human poverty index (HPI) for ten ESAP countries, showing a strong correlation between ICT and poverty. The higher the HPI, the lower the number of ISPs, telephone lines, personal computers and television sets per 1000 persons. Similarly, the higher the ICT indicator value, the lower the HPI—as in the case of Singapore, Brunei and Malaysia.

Research from Thailand, South Korea, India and China also indicates that growth rates of ICT consumption correlate well with growth in productivity and GDP (Mody and Dalman, 1992; Kraemer and Dedrick, 1998). Research for India in particular—which showed an average growth rate in IT-related investment of 22.21% per year—shows a strong correlation between growth in ICT consumption and productivity and GDP between 1984 and 1990 (Kraemer and Dedrick, 1998). ICT can have an impact on poverty alleviation and meet development and sustainability goals by educating rural and disenfranchised communities about their circumstances and rights and by providing access to health-related and agricultural and natural resource management information, among other issues (Pigato, 2001).

Knowledge management offers great promise to poverty alleviation initiatives in ESAP countries, particularly in the realm of policy, strategic planning and monitoring and evaluation. For instance, Southeast Asia, home to the two largest archipelagos in the world, Indonesia and the Philippines, is rich in marine biodiversity. A regional knowledge network enabled by information and communication technology is now being proposed to share best practices and lessons learned from these projects and more effectively address marine resource depletion.

International organizations have forged linkages to facilitate database information collection and sharing within the ESAP region through the creation of decision support systems, expert systems and similar initiatives to help farmers and other users make informed decisions. Two prominent examples of such efforts are the International Rice Research Institute's internet or computer-based "Rice Doctor", a field diagnostic to identify factors limiting rice crop growth in the Tropics. Another is the multi-agency (GTZ, DFID, ACIAR, CSIRO, CIAT, ILRI and Queensland Government) "Tropical Forages"—an interactive, computer-based selection tool that brings together in one package accumulated information on the adaptation, use and management of tropical and subtropical forage species.

Yet another notable knowledge initiative to improve the lives of Filipino farmers and fisherfolk is the K-AgriNet program (Knowledge Networking Towards Enterprising Communities), jointly implemented by a number of government agencies. It aims to modernize the agricultural sector by utilizing ICT to access information, modern technologies and indigenous knowledge. The Philippine Council for Agriculture, Forestry and Natural Resources Research and Development (PCARRD), with its e-Consortia and e-Farm components establishes a digital community by interconnecting the Farmers Information and Technology Service (FITS) centers that serve as the ICT/ICE kiosks for agricultural information, technology and other interventions.

ICT has allowed women and disadvantaged groups in the ESAP region more scope to project themselves. For example, the organization IT for Change, in partnership with Mahila Samkhya (women's collectives) is implementing a development strategy called Mahiti Manthana to institutionalize ICT-based innovations in the state of Karnataka (India) using radio, video and telecenter-based initiatives. This initiative focuses on adult learning, health care, the responsibilities of Panchayats (village governance bodies) and gender mainstreaming in local development processes (IT for Change Report, 2006). However, despite such benefits, uneven distribution and access to ICT has given rise to the "digital divide," a widening gap between the information-rich and information-poor, creating yet another form of social exclusion.

Despite the digital divide being a concern, there has been some success in ensuring equitable access to ICT within the region. The founder of Bangladesh's Grameen Bank, Professor Muhammad Yunus, embarked on an innovative undertaking based on a simple idea (Flor, 2001). The bank has initiated a cellular phone project, through which the Grameen Phone Company has put mobile phones in approximately 45,000 Bangladeshi villages, greatly increasing villagers' access to ICT. Each mobile phone is acquired through a small bank loan and becomes a village telephone service provider, earning income for the owner and providing a much-needed utility to the community. This endeavor is being followed with an experimental Village Computer and Internet Program or VCIP, which would provide a very low-cost email and Internet service to villagers. A simple form of e-commerce will also be initiated by this system, allowing farmers to check market prices and study the list of wholesalers in Dhaka by surfing the Web.

In a similar development, the Philippines government established a Short Message Service (SMS) using cellular phones to gather and disseminate knowledge about agriculture. The Philippine Rice Research Institute established the

"Open Academy for Philippine Agriculture" and embarked on e-learning modules for basic and advanced agricultural practices. Lifelines India (Soochana se Samadhan), uses a phone-based platform and a web-based interface to respond to the agricultural information needs of small-scale Indian farmers. By so doing, it bridges the information gap that currently prevents them from leveraging sustainable and efficient farming methods, integrating with global markets and improving the quality of their lives.

Another innovative program is the Village Knowledge Centers set up in 1998 by the M.S. Swaminathan Research Foundation in 22 villages in southern India. Two major features of the project were the development of Tamil language Software and gender sensitivity in assessing the information needs of local people. The National E-Governance Action Plan of the Government of India has emphasized community computer and internet kiosks in rural areas. For example, in the state of Tamil Nadu alone, 600 rural kiosks are functioning to deliver services such as health care, education, agriculture and communication (voice mail and e-mail). Most of these kiosks are run by self-employed entrepreneurs and NGOs; but women SHGs are increasingly beginning to manage them. However, existing social and cultural norms can be a major obstacle to the diffusion of such IT knowledge. Recent research in Tamil Nadu noted that although the technology itself may be gender-neutral, women often lack the literacy, time, decision-making power and financial resources to fully avail themselves of the kiosk services (Best and Maier, 2007).

A last example from Thailand involves the struggle against the forced relocation of Akha villagers in Huay Mahk village of Northern Thailand in 2000. The problems of this isolated people were communicated via internet to mobilize support and the villagers were able to retain their homes, lands and forests, thereby avoiding the drugs, crime and poverty cycle that has affected other relocated communities in the region (Satyawadhana, 2001).

5.3.7.2 Challenges

The biggest barrier to regional knowledge networking is the lack of de facto standards for information sharing and data exchange. This barrier will need to be eliminated if technology transfer and sharing of best management practices across the region is to be facilitated for the common good at a relatively low investment from any given country. Knowledge-based products, services and systems will need to be coordinated and integrated into usable formats such as: information, education and communications materials (publications and instructional materials at the community level—flyers, primers, technology bulletin); information management services (database and subject information networks); and documented records of local knowledge in the form of stories, case studies and plays.

In pursuing the fruits that ICT offers, it is important that farmers' information needs are not overshadowed by technological enthusiasm (Meera, 2004). Although they function within the dominant interests of the market and the state, ICT can help redefine traditional, social, cultural and gender norms within the region and support a media of information, understanding and knowledge in which interests, voices and rights of the rural poor, women and excluded social groups are taken into account, but only if they are equitably and widely disseminated (Kelkar et al., 2005; Ng and Mitter, 2005; Patel, 2006). Initiatives such as those described above provide models which are well worth pursuing throughout the ESAP region.

5.4 Technologies: Options for Achieving Development and Sustainability Goals

It is important to recognize that there is no one ideal agronomic/forestry/aquaculture system. For instance, modern best practice guidelines for conventional production systems advise the full use of all indigenous fertility sources (composts, crop residues and animal manures), with mineral fertilizers employed as a complement to bridge deficits between crop needs and indigenous supplies (http://www.knowledgebank.irri.org/ssnm/). The ideal agricultural system may be context-dependent, combining elements of traditional, organic, conventional and emerging practices and technologies in a locality and constraint-specific manner to maintain food security and improve micronutrient content of food with minimal adverse environmental consequences and maximum improvement in social inclusiveness. One mitigation strategy to cope with climate change-induced variability and extreme events is to increase the resilience of agricultural systems (Scheffer et al., 2001) through crop diversification, shifts to low input, no-till or organic agriculture where possible and the use of crop species and varieties that can withstand extreme weather conditions. The integrated use of inorganic fertilizers, green manures and organic materials has been proven to increase productivity and yield stability for a variety of vegetable and grain crops over that attained from using only inorganic or only organic inputs (Kanwar and Rego, 1995; Tandon, 1995; Satyanarayana et al., 2002; Paul et al., 2005; Chatterjee et al., 2006; Singh, 2006; The et al., 2006).

While organic inputs are desirable, significant trade-offs need to be contended with and addressed when considering policy options that favor such inputs. For example, replacing inorganic fertilizer with manure or a combination of manure and compost at recommended rates for rice generally presupposes the availability of close to 10 tonnes ha^{-1} of the manure and compost. Further, manure is generally subject to competing uses throughout much of the rural ESAP region: more used in the field generally means less available as cooking fuel. Similarly, green manures are an excellent way to add organic matter and nutrients to the soil; however, such a cover crop requires sufficient land holdings to dedicate to a non-food or cash crop and is too often not a practicality for small holders or subsistence farmers.

The integration of local knowledge and socioecological context in the design of agricultural technology are critical if sustainability and efficiency are to be attained and technological interventions are to succeed. A clearer understanding of various traditional or organic systems would be valuable, enabling the utilization of this knowledge to develop more sustainable production systems for use by future farmers. In contrast with the knowledge systems generated by universities, research institutions and private firms, indigenous knowledge is usually seen as residing locally within the farming community where it is applicable, with transfer of knowledge from farmer to farmer rather than scientist and extension agent to farmer. New paradigms in agricultural extension programs have recognized that local people conduct research on their own farms; it is even argued that their experiential knowledge is derived from their skills as experimenters (Stanley and Rice, 2003).

Although the absence of formal and rigorous methodology employed by formal institutions such as robust statistical design make analysis and interpretation of farmer experimentation difficult, such paradigms are increasingly becoming acceptable to the scientific community. An example is the agroforestry project of the International Institute of Rural Reconstruction (IIRR) in the Philippines, in which scientists worked with village farmers after a failed nursery operation that relied on exotic species. Local people were then asked to identify locally growing species (indigenous and introduced) according to criteria considered by the community as important—such as hardiness, fire resistance, general utility and seed availability. The exercise resulted in the formulation of community action plans for reforestation (IIRR, 1996).

Achieving sustainable agriculture in the region will require the integration of agricultural knowledge systems, practices and technologies through the provision of financial and infrastructural support to facilitate research, dissemination and constraint-specific utilization of all available technologies. Specifically, this will include but not be limited to:

- Increasing support for and applying traditional and low-input systems that function productively and in a socially inclusive manner, particularly in low productivity areas throughout the ESAP region.
- Augmenting indigenous knowledge with appropriate modern practices that can enhance the system, such as microbial inoculations, appropriate scale mechanization and small-scale technology—such as gravity-fed technology for sprinkler and drip irrigation for vegetable and fruit cultivation.
- Integrating elements of traditional and organic practices (such as rotational, trap and intercropping and agroforestry) in modern agricultural settings to maximize system productivity and natural resource sustainability.
- The application of emerging technologies such as bio and nanotechnology that have been rigorously evaluated in comparison with existing or developing technologies not only on the basis of potential gains in productivity, but on their ability to maintain ecosystem integrity, human and animal health and social and economic wellbeing in adopting communities and countries.

It is critical that unbiased science precedes, rather than follows the commercialization of new agricultural technologies such as genetic engineering and nanotechnology applications. Each new transgenic product has potential for unintended impact; hence each new technology should be evaluated on a case by case basis (NAS, 2000), acknowledging farmer needs and conditions. A stringent biosafety framework, enforcement, and rigorous site-specific scientific and social monitoring protocol are needed. ESAP countries intending to implement new technologies will need to ensure that their infrastructure is sufficient to support the safe development, transfer and application of the technologies with special attention paid to developing relevant policies, information systems and training in biotechnology risk assessment and biosafety procedures. Gene flow and resistance management issues in particular, among others, still warrant caution and long-term monitoring in the field, along with a holistic assessment that includes analysis of the social and economic consequence of biotechnology adoption by farming communities.

Methods and results of environmental risk assessments could be shared between countries that have similar agricultural environments, thus reducing the burden of proof for any one country (Nuffield Council on Bioethics, 2003). Adoption of such a model in the ESAP region may include data sharing to satisfy regulatory needs and similar extensions of prior findings, all of which could substantially reduce unwarranted restrictions and improve the benefits of these technologies for resource-poor farmers, as long as their socioeconomic and cultural environments are also considered alongside the agricultural. While there may be reason to be optimistic about the potential for a variety of bio and nanotechnologies to be beneficial in increasing agricultural productivity while reducing some inputs, there is also an imperative to exert the highest scientific, regulatory and policy standards to ensure negligible long-term ecological and human impact prior to their deployment.

Facilitating the access of rural communities to information through a range of ICTs and knowledge bases can lead to an increased understanding of the consequences of various management practices and the adoption of appropriate and sustainable agricultural and aquaculture management practices. The use of ICTs can improve the reliability of climate forecasts and the prediction of extreme weather events as well as their likely effects on agricultural ecosystems and rural livelihoods—both of which are invaluable in devising coping strategies. The adoption of common regional standards for sharing information will facilitate technology transfer and proliferate best management practices across the ESAP region for a relatively low R&D investment on the part of any given country.

The ESAP region accounts for about two-thirds of the world's rural poor, primarily concentrated in sparsely populated arid or marginal lands and forests of South Asia (CGIAR, 1999; Fan and Kang, 2004). Data on poverty distribution by land type for India and China indicates that in the 1990s, over 80% and 60% of the rural poor lived in low potential or rainfed areas of India and China, respectively (Fan and Hazell, 1999; Fan et al., 2002). Despite high rates of out-migration from these areas, populations continue to grow, often with increasing poverty and degradation of natural resources (Hazell et al., 2000). Government and donor investment in the past favored areas of high yield potential, leading to better infrastructure, schools, health facilities, credit programs, production pricing policies and access to agricultural technologies and services in these areas (Fan and Kang, 2004). Funding for public agricultural research and development grew at an annual rate of about 8.7% in the 1970s; this slowed to 6.2% annually in the 1980s and has likely been decreasing since, as have investments in new irrigation infrastructure (Gruhn et al., 2000).

Improving agricultural productivity and halting the degradation of the region's natural resource base will require that appropriate institutional arrangements, policies and investments be put in place to address the combined utilization of a variety of tools and technologies in optimal as well as marginal areas. Both the rates and levels of investments will need to increase if technologies are to address development and sustainability goals in the region. Further, enabling a more integrated approach to the use of the suite of agricultural systems and technologies described above to improve agricultural productivity at minimal cost to human health and the natural resource base will require a clear understanding of the complexities and regional variations that govern agriculture in the ESAP region. Thus, to truly address development and sustainability goals, investments must target not only agricultural research, technologies, services and capacities, but also infrastructure interventions that improve credit availability, health facilities and programs, roads, markets and electricity and fuel access and affordability.

5.5 Gender Inequality and Social Exclusion in ESAP Agriculture

Social exclusion derives from exclusionary relationships based on power and hierarchy and intersects with other aspects of social disadvantage, such as gender, caste, ethnicity, religious minority status. It is a multidimensional process that prevents individuals or groups from access to institutions of governance, public services like health care and education and economic resources as well as factors of production. Given increasing feminization of agriculture, the widespread gender-based inequalities in access to and control of productive resources throughout the ESAP region will be discussed here. These inequalities are linked with women's lower access to employment opportunities, social structures and institutions of governance and public services, such as healthcare, education and training for skill development.

Many labor market inequalities, the process of informalization, the lack of security and voice and the discriminations to which particular groups are subject are aspects of social exclusion. Thus, effectively addressing social exclusion requires action within the labor market. Exclusion from formal employment may lead to open unemployment or to different forms of informal work and under-employment, denying the dignity of livelihood in numerous cases.

5.5.1 Feminization of agriculture

Although the phenomenon of increasing feminization of agriculture within the ESAP region has drawn policy attention in recent years, its causes, extent and impact on women and productivity have not been sufficiently considered in policy implementation. Insufficient attention has been paid to areas where women are most active, such as crops and vegetable cultivation, forest regeneration and wasteland and watershed development, resulting in women's contributions and concerns remaining invisible in planning and ignored in AKST institutions (Sujaya, 2006). Further, the stress on self-employment and dependence on credit in most land-based economic activities has meant that women, who are mostly landless in many ESAP countries, are generally not eligible for assistance.

Rural women have an important role in the livestock sector (such as animal care, grazing, fodder collection, cleaning of animal shed, processing of milk and sale of livestock products). However, their control over livestock and products is minimal and these activities have been conventionally viewed as an extension of domestic work. With some regional variations, women account for 93% of employment in dairy production in India, yet 75% of dairy cooperative membership is male (World Bank, 1991; Sujaya, 2006). Introducing taxes and limits to overgrazing and uncontrolled burning and deforestation (largely caused by the expansion of livestock sector) are likely to prove effective steps for increased productivity and sustainability of this sector.

The Gender Assessment Report of China (IFAD, 2005b) indicated that women constitute about 70% of the agricultural labor force and perform more than 70% of farm labor in China. A general pattern is that the poorer the area, the higher women's contribution, largely as subsistence farmers, who farm small pieces of land, often less than 0.2 hectares. In India, close to 33% of cultivators and nearly 47% of agricultural workers are women (Vepa, 2005) (Table 5-1). This feminization of agriculture is caused by increased casualization of work, unprofitable crop production and distress migration of men "for higher casual work in agriculture and non-agriculture sectors", leaving women to take up low paid casual work in agriculture (Sujaya, 2006). Throughout the region women are more likely than men to work in agriculture. Manufacturing tends to employ a fairly large number of women followed by trade, hotel and restaurant businesses.

The feminization of agriculture model in ESAP is determined by two major factors. First, compared to men, women have much poorer access to and control over productive resources and they have inadequate access to public services, such as training, extension and credit. Technologies are often designed for irrigated land in favorable areas where male farmers predominate, with poor farmers, mainly women, lacking access to credit and appropriate technologies. Second, rural society structure makes it difficult for all members of the household to migrate, since cities have even more limited resources for masses of asset-poor, who lack not only income but production-related assets, human capabilities, social capital and physical assets. Women constitute the majority of this group and when men leave to become temporary laborers in cities, they are left behind to take care of the land, children and elderly. Thus, they have the compounded burden of productive and reproductive work. Although its impact on agricultural productivity is unclear, increasing feminization of agricultural labor is likely to have deep and wide ranging effects. It may rank as one of the leading foci for AKST policies centered on capacity development of (women) farmers, extension outreach, training in agricultural technologies and women's effective rights to land, trees, water bodies and other assets.

5.5.2 Gender wage differentials

Gender wage differentials in agriculture and related industries are generally but not always due to differences in educational attainments and work skills between women and men (Zhao and Zhang, 1999; Gustafsson and Li, 2000).

According to the National Census of China in 2000, women have an average of 1.1 years less schooling than men, but the unexplained gender wage differential against women in rural areas has been relatively constant, with some decline in the period 1988-95 (Rozelle et al., 2002). The major share (93.5%) of the wage differential between women and men is attributed to discrimination rather than to capital differences between the genders (Wang and Cai, 2006). The wage differential is largely due to gender discrimination which encourages women's engagement in low levels of occupation, like unskilled and semi-skilled work, low level management work and other related productive work (Hirway, 2006). The wage differentials between women and men agricultural workers therefore appear to be based on a pre-assumed gender character. Employers or contracts simply lower women's wages regardless of job performance. In a pervasive climate of social and economic neglect, women have no better options and work longer and harder to make ends meet, leading to exhaustion and injuries from stress and overwork.

A large proportion of women in ESAP countries are not able to retain their earned income: over 40% in Bangladesh and Gujarat, India and over 70% in Indonesia. In China only 53% of women said that they alone decided how their income is spent; this proportion was far less in Bangladesh and India (IFAD, 2005b). Thus, it is not sufficient to stop analysis at the point of household income level, but is necessary to examine how much control women have over that income and work out measures to increase their control over it.

5.5.3 Microfinance groups

Microfinance reaches over 10 million members of savings and credit groups in the region, nearly 90% of whom are women (Donaghue, 2004). Whether or not microfinance has increased the economic agency of members is debatable (Goetz and Sen Gupta, 1996). There is, however, a continuous creation of new norms and social contexts in favour of women as income earners and their access to resources, as evident in a study of four districts in rural Bangladesh (Kelkar et. al., 2004). The rise of women's Self-Help Groups (SHGs) or women's microcredit and microfinance groups, in India and other ESAP countries, has made women's income a permanent component of household income and weakened patriarchal gender relations, reducing women's dependency on the male provider.

The weakening of discriminatory gender relations has been noted in a number of ways: (1) women's greater presence in the market as buyers, though very restricted (in South Asia) as sellers; (2) women's participation in various types of agricultural field work; and (3) women's unescorted movement, though often in a group and not alone, to markets, schools and training. Although women's production activities are still largely confined to the homestead or hamlet, when women engage in activities outside their households and/or villages, the increased interaction they have with the outside world goes a long way towards establishing social and economic equality.

The roles that women in microfinance groups can play in development need not end with instilling new cultures of savings and repayment. In India, SHGs have gone beyond savings and individual loans to take up management of community-based projects—contracting to construct minor irrigation works or undertake soil conservation. Unlike men's groups doing the same tasks, they have saved considerable amounts of capital and used their savings to invest in tractors and other forms of mechanization. In a village in Andhra Pradesh, India, SHGs have even invested in electricity generation with diesel saving pongamia seed oil and sold the carbon saved in the international market. Women's microfinance groups can thus serve as agencies to introduce new and advanced energy technologies to villages and as links to the world.

The SHG concept can be used throughout the region to foster women's agency and extend to community-based organizations of women whose members often struggle for adequate supplies of food, water, housing and employment.

Table 5-1. *Sectoral composition of employment by gender (%).*

	Bangladesh		India		Nepal		Pakistan	
	M	F	M	F	M	F	M	F
Agriculture/Fisheries	54.3	75.7	53.1	74.8	67.1	85.2	36.0	64.2
Mining	0.4	1.1	0.7	0.3	0.1	0.0	< 0.01	< 0.01
Manufacturing	7.2	7.7	11.5	10.1	7.7	3.9	14.0	14.6
Utility	0.3	0.2	0.4	0.0	0.5	0.0	1.0	< 0.01
Construction	2.9	0.5	5.7	1.7	6.2	1.1	7.5	0.3
Trade, Hotel, Restaurant	18.0	2.5	13.1	4.3	7.3	3.7	17.3	1.9
Transport, Storage and Communications	7.2	0.4	5.2	0.4	2.7	0.1	7.3	0.4
Finance and Business	1.0	0.2	1.6	0.5	0.9	0.2	1.1	< 0.01
Community, Social and Personal Services	8.8	11.9	8.7	7.9	7.5	5.6	15.7	18.4
Total %	100	100	100	100	100	100	100	100

Source: Das, 2006.

An innovative model in this regard is the Rajasthan Ekta Nari Sangsthan, with a membership of 16,000 low-income single women who help each other to reclaim land rights and stop domestic and social violence (Planning Commission, 2005).

5.6 Institutions and Policies

Agricultural development is dependent upon the performance of a large number of actors/organizations. It includes not only those involved in Research, Development, Training and Extension (RDTE), but also those involved in a range of other activities, such as input generation and distribution, credit supply, value addition and marketing and policy development and implementation. It also depends on the overall institutional context shaping the interaction among these different actors/organizations. Though many of these actors are present in all ESAP countries, there is a wide diversity in the number, capability and performance. This diversity stems from the historical pattern of governance (colonization and independence), ideologies (role of the state and other actors), stage of development, distribution of holdings and share of population involved in farming.

This diversity has several important implications for planning agricultural development interventions, including agricultural science and technology. Firstly, importing models of technological change, which might have been successful elsewhere, is not the way to address agricultural development in this region. In other words, country and region-specific approaches for agricultural development need to be designed. Secondly, development or application of new technology need not necessarily be the starting point for agricultural development. While technologies do play an important role, there could be other areas for intervention (institutional innovations) that may better address agricultural development and sustainability and these need to be explored.

5.6.1 Institutions

There has been an increasing realization that institutions—the rules, norms, habits, practices and routines that determine how different actors interact with each other and respond to new challenges and opportunities—influence the performance of the agricultural innovation system. An innovation system could be defined as a network of organizations, enterprises and individuals focused on bringing new products, new processes and new forms of organization into social and economic use, together with the institutions and policies that affect their behavior and performance. The innovation system concept embraces not only the science suppliers but also the totality and interaction of actors involved (World Bank, 2006). However, the different actors in this system often do not sufficiently interact, collaborate, or share knowledge in most ESAP nations unless policy and practice address the institutional and related issues underpinning this situation. Several policies dealing with agriculture and allied sectors potentially affect agriculture and how AKST is deployed for agricultural and socioeconomic development. Policies influence or shape how programs are designed and operated. Exploring institutions and policies therefore assumes critical importance in strengthening AKST arrangements in the ESAP region.

Finding new ways of working and collaborating among large numbers of organizations is absolutely essential if sustainability and development challenges in the region are to be effectively addressed. Development of appropriate institutions will therefore assume great importance as these will facilitate the ability of the various actors to link with other sources of expertise and knowledge and enable timely and successful responses to new challenges and opportunities in the region. Many of the previous efforts in improving the functioning of AKST arrangements focused only on improving the links between research and extension. Though the research-extension linkage will continue to be important, organizations involved in RDTE will need to develop partnerships with a large number of other actors including farmers, NGOs, producer organizations, input agencies, agroprocessors, agribusiness houses, traders, retailers and even consumers (van Mele et al., 2005; Hall, 2006). Developing wider links will be essential not only to improve the performance of organizations involved in RDTE, but also to facilitate rural innovation—where new knowledge, information and technologies may be made available and put to socially and economically productive use.

Several institutional barriers currently constrain the development of appropriate working arrangements. There is an increasing realization that the research-extension-farmer paradigm of agricultural development is insufficient to address the new and rapidly evolving challenges to agricultural development in the ESAP region. Attempts to refine this linear paradigm started with ensuring farmer participation at different stages of technology development and promotion. Though it brought farmer perspectives into the process of agricultural technology development, several other important actors whose decisions also influence technology demand, promotion and uptake were left out. Moreover, most of the decisions on technologies were made by researchers and there has not been any change in the way science is organized, funded, managed or evaluated. Organizational reforms within public sector research and extension organizations—such as decentralization and interface meetings with a wide range of stakeholders—will need to go further if underlying paradigms governing the way research or extension is implemented in the region are to change.

Many organizations have narrow mandates that prevent them from working with others. For instance, the agenda, constituency and training opportunities available to those in the extension sector need to expand if it is to support the producers who need more diverse support. Public sector extension in ESAP countries is focused only on the dissemination of technologies to farmers. It will need to move beyond its restricted mandate of technology dissemination to helping producers cope with new challenges, including the provision of organizational, managerial and entrepreneurial support (Sulaiman and Hall, 2003). Its client base will also need to expand to include NGOs, producer associations, rural entrepreneurs, agricultural labor and women. If extension is to play these roles, it must develop new partnerships and capacities, including technological (new knowledge and skills) and institutional (new patterns of collaboration, new habits and practices) capacities.

Agricultural science and technology arrangements in the ESAP region need to be assessed not only in terms of the number of research institutions, technologies released, or number of scientists or extension workers, but also in terms of how they relate to other actors in the wider innovation system. Evaluation parameters in research and extension agencies also need re-examination. Evaluating performance based on number of technologies released has restricted scientists from engaging in other equally important aspects such as technology adaptation and problem-solving. Similarly exclusive focus on technology dissemination has restricted extension from engaging in other important institutional innovations that are required for raising farm incomes. The role of social scientists also needs to change from measuring impacts to experimenting with new institutional arrangements and learning from them.

The need for partnering with the various other organizations involved in agricultural development has been evident in many ESAP nations since the 1990s. There have been increasing calls for public-private partnerships in agricultural development in the last decade and several efforts were made to promote this approach. Several innovative institutional arrangements involving a wide range of partners emerged in response to the realization that agricultural development involves interaction among a wide range of actors. Fostering such interaction and increased collaboration among multiple partners will require the identification and elimination of the mistrust between potential partners and organizations in both, public and private sectors (Box 5-1).

The increasing complexity of agricultural development and the rapidly changing external environment necessitates that all actors in the agricultural innovation system including those directly dealing with AKST embrace partnerships as an organizational principle. It is increasingly clear that there cannot be a blueprint for promoting partnerships, but development of partnership arrangements could be facilitated through funding arrangements designed to promote and support stakeholder meetings and handholding development of joint collaborative activities. These need to be supplemented with efforts to reflect on partnership progress and lesson-learning to direct the much-needed institutional changes among different actors in the innovation system. Some of the key recommendations that have emerged through a joint analysis by different stakeholders who have participated in four NRM projects in India are relevant to those interested in promoting partnerships in RDTE (Box 5-2). The projects examined include: (1) integrated management of land and water resources (DFID/NRSP-ICAR); (2) improved livelihoods through a consortium approach (ICRISAT); (3) promotion of zero-tillage (Rice-Wheat Consortium) and (4) community development (Aga Khan Rural Support Project).

Agricultural innovation occurs when different actors in the innovation systems interact and share knowledge and work in partnerships (Figure 5-1). While understanding and planning agricultural development interventions, it is worthwhile to use the conceptual framework of "innovation system". Its attraction is that it recognizes that innovation is not a research driven process simply relying on technology transfer. Instead innovation is a process of generating, accessing and putting knowledge into use and is complicated and context-embedded. Consequently, its main determinants are the interactions of different people, the ideas they have and the social setting of these interactions and relationships. Its other important insight, which is now widely recognized in the development sector, is that institutions really matter. Thus, the attitudes, habits, practices and ways of working that shape how individuals behave have an enormous impact on whether or not innovation takes place (Hall, 2006). Addressing these issues related to governance and partnerships in AKST assumes primary importance in programs aimed at strengthening AKST in the ESAP region.

Conventional approaches to strengthening capacity in agriculture focused only on science and technology. This is important and will continue to be important especially for countries with limited science and technology capacity. Emerging frontiers of new knowledge will necessitate organizing special training programs in such select areas. Knowledge and information exchange among different countries is required to bridge the gaps in capacity to develop and apply new knowledge. CGIAR centers and international and regional donors play important roles and their efforts will need to be strengthened. However, science and technology capacity alone will not enough to bring about better knowledge uptake and use; applying new knowledge will necessitate the development of several kinds of capacities among several actors. Capacity to develop and implement policies, experiment and evaluate new approaches and address issues related to quality, standards and markets will all need to be upgraded throughout the ESAP region.

To attain the development and sustainability goals of AKST, organizations require a wide range of capacity—broadly called innovation capacity. Innovation capacity is the ability of the network of actors in an innovation system to address problems and to identify, test and implement solutions—in other words, to innovate. It comprises the context-specific range of scientific and other skills and information held by individuals and organizations, practices and routines (institutions) and the patterns of interaction and policies needed to create and put knowledge into productive use in response to an evolving set of challenges and opportunities (Hall, 2007).

Options for action are as follows:

- Capacity development will involve diagnosing the existing innovation system, including exploring the actors, their knowledge and skills, roles, patterns of interaction, habits and practices and the policy environment. An innovation systems framework could be used as a diagnostic tool to understand the existing innovation system and also as a framework for planning interventions (World Bank, 2006). Learning from the emerging institutional arrangements in the region necessitates a detailed analysis of cases where the various actors in specific contexts come together and collaborate to solve particular problems or address new challenges. What kinds of changes were made, how were they implemented, were they sustained at the end of the specific initiative and why?
- Not many organizations have a culture of learning. Opportunities will need to be created and specifically funded to bring in this change of culture. It will be useful to bring staff together to reflect on lessons learned and discuss how goals could be better achieved. The concept

Box 5-1. Lessons on partnership experiences in South Asia.

- Partnering is a pragmatic response to the need for accomplishing complex tasks that cut across disciplinary, organizational and sectoral mandates. Joint task identification and definition builds partnership. Forced partnerships and ritualistic partnerships have no value and will not be sustained.
- Partnership should last as along as there is a shared task to be accomplished and should not be viewed as a permanent linkage.
- Not all organizations have the appropriate skill to be good partners.
- While clear definition of roles of all partners is important, it also needs to be recognized that the roles of partners change during the innovation process, with different partners assuming greater importance at certain times.
- Partnering facilitates sharing of resources, skills and knowledge and is crucial to learning and innovation. Not all organizations have a culture of learning, restricting their ability to partner and generate institutional innovations.
- Rigid institutional and organizational structures, particularly those with hierarchical designs tend to stifle learning and the development of iterative relationships with broader set of partners.
- While it is easy to stereotype public, private and NGO organizations, and the organizational culture that goes with them, there is a need to examine these more closely in the analysis of project partnership viability.
- Successful partners have intuitive ways of identifying each other that relate to shared values of trust and complementarity; shared history built up over the previous partnerships contributes to this.
- Partnership skills are a range of capabilities that help organizations innovate, and that are learnt through interaction with partners and networks.
- How organizations learn and build these skills is not yet entirely clear.
- The strength of the learning process in project partners appears to be a key area of capacity development.
- Activities that widen the interaction of organizations with other partners and networks are likely to be an important way of building up innovation capabilities, both in individual organizations and in wider national systems.

Source: Hall et al., 2004.

of institutional learning concerns the process through which new ways of working emerge. It specifically asks the questions, what rules, habits and conventions have to be changed to do a new task or to do an old one better? (Hall et al., 2005)

Box 5-2. Encouraging effective R&D partnerships: Lessons learned from the Indian experience.

- Time: Donors and partners should allow at least one to two years before expecting R&D partnerships to begin to deliver results and achieve impacts; where partnerships already exist, it may be more efficient and effective to invest in those to leverage previous investments rather than establishing new ventures.
- Flexibility: Management systems need to provide sufficient flexibility, allowing new partners to join over time and others to leave once it is clear that their role has changed or been fulfilled.
- Leadership: Policy makers need to create an environment that allows, indeed encourages, the delegation of both responsibility and authority to those most closely involved in carrying out the work. This can be done using broad accountability frameworks to monitor impacts and ensure the delivery of results.
- Monitoring and evaluation: Partnerships require internal monitoring and evaluation mechanisms that allow them to respond effectively to changing needs and opportunities.
- Responsibility and authority for implementing this continuing activity should be vested with project leaders and be seen as complementary to formal mid-terms and end-of project monitoring and evaluation activities.
- Livelihoods: Project leaders should be formally encouraged to seek innovative ways of empowering local communities. The work of researchers and development specialists from outside the community is all too often guided by predetermined or assumed development priorities, and such deterministic community development activities should be avoided.

Source: Authors' elaboration.

- Opportunities will need to be created that bring different actors together and develop joint activities and long-term relationships. These will need to be mentored and have funding and other resources.
- Capacity development workshops with actors within the innovation system will need to be organized to enhance the ability of all the actors to think and act in a more systemic sense. These could also be used as a platform to share results of the diagnosis and identify the nature of interventions that are required to strengthen innovation capacity.
- Implementing a series of institutional changes (i.e., changes in rules, norms, conventions and habits within these organizations and the way they relate to other stakeholders) in the RDTE system and others related to RDTE will be necessary if ESAP governments want to improve the performance of this system. This has to be a learning-based approach appropriate to the specific institutional context and this process needs to be facilitated.

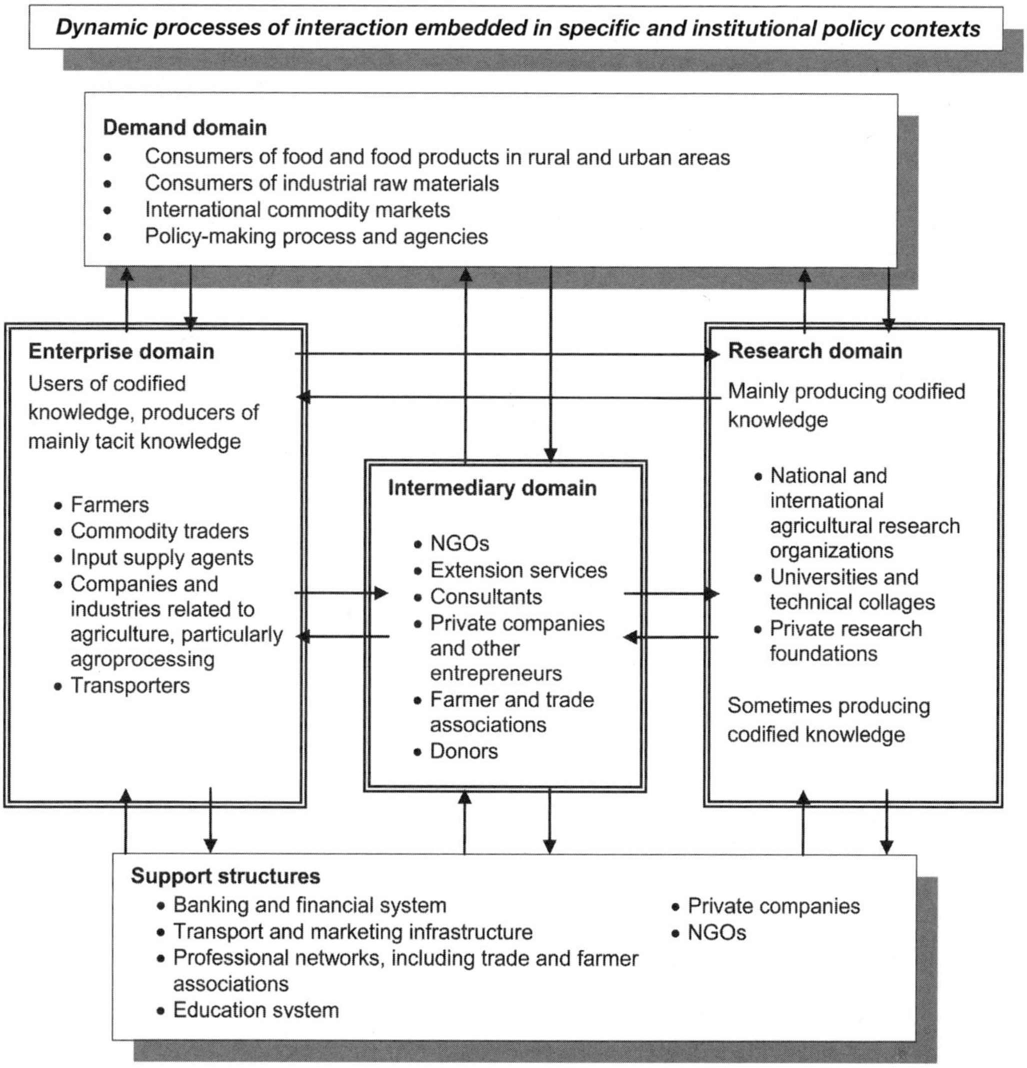

Figure 5-1. *Elements of an agricultural innovation system. Source: Adapted from Arnold and Bell, 2001; World Bank, 2006.*

5.6.2 Policies

A large number of policies affect agriculture and how AKST is deployed for agricultural and socioeconomic development in the face of stagnating grain yields, declining water and land availability, new threats and opportunities from WTO, emergence of supermarkets, increasing private sector participation, emerging concerns on food safety and the need for standards in production and processing. Policies influence the ways programs are designed and operated. In particular, policies related to agricultural R&D (IPR, biosafety), agriculture and allied sectors (livestock, fisheries, etc.), natural resource management, input use (seeds, fertilizer, etc.), trade, gender, conservation and utilization of genetic resources, biodiversity, etc., are critical for attaining the sustainability and development goals of AKST in the region and need to be developed and implemented by national governments (and inter-governmental organizations wherever necessary).

Having a sound policy does not ensure better compliance to guidelines or better performance of the system. Firstly, the countries in the region vary in their capacity to implement policies. There are significant gaps in the capacity of several countries in the ESAP region to implement policies related to biosafety, IPR and food quality standards. For instance, countries intending to implement transgenic and other developing technologies will need to ensure that their infrastructure is sufficient to support the safe development, transfer and application of these technologies with special attention to developing relevant policies, information systems and training in biotechnology risk assessment and biosafety procedures. Methods and results of environmental risk assessments, as well as a model policy framework could be shared between countries that have similar agricultural environments, thus reducing the burden of proof for any one country.

Secondly, quite often the policy is only prescriptive, without taking into account everything needed to get it implemented. To be effective, a policy should also facilitate change, through a process of experimentation, reflection and learning so that it develops the capacity of the various stakeholders to identify bottlenecks, experiment with alternative ways of working and evaluate performance. The actors in the policy system thereby learn what needs to be changed or modified and how to develop better policies. Thirdly, implementation of good policies and programs requires collaboration among a large number of organizations. This is especially so since most of the innovations needed in agriculture today have collective dimensions; i.e., they require new forms of interaction, organization and agreement between multiple actors (Leeuwis and van den Ban, 2004).

This essentially means that all organizations involved in the agricultural innovation system need to operate within a policy framework and have the capacity to produce and integrate new knowledge and apply it in their specific contexts to deal with challenges. For instance, the agenda and constituency of extension will need to expand beyond its current mandate of technology dissemination to help producers cope with new challenges and expand its client base to include NGOs, producer associations, rural entrepreneurs, agricultural labor and women. For extension to play these roles, it will need to develop new capacities, including technological (new knowledge and skills) and institutional (new patterns of collaboration, new habits and practices) ones.

5.7 IPR

Governments extend legal measures to protect intellectual rights over AKST to protect the rights of innovators from misappropriation and reward and encourage innovations. Intellectual rights protection for AKST in the ESAP region generally falls under intellectual property rights (IPR) systems such as patents and the more plant-specific plant variety protection. There is emerging consensus in international discussions that the current IPR regime may only be appropriate for innovations generated by formal institutions, particularly involving conventional AKST and emerging frontiers in AKST and is not an appropriate system to protect traditional and informal knowledge systems. There are ongoing efforts to address this issue, such as the multilateral platforms of the World Intellectual Property Organization (WIPO) to develop an appropriate system of intellectual rights protection over these knowledge systems (CIPR, 2002; Tauli-Corpuz, 2003; Khor, 2004).

Some governments in the ESAP have adopted policies and laws that specifically provide for protection of traditional knowledge and resources. Thailand enacted its Farmers Rights and Plant Variety Protection Law in 2000, providing protection to farmers' traditional varieties through registration under the name of the local community and mandating benefit-sharing on new varieties derived from endemic varieties. India has a Farmers' Rights and Plant Breeders' Rights Law (1999) which exempts farmers' varieties from plant breeders' rights protection and allows farmers to use, exchange and even sell protected varieties. Malaysia's Plant Variety Protection Act was passed in 2004 and provides for separate category and criteria for farmers' varieties.

There is contentious debate on whether or not the IPR regime promotes innovations. The UK Commission on Intellectual Property Rights (CIPR), comprised of international IPR experts who looked into the role and impacts of the IPR on developing countries, concluded that patents in particular are not considered important determinants of innovation even in developed countries, except in such sectors as pharmaceuticals (CIPR, 2002). There are concerns among international and national public research institutions that the stringent application of IPR has impeded free exchange and flow of germplasm needed for research and development efforts. This has resulted in calls to review the appropriateness of the current IPR systems in protecting innovations while at the same time ensuring continuing innovation and exchange of germplasm. ESAP governments need to strike a balance between intellectual protection over innovations and ensuring that innovations and genetic materials are continuously available for further research and development for public interest.

Over the years, concerns have been raised by various sectors in the region about the threats posed by the current IPR regime in misappropriating traditional knowledge and resources through biopiracy. Controversial regional biopiracy cases include several from India: the neem tree, whose insecticidal properties have been patented by W.R. Grace; a traditional chickpea variety covered by plant breeders' rights in Australia; and turmeric, patented in the US for post-surgical wound-healing. Thailand's plau noi has been patented by a Japanese company. The Indian government won a petition for the revocation of the turmeric patent in the US in 1999 and Indian civil society organizations won a petition for the revocation of the W.R. Grace patents on the neem tree in 2005. Such biopiracy issues have been brought to the attention of the Convention on Biological Diversity (CBD) which serves as the major multilateral forum to address issues of access to biological and genetic resources. The CBD has been discussing mechanisms for the regulation of access to genetic resources and benefit-sharing arising from the commercial utilization of these resources.

In compliance with their commitment to the WTO Agreement on Trade Related Aspects of Intellectual Property Rights (TRIPS), some countries in the ESAP have already adopted or formulated their own patent laws or sui generis systems for plants and plant varieties. However, most of these countries still lack the capacity to address the complex issues surrounding this agreement (Hossain, 2004). As of 2006, only a few countries in the ESAP had actually adopted sui generis legislations for the protection of plant varieties (PVP) as mandated in TRIPS 27.3(b). There are debates that surround this system of plant variety protection, with some authors claiming that the sui generis PVP system is a form of IPR (Evenson, 2005) while others assert that it is not a form of IPR but actually a model to balance the monopolistic tendencies of IPR regimes (Khor, 2004). The most notable sui generis plant variety protection laws were enacted by India (1999), Thailand (2000), Pakistan (2002) and Malaysia (2004). While the Philippines (2002) and Indonesia (2002) have also adopted their own PVP laws which they claim as sui generis, their laws may be more akin to the conventional plant breeders' rights model promoted by the Union for the Protection of New Varieties of Plants (UPOV). There are systems of intellectual rights protection that are evolving outside of the conventional IPR regimes and even the sui generis system that developed from TRIPS Article 27.3b. One such concept is farmers' rights which developed under the FAO's International Treaty on Plant Genetic Resources for Food and Agriculture (ITPGRFA).

The ESAP has shown examples of civil society-led initiatives to evolve a specific system of intellectual rights protection for traditional knowledge and resources. Indian civil society organizations have pioneered the concept and practice of Community Registry involving the documentation and formal local registration of community knowledge systems and resources (Khor, 2004). The Community Registry model is replicated in Nepal by a formal institution which has expanded the model beyond AKST and covered traditional knowledge and resources in such areas as forestry. The model has been widely replicated and has since evolved in other parts of ESAP, such as Bangladesh, Thailand and the Philippines. There are also ongoing efforts that attempt to combine useful and workable traits of the conventional IPR regime while at the same time exploring the potentials of alternative systems of intellectual rights protection. Other efforts involve the extension of the provisions on geographic indication provided in the TRIPS to agriculture as a system of protecting traditional and conventional AKST. These national initiatives may be considered as "hybrid" IPR since they assert the rights of communities over specific AKST and products of AKST by maximizing the opportunities available in the current IPR regime.

5.8 Trade and Markets

5.8.1 Domestic regional markets and trade

The importance of trade in discussions of technological change lies in the relationship between efficiency and/or productivity change and export growth and vice versa. On the one hand the productivity gap between economies is posited as the basis for international trade, while on the other hand there is a contention that trade liberalization will enhance factor productivity. This debate has important implications for agricultural modernization, tariff policies and technological priorities. Economic arguments can be mustered to support causality in either direction. Further liberalization of world trade could improve the efficiency of the agricultural sector for countries which enter agricultural export markets for the first time. However, only in very few countries does increased productivity trigger sustained export growth (Arnade and Vasavada, 1995; Suhariyanto and Thirtle, 2001).

The ESAP region's development over the last two decades has been closely linked to economic reforms and trade opportunities. The policy reforms have been triggered by many changes or crises which have affected domestic markets as well as regional and international trade patterns. Additionally, transport and related infrastructure as well as non-transport logistics and market arrangements can influence how these economies react to policy changes. It has been estimated that in East Asia non-transport logistics impose heavier trade penalties than do transport inadequacies and that domestic market arrangements can constrain diversification and impede international trade (Carruthers, 2003). Thus, the pattern of development so far experienced in the region is dependent on continuing improvements in transport, storage, distribution and processing. The prevalence and quality of non-market measures which characterize the region's trade will also determine the nature and direction of its growth.

Although there remain marked differences in the policy regimes across the region, one factor that will influence ESAP's future policies and trade patterns is liberalization. China and Vietnam are in the process of opening their economies in response to their new WTO obligations. The liberalization which has already taken place in the region has been contemporaneous with both increased food production and greater food availability. In that sense at least, the liberalization strategy may be said to have worked so far. Nonetheless, food availability remains a major problem for hundreds of millions in the region and in such circumstances the dependence on an international market entails risks. Consequently, there is a need for public intervention by governments to address the risk of variations in availability, chronic poverty and household food insecurity. The policies undertaken in pursuit of these ends often compound the original problems; one study of the China grain market, for example, concluded that Government action destabilized rather than abetted the market. Since such action is costly and often leads to scarcity and food insecurity the search for more stable policies will be a feature of future action (Findlay, 2003).

Liberalization has itself contributed to extensive diversification of production away from the "old order" of growing and marketing cereal crops in a subsistence farming system. There has been considerable diversification of the region's production, stimulated by trade in high value commodities, particularly in horticulture for urban and periurban areas (rather than in the hinterlands and near-urban areas). This trend, the extent and impact of which differ across the region, is expected to continue in response to increasing urbanization, rising per capita incomes, further trade liberalization and supporting strategies and the removal of restrictions on foreign direct investment in the food sector.

Other developments with important implications for the region's future are changes in lifestyles, rapid changes in dietary preferences from cereal-based to high-value commodities and dramatic growth (10-90% in recent years) in the number of supermarkets and other food retail shops. The rate of growth of food through supermarkets varied from 5% in Bangladesh and similar low income states to 50% in others such as Thailand and the Philippines (Bayes, 2005).

The removal of restrictions on foreign direct investment in the food sector has been associated with important institutional changes. These have resulted in new developments involving bilateral cooperation in the reorganization of the supply chain in some states such as China. Such restructuring will continue to be critical if the potential of the various states and their endowments is to be fully exploited without income and food crises. In the process of this regional diversification and cross-border investment, China has been shifting production from grains, cotton and sugar to more labor-intensive crops such as fruit and vegetables in which it would appear to have a comparative advantage and is importing Thai cassava, logans durians and prawns. The specific cross-border arrangements in the case of China, Japan and Korea have been quite innovative, centering on linking producers with agribusinesses in the form of cooperatives, the domestic private sector and TNCs. They have been responsible for initiatives in niche areas and include contract-growing of perishables for competing NGOs, agroprocessors and supermarkets (Bayes, 2005). These arrangements have had a favorable impact on production by providing or enhancing access to assured markets and reliable information, reducing transaction costs, providing means to handle market risks and increasing producer profits. They may provide models to explore for other countries in the region (Bayes, 2005).

The emergence of such marketing arrangements has not been an unmitigated blessing for small farmers and distributors, however (Humphrey, 2006). Many studies have already been devoted to devising guidelines for policy makers in this context and Governments may wish to consider them (Reardon et al., 2006; Altenburg, 2006). Another problem arising from these developments in trade includes wildlife endangerment. For instance, the rapid transformation of the Chinese economy and the allied increase in consumer purchasing power implies growing pressure on individual species over the coming years (von Moltke, 2000). Increasingly, ESAP states will need to address domestic as well as international concerns on this front.

5.8.2 Poverty and the liberalization of international trade

The impact of liberalization on the levels and prevalence of poverty is a highly contentious issue in the region. The process of liberalization may involve higher risks for national producers and labor than for consumers. The analysis of poverty is especially difficult due to the simplifications of the theory, the fact that the category "poor" is both diverse and dynamic (i.e., changing over time, even seasonally) and the challenges of measurement.

The theory points to positive net results and many assessments confirm this although they acknowledge that for some groups such as small farmers and farmers in less favored areas incomes may deteriorate (CGPRT, 1999). There is, however, no shortage of studies which are unequivocally pessimistic about the overall outcomes of these reforms (Rodrik, 2001; SAPRIN, 2002; Patnaik, 2004).

"While theory may suggest that the liberalization of trade policies will result in net benefits to the liberalizing country and while there may be a growing collection of empirical studies to support the theory, it is also clear from the preceding discussion that the benefits of liberalization will not necessarily be achieved and even where they are, some groups of individuals within some countries are likely to be disadvantaged. In a concise and convincing paper, Winters . . . argues that although he believes that trade liberalization aids economic growth, it "may have some adverse consequences for some—including some poor people—that should be avoided or ameliorated to the greatest extent possible". He suggests that rather than using this as a reason for resisting reform, it should "stimulate the search for complementary policies to minimize adverse consequences and reduce the hurt that they cause". It is clear that there is no clear consensus that liberalization results in economic growth, despite a number of major research programmes investigating this relationship. It is therefore important to understand the types of reform that have had the greatest impact on economic growth in each country." (FAO, 2003)

In essence, the impact of trade reform on employment and hence on poverty is context specific. There are guidelines to the questions that need to be answered in order to accurately anticipate its net impact (Bussolo et al., 1999; Kanai, 2000).

5.8.3 Bilateral and regional agreements and their implications

Pascal Lamy, Director General of the WTO, is reported to have anticipated 400 preferential trade agreements by 2010. The reasons for the continuation of this trend have been well documented and analysed and are numerous (Sagar, 1997; Gilbert et al., 2001; Hilaire and Yang, 2003; Bhagwati et al., 2005; Menon, 2006). The Asia-Pacific region, including the smaller island states, has been particularly active in the drive to so-called "new regionalism" (Figure 5-2) (Ethier, 1998; Majluf, 2004). Bilateral and preferential agreements involving at least one Asia-Pacific state rose from 57 in 2002 to 176 in October 2006, about 70% of which have yet to be implemented. Furthermore, the region can be expected to be even more active in this regard as a result of post WTO-entry initiatives by China and Vietnam. The agreements have distinctive product coverage, time lines and varying rules, etc., so future harmonization of these agreements would be very difficult. Given the current activities of larger economies in the region including China, Japan and India, there is clearly the risk that a hub-and-spoke system will dominate, with these leading economies as the hubs. While ASEAN may also include hub contenders, this is muddied by individual ASEAN members also pursuing bilateral trade agreements (BTAs), especially with the USA and Japan. The resulting "spaghetti bowl" of agreements and rules can enormously complicate the life of international traders, so that an exporter can enter another market under different sets of preferences, multiple agreements may exist, preferences may be prohibited from being realized and MFN entry may appear to be the least costly action.

The negotiation of bilateral agreements is often politically easier than multilateral or regional approaches; however, some are sanguine about the prospects of such agreements eventually being aggregated to a wider grouping. Several options exist, including those of linking individual "spokes" under a single PTA to create de facto regionalism and reduce trade diversion and "open regionalism" (as per the APEC 1994 Bogor Declaration) in which BTA preferences are extended to non-members (Strutt and Lim, 2003).

The challenge for ESAP governments is to configure and phase the transition along the spectrum of bilateral/sub-regional to multilateral integration in ways that enable the development of a global, non-discriminatory trading system from which the agricultural sector can benefit. A comprehensive set of guidelines has been suggested by the Asian Development Bank whereby the adoption and effective implementation of ten basic principles would minimize the potential damage from bilateral agreements while allowing for trade and investment creation and efficient behind-the-border reforms. While this agenda would be best adopted at the multilateral WTO level, it will require strong political support. Asia could play a significant leadership role by adopting these principles and by incorporating them in its bilateral and regional trade agreements.

5.8.4 Agriculture in the liberalizing process

Since agriculture is among the most sensitive of the products in such arrangements its treatment can be problematic. Although tariff barriers in the ESAP region are generally being rapidly dismantled, there has not been a parallel attempt to provide for timely removal of barriers to agricultural trade. Of the six preferential trading arrangements which are in force in ESAP only one has a significant number of eligible agricultural products. All but one of those provides for non-tariff barriers (NTBs) to be employed against agricultural imports and none of the agreements constrain domestic support subsidies (Parakrama and Thibbotuwawa, 2006). Many of the states with low tariff barriers in agriculture actually have NTBs against agricultural commodities. It comes as no surprise therefore to find that agricultural trade flows show relatively little buoyancy. For ASEAN, what agricultural export expansion there is has been mostly due to extra-ASEAN agricultural trade and is less than that of intra-ASEAN trade for industry.

ASEAN member states impose higher agricultural tariffs than they face abroad. Nevertheless, some agreements are relatively liberal and the Early Harvest Program—China ASEAN, for example—provides for substantial inclusion of agricultural products in the liberalization exercise. A majority of the agreements do make provision for the eventual inclusion of most agricultural products, albeit with long transitional periods. The major agreement, AFTA, is the least liberalized of all the trading arrangements listed in a multilateral liberalization index of agricultural measures (Table 5-2). Thus, AFTA is not considered to be as good a building block for agricultural trade as it is for trade as a whole.

5.8.5 The region and the WTO

The realization of minimal benefits in return for unprecedented intrusiveness into the domestic sovereignty of developing states has triggered an attack on the multilateral trading system and its legitimacy (Srinivasan, 2002; Aksoy et al., 2004). Despite this and the adverse consequences of WTO membership several states are waiting to join this body (Bello, 2003). Both China and Vietnam have sought to join the WTO in spite of their political philosophies and in recognition of the cost of staying out. Burma and most Pacific Island states have yet to join.

The rules governing global markets and the management of those markets are of great importance to the ESAP

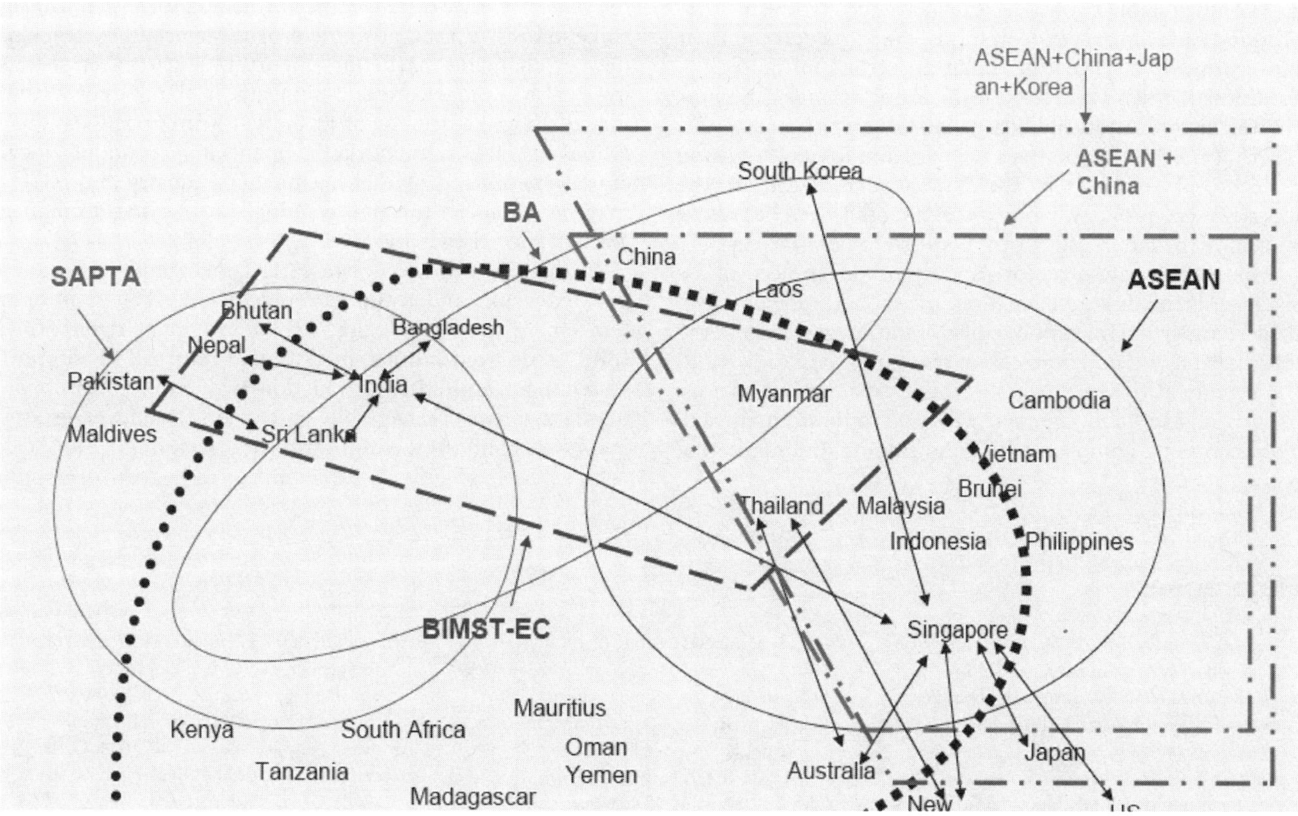

Figure 5-2. *BTAs, RTAs in force and potential RTAs in the Asia-Pacific region. Source: Samaratunga and Thibbotuwawa, 2006.*

region because they can affect the region's access to other markets and the sharing of the gains from expanded trade. At the same time, the price of late entry can be high, especially for small states which tend to be open and highly trade-dependent. Many developing countries need policy flexibility to support and promote their enterprises, investments in production and marketing and export expansion and diversification for this reason. Nonetheless, latecomers to WTO membership now face more stringent policy conditions than earlier ones. Under its accession terms, Tonga, for example, is committed to lowering trade barriers, expanding market access for foreign goods to bind all tariff lines at a level lower than most other developing countries. Moreover, extensive concessions were also made with regard to services in sectors such as health, education, finance and telecommunications. Oxfam described these accession concessions as "eye watering" and "the worst terms ever offered to any country" (Manduna, 2006). Concern about similarly harsh terms caused Vanuatu to baulk at signing the accession agreement after negotiating entry in 2001.

Apart from terms of entry, smaller states face many difficulties in spite of arrangements to render them technical assistance and in spite of the range of agreements which pay special attention to them such as EBA, the GSPs of the US and Japan, SAPTA (SAARC Preferential Trading Arrangement), the Bangkok Arrangement, the Thailand-Bangladesh Preferential Trade Arrangement and the AFTA. Writing on the challenges of the Rules of Origin in particular, the Centre for Policy Dialogue has suggested that ESAP LDCs and Bangladesh and Cambodia in particular, engage more proactively in the negotiations on agriculture.

For the region as a whole, leaders will need to ensure that changes, including the AoA reforms, take into account their impact on the divergent agricultural sector/s in the region. Additionally, ways will need to be found to make the process of liberalization politically palatable in circumstances where it inevitably generates winners and losers. Provision of technical assistance by the more advanced states to the less advanced is one way of approaching this need because it is usually the latter which can ill-afford to bear the costs.

However, states are not the only parties in need of assistance. If the desired changes in the global trade environment, namely liberalization, global trade integration and the information and communication revolution, are to work to the benefit of the rural economy, the capacities of various actors and especially the poor will need to be strengthened. The increasing complexity of these trading and marketing activities will challenge the resources of the traditional actors involved in trade and in economic policy in the region. In order to be in a position to successfully survive in such an environment, the capacities of all the stakeholders in the region will need to be enhanced, including those allowing stakeholders to take advantage of the market, to react to new or changing marketing opportunities, to meet changing health standards to communicate in the domain of the internet, etc. Capacities of farmers, researchers, local governments, extension workers, financial institutions, local entrepreneurs and market agents, agroindustry and NGOs may be enhanced through training, professional exchange and vocational education.

Efforts in the areas of research, policy and governance and extension and training could include: (1) traditional and emerging technologies, (2) international regulations, IPRs, trade negotiations, institutional reforms, (3) support systems not limited to production such as organizational, marketing, entrepreneurship to farmers, producer groups and NGOs and (4) the non-farm rural sector.

Table 5-2. *Ranking of selected regional trade agreements by degree of liberalization of trade.*

	All Trade	Agriculture	Industry
Singapore-NZ	1	1	1
EU	2	6	2
ANZCERTA	3	2	4
Chile-MERCOSUR	4	4	3
Chile-Mexico	5	3	6
NAFTA	6	11	10
EU-Poland	7	7	13
ANDEAN	8	5	5
MERCOSUR	9	8	7
Chile-Colombia	10	13	8
ASEAN-FTA	**11**	**16**	**14**
EFTA	12	9	8

Note: Most liberalizing = 1.
Source: Adams et al., 2003.

References

Abrol, I.P. and M.S. Gill. 1994. Sustainability of rice-wheat systems in India. p. 172-184. *In* R.S. Paroda et al. (ed) Sustainability of rice-wheat systems in Asia. RAPA Publ. 1994/11. FAO, Bangkok.

Acharya, S.S. 1982. Green revolution impact on farm employment and income. Green-revolution-impact-on-farm-employment-and-income. Sanghi Prakashan, Jaipur.

Adams, R. M., D.M. Adams, J.M. Callaway, C.C. Chang and B.A. McCarl. 1993. Sequestering carbon on agricultural land: Social cost and impacts on timber markets. Contemp. Policy Iss. 11:76-87.

Adams, D.M., R.J. Alig, B.A. McCarl, J.M. Callaway and S.M. Winnett. 1999. Minimum cost strategies for sequestering carbon in forests. Land Econ. 75:360-374.

Agarwal, B. 2003. Community forestry groups and gender inequality in South Asia. id21 Res. Highlight. Feb 26. Available at http://www.id21.org/id21ext/s2cba1g1.html.

Aksoy, M., and J.C. Beghin (ed) 2004. Global agricultural trade and developing countries. World Bank, Washington.

Al-Deeb, M.A., G.E. Wilde, J.M. Blair and T.C. Todd. 2003. Effect of Bt corn for corn rootworm control on nontarget soil microarthropods and nematodes. Environ. Entomol. 32:859-865.

Altenburg, T. 2006. Governance patterns in value chains and their development impact. Eur. J. Dev. Res. 18:498-521.

Apu, N.A. and H. Middendorp. 1998. Establishing fishers groups for self management of enhanced fisheries in semi-closed water bodies in Western Bangladesh: The experience of the oxbow lakes small-scale fisherman project (OLP II). Fish. Tech. Pap. 374. FAO, Rome.

Arnade, C and U. Vasavada. 1995. Causality between productivity and exports in agriculture: evidence from Asia and Latin America. J. Agric. Econ. 46:174-186.

Arnason, J.T., B.J.R. Philogène, N. Donskov, H.M.C. McDougall, G. Fortier, P. Morand et al. 1985. Antifeedant and insecticidal properties of azadirachtin to the European Corn Borer, Ostrinia Nubilalis. Entomol. Exp. Appl. 38:29-34.

Arnold, E., and M. Bell. 2001. Some new ideas about research for development. p. 279-316. *In* Partnership at the leading edge: A Danish vision for knowledge, research, and development. Danish Ministry of Foreign Affairs, Copenhagen.

ASEAN. 2006. Towards an environmentally sustainable ASEAN community. Third Association of Southeast Asian Nations (ASEAN) state of the environment report. ASEAN Sec., Jakarta.

Ash, J., C. Novak and S.E. Scheideler. 2003. The fate of genetically modified protein from Roundup-Ready soybeans in laying hens. J. Applied Poultry Res. 12:242-245.

Auernhammer, H. 2001. Precision farming—the environmental challenge. Comput. Electron. Agric. 30:31-43.

Balmford, A., P. Gravestock, N. Hockley, C.J. McClean, and C.M. Roberts. 2004. The worldwide costs of marine protected areas. PNAS 101:9694-9697.

Barker, T., M. Lesnick, T. Mealey, R. Raimond, S. Walker, D. Rejeski, and L. Timberlake. 2005. Nanotechnology and the poor: Opportunities and risks; Closing the gaps within and between sectors of society. Available at http://www.meridian-nano.org/gdnp/NanoandPoor.pdf. Meridian Inst., Dillon CO.

Bautista, R.M. 1997. Income and equity effects of the Green Revolution in the Philippines: a macroeconomic perspective. J. Int. Dev. 9:151-168.

Bayes, A. 2005. Agriculture in South Asia: Destined for a new deal. p. 19-23. In South Asia: Agricultural and rural development. Proc. Seminar. IFPRI, Washington DC.

Beall, J. and L.H. Piron. 2005. Social exclusion review. DFID, London.

Bello, W. 2003. Multilateral punishment: the Philippines in the WTO, 1995-2003. Focus on the Global South. Available at http://www.apl.org.ph/ps/multilateral-punishment .pdf.

Bengtsson, H., I. Öborn, S. Jonsson, I. Nilsson, and A. Andersson. 2003. Field balances of some mineral nutrients and trace elements in organic and conventional dairy farming—a case study at Öjebyn, Sweden. Eur. J. Agron. 20:101-116.

Best, M.L. and S.G. Maier. 2007. Gender, culture and ICT use in rural South India. Gender Tech. Dev. 11:137-155.

Blackwood, C. and J. Buyer. 2004. Soil microbial communities associated with Bt and non-Bt corn in three soils. J. Environ. Qual. 33:832-836.

Bhatla, N., H. Swaminathan and S. Chakraborty. 2007. Women's property rights as an AIDS response. Int. Center Res. Women, Washington DC.

Borlaug, N.E. and C.R. Dowswell. 2005. Feeding a world of ten billion people: A 21st century challenge. p. 11-24. *In* R. Tuberosa et al. (ed) The wake of the double helix: from the green revolution to the gene revolution. Proc. Int. Congr., Bologna, 27-31 May 2003.

Bouis, H.E. 1993. Measuring the sources of growth in rice yields: Are growth rates declining in Asia? Food Res. Inst. Studies 22(3). Stanford CA.

Bouis, H.E. 2000. Enrichment of food staples through plant breeding: a new strategy for fighting micronutrient malnutrition. Nutrition 16:701-704.

Bourguet, D., J. Chaufaux, A. Micoud, M. Delos, B. Naibo, F. Bombarde et al. 2002. Ostrina nubilalis parasitism and the field abundance of non-target insects in transgenic Bacillus thuringiensis corn (Zea mays). Environ. Biosafety Res. 1:49-60.

Bovy, A., R. deVos, M. Kemper, E. Schijlen, M.A. Pertejo, S. Muir et al. 2002. High-flavonol tomatoes resulting from the heterologous expression of the maize transcription factor genes LC and C1. Plant Cell 14:2509-2526.

Brown, S. 1998. Present and future role of forests in global climate change. p. 59-74. *In* B. Gopal et al. (ed) Ecology today: An anthology of contemporary ecological research. Int. Sci. Publ., New Delhi.

Brown, L. 2006. Ethanol could leave the world hungry. Fortune 17 Aug 2006.

Bussolo, M and H.-B.S. Lecompte.1999. Trade liberalisation and poverty. Poverty Briefing 6. Available at http://www.odi.org .uk/publications/briefing/pov6.pdf. ODI, London.

Byerlee, D. and P. Moya. 1993. Impacts of international wheat breeding research in the developing world, 1966-1990. CIMMYT, Mexico.

Carruthers, I., M. W. Rosegrant and D. Seckler. 1997. Irrigation and food security in the 21st Century. Irrig. Drain. Syst. 11:83-101.

Cassman, K.G., S.K. De Datta, D.C. Olk, J. Alcantara, M. Samson, J. Descalsota et al. 1995. Yield decline and the nitrogen economy of long term experiments on continuous, irrigated rice systems in the tropics. p. 181-222. *In* R. Lal and B. Stewart (ed) Soil management: Experimental basis for sustainability and environmental quality. CRC Press, Boca Raton FL.

Cassman, K.G. 1999. Ecological intensification of cereal production systems: Yield potential, soil quality, and precision agriculture. PNAS 96:5952-5959.

Castells, M. 1999. Information technology, globalization and docial development. UNRISD Disc. Pap. 114. United Nations, NY.

CFAN. 2004. Forest, trees, and climate change: Recent experiences in Asia. CIDA Forestry Advisers Network.

Chatterjee, R., P. Choudhuri and S. Bandyopadhyay. 2006. Influence of integrated use of organic and inorganic sources of nutrients on green yield of palak (Beta vulgaris L. var bengalensis Hort.) cv Pusa Bharati in terai zone of West Bengal. Environ. Ecol. 24S (Spec. 3):722-723.

Chrispeels, M.J. 2000. Biotechnology and the poor. Plant Physiol. 124:3-6.

Christensen, V. and C. Saunders. 2003. Economic analysis of issues concerning organic dairy farming. Res. Report No. 257. Agribusiness Econ. Res. Unit, Lincoln Univ. Canterbury, New Zealand.

CIPR. 2002. Integrating intellectual property rights and development policy. UK CIPR, London. Available at http://www .iprcommission.org/papers/pdfs/final_report/reportpdf.pdf.Cloin, J. 2007. Coconut oil as a fuel in the Pacific Islands. Nat. Res. Forum 31:119-127.

Clover, C. 2004. The end of the line: How overfishing is changing the world and what we eat. Ebury Press. London, UK.

Combs, Jr. J.F., R.M. Welch, J.M. Duxbury, N.T. Uphoff and M.C. Nesheim. 1996. Food-based approaches to preventing micronutrient malnutrition: An international research agenda. CIIFAD, Cornell Univ., Ithaca NY.

Conway, G. 1999. The doubly-green revolution: Food for all in the twenty-first century. Cornell Univ. Press. Ithaca NY.

Conway, G.R. and J.N. Pretty. 1991. Unwelcome harvest: Agriculture and pollution. Earthscan, London.

Conway, G., and G. Toenniessen. 1999. Feeding the world in the twenty-first century. Nature 402:C55-C58.

Cook, S.E., M.L. Adams and R.G. Bramley. 2000. What is obstructing the wider adoption of precision agriculture technology? Proc. Fifth Int. Conf. Precision Agriculture, Bloomington MN 16-19 July.

Correa, C.M. 2002. Protection and promotion of traditional medicine: Implications for public health in developing countries. WHO, South Centre, Geneva.

Cowgill, S.E., and H.J. Atkinson. 2003. A sequential approach to risk assessment of transgenic plants expressing protease inhibitors: effects on nontarget herbivorous insects. Transgenic Res. 12(4):439-444.

Cowgill, S.E., R.D. Bardgett, D.T. Kiezebrink and H.J. Atkinson. 2002. The effect of transgenic nematode resistance on non-target organisms in the potato rhizosphere. J. Appl. Ecol. 39(6):915-923.

Cox, S. 2002. Information technology: the global key to precision agriculture and sustainability. Comput. Electron. Agric. 36:93-111.

Craswell, E.T., A. Sajjapongse, D.J.B. Howlett, and A.J. Dowling. 1998. Agroforestry in the management of sloping lands in Asia and the Pacific. Agrofor. Syst. 38:121-137.

Crosson, P., and J.R. Anderson. 1992. Resources and global food prospects: Supply and demand for cereals to 2030. Tech. Pap. 184. World Bank, Washington DC.

Dalgaard, T., N. Halberg, and R.P. Porter. 2001. A model for fossil energy use in Danish agriculture used to compare organic and conventional farming. Agric. Ecosyst. Environ. 87:51-65.

Dalke. K. 2003. Inside information: Nanofibers deliver DNA to cells. Available at http://www.genomenewsnetwork.org/articles/06_03/nano.shtml. Genome News Network, June 13.

Das, M.B. 2006. Do traditional axes of exclusion affect labor market outcomes in India? Social Dev. Pap. South Asia Ser. 97. World Bank, Washington DC.

David, C.C., and K. Otsuka. 1993. Modern rice technology and income distribution in Asia. IRRI, Manila.

De, A., H.P.S. Arya, B. Tudu, and A. Goswami. 2004. Indigenous technical knowledge in animal husbandry. Livestock Res. Rural Dev. 16(8):Art. 59. Available at http://www.cipav.org.co/lrrd/lrrd16/8/arun16059.htm.

De la Fuente-Martinez, J., V. Ramirez-Rodriguez, L. Cabrera-Ponce, and L. Herrera-Estrella. 1997. Aluminum tolerance in transgenic plants by alteration of citrate synthesis. Science 276:1566-1568.

De la Fuente-Martinez, J.M., and L. Herrara-Estrella 1999. Advances in the understanding of aluminum toxicity and the development of aluminum-tolerant transgenic plants. Adv. Agron. 66:103-120.

Dene, C. 2005. The good the bad and the ugly: discourse, policy controversies and the role of science in the politics of shrimp farming development. Dev. Policy Rev. 23:585-614.

Devare, M.H., C.M. Jones, and J.E. Thies. 2004. Effect of Cry3Bb transgenic corn and tefluthrin on the soil microbial community: Biomass, activity, and diversity. J. Environ. Qual. 33:837-843.

Devare, M.H., C.M. Jones, and J.E. Thies. 2007. Neither transgenic Bt maize (MON863) nor tefluthrin insecticide adversely affect soil microbial activity or biomass: A 3-year field analysis. Soil Biol. Biochem. 39:2038-2047.

Di Giovanni, G., L.S. Watrud, R.J. Seidler and F. Widmer. 1999. Comparison of parental and transgenic alfalfa rhizosphere bacterial communities using BIOLOG GN metabolic fingerprinting and enterobacterial repetitive intergenic consensus sequence-PCR (ERIC-PCR). Microbial Ecol. 37:129-139.

Doane, D.L. 1999. Indigenous knowledge, technology blending and gender implications. Gender Tech. Dev.3:235-257.

Donaghue, K. 2004. Microfinance in the Asia Pacific. Asian-Pacific Econ. Lit. 18:41-61.

Donegan, K.K., C.J. Palm, V.J. Fieland, L.A. Porteous, L.M. Ganio, D.L. Schaller et al. 1995. Changes in levels, species and DNA fingerprints of soil microorganisms associated with cotton expressing the Bacillus thuringiensis var. kurstaki endotoxin. Appl. Soil Ecol. 2:111-124.

Donegan, K.K., R.J. Seidler, J.D. Doyle, L.A. Porteous, G. di Giovanni, F. Widmer et al. 1999. A field study with genetically engineered alfalfa inoculated with recombinant Sinorhizobium meliloti: Effects on the soil ecosystem. J. Appl. Ecol. 36:920-936.

Donkin S.S., J.C.Velez, A.K. Totten, E.P. Stanisiewski and G.F. Hartnell 2000. Effect of feeding Roundup Ready corn silage and grain on feed intake, milk production and milk composition in lactating dairy cattle. J. Dairy Sci. 83(Suppl 1):273.

Duan, K.Q., T.D. Yao, and L.G. Thompson. 2006. Response of monsoon precipitation in the Himalayas to global warming. J. Geophys. Res. 111:D19110.

Dutton, A., J. Romeis, and F. Bigler. 2003. Assessing the risks of insect resistant transgenic plants on entomophagous arthropods: Bt-maize expressing Cry1Ab as a case study. BioControl 48:611-636.

Edmundson, W.C. 1994. Do the rich get richer, do the poor get poorer? East Java, two decades, three villages, 46 people. Bull. Indonesian Econ. Studies 30:133-148.

Edwards, P. 2000. Aquaculture, poverty impacts and livelihoods. Nat. Resourc. Perspect. 56. ODI, London.

Ellstrand, N.C. 2003. Current knowledge of gene flow in plants. Phil. Trans. R. Soc. London B 358:1163-1170.

Estudillo, J.P., Y. Sawada, and K. Otsuka. 2006. The Green Revolution, development of labor markets, and poverty reduction in the rural Philippines, 1985-2004. Agric. Econ. 35(Suppl. 3):399-407.

ETC Group. 2002. No small matter! Nanotech particles penetrate living cells and accumulate in animal organs. Communique 76, May/June. Available at www.etcgroup .com.

ETC Group. 2004. Down on the farm: The impact of nano-scale technologies on food and agriculture. Available at www.etcgroup.com.

ETC Group. 2005a. NanoGeopolitics: ETC Group surveys the political landscape. ETC Group Report/Communique 89, July/August. Available at www.etcgroup.com.

ETC Group. 2005b. The potential impacts of nano-scale technologies on commodity markets: The implications for commodity dependent developing Countries. Trade-Related Agenda, Development and Equity (TRADE) Res. Papers 4. South Center, Nov 2005. Available at www.etcgroup.com.

Ethier, W.J. 1998. The new regionalism. Economic J. 108:1149-1161.

Fan, S. and C.C. Kang. 2004. Returns to investment in less-favored areas in developing countries: A synthesis of evidence and implications for Africa. Food Policy 29:431-444.

Fan, S. and P. Hazell. 1999. Are returns to public investment lower in less-favored rural areas? An empirical analysis of India. EPTD Disc. Pap. No. 43. IFPRI, Washington DC.

Fan, S., L. Zhang, and X. Zhang. 2002. Growth, inequality, and poverty in rural China. Res. Rep. 125. IFPRI, Washington DC.

Fang, M., R.J. Kremer, P.P. Motavalli, and G. Davis. 2005. Bacterial diversity in rhizospheres of nontransgenic and transgenic corn. Appl. Environ. Microbiol. 71:4132-4136.

FAO. 1999. State of the world's forests, 1999. FAO, Rome.

FAO. 2001. Potential impacts of genetic use restriction technologies (GURTs) on agricultural biodiversity and agricultural production systems. Item 6 Draft Provis. Agenda Working Group on Plant Genetic Resourc. Food Agric. Rome, 2-4 July 2001. FAO, Rome.

FAO. 2002. World agriculture: Towards 2015/2030. Summary report. Available at http://www.fao.org/docrep/004/y3557e/y3557e00.HTM. FAO, Rome.

FAO. 2003a. The state of the world's forests. FAO. Rome.

FAO. 2003b. Trade reforms and food security: Conceptualizing the linkages. Expert consultation, 11-12 Jul 2003. FAO, Rome.

FAO. 2004a. Guiding principles for highly pathogenic avian influenza surveillance and diagnostic networks in Asia. Experts meeting on surveillance and diagnosis of avian influenza in Asia. Bangkok, 21-24 Jul 2004. FAO, Rome.

FAO. 2004b. State of food and agriculture 2003-04. FAO, Rome.

FAO. 2006. World agriculture: Towards 2030/2050. Interim Report. Projections of food, nutrition, agriculture and major commodity groups. FAO, Rome.

Fernández, P.G. 1994. Indigenous seed practices for sustainable agriculture. Indigen. Knowl. Dev. Monitor 2:9-12.

FFTC. 2007. Indigenous feed resources for Asian livestock. Issues in Asian agriculture. Available at http://www.agnet.org/library/ac/1995b/. Food Fertilizer Tech. Center Asian and Pacific Region.

Findlay, R. 2003. Chinese grain market policy with special emphasis on the domestic grain trade. ACIAR Proj. ADP/1997/021. ACIAR, Australia.

Flor, A.G. 2001. ICT and poverty: The indisputable link. Third Asia Dev. Forum Regional Economic Cooperation in Asia and the Pacific. Bangkok. 11-14 June, 2001. Asian Dev. Bank.

Folmer, J.D., G.E. Erickson, C.T. Milton, T.J. Klopfenstein and J.F. Beck. 2000. Utilization of Bt corn residue and corn silage for growing beef steers. J. Anim. Sci. 80:1352-1361.

Fox, J.L. 2003. Resistance to Bt toxin surprisingly absent from pests. Nature Biol. 21:958-959.

Franke-Snyder, M., D.D. Douds, Jr., L. Galvez, J.G. Phillips, P. Wagoner, L. Drinkwater et al. 2001. Diversity of communities of arbuscular mycorrhizal (AM) fungi present in conventional versus low-input agricultural sites in eastern Pennsylvania, USA. Appl. Soil Ecol. 16:35-48.

Fromm, M.E., D.M. Stark, G.D. Austin and F.J. Perlak. 1993. Improved agronomic and quality traits in transgenic crops: Recent advances. Phil. Trans. R. Soc. London B 339:233-237.

Gajalakshmi, S. and S.A. Abbasi. 2004. Neem leaves as a source of fertilizer-cum-pesticide vermicompost. Bioresour. Tech. 92:291-296.

Garg, A.K., J. Kim, T.G. Owens, A.P. Ranwala, Y.D. Choi, L.V. Kochian et al. 2002. Trehalose accumulation in rice plants confers high tolerance levels to different abiotic stresses. PNAS 99:15898-15903.

Goto, F., T. Yoshihara, N. Shigemoto, S. Toki and F. Takaiwa. 1999. Iron fortification of rice seed by the soybean ferritin gene. Nature Biotech. 17:282-286.

Gray, M.E. 2000. Prescriptive use of transgenic hybrids for corn rootworms: an ominous cloud on the horizon? Crop Prot. Tech. Conf., 5-6, Jan 2000, Univ. Illinois, Champaign-Urbana.

Gray, M.E., K.L.Steffey and H. Oloumi-Sadeghi. 1993. Participatory on-farm research in Illinois cornfields: an evaluation of established soil insecticide rates and prevalence of corn rootworm (Coleoptera: Chrysomelidae). J. Econ. Entomol. 86:1473-1482.

Greiner, S., T. Rausch, U. Sonnewald, and K. Herbers.1999 Ectopic expression of a tobacco invertase inhibitor homolog prevents cold-induced sweetening of potato tubers. Nature Biotech. 17:708-711.

Grenier, L. 1998. Working with indigenous knowledge: A guide for researchers. IDRC, Ottawa.

Grierson, D. and R. Fray. 1994. Control of ripening in transgenic tomatoes. Euphytica 79:251-263.

Griffiths, B.S., I.E. Geoghegan and W.M. Robertson. 2000. Testing genetically engineered potato, producing the lectins GNA and Con A, on non-target soil organisms and processes. J. Appl. Ecol. 37:159-170.

Gruhn, P., F. Goletti and M. Yudelman. 2000. Integrated nutrient management, soil fertility, and sustainable agriculture: Current issues and future challenges. FAE Disc. Pap. 2. IFPRI, Washington DC.

Gupta, R. and A. Seth. 2007. A review of resource conserving technologies for sustainable management of the rice–wheat cropping systems of the Indo-Gangetic plains (IGP). Crop Prot. 26:436-447.

Gustafsson, B. and S. Li, 2000. Economic transformation and gender earnings gap in urban China. J. Popul. Econ. 13:305-329.

Haigler, C.H. and A.S. Holaday. 2001. Transgenic fiber-producing plants with increased expression of sucrose phosphate synthase. Patent Application WO 01/17333.

Hall, A. J. 2006. Public-private sector partnerships in a system of agricultural innovation: Concepts and challenges. Int. J. Tech. Manage. Sustainable Dev. 5:3-20.

Hall, A.J., B. Yoganand, R. Sulaiman, R. Raina, S. Prasad, G. Naik et al. 2004. Innovations in innovation: Reflections on partnership and learning. ICRISAT, Patancheru, and NCAP, New Delhi.

Hammond, B.G., J.L. Vicini, G.F. Hartnell, M.W. Naylor, C.D. Knight, E.H. Robinson et al. 1996. The feeding value of soybeans fed to rats, chickens, catfish and dairy cattle is not altered by genetic incorporation of glyphosate tolerance. J. Nutr. 126:717-727.

Hammond, B., R. Dudek, J. Lemen, and M. Nemeth. 2004. Results of a 13 week safety assurance study with rats fed grain from glyphosate tolerant corn. Food Chem. Toxicol. 42:1003-1014.

Han, F.X., J.S. Lindner and C. Wang. 2007. Making carbon sequestration a paying proposition. Naturwissenschaften 94:170-182.

Harrington, L.W., S. Fujisaka, P.R. Hobbs, C. Adhikary, G.S Giri and K. Cassaday. 1989. Wheat and rice in the hills: Farming systems, production techniques, and research issues for rice-wheat cropping patterns in the mid-hills of Nepal. Nepal Agric. Res. Council, CIMMYT, and IRRI.

Harrison, L.A., M.R. Bailey, M.W. Naylor, J.E. Ream, B.G. Hammond, D.L. Nida et al. 1996. The expressed protein in glyphosate-tolerant soybean, 5-enolypyruvylshikimate-3-phosphate synthase from Agrobacterium sp. Strain CP4, is rapidly digested in vitro and is not toxic to acutely gavaged mice. J. Nutr. 126:728-740.

Hashimoto, W., K. Momma, H.J. Yoon, S. Ozawa, Y. Ohkawa, T. Ishige et al. 1999. Safety assessment of transgenic potatoes with soybean glycinin by feeding studies in rats. Biosci. Biotech. Biochem. 63:1942-1946.

Haswidi, A. 2006. Lack of land and incentives hamper biofuel investment. The Jakarta Post 17 Nov 2006.

Hazell, P., P. Jagger and A. Knox. 2000. Technology, natural resources and the poor. IFPRI, Washington DC.

He, Y.K., J.G. Sun, X.Z. Feng, M. Czako and L. Marton. 2001. Differential mercury volatilization by tobacco organs expressing a modified merA gene. Cell Res. 11:231-236.

Heisey, P.W., M.A. Lantican and H.J. Dubin. 2002. Impacts of inizernational wheat breeding research in developing countries, 1966-1997. CIMMYT, Mexico.

Hellmich, R.L., B.D. Siegfried, M.K. Sears, D.E. Stanley-Horn, M.J. Daniels, H.R. Mattila et al. 2001. Monarch larvae sensitivity to Bacillus thuringiensis purified proteins and pollen. PNAS 98:11925-11930.

Henao, J. and C. Baanante. 1999. Nutrient depletion in the agricultural soils of Africa. 2020 Vision Brief 62. IFPRI, Washington DC.

Hilaire, A. and Y.Yang. 2003. The United States and the New Regionalism/Bilateralism. Working Pap. WP/03/206. IMF, Washington DC.

Hilbeck, A., M. Baumgartner, P.M. Fried and F. Bigler. 1998a. Effects of transgenic Bacillus thuringiensis corn-fed prey on mortality and development time of immature Chrysoperla carnea (Neuroptera: Chrysopidae). Environ. Entomol. 27:481-487.

Hilbeck, A., W.J. Moar, M. Pusztai-Carey, A. Filippini and F. Bigler. 1998b. Toxicity of Bacillus thuringiensis Cry1Ab toxin to the predator Chrysoperla carnea (Neuroptera: Chrysopidae). Environ. Entomol. 27:1255-1263. Available at http://www.biotech-info.net/rootworm.html.

Ho, M.W. 2007. How to stop bird flu instead of the vaccine-antiviral model. Press Inst. Sci. Society (I-SIS) Press Release, 14 May 2007. www.i-sis.org.uk

Hossain, F., C. Pray, Y. Lu, J. Huang, C. Fan and R. Hu. 2004. Genetically modified cotton and farmers' health in China. Int. J. Occup. Environ. Health 10: 296-303.

Huang, J., C. Pray, and S. Rozelle. 2002. Enhancing the crops to feed the poor. Nature 418:678-684.

Huang, J., S. Rozelle, C. Pray, and Q. Wang. 2002. Science 295:674-677.

Hulme, D. and A. McKay. 2005. Identifying and understanding chronic poverty—beyond monetary measures. UNDP-Int. Poverty Centre Conf. Many Dimensions of Poverty, Brasilia, 29-31 Aug. UNDP, NY.

Humphrey, J. 2006. Policy implications of trends in agribusiness value chains. Eur. J. Dev. Res. 18:572-592.

Hunter, M.D. 2000. Between hyperbole and hysteria. Entomological issues and the deployment of transgenic plants. Agric. Forest Entomol. 2:77-84.

Husain, H.J., and A.K. Raina. 2004. Adoption of agriculture technology before and after the introduction of watershed management project. Indian J. Forestry 27:201-205.

Hussain, I., L. Raschid, M.A. Hanjra, F. Marikar and W. van der Hoek. 2002. Wastewater use in agriculture: Review of impacts and methodological issues in valuing impacts. Working Pap. 37. IWMI, Colombo.

IFAD. 2002. Regional strategy paper-Asia and the Pacific. IFAD, Rome.

IFAD. 2005a. Organic agriculture and poverty reduction in Asia: China and India focus. Thematic evaluation. Rep. 1664. IFAD, Rome.

IFAD. 2005b. Gender assessment report: Impact of IFAD's commitment to women in China and insights for gender mainstreaming. Rep. 640. IFAD, Rome.

IPCC. 1996. Scientific-technical analysis of impacts, adaptations, and mitigation of climate change. Report of Working Group II. Climate change 1995: Intergovernmental Panel on Climate Change second assessment report. Cambridge Univ. Press, UK.

IPCC. 2000. Intergovernmental Panel on Climate Change (IPCC) special report: Land use, land-use change, and forestry. Cambridge Univ. Press, UK.

IRRI. 1994. Filling the world's rice bowl. IRRI, Los Baños.

Jackson, W.J. and P.F. Moore. (1998). The role of indigenous use of fire in forest management and conservation. Int. Seminar. Pap. Cultivating forests: Alternative forest management practices and techniques for community forestry. Reg. Community Forestry Training Centre, Bangkok.

James, C. 2005. Preview: Global status of commercialized biotech/GM crops: 2005. ISAAA Brief 34. Ithaca NY.

Janzen, H.H. 2004. Carbon cycling in earth systems—a soil science perspective. Agric. Ecosyst. Environ. 104:399-417.

Jesse, L.C.H. and J.J. Obrycki. 2000. Field deposition of Bt transgenic corn pollen: lethal effects on the monarch butterfly. Oecologia 1125:241-248.

Jeyaratnam, J. 1990. Acute pesticide poisoning: a major global health problem. World Health Stat. Q. 43:139-144.

Johnson, M. 1992. LORE: Capturing traditional environmental knowledge. Dene Cultural Inst. and IDRC, Ottawa.

Jorg, R., M. Battini and F. Bigler. 2003. Transgenic wheat with enhanced fungal resistance causes no effects on Folsomia candida (Collembola: Isotomidae). Pedobiolgia 47:141-147.

Kalamaki, M.S., M.H. Harpster, J.M. Palys, J.M. Labavitch, D.S. Reid and D.A. Brummell. 2003a. Simultaneous transgenic suppression of LePG and LeExp1 influences rheological properties of juice and concentrates from a processing tomato variety. J. Agric. Food Chem. 51:7459-7464.

Kalamaki, M.S., A.L.T. Powell, K. Struijs, J.M. Labavitch, D.S. Reid and A.B. Bennett. 2003b. Transgenic overexpression of expansin influences particle size distribution and improves viscosity of tomato juice and paste. J. Agric. Food Chem. 51:7465-7471.

Kan, C.A. and G.F. Hartnell. 2004. Evaluation of broiler performance when fed Roundup-Ready wheat (Event MON 71800), control, and commercial wheat varieties. Poultry Sci. 83:1325-1334.

Kanai, M., B. Titapiwatanakun and D.R. Stolz. 1999. Effects of trade liberalization on agriculture in selected Asia. Proc. Workshop, Bogor, 5-8 Oct 1999. CGPRT NO. 38. Bogor, Indonesia.

Kaosa-ard, M.S. and B. Rerkasem. 2000. The growth and sustainability of Agriculture in Asia. Asian Development Bank's study of rural Asia: Beyond the green revolution. Vol. 2. Oxford Univ. Press, UK.

Kar, A. 1991. Measurement of risk premium: A case study of some areas in West Bengal. Indian J. Reg. Sci. 23:43-55.

Kelkar, G., D. Nathan, and P. Walter. 2003. Gender relations in forest societies in Asia: Patriarchy at odds. Sage, New Delhi.

Khush, G.S. 2005. What it will take to feed 5.0 billion rice consumers in 2030. Plant Mol. Biol. 59:1-6.

Kijne, J. 1994. Irrigation and drainage for the rice-wheat production system: management aspects. p. 112-126. *In* R.S. Paroda et al. (ed) Sustainability of rice-wheat systems in Asia. RAPA Publ. 1994/11. FAO, Bangkok.

Kirchmann, H. and L. Bergstroem. 2001. Do organic farming practices reduce nitrate leaching? Comm. Soil Sci. Plant Anal. 32:997-1028.

Kraemer, K. L. and J. Dedrick. 1998. The Asian computer challenge: Threat or opportunity for the U.S.? Oxford Univ. Press, NY.

Kramer, S.B., J.P. Reganold, J.D. Glover, B.J.M. Bohannan, and H.A. Mooney. Reduced nitrate leaching and enhanced denitrifier activity and efficiency in organically fertilized soils. PNAS 103:4522-4527.

Krattiger, A.F. 1998. The importance of ag-biotech to global prosperity. ISAAA Brief 6. ISAAA, Ithaca NY.

Krishna, S. 2004. Livelihood and gender equity in community resource management. Sage, New Delhi.

Kulkarni, A.V., I.M. Bahuguna, B.P. Rathore, S.K. Singh, S.S. Randhawa, R.K. Sood et al. 2007. Glacial retreat in Himalaya using Indian Remote Sensing satellite data. Current Sci. 92:69-74.

Lambrou, Y., and G. Piana. 2006a. Gender: The missing components of the response to climate change. FAO, Rome.

Lambrou, Y. and G. Piana. 2006b. Energy and gender issues in rural sustainable development. FAO, Rome.

Lamphear, B.J., J.M. Jilka, L. Kesl, M. Welter, J.A. Howard and S.J. Streatfield. 2004. A corn-based delivery system for animal vaccines: an oral transmissible gastroenteritis virus boosts lactogenic immunity in swine. Vaccine 22:2420-2424.

Lantican, M.A., P.L. Pingali, and S. Rajaram. 2003. Is research on marginal lands catching up? The case of unfavourable wheat growing environments. Agric. Econ. 29:353-361.

Leeuwis, C. and A.W. Van den Ban. 2004. Communication for rural innovation: Rethinking agricultural extension. 3rd ed. Blackwell Sci., Oxford, UK.

Lipton, M. and R. Longhurst. 1989. New seeds and poor people. Johns Hopkins Univ. Press, Baltimore.

Liu, Y., G. Wang, J. Liu, X. Peng, Y. Xie, J. Dai et al. 1999. Transfer of E. Coli gutD gene into maize and regeneration of salt-tolerant transgenic plants. Sci. China Ser. C: Life Sci. 42:90-95.

Loges, R., M.R. Kelm, and F. Taube. 2006. Nitrogen balances, nitrate leaching and energy efficiency of conventional and organic farming systems on fertile soils in northern Germany. Adv. Geoecol. 38:407-414.

Lopez, R. and D.N. Ferro. 1995. Larviposition response of Myioharus-Doryphorae (Diptera, Tachirridae) to Colorado Potato Beetle (Coleoptera, Chrysomelidae) larvae treated with lethal and sublethal doses of Bacillus thuringiensis Berliner subsp. plants. J. Econ. Entomol. 88:870-874.

Losey, J.E., L.S. Rayor, and M.E. Carter. 1999. Transgenic pollen harms monarch larvae. Nature 399:214.

Lucca, P. 1999. Genetic engineering approaches to improve the bioavailability and the level of iron in rice grains. Gen. Meeting Int. Prog. Rice Biotech., Phuket, Thailand. 20-24 Sep 1999.

Lukow, T., P.F. Dunfield and W. Liesack. 2000. Use of the T-RFLP technique to assess spatial and temporal changes in the bacterial community structure within an agricultural soil planted with transgenic and non-transgenic potato plants. FEMS Microbiol. Ecol. 32:241-247.

Majluf, L.A. 2004. Swimming in the spaghetti bowl: Challenges for developing countries under the "new regionalism". Policy issues in international trade and commodities study ser. 27. UNCTAD, Geneva.

Manduna, C. 2006. A review of the results of the 6th WTO Hong Kong ministerial conference—considerations for African, Caribbean and Pacific countries. Working Pap. No. 6/2006. Trade Law Center for Southern Africa (TRALAC), Cape Town.

Marroquin, L.D., D. Elyassnia, J.S. Griffiths, J.S. Feitelson and R.V. Aroian. 2000. Bacillus thuringiensis (Bt) toxin susceptibility and isolation of resistance mutants in the nematode Caenorhabditis elegans. Genetics 155:1693-1699.

Martin, V., A. Forman and J. Lubroth. 2006. Preparing for highly pathogenic avian

influenza. Anim. Prod. Health Manual 3. FAO, Rome.
Mason, A.D. and E.M. King. 2001. Engendering development: Through gender equality in rights, resources and voice. World Bank Policy Res. Rep. Oxford Univ. Press, UK.
Mehta, A.K. and S. Gupta. 2005. The impact of HIV/AIDS on women care givers in situations of poverty: Policy issues. UNIFEM, Delhi.
Meier, M.S., and A. Hilbeck. 2001. Influence of transgenic Bacillus thuringiensis corn-fed prey on prey preference of immature Chrysoperla carnea (Neuroptera: Chrysopidae). Basic Appl. Entomol. 2:35-44.
Melero, S., J.C.R. Porras, J.F. Herencia and E. Madejon. 2006. Chemical and biochemical properties in a silty loam soil under conventional and organic management. Soil Till. Res. 90:162-170.
Menon, J. 2007. Bilateral trade agreements. Asian-Pacific Econ. Lit. 21:29-47.
Mody, A. and C. Dalman. 1992. Performance and Potential of Information Technology: An international perspective. World Dev. 20:1703-1719.
Momma, K., W. Hashimoto, H.J. Yoon, S. Ozawa, Y. Fukuda, S. Kawai et al. 2000. Safety assessment of rice genetically modified with soybean glycinin by feeding studies on rats. Biosci. Biotech. Biochem. 64:1881-1886.
Morse, S., R. Bennett and Y. Ismael. 2004. Why Bt cotton pays for small-scale producers in South Africa. Nature Biotech. 22:379-380.
Nader, W. and N. Mateo. 1998. Biodiversity: Resource for new products, development, and self-reliance. p. 181-199. W. Barthlott and M. Winiger (ed) Biodiversity: A challenge for development research and policy. Springer Life Sci., NY.
Nagarajan, L., P.G. Pardey, and M. Smale. 2006. Local seed systems for millet crops in marginal environments of India: Industry and policy perspectives. EPT Disc. Pap. 151. IFPRI, Washington DC.
Narayan, D., R. Patel, K. Schafft, A. Rademacher and S. Koch-Schulte. 2000. Voices of the poor: Can anyone hear us? Oxford Univ. Press and World Bank, Washington DC.
NAS. 2000. Transgenic plants and world agriculture. Nat. Acad. Sci., Washington DC.
NAS. 2003. Environmental effects of transgenic crops: The scope and adequacy of regulation. Nat. Acad. Sci., Washington DC.
Nguyen, M.L. and R.J. Haynes. 1995. Energy and labour efficiency for three pairs of conventional and alternative mixed cropping (pasture-arable) farms in Canterbury, New Zealand. Agric. Ecosyst. Environ. 52:163-172.
Noteborn, H.P.J.M., M.E. Bienenmann-Ploum, J.H.J. Van Den Berg, G.M. Alink, L. Zolla, A. Reynaerts et al. 1995. Safety assessment of the Bacillus thuringiensis insecticidal crystal protein CRY1A(b) expressed in transgenic tomatoes. ACS Symp. Ser. 605:134-147.
NRC. 1989. Alternative agriculture. Nat. Res. Council, Nat. Acad. Press, Washington DC.
NRC. 1997. Precision agriculture in the 21st century: Geospatial and information technologies in crop management. Nat. Res. Council, Nat. Acad. Press, Washington DC.
Nuffield Council on Bioethics. 1999. Genetically modified crops: The ethical and social issues. Nuffield Council Bioethics, London.
Oberdörster, Eva. 2004. Manufactured nanomaterials (fullerenes, C_{60}) induce oxidative stress in the brain of juvenile large-mouth bass. Environ. Health Perspect. 112:1058-1062.
Ohga, K., and T. Koizumi. 2007. Biofuels policies in Asia: Trade effects on world agricultural and biofuels trade. 2 Mar 2007. Agric. Outlook Forum. Available at http://www.usda.gov/agency/oce/forum/2007%20Speeches/PDF%20PPT/K%20Ohga.pdf. USDA, Washington DC.
Olk, D.C., K.G. Cassman, G.C. Simbahan, P.C. Sta-Cruz, S. Abdulrachman, R. Nagarajan et al. 1999. Interpreting fertilizer use efficiency in relation to soil nutrient-supplying capacity, factor productivity, and agronomic efficiency. p. 45-55. *In* V. Balasubramanian et al. (ed) Resource management in rice systems: Nutrients. Kluwer Acad. Publ., NY.
Olsen, S., Y. Laosiritaworn, S. Pattanasin, P. Prapasiri and S.F. Dowell. 2005. Poultry-handling practices during avian flu outbreak, Thailand. Emerg. Infect. Dis. 11:1601-1603.
Orr, D.B., and D.A. Landis. 1997. Oviposition of European corn borer (Lepidoptera: Pyralidae) and impact of natural enemy populations in transgenic versus isogenic corn. J. Econ. Entomol. 90:905-909.
Osmani, S.R. 1998. Did the green revolution hurt the poor? A reexamination of the early critique. p. 193-212. *In* P.L. Pingali and M. Hossain (ed) Impact of rice research. IRRI, Manila.
Panagariya, A. and J. Bhagwati. 2005. Bilateral treaties are a sham. Financial Times. 13 July 2005. Available at http://www.cfr.org/publication/6118/bilateral_trade_treaties_are_a_sham.html.
Parakrama, S. and M. Thibbotuwawa. 2006. Customs mapping and analysis of South Asian agricultural trade liberalization effort. Working Pap. 2606, Asia-Pacific Res. and Training Network on Trade (ARTNeT). An initiative of UNESCAP and IDRC, Canada.
Patnaik, U. 2004. External trade domestic employment and food security: recent outcomes of neo-liberal economic reforms. Proc. Conf. The question of Asia and the new global order, Asia/Pacific Studies Inst., Duke Univ., 1-2 Oct 2004.
Paul, G.C., M. Rahman, N.U. Khan and A.B.M.M. Rahman. 2005. Contribution of rice mill ash and press mud with inorganic fertilizers to sugarcane production in Old Himalayan Piedmont Plain soils of Bangladesh. Korean J. Crop Sci. 50:108-111.
Paul, S. 1989. Green Revolution and income distribution among farm families in Haryana 1969-70 to 1982-83. Econ. Polit. Weekly 24:A154-A158.
Peng, J., E. Richards, N.M. Hartley, G.P. Murphy, K.M. Devos, J.E. Flintham et al. 1999. "Green revolution" genes encode mutant gibberellin response modulators. Nature 400:256-261.
Peng, S., G. S. Khush and K. G. Cassman. 1994. Evolution of the new plant ideotype for increased yield potential. p. 5-20. *In* K.G. Cassman (ed) Breaking the yield barrier: Proc. workshop on rice yield potential in favourable environments. IRRI, Manila.
Persley, G.J., and J.J. Doyle. 1999. Biotechnology for developing country agriculture: Problems and opportunities. Brief 1. 2020 Vision for food, agriculture and the environment Focus 2. IFPRI, Washington DC.
Phipps, R.H. and J.R. Park. 2002. Environmental benefits of genetically modified crops: Global and European perspectives on their ability to reduce pesticide use. J. Anim. Feed Sci. 11:1-18.
Picton, S., J.E. Gray and D. Grierson. 1995. The manipulation and modification of tomato fruit ripening by expression of antisense RNA in transgenic plants. Euphytica 85:193-202.
Pigato, M. 2001. Information and communication technology, poverty, and development in Sub-Saharan Africa and South Asia. Africa Region Working Pap. Series 20. World Bank, Washington DC.
Pilcher, C.D., J.J. Obrycki, M.E. Rice and L.C. Lewis. 1997. Preimaginal development, survival, and field abundance of insect predators on transgenic Bacillus thuringiensis corn. Environ. Entomol. 26:446-454.
Pingali, P.L. and M.W. Rosegrant. 1994. Confronting the environmental consequences of the Green Revolution in Asia. EPTD Disc. Pap. 2. IFPRI, Washington DC.
Pingali, P.L. and P.A. Roger. 1995. Impact of pesticides on farmer health and the rice environment. Kluwer Publ., Norwell MA and IRRI, Los Baños.
Poirier, Y. 2001. Production of polyesters in transgenic plants. Adv. Biochem. Engineer. Biotech. 71:209-240.
Ponsard-Sergine, A., P. Guitierrez and N.J. Mills. 2002. Effect of Bt toxin (Cry1Ac) in transgenic cotton on the adult longevity of four heteropteran predators. Environ. Entomol. 31:1197-1205. World

population datasheet. Population Reference Bureau. 2002. Washington DC.

Powell, A.L.T., M.S. Kalamaki, P.A. Kurien, S. Gurrieri and A.B. Bennett. 2003. Simultaneous transgenic expression of LePG and LeExp1 influences fruit texture and juice viscosity in a fresh market tomato variety. J. Agric. Food Chem. 51:7450-7455.

Pretty J. and R. Hine. 2000. The promising spread of sustainable agriculture in Asia. Nat. Resourc. Forum 24:107-121.

Prijono, D. and E. Hassan. 1993. Laboratory and field efficacy of neem (*Azadirachta indica* A: Juss) extracts against two broccoli pests. Zeitschrift für Pflantzenkrankheiten und Pflantzenschutz 100:354-570.

Principe, P.P. 1989. Valuing the biodiversity of medicinal plants. p. 79-124. *In* O. Akerele et al. (ed) The conservation of medicinal plants. Cambridge Univ. Press, UK.

Qaim, M. and D. Zilberman. 2003. Yield effects of genetically modified crops in developing countries. Science 299:900-902.

Rai, S.C. and A. Gurung. 2005. Raising awareness of the impacts of climate change: Initial steps in shaping policy in Nepal. Mount. Res. Dev. 25:316-320.

Rajitha K., C.K. Mukherjee and R. Vinu Chandran. 2007. Applications of remote sensing and GIS for sustainable management of shrimp culture in India. Aquacult. Engineer. 36:1-17.

Raloff, J. 2005. Nano hazards: Exposure to minute particles harms lungs, circulatory system. Available at http://www .sciencenews.org/articles/20050319/fob1. asp. Sci. News Online 167:179.

Ranjana, W. 2004. Thailand embarks on the nano path to better rice and silk. Bangkok Post, 21 Jan 2004. Available at http:// www.smalltimes.com/document_display .cfm?document_id=7266.

Raynolds, L. 2000. Re-embedding global agriculture: The international organic and fair trade movements. Agric. Human Values 17:297-309.

Reardon, T., and R. Hopkins. 2006. The supermarket revolution in developing countries: policies to address emerging tensions among supermarkets, suppliers and traditional retailers. Eur. J. Dev. Res. 18:522-545.

Regierer, B., A.R. Fernie, F. Springer, A. Perez-Melis, A. Leisse, K. Koehl et al. 2002. Starch content and yield increase as a result of altering adenylate pools in transgenic plants. Nature Biotech. 20:1256-1260.

Rerkasem, B., and K. Rerkasem. 2002. Agrodiversity for *in situ* conservation of Thailand's native rice germplasm. Chiang Mai University J. 1:129-148. Available at http://cmuj.chiangmai.ac.th/full/2002/ may2002-1f.pdf.

Reynolds, M.P. and N.E. Borlaug. 2006. Impacts of breeding on international collaborative wheat improvement. J. Agric. Sci. 144:3-17.

Riddick, E.W. and P. Barbosa. 1998. Impact of Cry3A-intoxicated Leptinotarsa decemlineata, (Coleoptera: Chrysomelidae) and pollen on consumption, development, and fecundity of Coleomagilla maculata (Coleoptera: Coccinellidae). Ann. Entomol. Soc. Am. 91:303-307.

Rodrik, D. 2001. The global governance of trade: as if development really mattered. Background paper, UNDP project on Trade and Sustainable Human Dev. UNDP, NY.

Rosegrant, M.W. 1997. Water resources in the twenty-first century: Challenges and implications for action. FAE Disc. Pap. 20. IFPRI, Washington DC.

Roush, R.T. 1994. Managing pests and their resistance to Bacillus thuringiensis—can transgenic crops be better than sprays. Biocontr. Sci. Tech. 4:501-516.

Rozelle, S., X.Y. Dong, L. Zhang, and A.S. Mason. 2002. Gender wage raps in post-reform rural China. Pacific Econ. Rev. 7:157-179.

Samaratunga, P., and M. Thibbotuwawa. 2006. Mapping and analysis of South Asian agricultural trade liberalization effort. Available at http://www.unescap.org/tid/ artnet/pub/wp2606.pdf. Working Paper 2606. Asia Pacific Research and Training Network on Trade (ARTNeT), UNESCAP, IDRC, Canada.

SAPRIN. 2002. The policy roots of economic crisis and poverty: A multicountry participatory assessment of structural adjustment. 1st ed. Structural Adjustment Participatory Rev. Int. Network (SAPRIN), Washington DC.

Satyanarayana, V., P.V.V. Prasad, V.R.K. Murthy and K.J. Boote. Influence of integrated use of farmyard manure and inorganic fertilizers on yield and yield components of irrigated lowland rice. J. Plant Nutr. 25:2081-2090.

Scheller, J., K.H. Guhrs, F. Grosse and U. Conrad. 2001. Production of spider silk proteins in tobacco and potato. Nature Biotech. 19:573-577.

Schuler, T.H., G.M. Poppy, B.R. Kerry and I. Denholm. 1999. Potential side effects of insect-resistant transgenic plants on arthropod natural enemies. Trends Biotech. 17:210-216.

Seguin, B., D. Arrouays, J. Balesdent, J-F. Soussana, A. Bondeau, P. Smith et al. 2007. Moderating the impact of agriculture on climate. Agric. Forest Meteorol. 142:278-287.

Serageldin, I. 1999. Biotechnology and food security in the 21st century. Science 285:387-389.

Sidhu, R.S., B.G. Hammond, R.L. Fuchs, J.N. Mutz, L.R. Holden, B. George et al. 2000. Glyphosate-tolerant corn: The composition and feeding value of grain from glyphosate-tolerant corn is equivalent to that of conventional corn (Zea mays L.). J. Agric. Food Chem. 48:2305-2312.

Singh, P., M. Arora and N.K. Goel. 2006. Effect of climate change on runoff of a glacierized Himalayan basin. Hydrol. Process. 20:1979-1992.

Singh, R.B. and R.S. Paroda. 1994. Sustainability and productivity of rice-wheat systems in the Asia-Pacific region: research and technology development needs. p. 1-36. *In* R.S. Paroda et al. (ed) Sustainability of rice-wheat systems in Asia. RAPA Publ. 1994/11. FAO, Rome.

Singh, V. 2006. Productivity and economics of rice (Oryza sativa)-wheat (Triticum aestivum) cropping system under integrated nutrient-supply system in recently reclaimed sodic soil. Indian J. Agron. 51:81-84.

Singla-Pareek, S.L., M.K. Reddy, and S.K. Sopory. 2003. Genetic engineering of the glyoxalase pathway in tobacco leads to enhanced salinity tolerance. PNAS 100:14672-14677.

Siritunga, D. and R.T. Sayre. 2003. Generation of cyanogens-free cassava. Planta 217:367-373.

Srinivasan, T.N. 2002. Developing countries and the multilateral trading system after Doha. Working Pap. 842. Econ. Growth Center, Yale University.

Stanford, K., J.L. Aalhus, M.E.R. Dugan, G.L. Wallins, R. Sharma and T.A. McAllister. 2003. Effects of feeding transgenic canola on apparent digestibility, growth performance and carcass characteristics of lambs. Can. J. Anim. Sci. 83:299-305.

Stanley-Horn, D.E., G.P. Dively, R.L. Hellmich, H.R. Mattila, M.K. Sears, R. Rose et al. 2001. Assessing the impact of Cry1Ab-expressing corn pollen on monarch butterfly larvae in field studies. PNAS 98:11921-11936.

Stoorvogel, J.J. and E.M.E. Smaling. 1990. Assessment of soil nutrient depletion in sub-Saharan Africa. Rep. 28, Vol. 1-4. The Winand Staring Centre, Wageningen.

Strutt, A., and S. Lim. 2003. Trade liberalisation and rural poverty in Asia. Inst. Social Sci. Bull. 2:1-21.

Suhariyanto, K. and C. Thirtle. 2001. Asian agricultural productivity and convergence. J. Agric. Econ. 52:96-110.

Sujaya, C.P. 2006. Climbing a long road: Women in agriculture in India, ten years after Beijing. M.S. Swaminathan Res. Foundation, Chennai.

Sulaiman, V.R. and A.J. Hall. 2003. The emergence of extension plus in India: A future for extension beyond technology transfer? p. 19-30. *In* G. Alex and W. Rivera (ed) Extension reforms for development. Vol. 1: Decentralized systems; case studies of international initiatives. ARD Disc. Pap. 8. The World Bank, Washington DC.

Sun, X. 2007. Crops: A new source for ethanol. China Daily 13 June 2007.

Tabashnik, B.E. 1994. Delaying insect adaptation to transgenic plants: Seed mixtures and refugia reconsidered. Phil. Trans. R. Soc. London B 255:7-12.

Tabashnik, B.E., Y. Carriere, T.J. Dennehy, S. Morin, M.S. Sisterson, R.T. Roush et al. 2003. Insect resistance to transgenic Bt crops: Lessons from the laboratory and field. J. Econ. Entomol. 96:1031-1038.

Tamang, D. 1993. How hill farmers manage their soils. p. 164-181. *In* D. Tamang et al (ed) Indigenous management of natural resources in Nepal. Winrock Int., Kathmandu.

Tandon, H.L.S. 1998. Use of external inputs and the state of efficiency of plant nutrient supplies in irrigated cropping systems in Uttar Pradesh, India. p. 199-235. *In* P. Gruhn et al. (ed) Proc. IFPRI/FAO workshop on soil fertility, plant nutrient management, and sustainable agriculture: The future through 2020. IFPRI, Washington DC and FAO, Rome.

Tandon, H.L.S. 1995. Major nutritional constraints to crop production and the soil fertility management strategies in different agroclimatic regions of Asia. p. 43-72. *In* Proc. Int. Potash Inst. colloquium on potassium in Asia: Balanced fertilization to increase and sustain agricultural production. 21-24 Feb 1995. Chiang Mai, Thailand. Int. Potash Inst., Basel.

Taylor, D.C., V. Katavic, J. Zou, S.L. MacKenzie, W.A. Keller, J. An et al. 2001. Field testing of transgenic rapeseed cv. Hero transformed with a yeast sn-2 acyltransferase results in increased oil content, erucic acid content and seed yield. Mol. Breed. 8:317-322.

Teshima, R., H. Akiyama, H. Okunuki, J.I. Sakushima, Y. Goda, H. Onodera et al. 2000. Effect of GM and non-GM soybeans on the immune system of BN rats and B10A mice. J. Food Hyg. Soc. Japan 41:188-193.

Thanavala, Y., Y.F. Yang, P. Lyons, H.S. Mason and C. Arntzen. 1995. Immunogenicity of transgenic plant-derived hepatitis B surface antigen. PNAS 92:3358-3361.

Thanh, T.D., T.D. Lan and P.V. Luong. 2005. Protecting the marine environment: International assistance and the Vietnam. p. 183-200. *In* P.G. Harris (ed) Confronting environmental change in East and Southeast Asia. UNU Press, Tokyo.

The, C., H. Calba, C. Zonkeng, E.L.M. Ngonkeu, V.O. Adetimirin, H.A. Mafouasson et al. 2006. Responses of maize grain yield to changes in acid soil characteristics after soil amendments. Plant Soil 284:45-57.

Tiensin, T., P. Chaitaweesub, T. Songserm, A. Chaisingh, W. Hoonsuwan, C. Buranathai et al. 2005. Highly pathogenic avian influenza H5N1, Thailand, 2004. Emerg. Infect. Dis. 11:1664-1672.

Timsina, N.P. 2000. Seed management system in Nepal: Where does it stand? An assessment of the existing seed situation and related policies in Nepal. Action Aid Nepal, Kathmandu.

Torstensson, G., H. Aronsson and L. Bergström. 2006. Nutrient use efficiencies and leaching of organic and conventional cropping systems in Sweden. Agron. J. 99:960-972.

UK Royal Society and the Royal Academy of Engineering. 2004. Nanoscience and nanotechnologies: Opportunities and uncertainties. Available at http://www .nanotec.org.uk/report/Nano%20 report%202004%20fin.pdf. Clyvedon Press, Cardiff.

Umetsu, C., T. Lekprichakul, and U. Chakravorty. 2003. Efficiency and technical change in the Philippine rice sector: A Malmquist total factor productivity analysis. Am. J. Agric. Econ. 85:943-963.

UN. 2007. The millennium development goals report. United Nations, NY.

UNDP. 1999. Human development report. Globalization with a human face. UNDP, Oxford Univ. Press, UK.

UNDP. 2006. Human development report. Beyond scarcity: Power, poverty and the global water crisis. UNDP, Palgrave Macmillan, NY.

UNESCAP. 1995. State of the environment in the Asia-Pacific report 1995. UNESCAP, NY.

UNESCAP. 2002. Organic agriculture and rural poverty alleviation: Potential and best practices in Asia. UNESCAP, NY.

UNESCAP. 2002b. International migration: An emerging opportunity for the socio-economic development of the ESCAP Region. Social Policy Pap. 6. UNESCAP, NY.

Upreti, B.R. and Y.G. Upreti. 2002. Factors leading to agro-biodiversity loss in developing countries: The case of Nepal. Biodivers. Conserv. 11:1607-1621.

Van Mele, P., A. Salahuddin and N.P. Magor. 2005. Innovations in rural extension: Case studies from Bangladesh. CABI Publ., UK.

Van Veenhuizen, R. 2006. Cities farming for the future—urban agriculture for green and productive cities. RUAF Foundation, Leusden, The Netherlands.

Vasil, I.K. 2003. The science and politics of plant biotechnology—a personal perspective. Nature Biotech. 21:849-851.

Vepa, S.S. 2004. Feminization of agriculture and marginalization of their economic stake. Econ. Polit. Weekly 40:2563-2568.

Verdin J., C. Funk, G. Senay, and R. Choularton. 2005. Climate science and famine early warning. Phil. Trans. R. Soc. B 360:2155-2168.

Vermani, K. and S. Garg. 2002. Herbal medicines for sexually transmitted diseases and AIDS. J. Ethnopharmacol 80:49-66.

Volkmar, C., T. Kreuter, L. Richter, M.L. Al-Hussein, D. Jany, K. Schmutzler et al. 2000. Ecological studies accompanying the cultivation of transgenic and conventional rape plants in the Central German region from 1996 to 1998. Zeitschrift fuer Pflanzenkrankheiten und Pflanzenschutz. J. Plant Dis. Prot. 27:337-345.

Volkmar, C.M., M.L. Al-Hussein, D. Jany, I. Hunold, L. Richter, T. Kreuter et al. 2003. Ecological studies on epigeous arthropod populations of transgenic sugar beet at Friemar (Thuringia, Germany). Agric. Ecosyst. Environ. 95(1):37-47.

Von Moltke, K. and F. Spaninks. 2000. Traditional Chinese medicine and species endangerment: an economic research agenda. Working Pap. 32. IIED, London.

Wang, M. 2001. Possible adoption of precision agriculture for developing countries at the threshold of the new millennium. Comput. Electron. Agric. 30:45-50.

Wang, M. and F. Cai. 2006. Gender sage differentials in China's urban labour market. Res. Pap. No. 2006/146. UN Univ., World Inst. Dev. Econ. Res., Tokyo.

Warren, D.M. Using indigenous knowledge in agricultural development. Disc. Pap. 127. World Bank, Washington DC.

Way, M.J., and K.L. Heong. 1994. The role of biodiversity in the dynamics and management of insect pests of tropical rice—a review. Bull. Entomol. Res. 84:567-587.

Webster, R.G., M. Peiris, H. Chen and Y. Guan. 2006. H5N1 outbreaks and enzootic influenza. Emerg. Infect. Dis. 12:3-8.

Whitten, M. and W.H. Settle. 1998. The role of the small-scale farmer in preserving the link between biodiversity and sustainable agriculture. Frontiers in biology: the challenges of biodiversity, biotechnology and sustainable agriculture. p. 187-207. *In* Proc. IUBS Symp. Taipei, 16-22 Nov 1997.

WHO Global Influenza Program Surveillance Network. 2005. Evolution of H5N1 avian influenza viruses in Asia. Emerging Infect. Dis. 11:1515-1521.

World Bank. 1991. A World Bank country report: Gender and poverty in India. World Bank, Washington DC.

World Bank. 2006. Enhancing agricultural innovation: How to go beyond strengthening agricultural research systems. World Bank, Washington DC.

Wu, W., J.E. Thies, Q. Ye and H. Min. 2003. Effect of Bt transgenic rice straw on the bacterial community diversity in a flooded soil. Ann. Meetings Abstr.. ASA, CSSA, and SSSA, Madison WI.

Wu, W., Q. Ye, H. Min, and J.E. Thies. 2003. Microbial enzyme activities and bacterial community composition in the rhizosphere of Bt-transgenic rice as compared to its non-transgenic cultivar. Abstr. Ann. Meeting Am. Soc. Microbiology, Washington DC.

Wu, K., and Y. Guo. 2005. The evolution of cotton pest management practices in China.

Ann. Rev. Entomol. 50:31-52.
Ye, X., S. Al-Babili, A. Kloti, J. Zhang, P. Lucca, P. Beyer et al. 2000. Engineering the provitamin A (beta-carotene) biosynthetic pathway into (carotenoid-free) rice endosperm. Science 287:303-305.
Yussefi, M. and H. Willer. 2003. The world of organic agriculture—statistics and future prospects. Available at http://www.ionenviro.com/pdf/ORGANIC%20STUDY%202003.pdf. IFOAM, Bonn.
Zangerl, A.R., D. McKenna, C.L. Wraight, M. Carroll, P. Ficarello, R. Warner et al. 2001. Effects of exposure of exposure to event 176 Bacillus thuringiensis corn pollen on monarch and black swallowtail caterpillars under field conditions. PNAS 98:11908-11912.
Zhang, H.X., and E. Blumwald. 2001. Transgenic salt tolerant tomato plants accumulate salt in foliage but not in fruit. Nature Biotech. 19:765-768.
Zhang, H.X., J.N. Hodson, J.P. Williams and E. Blumwald. 2001. Engineering salt-tolerant Brassica plants: Characterization of yield and seed oil quality in transgenic plants with increased vacuolar sodium accumulation. PNAS 98:12832-12836.
Zwahlen, C., W. Nentwig, F. Bigler, and A. Hilbeck. 2000. Tritrophic interactions of transgenic Bacillus thuringiensis corn, Anaphothrips obscurus (Thysanoptera: Thripidae), and the predator Orius majusculus (Heteroptera: Anthocoridae). Environ. Entomol. 29:846-850.

Annex A
ESAP Authors and Review Editors

Australia
David J. Connor • University of Melbourne
Anna Matysek • Concept Economics
Girija Shrestha • Monash Asia Institute, Monash University

Bangladesh
Wais Kabir • Bangladesh Agricultural Research Council (BARC)
Karim Mahmudul • Bangladesh Shrimp and Fish Foundation

Barbados
Carl B. Greenidge • CFTC and Caribbean Regional Negotiating Machinery

Canada
M. Monirul Qader Mirza • Environment Canada and University of Toronto, Scarborough

China
Fu Quin • Chinese Academy of Agricultural Sciences (CAAS)
Ma Shiming • Institute for Agro-environment and Sustainable Development, CAAS
Li Xiande • Institute of Agricultural Economics and Development, CAAS
Zhu Xiaoman • China National Institute for Educational Research

France
Pascal Bergeret • Ministry of Agriculture

India
Satinder Baja • Eastern Institute for Integrated Learning in Management University
Indu Grover • CCS Haryana Agricultural University
Govind Kelkar • UNIFEM
Dev Nathan • Institute for Human Development
Rajeswari Sarala Raina • Centre for Policy Research
Vanaja Ramprasad • Green Foundation
Sukhpal Singh • Centre for Management in Agriculture, Indian Institute of Management
Rasheed Sulaiman V. • Centre for Research on Innovation and Science Policy (CRISP)

Indonesia
Hira Jhamtani • Third World Network

Malaysia
Lim Li Ching • Third World Network

Nepal
Rajendra Shrestha • AFORDA

New Zealand
Meriel Watts • Pesticide Action Network Aotearoa

Philippines
Arturo S. Arganosa • Philippine Council for Agriculture, Forestry and Natural Resources Research and Development
Danilo C. Cardenas • Philippine Council for Agriculture, Forestry and Natural Resources Research and Development
Richard B. Daite • Philippine Council for Agriculture, Forestry and Natural Resources Research and Development
Elenita C. Dano • Participatory Enhancement and Development of Genetic Resources in Asia (PEDIGREA)
Fezoil Luz C. Decena • Philippine Council for Agriculture, Forestry and Natural Resources Research and Development
Digna Manzanilla • Philippine Council for Agriculture, Forestry and Natural Resources Research and Development
Charito P. Medina • MASIPAG (Farmer-Scientist Partnership for Development, Inc.)
Thelma Paris • International Rice Research Institute

Switzerland
Hong Yang • Swiss Federal Institute for Aquatic Science and Technology.
Yuan Zhou • Swiss Federal Institute for Aquatic Science and Technology.

USA
Revathi Balakrishnan • Independent
Medha Devare • Cornell University
Shelley Feldman • Cornell University
JB Friday • University of Hawaii
Marcia Ishii-Eiteman • Pesticide Action Network, North America
Harold J. McArthur • University of Hawaii at Manoa
Douglas L. Vincent • University of Hawaii at Manoa
Dale Wen Jiajun • International Forum on Globalization

Viet Nam
Duong Van Chin • The Cuulong Delta Rice Research Institute

Annex B
Peer Reviewers

Australia
Les Copeland • University of Sydney
Tony Fischer • CSIRO Government of Australia
James Hardcastle • The Nature Conservancy, Indo-Pacific Resource Centre
Stuart B. Hill • University of Western Sydney
Lindsay Falvey • University of Melbourne
Geoff Lipsett-Moore • The Nature Conservancy, Indo-Pacific Resource Centre
David F. Smith • University of Melbourne
David Vincent • Australian Centre for International Agricultural Research
Sarah Withers • Department of Foreign Affairs and Trade
Jacqui Wright • Australian Centre for International Agricultural Research (ACIAR)

Bangladesh
Karim Mahmudul • Bangladesh Shrimp and Fish Foundation

Benin
Peter Neuenschwander • International Institute for Tropical Agriculture

Canada
Michael Cohen • Independent

Fiji
Anjeela Jokhan • University of South Pacific

France
Pascal Bergeret • Ministry of Agriculture and Fisheries
Patrick Safran • CIRAD

India
Mridul Eapen • Centre for Development Studies
Indu Grover • CCS Haryana Agricultural University
Pramod K. Joshi • NCAP
Sushil Kumar • NDRI
Mangesh V. Nadkarni • Institute for Social & Economic Change
Chand Ramesh • NCAP
V.N. Sharda • Central Soil & Water Conservation Research & Training Institute
Paul Singh Sidhu • The World Bank
Surijit Singh • Institute of Development Studies
Pal Suresh • NCAP
Vijay Taneja • Indian Council of Agricultural Research
Swarna S. Vepa • M.S. Swaminathan Research Foundation

Indonesia
Russell Dilts • FIELD Indonesia
Shobha Shetty • The World Bank

Italy
Barbara Gemmill Herren • FAO

Japan
Kiyoshi Takahashi • National Institute for Environmental Studies

Madagascar
Xavier Rakotonjanahary • FOFIFA

Malaysia
Lim Li Ching • Third World Network
Percy Sajise • Biodiversity International

New Zealand
Elvira Dommisse • National Council of NZ Soil & Health
Neil Macgregor • Journal of Organic Systems

Papua New Guinea
Paul Lokani • The Nature Conservancy, Indo-Pacific Resource Centre

Philippines
Beatriz Del Rosario • APAARI
Jacqueline Haessig Alleje • IFOAM
Rodel Lacso • World Agroforestry Centre (ICRAF)
Teodoro C. Mendoza • University of the Philippines
Priscilla Rubio • National Agricultural Development Plan, PNG
Jagadish Timsina • International Rice Research Institute

Poland
Ursula Soltysiak • AgroBio Test

Thailand
Jinhua Zhang • United Nations Environment Programme

United Kingdom
Zoe Cokeliss • United Nations Environment Programme
Lera Miles • United Nations Environment Programme

United States
Patrick Avato • The World Bank
Philip Bereano • University of Washington
Richard Chisholm • The World Bank
Ralph Crawford • U.S. Forest Service

Habiba Gitay • World Bank Institute
Doug Gurian-Sherman • Union of Concerned Scientists
Marcia Ishi-Eitemann • Pesticide Action Network North America
Willem Janssen • The World Bank
Susan J. Owens • U.S. Department of Agriculture
Ivette Perfecto • University of Michigan
Margaret Reeves • Pesticide Action Network North America
Doreen Stabinsky • College of the Atlantic
Goro Uehara • University of Hawaii
Jay Vroom • CropLife America

Annex C
Glossary

Agriculture A linked, dynamic social-ecological system based on the extraction of biological products and services from an ecosystem, innovated and managed by people. It thus includes cropping, animal husbandry, fishing, forestry, biofuel and bioproducts industries, and the production of pharmaceuticals or tissue for transplant in crops and livestock through genetic engineering. It encompasses all stages of production, processing, distribution, marketing, retail, consumption and waste disposal.

Agricultural biodiversity Encompasses the variety and variability of animals, plants and microorganisms necessary to sustain key functions of the agroecosystem, its structure and processes for, and in support of, food production and food security.

Agricultural extension Agricultural extension deals with the creation, transmission and application of knowledge and skills designed to bring desirable behavioral changes among people so that they improve their agricultural vocations and enterprises and, therefore, realize higher incomes and better standards of living.

Agricultural innovation Agricultural innovation is a socially constructed process. Innovation is the result of the interaction of a multitude of actors, agents and stakeholders within particular institutional contexts. If agricultural research and extension are important to agricultural innovation, so are markets, systems of government, relations along entire value chains, social norms, and, in general, a host of factors that create the incentives for a farmer to decide to change the way in which he or she works, and that reward or frustrate his or her decision.

Agricultural population The agricultural population is defined as all persons depending for their livelihood on agriculture, hunting, fishing or forestry. This estimate comprises all persons actively engaged in agriculture and their non-working dependants.

Agricultural subsidies Agricultural subsidies can take many forms, but a common feature is an economic transfer, often in direct cash form, from government to farmers. These transfers may aim to reduce the costs of production in the form of an input subsidy, e.g., for inorganic fertilizers or pesticides, or to make up the difference between the actual market price for farm output and a higher guaranteed price. Subsidies shield sectors or products from international competition.

Agricultural waste Farming wastes, including runoff and leaching of pesticides and fertilizers, erosion and dust from plowing, improper disposal of animal manure and carcasses, crop residues and debris.

Agroecological Zone A geographically delimited area with similar climatic and ecological characteristics suitable for specific agricultural uses.

Agroecology The science of applying ecological concepts and principles to the design and management of sustainable agroecosystems. It includes the study of the ecological processes in farming systems and processes such as: nutrient cycling, carbon cycling/sequestration, water cycling, food chains within and between trophic groups (microbes to top predators), lifecycles, herbivore/predator/prey/host interactions, pollination etc. Agroecological functions are generally maximized when there is high species diversity/perennial forest-like habitats.

Agroecosystem A biological and biophysical natural resource system managed by humans for the primary purpose of producing food as well as other socially valuable nonfood goods and environmental services. Agroecosystem function can be enhanced by increasing the planned biodiversity (mixed species and mosaics), which creates niches for unplanned biodiversity.

Agroforestry A dynamic, ecologically based, natural resources management system that through the integration of trees in farms and in the landscape diversifies and sustains production for increased social, economic and environmental benefits for land users at all levels. Agroforestry focuses on the wide range of work with trees grown on farms and in rural landscapes. Among these are fertilizer trees for land regeneration, soil health and food security; fruit trees for nutrition; fodder trees that improve smallholder livestock production; timber and fuelwood trees for shelter and energy; medicinal trees to combat disease; and trees that produce gums, resins or latex products. Many of these trees are multipurpose, providing a range of social, economic and environmental benefits.

AKST Agricultural Knowledge, Science and Technology (AKST) is a term encompassing the ways and means used to practice the different types of agricultural activities, and including both formal and informal knowledge and technology.

Alien Species A species occurring in an area outside of its historically known natural range as a result of intentional or accidental dispersal by human activities. Also referred to as introduced species or exotic species.

Aquaculture The farming of aquatic organisms in inland and coastal areas, involving intervention in the rearing process to enhance production and the individual or corporate ownership of the stock being cultivated. Aquaculture practiced in a marine environment is called mariculture.

Average Rate of Return Average rate of return takes the whole expenditure as given and calculates the rate of return to the global set of expenditures. It indicates whether or not the entire investment package was successful, but it does not indicate whether the allocation of resources between investment components was optimal.

Biodiversity The variability among living organisms from all sources, including inter alia, terrestrial, marine and other aquatic ecosystems and the ecological complexes of which they are part; including diversity within species and gene diversity among species, between species and of ecosystems.

Bioelectricity Electricity derived from the combustion of biomass, either directly or co-fired with fossil fuels such as coal and natural gas. Higher levels of conversion efficiency can be attained when biomass is gasified before combustion.

Bioenergy (biomass energy) Bioenergy is comprised of bioelectricity, bioheat and biofuels. Such energy carriers can be produced from energy crops (e.g., sugar cane, maize, oil palm), natural vegetation (e.g., woods, grasses) and organic wastes and residues (e.g., from forestry and agriculture). Bioenergy refers also to the direct combustion of biomass, mostly for heating and cooking purposes.

Biofuel Liquid fuels derived from biomass and predominantly used in transportation. The dominant biofuels are ethanol and biodiesel. Ethanol is produced by fermenting starch contained in plants such as sugar cane, sugar beet, maize, cassava, sweet sorghum or beetroot. Biodiesel is typically produced through a chemical process called trans-esterification, whereby oily biomass such as rapeseed, soybeans, palm oil, jatropha seeds, waste cooking oils or vegetable oils is combined with methanol to form methyl esters (sometimes called "fatty acid methyl ester" or FAME).

Bioheat Heat produced from the combustion of biomass, mostly as industrial process heat and heating for buildings.

Biological Control The use of living organisms as control agents for pests, (arthropods, nematodes mammals, weeds and pathogens) in agriiculture. There are three types of biological control:

Conservation biocontrol: The protection and encouragement of local natural enemy populations by crop and habitat management measures that enhance their survival, efficiency and growth.

Augmentative biocontrol: The release of natural enemies into crops to suppress specific populations of pests over one or a few generations, often involving the mass production and regular release of natural enemies.

Classical biocontrol: The local introduction of new species of natural enemies with the intention that they establish and build populations that suppress particular pests, often introduced alien pests to which they are specific.

Biological Resources Include genetic resources, organisms or parts thereof, populations, or any other biotic component of ecosystems with actual or potential use or value for humanity.

Biotechnology The IAASTD definition of biotechnology is based on that in the Convention on Biological Diversity and the Cartagena Protocol on Biosafety. It is a broad term embracing the manipulation of living organisms and spans the large range of activities from conventional techniques for fermentation and plant and animal breeding to recent innovations in tissue culture, irradiation, genomics and marker-assisted breeding (MAB) or marker assisted selection (MAS) to augment natural breeding. Some of the latest biotechnologies, called "modern biotechnology", include the use of *in vitro* modified DNA or RNA and the fusion of cells from different taxonomic families, techniques that overcome natural physiological reproductive or recombination barriers.

Biosafety Referring to the avoidance of risk to human health and safety, and to the conservation of the environment, as a result of the use for research and commerce of infectious or genetically modified organisms.

Blue Water The water in rivers, lakes, reservoirs, ponds and aquifers. Dryland production only uses green water, while irrigated production uses blue water in addition to green water.

BLCAs Brookered Long-term Contractual Arrangements (BLCAs) are institutional arrangements often involving a farmer cooperative, or a private commercial, parastatal or a state trading enterprise and a package (inputs, services, credit, knowledge) that allows small-scale farmers to engage in the production of a marketable commodity, such as cocoa or other product that farmers cannot easily sell elsewhere.

Catchment An area that collects and drains rainwater.

Capacity Development Any action or process which assists individuals, groups, organizations and communities in strengthening or developing their resources.

Capture Fisheries The sum (or range) of all activities to harvest a given fish resource from the "wild". It may refer to the location (e.g., Morocco, Gearges Bank), the target resource (e.g., hake), the technology used (e.g., trawl or beach seine), the social characteristics (e.g., artisanal, industrial), the purpose (e.g., commercial, subsistence, or recreational) as well as the season (e.g., winter).

Carbon Sequestration The process that removes carbon dioxide from the atmosphere.

Cellulosic Ethanol Next generation biofuel that allows converting not only glucose but also cellulose and hemi-cellulose—the main building blocks of most biomass—into ethanol, usually using acid-based catalysis or enzyme-based reactions to break down plant fibers into sugar, which is then fermented into ethanol.

Climate Change Refers to a statistically significant variation in either the mean state of the climate or in its variability, persisting for an extended period (typically decades or longer). Climate change may be due to natural internal processes or external forcing, or to persistent anthropogenic changes in the composition of the atmosphere or in land use.

Clone A group of genetically identical cells or individuals that are all derived from one selected individual by vegetative propagation or by asexual reproduction, breeding of completely inbred organisms, or forming genetically identical organisms by nuclear transplantation.

Commercialization The process of increasing the share of income that is earned in cash (e.g., wage income, surplus production for marketing) and reducing the share that is earned in kind (e.g., growing food for consumption by the same household).

Cultivar A cultivated variety, a population of plants within a species of plant. Each cultivar or variety is genetically different.

Deforestation The action or process of changing forest land to non-forested land uses.

Degradation The result of processes that alter the ecological characteristics of terrestrial or aquatic (agro)ecosystems so that the net services that they provide are reduced. Continued degradation leads to zero or negative economic agricultural productivity.

For loss of *land* in quantitative or qualitative ways, the term *degradation* is used. For water resources rendered unavailable for agricultural and non-agricultural uses, we employ the terms *depletion* and *pollution*. *Soil* degradation refers to the processes that reduce the capacity of the soil to support agriculture.

Desertification Land degradation in drylands resulting from various factors, including climatic variations and human activities.

Domesticated or Cultivated Species Species in which the evolutionary process has been influenced by humans to meet their needs.

Domestication The process to accustom animals to live with people as well as to selectively cultivate plants or raise animals in order to increase their suitability and compatibility to human requirements.

Driver Any natural or human-induced factor that directly or indirectly causes a change in a system.

Driver, direct A driver that unequivocally influences ecosystem processes and can therefore be identified and measured to different degrees of accuracy.

Driver, endogenous A driver whose magnitude can be influenced by the decision-maker. The endogenous or exogenous characteristic of a driver depends on the organizational scale. Some drivers (e.g., prices) are exogenous to a decision-maker at one level (a farmer) but endogenous at other levels (the nation-state).

Driver, exogenous A driver that cannot be altered by the decision-maker.

Driver, indirect A driver that operates by altering the level or rate of change of one or more direct drivers.

Ecoagriculture A management approach that provides fair balance between production of food, feed, fuel, fiber, and biodiversity conservation or protection of the ecosystem.

Ecological Pest Management (EPM) A strategy to manage pests that focuses on strengthening the health and resilience of the entire agro-ecosystem. EPM relies on scientific advances in the ecological and entomological fields of population dynamics, community and landscape ecology, multi-trophic interactions, and plant and habitat diversity.

Economic Rate of Return The net benefits to all members of society as a percentage of cost, taking into account externalities and other market imperfections.

Ecosystem A dynamic complex of plant, animal, and microorganism communities and their nonliving environment interacting as a functional unit.

Ecosystem Approach A strategy for the integrated management of land, water, and living resources that promotes conservation and sustainable use in an equitable way.

An ecosystem approach is based on the application of appropriate scientific methodologies focused on levels of biological organization, which encompass the essential structure, processes, functions, and interactions among organisms and their environment. It recognizes that humans, with their cultural diversity, are an integral component and managers of many ecosystems.

Ecosystem Function An intrinsic ecosystem characteristic related to the set of conditions and processes whereby an ecosystem maintains its integrity (such as primary productivity, food chain biogeochemical cycles). Ecosystem functions include such processes as decomposition, production, pollination, predation, parasitism, nutrient cycling, and fluxes of nutrients and energy.

Ecosystem Management An approach to maintaining or restoring the composition, structure, function, and delivery of services of natural and modified ecosystems for the goal of achieving sustainability. It is based on an adaptive, collaboratively developed vision of desired future conditions that integrates ecological, socioeconomic, and institutional perspectives, applied within a geographic framework, and defined primarily by natural ecological boundaries.

Ecosystem Properties The size, biodiversity, stability, degree of organization, internal exchanges of material and energy among different pools, and other properties that characterize an ecosystem.

Ecosystem Services The benefits people obtain from ecosystems. These include provisioning services such as food and water; regulating services such as flood and disease control; cultural services such as spiritual, recreational, and cultural benefits; and supporting services such as nutrient cycling that maintain the conditions for life on Earth. The concept "ecosystem goods and services" is synonymous with ecosystem services.

Ecosystem Stability A description of the dynamic properties of an ecosystem. An ecosystem is considered stable if it returns to its original state shortly after a perturbation (resilience), exhibits low temporal variability (constancy), or does not change dramatically in the face of a perturbation (resistance).

Eutrophication Excessive enrichment of waters with nutrients, and the associated adverse biological effects.

Ex-ante The analysis of the effects of a policy or a project based only on information available before the policy or project is undertaken.

Ex-post The analysis of the effects of a policy or project based on information available after the policy or project has been implemented and its performance is observed.

Ex-situ Conservation The conservation of components of biological diversity outside their natural habitats.

Externalities Effects of a person's or firm's activities on others which are not compensated. Externalities can either hurt or benefit others—they can be negative or positive. One negative externality arises when a company pollutes the local environment to produce its goods and does not compensate the negatively affected local residents. Positive externalities can be produced through primary education, which benefits not only primary school students but also society at large. Governments can reduce negative externalities by regulating and taxing goods with negative externalities. Governments can increase positive externalities by subsidizing goods with positive externalities or by directly providing those goods.

Fallow Cropland left idle from harvest to planting or during the growing season.

Farmer-led Participatory Plant Breeding Researchers and/or development workers interact with farmer-controlled, managed and executed PPB activities, and build on farmers' own varietal development and seed systems.

Feminization The increase in the share of women in an activity, sector or process.

Fishery Generally, a fishery is an activity leading to harvesting of fish. It may involve capture of wild fish or the raising of fish through aquaculture.

Food Security Food security exists when all people of a given spatial unit, at all times, have physical and economic access to sufficient, safe and nutritious food to meet their dietary needs and food preferences for an active and healthy life, and that is obtained in a socially acceptable and ecologically sustainable manner.

Food Sovereignty The right of peoples and sovereign states to democratically determine their own agricultural and food policies.

Food System A food system encompasses the whole range of food production and consumption activities. The food system includes farm input supply, farm production, food processing, wholesale and retail distribution, marketing, and consumption.

Forestry The human utilization of a piece of forest for a certain purpose, such as timber or recreation.

Forest Systems Forest systems are lands dominated by trees; they are often used for timber, fuelwood, and non-wood forest products.

Gender Refers to the socially constructed roles and behaviors of, and relations between, men and women, as opposed to sex, which refers to biological differences. Societies assign specific entitlements, responsibilities and values to men and women of different social strata and sub-groups.

Worldwide, systems of relation between men and women tend to disadvantage women, within the family as well as in public life. Like the hierarchical framework of a society, gender roles and relations vary according to context and are constantly subject to changes.

Genetic Engineering Modifying genotype, and hence phenotype, by transgenesis.

Genetic Material Any material of plant, animal, microbial or other origin containing functional units of heredity.

Genomics The research strategy that uses molecular characterization and cloning of whole genomes to understand the structure, function and evolution of genes and to answer fundamental biological questions.

Globalization Increasing interlinking of political, economic, institutional, social, cultural, technical, and ecological issues at the global level.

GMO (Genetically Modified Organism) An organism in which the genetic material has been altered anthropogenically by means of gene or cell technologies.

Governance The framework of social and economic systems and legal and political structures through which humanity manages itself. In general, governance comprises the traditions, institutions and processes that determine how power is exercised, how citizens are given a voice, and how decisions are made on issues of public concern.

Global Environmental Governance The global biosphere behaves as a single system, where the environmental impacts of each nation ultimately affect the whole. That makes a coordinated response from the community of nations a necessity for reversing today's environmental decline.

Global Warming Refers to an increase in the globally-averaged surface temperature in response to the increase of well-mixed greenhouse gases, particularly CO_2.

Global Warming Potential An index, describing the radiative characteristics of well-mixed greenhouse gases, that represents the combined effect of the differing times these gases remain in the atmosphere and their relative effectiveness in absorbing outgoing infrared radiation. This index approximates the time-integrated warming effect of a unit mass of a given greenhouse gas in today's atmosphere, relative to that of carbon dioxide.

Green Revolution An aggressive effort since 1950 in which agricultural researchers applied scientific principles of genetics and breeding to improve crops grown primarily in less-developed countries. The effort typically was accompanied by collateral investments to develop or strengthen the delivery of extension services, production inputs and markets and develop physical infrastructures such as roads and irrigation.

Green Water Green water refers to the water that comes from precipitation and is stored in unsaturated soil. Green water is typically taken up by plants as evapotranspiration.

Ground Water Water stored underground in rock crevices and in the pores of geologic materials that make up the Earth's crust. The upper surface of the saturate zone is called the water table.

Growth Rate The change (increase, decrease, or no change) in an indicator over a period of time, expressed as a percentage of the indicator at the start of the period. Growth rates contain several sets of information. The first is whether there is any change at all; the second is what direction the change is going in (increasing or decreasing); and the third is how rapidly that change is occurring.

Habitat Area occupied by and supporting living organisms. It is also used to mean the environmental attributes required by a particular species or its ecological niche.

Hazard A potentially damaging physical event, phenomenon and/or human activity, which my cause injury, property damage, social and economic disruption or environmental degradation.

Hazards can include latent conditions that may represent future threats and can have different origins.

Household All the persons, kin and non-kin, who live in the same or in a series of related dwellings and who share income, expenses and daily subsistence tasks. A basic unit for socio-cultural and economic analysis, a household may consist of persons (sometimes one but generally two or more) living together and jointly making provision for food or other essential elements of the livelihood.

Industrial Agriculture Form of agriculture that is capital-intensive, substituting machinery and purchased inputs for human and animal labor.

Infrastructure The facilities, structures, and associated equipment and services that facilitate the flows of goods and services between individuals, firms, and governments. It includes public utilities (electric power, telecommunications, water supply, sanitation and sewerage, and waste disposal); public works (irrigation systems, schools, housing, and hospitals); transport services (roads, railways, ports, waterways, and airports); and R&D facilities.

Innovation The use of a new idea, social process or institutional arrangement, material, or technology to change an activity, development, good, or service or the way goods and services are produced, distributed, or disposed of.

Innovation system Institutions, enterprises, and individuals that together demand and supply information and technology, and the rules and mechanisms by which these different agents interact.

In recent development discourse agricultural innovation is conceptualized as part and parcel of social and ecological organization, drawing on disciplinary evidence and understanding of how knowledge is generated and innovations occur.

In-situ Conservation The conservation of ecosystems and natural habitats and the maintenance and recovery of viable populations of species in their natural habitats and surroundings and, in the case of domesticated or cultivated species, in the surroundings where they have developed their distinctive properties and were managed by local groups of farmers, fishers or foresters.

Institutions The rules, norms and procedures that guide how people within societies live, work, and interact with each other. Formal institutions are written or codified rules, norms and procedures. Examples of formal institutions are the Constitution, the judiciary laws, the organized market, and property rights. Informal institutions are rules governed by social and behavioral norms of the society, family, or community. Cf. Organization.

Integrated Approaches Approaches that search for the best use of the functional relations among living organisms in relation to the environment without excluding the use of external inputs. Integrated approaches aim at the achievement of multiple goals (productivity increase, environmental sustainability and social welfare) using a variety of methods.

Integrated Assessment A method of analysis that combines results and models from the physical, biological, economic, and social sciences, and the interactions between these components in a consistent framework to evaluate the status and the consequences of environmental change and the policy responses to it.

Integrated Natural Resources Management (INRM) An approach that integrates research of different types of natural resources into stakeholder-driven processes of adaptive management and innovation to improve livelihoods, agroecosystem resilience, agricultural productivity and environmental services at community, eco-regional and global scales of intervention and impact. INRM thus aims to help to solve complex real-world problems affecting natural resources in agroecosystems.

Integrated Pest Management The procedure of integrating and applying practical management methods to manage insect populations so as to keep pest species from reaching damaging levels while avoiding or minimizing the potentially harmful effects of pest management measures on humans, non-target species, and the environment. IPM tends to incorporate assessment methods to guide management decisions.

Intellectual Property Rights (IPRs) Legal rights granted by governmental authorities to control and reward certain products of human intellectual effort and ingenuity.

Internal Rate of Return The discount rate that sets the net present value of the stream of the net benefits equal to zero. The internal rate of return may have multiple values when the stream of net benefits alternates from negative to positive more than once.

International Dollars Agricultural R&D investments in local currency units have been converted into international dollars by deflating the local currency amounts with each country's inflation ration (GDP deflator) of base year 2000. Next, they were converted to US dollars with a 2000 purchasing power parity (PPP) index. PPPs are synthetic exchange rates used to reflect the purchasing power of currencies.

Knowledge The way people understand the world, the way in which they interpret and apply meaning to their experiences. Knowledge is not about the discovery of some final objective "truth" but about the grasping of subjective culturally-conditioned products emerging from complex and ongoing processes involving selection, rejection, creation, development and transformation of information. These processes, and hence knowledge, are inextricably linked to the social, environmental and institutional context within which they are found.

Scientific knowledge: Knowledge that has been legitimized and validated by a formalized process of data gathering, analysis and documentation.

Explicit knowledge: Information about knowledge that has been or can be articulated, codified, and stored and exchanged. The most common forms of explicit knowledge are manuals, documents, procedures, cultural artifacts and stories. The information about explicit knowledge also can be audio-visual. Works of art and product design can be seen as other forms of explicit knowledge where human skills, motives and knowledge are externalized.

Empirical knowledge: Knowledge derived from and constituted in interaction with a person's environment. Modern communication and information technologies, and scientific instrumentation, can extend the "empirical environment" in which empirical knowledge is generated.

Local knowledge: The knowledge that is constituted in a given culture or society.

Traditional (ecological) knowledge: The cumulative body of knowledge, practices, and beliefs evolved by adaptive processes and handed down through generations. It may not be indigenous or local, but it is distinguished by the way in which it is acquired and used, through the social process of learning and sharing knowledge.

Knowledge Management A systematic discipline of policies, processes, and activities for the management of all processes of knowledge generation, codification, application and sharing of information about knowledge.

Knowledge Society A society in which the production and dissemination of scientific information and knowledge function well, and in which the transmission and use of valuable experiential knowledge is optimized; a society in which the information of those with experiential knowledge is used together with that of scientific and technical experts to inform decision-making.

Land Cover The physical coverage of land, usually expressed in terms of vegetation cover or lack of it. Influenced by but not synonymous with land use.

Land Degradation The reduction in the capability of the land to produce benefits from a particular land use under a specific form of land management.

Landscape An area of land that contains a mosaic of ecosystems, including human-dominated ecosystems. The term cultural landscape is often used when referring to landscapes containing significant human populations.

Land Tenure The relationship, whether legally or customarily defined, among people, as individuals or groups, with respect to land and associated natural resources (water, trees, minerals, wildlife, and so on).

Rules of tenure define how property rights in land are to be allocated within societies. Land tenure systems determine who can use what resources for how long, and under what conditions.

Land Use The human utilization of a piece of land for a certain purpose (such as irrigated agriculture or recreation). Land use is influenced by, but not synonymous with, land cover.

Leguminous Cultivated or spontaneous plants which fix atmospheric nitrogen.

Malnutrition Failure to achieve nutrient requirements, which can impair physical and/or mental health. It may result from consuming too little food or a shortage or imbalance of key nutrients (eg, micronutrient deficiencies or excess consumption of refined sugar and fat).

Marginal Rates of Return Calculates the returns to the last dollar invested on a certain activity. It is usually estimated through econometric estimation.

Marker Assisted Selection (MAS) The use of DNA markers to improve response to selection in a population. The markers will be closely linked to one or more target loci, which may often be quantitative trait loci.

Minimum Tillage The least amount possible of cultivation or soil disturbance done to prepare a suitable seedbed. The main purposes of minimum tillage are to reduce tillage energy consumption, to conserve moisture, and to retain plant cover to minimize erosion.

Model A simplified representation of reality used to simulate a process, understand a situation, predict an outcome or analyze a problem. A model can be viewed as a selective approximation, which by elimination of incidental detail, allows hypothesized or quantified aspects of the real world to appear manipulated or tested.

Multifunctionality In IAASTD, multifunctionality is used solely to express the inescapable interconnectedness of agriculture's different roles and functions. The concept of multifunctionality recognizes agriculture as a multi-output activity producing not only commodities (food, feed, fibers, agrofuels, medicinal products and ornamentals), but also non-commodity outputs such as environmental services, landscape amenities and cultural heritages (See Global SDM Text Box)

Natural Resources Management Includes all functions and services of nature that are directly or indirectly significant to humankind, i.e. economic functions, as well as other cultural and ecological functions or social services that are not taken into account in economic models or not entirely known.

Nanotechnology The engineering of functional systems at the atomic or molecular scale.

Net Present Value (NPV) Net present value is used to analyze the profitability of an investment or project, representing the difference between the discounted present value of benefits and the discounted present value of costs. If NPV of a prospective project is positive, then the project should be accepted. The analysis of NPV is sensitive to the reliability of future cash inflows that an investment or project will yield.

No-Till Planting without tillage. In most systems, planter-mounted coulters till a narrow seedbed assisting in the placement of fertilizer and seed. The tillage effect on weed control is replaced by herbicide use.

Obesity A chronic physical condition characterized by too much body fat, which results in higher risk for health problems such as high blood pressure, high blood cholesterol, diabetes, heart disease and stroke. Commonly it is defined as a Body Mass Index (BMI) equal to or more than 30, while overweight is equal to or more than 25. The BMI is an index of weight-for-height and is defined as the weight in kilograms divided by the square of the height in meters (kg/m^2).

Organic Agriculture An ecological production management system that promotes and enhances biological cycles and soil biological activity. It is based on minimal use of off-farm inputs and on management practices that restore, maintain and enhance ecological harmony.

Organization Organizations can be formal or informal. Examples of organizations are government agencies (e.g., police force, ministries, etc.), administrative bodies (e.g., local government), nongovernmental organizations, associations (e.g., farmers' associations) and private companies (firms). Cf. with Institutions.

Orphan Crops Crops such as tef, finger millet, yam, roots and tubers that tend to be regionally or locally important for income and nutrition, but which are not traded globally and receive minimal attention by research networks.

Participatory Development A process that involves people (population groups, organizations, associations, political parties) actively and significantly in all decisions affecting their lives.

Participatory Domestication The process of domestication that involves agriculturalists and other community members actively and significantly in making decisions, taking action and sharing benefits.

Participatory Plant Breeding (PPB) Involvement of a range of actors, including scientists, farmers, consumers, extension agents, vendors, processors and other industry stakeholders—as well as farmer and community-based organizations and non-government organization (NGOs) in plant breeding research and development.

Participatory Varietal Selection (PVS) A process by which farmers and other stakeholders along the food chain are involved with researchers in the selection of varieties from formal and farmer-based collections and trials, to determine which are best suited to their own agroecosystems' needs, uses and preferences, and which should go ahead for finishing, wider release and dissemination. The information gathered may in turn be fed back into formal-led breeding programs.

Pesticide A toxic chemical or biological product that kills organisms (e.g., insecticides, fungicides, weedicides, rodenticides).

Poverty There are many definitions of poverty.

Absolute Poverty: According to a UN declaration that resulted from the World Summit on Social Development in 1995, absolute poverty is a condition characterized by severe deprivation of basic human needs, including food, safe drinking water, sanitation facilities, health, shelter, education and information. It depends not only on income but also on access to services.

Dimensions of Poverty: The individual and social characteristics of poverty such as lack of access to health and education, powerlessness or lack of dignity. Such aspects of deprivation experienced by the individual or group are not captured by measures of income or expenditure.

Extreme Poverty: Persons who fall below the defined poverty line of US$1 income per day. The measure is converted into local currencies using purchasing power parity (PPP) exchange rates. Other definitions of this concept have identified minimum subsistence requirements, the denial of basic human rights or the experience of exclusion.

Poverty Line: A minimum requirement of welfare, usually defined in relation to income or expenditure, used to identify the poor. Individuals or households with incomes or expenditure below the poverty line are poor. Those with incomes or expenditure equal to or above the line are not poor. It is common practice to draw more than one poverty line to distinguish different categories of poor, for example, the extreme poor.

Private Rate of Return The gain in net revenue to the private firm/business divided by the cost of an investment expressed in percentage.

Processes A series of actions, motions, occurrences, a method, mode, or operation, whereby a result or effect is produced.

Production Technology All methods that farmers, market agents and consumers use to cultivate, harvest, store, process, handle, transport and prepare food crops, cash crops, livestock, etc. for consumption.

Protected Area A geographically defined area which is designated or regulated and managed to achieve specific conservation objectives as defined by society.

Public Goods A good or service in which the benefit received by any one party does not diminish the availability of the benefits to others, and/or where access to the good cannot be restricted. Public goods have the properties of non-rivalry in consumption and non-excludability.

Public R&D Investment Includes R&D investments done by government agencies, nonprofit institutions, and higher-education agencies. It excludes the private for-profit enterprises.

Research and Development (R&D) Organizational strategies and methods used by research and extension program to conduct their work including scientific procedures, organizational modes, institutional strategies, interdisciplinary team research, etc.

Scenario A plausible and often simplified description of how the future may develop based on explicit and coherent and internally consistent set of assumptions about key driving forces (e.g., rate of technology change, prices) and relationships. Scenarios are neither predictions nor projections and sometimes may be based on a "narrative storyline". Scenarios may be derived from projections but are often based on additional information from other sources.

Science, Technology and Innovation Includes all forms of useful knowledge (codified and tacit) derived from diverse branches of learning and practice, ranging from basic scientific research to engineering to local knowledge. It also includes the policies used to promote scientific advance, technology development, and the commercialization of products, as well as the associated institutional innovations. *Science* refers to both basic and applied sciences. *Technology* refers to the application of science, engineering, and other fields, such as medicine. *Innovation* includes all of the processes, including business activities that bring a technology to market.

Shifting Cultivation Found mainly in the tropics, especially in humid and subhumid regions. There are different kinds; for example, in some cases a settlement is permanent, but certain fields are fallowed and cropped alternately ("rotational agriculture"). In other cases, new land is cleared when the old is no longer productive.

Slash and Burn Agriculture A pattern of agriculture in which existing vegetation is cleared and burned to provide space and nutrients for cropping.

Social Rate of Return The gain to society of a project or investment in net revenue divided by cost of the investment, expressed by percentage.

Soil and Water Conservation (SWC) A combination of appropriate technology and successful approach. Technologies promote the sustainable use of agricultural soils by minimizing soil erosion, maintaining and/or enhancing soil properties, managing water, and controlling temperature. Approaches explain the ways and means which are used to realize SWC in a given ecological and socio-economic environment.

Soil Erosion The detachment and movement of soil from the land surface by wind and water in conditions influenced by human activities.

Soil Function Any service, role, or task that a soil performs, especially: (a) sustaining biological activity, diversity, and productivity; (b) regulating and partitioning water and solute flow; (c) filtering, buffering, degrading, and detoxifying potential pollutants; (d) storing and cycling nutrients; (e) providing support for buildings and other structures and to protect archaeological treasures.

Staple Food (Crops) Food that is eaten as daily diet.

Soil Quality The capacity of a specific kind of soil to function, within natural or managed ecosystem boundaries, to sustain plant and animal productivity, maintain or enhance water and air quality, and support human health and habitation. In short, the capacity of the soil to function.

Subsidy Transfer of resources to an entity, which either reduces the operating costs or increases the revenues of such entity for the purpose of achieving some objective.

Subsistence Agriculture Agriculture carried out for the use of the individual person or their family with few or no outputs available for sale.

Sustainable Development Development that meets the needs of the present without compromising the ability of future generations to meet their own needs.

Sustainable Land Management (SLM) A system of technologies and/or planning that aims to integrate ecological with socioeconomic and political principles in the management of land for agricultural and other purposes to achieve intra- and intergenerational equity.

Sustainable Use of Natural Resources Natural resource use is sustainable if specific types of use in a particular ecosystem are considered reasonable in the light of both the internal and the external perspective on natural resources. "Reasonable" in this context means that all actors agree that resource use fulfils productive, physical, and cultural functions in ways that will meet the long-term needs of the affected population.

Technology Transfer The broad set of deliberate and spontaneous processes that give rise to the exchange and dissemination of information and technologies among different stakeholders. As a generic concept, the term is used to encompass both diffusion of technologies and technological cooperation across and within countries.

Terms of Trade The *international terms* of trade measures a relationship between the prices of exports and the prices of imports, this being known strictly as the barter terms of trade. In this sense, deterioration in the terms of trade could have resulted if unit prices of exports had risen less than unit prices for imports. The *inter-sectoral terms of trade* refers to the terms of trade between sectors of the economy, e.g., rural and urban, agriculture and industry.

Total Factor Productivity A measure of the increase in total output which is not accounted for by increases in total inputs. The total factor productivity index is computed as the ratio of an index of aggregate output to an index of aggregate inputs.

Tradeoff Management choices that intentionally or otherwise change the type, magnitude, and relative mix of services provided by ecosystems.

Transgene An isolated gene sequence used to transform an organism. Often, but not always, the transgene has been derived from a different species than that of the recipient.

Transgenic An organism that has incorporated a functional foreign gene through recombinant DNA technology. The novel gene exists in all of its cells and is passed through to progeny.

Undernourishment Food intake that is continuously inadequate to meet dietary energy requirement.

Undernutrition The result of food intake that is insufficient to meet dietary energy requirements continuously, poor absorption, and/or poor biological use of nutrients consumed.

Urban and Peri-Urban Agriculture Agriculture occurring within and surrounding the boundaries of cities throughout the world and includes crop and livestock production, fisheries and forestry, as well as the ecological services they provide. Often multiple farming and gardening systems exist in and near a single city.

Value Chain A set of value-adding activities through which a product passes from the initial production or design stage to final delivery to the consumer.

Virtual Water The volume of water used to produce a commodity. The adjective "virtual" refers to the fact that most of the water used to produce a product is not contained in the product. In accounting virtual water flows we keep track of which parts of these flows refer to green, blue and grey water, respectively.

The real-water content of products is generally negligible if compared to the virtual-water content.

Waste Water "Grey" water that has been used in homes, agriculture, industries and businesses that is not for reuse unless it is treated.

Watershed The area which supplies water by surface and subsurface flow from precipitation to a given point in the drainage system.

Watershed Management Use, regulation and treatment of water and land resources of a watershed to accomplish stated objectives.

Water Productivity An efficiency term quantified as a ration of product output (goods and services) over water input.

Expressions of water productivity. Three major expressions of water productivity can be identified: 1) the amount of carbon gain per unit of water transpired by the leaf or by the canopy (photosynthetic water productivity); 2) the amount of water transpired by the crop (biomass water productivity); or 3) the yield obtained per unit amount of water transpired by the crop (yield water productivity).

Agricultural water productivity relates net benefits gained through the use of water in crop, forestry, fishery, livestock and mixed agricultural systems. In its broadest sense, it reflects the objectives of producing more food, income, livelihood and ecological benefits at less social and environmental cost per unit of water in agriculture.

Physical water productivity relates agricultural production to water use—more crop per drop. Water use is expressed either in terms of delivery to a use, or depletion by a use through evapotranspiration, pollution, or directing water to a sink where it cannot be reused. Improving physical water productivity is important to reduce future water needs in agriculture.

Economic water productivity relates the value of agricultural production to agricultural water use. A holistic assessment should account for the benefits and costs of water, including less tangible livelihood benefits, but this is rarely done. Improving economic water productivity is important for economic growth and poverty reduction.

Annex D

Acronyms, Abbreviations and Units

ADB Asian Development Bank
ACIAR Australian Centre International Agricultural Research
AI Artificial Insemination
AIDS Acquired immune deficiency syndrome
AKF Aga Khan Foundation
AKST Agricultural Knowledge, Science, and Technology
AMS Aggregate measure of support
AoA Agreement on Agriculture
APAARI Asia Pacific Association of Agricultural Research Institutions
APEC Asia Pacific Economic Cooperation
APOP Agricultural Products Options Program
APRN Asia Pacific Research Network
ASEAN Association of Southeast Asian Nations
AVRDC Asian Vegetable Research and Development Centre
billion one thousand million
BIMSTEC Bengal Initiative for Multisectoral Technical and Economic Cooperation
BPL Below the Poverty Line
BRAC Bangladesh Rural Advancement Committee
Bt soil bacterium (usually refers to plants made insecticidal using a variant of various toxin genes sourced from plasmids of these bacteria)
BTA bilateral trade agreement
C carbon
Ca calcium
CA Comprehensive Assessment of Water Management in Agriculture
CAC Codex Alimentarius Commission
CAFTA Central American Free Trade Agreement
CAP European Common Agricultural Policy
CAPHP commercial animal and plant health protection
CBD Convention on Biological Diversity
CDM Clean Development Mechanism
CGIAR Consultative Group on International Agricultural Research
CIAT Centro Internacional de Agricultura Tropical
CIMMYT Centro Internacional de Mejoramiento de Maíz y Trigo
CIP International Potato Center
CIPR Commission on Intellectual Property Rights
CITES Convention on International Trade in Endangered Species of Wild Fauna and Flora
CPB Cartagena Protocol on Biosafety
CSO Civil Society Organization
CWANA Central and West Asia and North Africa
DFID UK Department for International Development
DRC Domestic Resource Cost
DSS Decision Support System
EDV essentially derived variety
EJ Exajoules
EPA US Environmental Protection Agency
ESAP East and South Asia and Pacific
ESCAP Economic and Social Commission for Asia and the Pacific
ET embryo transfer
EU European Union
HACCP Hazard Analysis Critical Control Point
FAO Food and Agriculture Organization of the United Nations
FDI foreign direct investment
FITS Farmers Information and Technology Service
FFS Farmer Field School
FSC Forest Stewardship Council
FTA Free trade agreement
FTO Freedom to operate
GATT General Agreement on Trade and Tariffs
GDP Gross Domestic Product
GE genetically engineered
GEF Global Environment Facility
GFAR Global Forum on Agricultural Research
GHG greenhouse gas
GIFT Genetically Improved Fish Tilapia
GIS Geographic information systems
GLASOD Global Assessment of Human Induced Soil Degradation
GMO Genetically Modified Organisms
GNI gross national income
GTZ German Agency for Technical Cooperation
GURT Genetic Use Restriction Technologies
ha hectare
HDI human development index
HIV human immunodeficiency virus
HPI human poverty index
HYV high yielding variety
IAASTD International Assessment of Agricultural Knowledge, Science and Technology for Development
IARC International Agricultural Research Center
ICA International Coffee Agreement
ICAR Indian Council for Agricultural Research
ICT Information and communication technologies
IDRC International Development Research Centre

IFAD International Fund for Agricultural Development
IFPRI International Food Policy Research Institute
IIRR International Institute of Rural Reconstruction
IIT Indian Institute of Technology
ILO International Labour Organization
ILRI International Livestock Research Institute
IMF International Monetary Fund
INM Integrated nutrient management
IPM Integrated pest management
IPCC Intergovernmental Panel on Climate Change
IPPC International Plant Protection Convention
IPR Intellectual property rights
IR insect resistant
IRRI International Rice Research Institute
ISP internet service provider
ITPGRFA International Treaty on Plant Genetic Resources for Food and Agriculture
IWMI International Water Management Institute
K potassium
kcal kilocalorie
kg kilogram
km kilometer
kW kilowatt (10^3 watts)
LAC Latin America and Caribbean
LDC least developed countries
LLP livestock and livestock products
m meter (10^2 cm)
MA Millennium Ecosystem Assessment
MEA multilateral environment agreement
mm millimeter
Mn manganese
MNC multinational corporation
Mo molybdenum
MTA material transfer agreement
MWe megawatt electricity
MWh megawatt hours
NAE North America and Europe
NAFTA North American Free Trade Agreement
NARA Nippon Australia Relations Agreement
NARS National Agricultural Research Systems
NASA National Aeronautic and Space Administration
NBSAP Implementing National Biodiversity Strategy and Action Plans
NDDB National Dairy Development Board
NFE non-formal education
NGO Nongovernmental organizations
NOx generic term for mononitrogen oxides (NO and NO_2)
nPAH nitrated polycyclic aromatic hydrocarbons
NPM non-pesticide management
NREGA National Rural Employment Guarantee Act
NTB non-tariff barriers
NTFP Non-Timber Forest Products
OECD Organization for Economic Cooperation and Development
OMC Organización Mundial del Comercio
P phosphorus
PAH polycyclic aromatic hydrocarbons
PBR plant breeder rights
PCARRD Philippine Council for Agriculture, Forestry and Natural Resources Research and Development
PIPRA Public-Sector Intellectual Property Resource for Agriculture
PGR plant genetic resource
PVO Private Voluntary Organization
PVP plant variety protection
R&D research and development
RDE research, development, and extension
RDTE research, development, training and extension
RMP Residue Monitoring Plan
S&T science and technology
SAARC South Asian Association for Regional Cooperation
SARS Severe Acute Respiratory Syndrome
SDR Special Drawing Right
SEWA Self-Employed Women's Organization
SHG self help group
SPS Sanitary and phytosanitary
SSA Sub-Saharan Africa
TBT Technical Barriers to Trade
Tg teragram (megatonne, 10^{12} grams)
TK traditional knowledge
TNC transnational corporation
tonne 10^3 kg (metric ton)
TRIPS Trade-Related Aspects of Intellectual Property Rights
UAE United Arab Emirates
UNAIDS United Nations Joint Program on HIV/AIDS
UNCBD UN Convention on Biodiversity
UNCCD United Nations Convention to Combat Desertification
UNCED United Nations Conference on Environment and Development
UNCTAD United Nations Conference on Trade and Development
UNDESA United Nations Department of Economic and Social Affairs
UNDP United Nations Development Program
UNESCO United Nations Educational, Scientific and Cultural Organization
UNEP United Nations Environment Program
UNFPA United Nations Population Fund
UNIFEM United Nations Development Fund for Women
UNEP United Nations Environment Programme
UNESCO United Nations Educational, Scientific and Cultural Organization
UNFCCC United Nations Framework Convention on Climate Change
UPA Urban and periurban agriculture
UPOV International Union for the Protection of New Varieties of Plants
USDA U.S. Department of Agriculture
VCIP Village Computer and Internet Program
WCSDG World Commission on the Social Dimension of Globalization
WHO World Health Organization
WIPO World Intellectual Property Organization
WTO World Trade Organization
Zn zinc

Annex E

Steering Committee for Consultative Process and Advisory Bureau for Assessment

Steering Committee

The Steering Committee was established to oversee the consultative process and recommend whether an international assessment was needed, and if so, what was the goal, the scope, the expected outputs and outcomes, governance and management structure, location of the secretariat and funding strategy.

Co-chairs

Louise Fresco, Assistant Director General for Agriculture, FAO
Seyfu Ketema, Executive Secretary, Association for Strengthening Agricultural Research in East and Central Africa (ASARECA)
Claudia Martinez Zuleta, Former Deputy Minister of the Environment, Colombia
Rita Sharma, Principal Secretary and Rural Infrastructure Commissioner, Government of Uttar Pradesh, India
Robert T. Watson, Chief Scientist, The World Bank

Nongovernmental Organizations

Benny Haerlin, Advisor, Greenpeace International
Marcia Ishii-Eiteman, Senior Scientist, Pesticide Action Network North America Regional Center (PANNA)
Monica Kapiriri, Regional Program Officer for NGO Enhancement and Rural Development, Aga Khan
Raymond C. Offenheiser, President, Oxfam America
Daniel Rodriguez, International Technology Development Group (ITDG), Latin America Regional Office, Peru

UN Bodies

Ivar Baste, Chief, Environment Assessment Branch, UN Environment Programme
Wim van Eck, Senior Advisor, Sustainable Development and Healthy Environments, World Health Organization
Joke Waller-Hunter, Executive Secretary, UN Framework Convention on Climate Change
Hamdallah Zedan, Executive Secretary, UN Convention on Biological Diversity

At-large Scientists

Adrienne Clarke, Laureate Professor, School of Botany, University of Melbourne, Australia
Denis Lucey, Professor of Food Economics, Dept. of Food Business & Development, University College Cork, Ireland, and Vice-President NATURA
Vo-tong Xuan, Rector, Angiang University, Vietnam

Private Sector

Momtaz Faruki Chowdhury, Director, Agribusiness Center for Competitiveness and Enterprise Development, Bangladesh
Sam Dryden, Managing Director, Emergent Genetics
David Evans, Former Head of Research and Technology, Syngenta International
Steve Parry, Sustainable Agriculture Research and Development Program Leader, Unilever
Mumeka M. Wright, Director, Bimzi Ltd., Zambia

Consumer Groups

Michael Hansen, Consumers International
Greg Jaffe, Director, Biotechnology Project, Center for Science in the Public Interest
Samuel Ochieng, Chief Executive, Consumer Information Network

Producer Groups

Mercy Karanja, Chief Executive Officer, Kenya National Farmers' Union
Prabha Mahale, World Board, International Federation Organic Agriculture Movements (IFOAM)
Tsakani Ngomane, Director Agricultural Extension Services, Department of Agriculture, Limpopo Province, Republic of South Africa
Armando Paredes, Presidente, Consejo Nacional Agropecuario (CNA)

Scientific Organizations

Jorge Ardila Vásquez, Director Area of Technology and Innovation, Inter-American Institute for Cooperation on Agriculture (IICA)
Samuel Bruce-Oliver, NARS Senior Fellow, Global Forum for Agricultural Research Secretariat
Adel El-Beltagy, Chair, Center Directors Committee, Consultative Group on International Agricultural Research (CGIAR)
Carl Greenidge, Director, Center for Rural and Technical Cooperation, Netherlands
Mohamed Hassan, Executive Director, Third World Academy of Sciences (TWAS)
Mark Holderness, Head Crop and Pest Management, CAB International
Charlotte Johnson-Welch, Public Health and Gender Specialist and Nata Duvvury, Director Social Conflict and Transformation Team, International Center for Research on Women (ICRW)
Thomas Rosswall, Executive Director, International Council for Science (ICSU)
Judi Wakhungu, Executive Director, African Center for Technology Studies

Governments

Peter Core, Director, Australian Centre for International Agricultural Research

Keming Qian, Director General Inst. Agricultural Economics, Dept. of International Cooperation, Chinese Academy of Agricultural Science

Tiina Huvio, Senior Advisor, Agriculture and Rural Development, Ministry of Foreign Affairs

Alain Derevier, Senior Advisor, Research for Sustainable Development, Ministry of Foreign Affairs

Germany: Hans-Jochen de Haas, Head, Agricultural and Rural Development, Federal Ministry of Economic Cooperation and Development (BMZ)

Zoltan Bedo, Director, Agricultural Research Institute, Hungarian Academy of Sciences

Aidan O'Driscoll, Assistant Secretary General, Department of Agriculture and Food

Hamid Narjisse, Director General, INRA

Eugenia Serova, Head, Agrarian Policy Division, Institute for Economy in Transition

Grace Akello, Minister of State for Northern Uganda Rehabilitation

Paul Spray, Head of Research, DFID

Rodney Brown, Deputy Under Secretary of Agriculture and Hans Klemm, Director of the Office of Agriculture, Biotechnology and Textile Trade Affairs, Department of State

Foundations and Unions

Susan Sechler, Senior Advisor on Biotechnology Policy, Rockefeller Foundation

Achim Steiner, Director General, The World Conservation Union (IUCN)

Eugene Terry, Director, African Agricultural Technology Foundation

Advisory Bureau

Non-government Representatives

Consumer Groups

Jaime Delgado • Asociación Peruana de Consumidores y Usuarios
Greg Jaffe • Center for Science in the Public Interest
Catherine Rutivi • Consumers International
Indrani Thuraisingham • Southeast Asia Council for Food Security and Trade
Jose Vargas Niello • Consumers International Chile

International organizations

Nata Duvvury • International Center for Research on Women
Emile Frison • CGIAR
Dennis Rangi • CAB International
Mohamed Hassan • Third World Academy of Sciences
Jeffrey McNeely • World Conservation Union (IUCN)
John Stewart • International Council of Science (ICSU)

NGOs

Kevin Akoyi • Vredeseilanden
Hedia Baccar • Association pour la Protection de l'Environment de Kairouan
Benedikt Haerlin • Greenpeace International
Juan Lopez • Friends of the Earth International
Khadouja Mellouli • Women for Sustainable Development
Patrick Mulvaney • Practical Action
Romeo Quihano • Pesticide Action Network
Maryam Rahmaniam • CENESTA
Daniel Rodriguez • International Technology Development Group

Private Sector

Momtaz Chowdhury • Agrobased Technology and Industry Development
Giselle L. D'Almeida • Interface
Eva Maria Erisgen • BASF
Armando Paredes • Consejo Nacional Agropecuario
Steve Parry • Unilever
Harry Swaine • Syngenta (resigned)

Producer Groups

Shoaib Aziz • Sustainable Agriculture Action Group of Pakistan
Philip Kiriro • East African Farmers Federation
Kristie Knoll • Knoll Farms
Prabha Mahale • International Federation of Organic Agriculture Movements
Anita Morales • Apit Tako
Nizam Selim • Pioneer Hatchery

Government Representatives

Central and West Asia and North Africa

Egypt • Ahlam Al Naggar
Iran • Hossein Askari
Kyrgyz Republic • Djamin Akimaliev
Saudi Arabia • Abdu Al Assiri, Taqi Elldeen Adar, Khalid Al Ghandi
Turkey • Yalcin Kaya, Mesut Keser

East and South Asia and the Pacific

Australia • Simon Hearn
China • Puyun Yang
India • PK Joshi
Japan • Ryuko Inoue
Philippines • William Medrano

Latin America and Caribbean

Brazil • Sebastiao Barbosa, Alexandre Cardoso, Paulo Roberto Galerani, Rubens Nodari
Dominican Republic • Rafael Perez Duvergé
Honduras • Arturo Galo, Roberto Villeda Toledo
Uruguay • Mario Allegri

North America and Europe

Austria • Hedwig Woegerbauer
Canada • Iain MacGillivray
Finland • Marja-Liisa Tapio-Bistrom
France • Michel Dodet
Ireland • Aidan O'Driscoll, Tony Smith
Russia • Eugenia Serova, Sergey Alexanian
United Kingdom • Jim Harvey, David Howlett, John Barret
United States • Christian Foster

Sub-Saharan Africa

Benin • Jean Claude Codjia
Gambia • Sulayman Trawally
Kenya • Evans Mwangi
Mozambique • Alsácia Atanásio, Júlio Mchola
Namibia • Gillian Maggs-Kölling
Senegal • Ibrahim Diouck

Annex F

Secretariat and Cosponsor Focal Points

Secretariat

World Bank

Marianne Cabraal, Leonila Castillo, Jodi Horton, Betsi Isay, Pekka Jamsen, Pedro Marques, Beverly McIntyre, Wubi Mekonnen, June Remy

UNEP

Marcus Lee, Nalini Sharma, Anna Stabrawa

UNESCO

Guillen Calvo

With special thanks to the Publications team: Audrey Ringler (logo design), Pedro Marques (proofing and graphics), Ketill Berger and Eric Fuller (graphic design)

Regional Institutes

Sub-Saharan Africa – African Centre for Technology Studies (ACTS)

Ronald Ajengo, Elvin Nyukuri, Judi Wakhungu

Central and West Asia and North Africa – International Center for Agricultural Research in the Dry Areas (ICARDA)

Mustapha Guellouz, Lamis Makhoul, Caroline Msrieh-Seropian, Ahmed Sidahmed, Cathy Farnworth

Latin America and the Caribbean – Inter-American Institute for Cooperation on Agriculture (IICA)

Enrique Alarcon, Jorge Ardila Vásquez, Viviana Chacon, Johana Rodríguez, Gustavo Sain

East and South Asia and the Pacific – WorldFish Center

Karen Khoo, Siew Hua Koh, Li Ping Ng, Jamie Oliver, Prem Chandran Venugopalan

Cosponsor Focal Points

GEF	Mark Zimsky
UNDP	Philip Dobie
UNEP	Ivar Baste
UNESCO	Salvatore Arico, Walter Erdelen
WHO	Jorgen Schlundt
World Bank	Mark Cackler, Kevin Cleaver, Eija Pehu, Juergen Voegele

Index

Page references followed by *f*, *t*, and *b* indicate figures, tables, and boxes, respectively.

D

E

About Island Press

Since 1984, the nonprofit Island Press has been stimulating, shaping, and communicating the ideas that are essential for solving environmental problems worldwide. With more than 800 titles in print and some 40 new releases each year, we are the nation's leading publisher on environmental issues. We identify innovative thinkers and emerging trends in the environmental field. We work with world-renowned experts and authors to develop cross-disciplinary solutions to environmental challenges.

Island Press designs and implements coordinated book publication campaigns in order to communicate our critical messages in print, in person, and online using the latest technologies, programs, and the media. Our goal: to reach targeted audiences—scientists, policymakers, environmental advocates, the media, and concerned citizens—who can and will take action to protect the plants and animals that enrich our world, the ecosystems we need to survive, the water we drink, and the air we breathe.

Island Press gratefully acknowledges the support of its work by the Agua Fund, Inc., Annenberg Foundation, The Christensen Fund, The Nathan Cummings Foundation, The Geraldine R. Dodge Foundation, Doris Duke Charitable Foundation, The Educational Foundation of America, Betsy and Jesse Fink Foundation, The William and Flora Hewlett Foundation, The Kendeda Fund, The Andrew W. Mellon Foundation, The Curtis and Edith Munson Foundation, Oak Foundation, The Overbrook Foundation, the David and Lucile Packard Foundation, The Summit Fund of Washington, Trust for Architectural Easements, Wallace Global Fund, The Winslow Foundation, and other generous donors.